Food, Energy, and Water

Food, Energy, and Water

The Chemistry Connection

Edited by

Satinder Ahuja
Ahuja Consulting
Calabash, NC

ELSEVIER

Amsterdam • Boston • Heidelberg • London • New York • Oxford
Paris • San Diego • San Francisco • Singapore • Sydney • Tokyo

Elsevier
Radarweg 29, PO Box 211, 1000 AE Amsterdam, Netherlands
The Boulevard, Langford Lane, Kidlington, Oxford OX5 1GB, UK
225 Wyman Street, Waltham, MA 02451, USA

ISBN: 978-0-12-800211-7

British Library Cataloguing in Publication Data
A catalogue record for this book is available from the British Library

Library of Congress Cataloging-in-Publication Data
A catalog record for this book is available from the Library of Congress

For Information on all Elsevier publications
visit our website at http://store.elsevier.com/

Contents

List of Contributors

Satinder Ahuja Ahuja Consulting, Calabash, NC, USA

Mark A. Benvenuto Department of Chemistry and Biochemistry, University of Detroit Mercy, Detroit, MI, USA

Carleton R. Bern U.S. Geological Survey, Crustal Geophysics and Geochemistry Science Center, Denver, CO, USA

Justin E. Birdwell U.S. Geological Survey, Central Energy Resources Science Center, Denver Federal Center, Denver, CO, USA

David Bromley President DBE Engineering Ltd., Vancouver, British Columbia, Short Course Instructor – Texas A and M University

Lawrence B. Cahoon Department of Biology and Marine Biology, UNC Wilmington, Wilmington, NC, USA

Neera Chawla Consultant, Energy & Petroleum, Chicago, IL

Franz S. Ehrenhauser Cain Department of Chemical Engineering, Louisiana State University, Baton Rouge, LA, USA

Mark Engle U.S. Geological Survey, Department of Geological Sciences, University of Texas at El Paso, El Paso, TX, USA

John W. Finley School of Nutrition and Food Sciences, Louisiana State University, Baton Rouge, LA, USA

Mark Fitch Civil, Architectural and Environmental Engineering, Missouri University of Science and Technology

Tanya J. Gallegos U.S. Geological Survey, Eastern Energy Resources Science Center, Reston, VA, USA

Seth S. Haines U.S. Geological Survey, Central Energy Resources Science Center, Denver Federal Center, Denver, CO, USA

Aubrey A. Heath Cain Department of Chemical Engineering, Louisiana State University, Baton Rouge, LA, USA

Neha Mehta Civil and Environmental Engineering, Massachusetts Institute of Technology, Cambridge, MA, USA

Stephen P. Mezyk Department of Chemistry and Biochemistry, California State University at Long Beach, Long Beach, CA, USA

Parimita Mohanty Fellow and Team Leader, Teri, New Delhi, India

Donna N. Myers U.S. Geological Survey, Reston, VA, USA

Shauna C. Otto Department of Chemistry and Biochemistry, California State University at Long Beach, Long Beach, CA, USA

Nikul K. Patel Mechanical Engineering Department, Faculty of Technology & Engineering, The M S University of Baroda, Vadodara, India

Elizabeth S. Roberts-Kirchhoff Department of Chemistry and Biochemistry, University of Detroit Mercy, Detroit, MI, USA

Alexandra E. Schroeder Department of Chemistry and Biochemistry, University of Detroit Mercy, Detroit, MI, USA

James N. Seiber Department of Environmental Toxicology, University of California, Oakland, CA, USA

Shailesh N. Shah Chemistry Department, Faculty of Science, The M S University of Baroda, Vadodara, India

Zachary R. Smith Department of Chemistry and Biochemistry, University of Detroit Mercy, Detroit, MI, USA

Francis O' Sullivan MIT Energy Initiative, Massachusetts Institute of Technology, Cambridge, MA, USA

Hessy L. Taft St. John's University, New York, NY

Jim Thebaut The Chronicles Group, Redondo Beach, CA

Tsanangurayi Tongesayi Department of Chemistry, Medical Technology and Physics, Monmouth University, West Long Branch, NJ, USA

Sunungurai Tongesayi Walden University Public Health, Minneapolis, MN, USA

Akshay Tyagi Research Associate Trainee, Teri, New Delhi, India

Mickael Vaitilingom Cain Department of Chemical Engineering, Louisiana State University, Baton Rouge, LA, USA

Kalliat T. Valsaraj Cain Department of Chemical Engineering, Louisiana State University, Baton Rouge, LA, USA

Kathrin D. Zimmerman Department of Chemistry and Biochemistry, California State University at Long Beach, Long Beach, CA, USA

Preface

Food, energy, and water are essential for our existence; however, which one of these is most important can be a subject of some debate. The interrelationship of food, energy, and water, i.e., the chemistry connection, presents us with a multidimensional problem that links resources, national policies, politics, and quality of life for all of us on the planet now, and for the 9 billion people in the next 40 years. The nexus of food, water, and energy can help us determine how we deploy these resources to address our needs—collectively rather than individually. Addressing these complex issues requires commitment and also collaborative efforts among federal and state agencies, private enterprise, and nongovernment organizations. More investment is necessary to develop sound science and policies which address these complex issues that affect food security, sustainable energy, and water.

This book discusses how we can utilize these resources without depleting them or decreasing their quality. Most importantly, we need to know how to use these resources judiciously. For example, coal and fossil fuels have served our needs well; however, they can do significant harm to our environment including water and air. We need to minimize or eliminate that damage. Can natural gas help us achieve that goal, or are we likely to harm our water supplies? We need to provide more clean water and raise more food crops to feed the growing world population. This means we need to consider these sources in terms of links where excess zeal for one does not harm the other, i.e., they need to be in balance and harmony.

The book addresses these objectives and provides valuable information on these resources in terms of issues and solutions. Chapter 1 offers a broad overview of the subject. The need to regulate water quality is discussed at some length in Chapter 2. The interactions at water–air interface are discussed in Chapter 3. Two chapters (4 and 5) provide the means to purify water contaminated with pharmaceuticals or metals. Chapter 6 covers at some length how natural gas can fill the gap of our energy needs with minimum pollution. Three chapters (7–9) discuss the role of water in energy extraction and water management in oil and gas development. Chapter 10 alerts us to the fact that coal energy can cause water contamination. Clearly, we need to find ways to minimize these effects.

Chapters 11 and 12 deal with alternate sources of energy, e.g., plant-based biodiesel and also photovoltaic technology. These sources are less polluting; however, the cost of solar energy needs further improvement.

Chapter 13 deals with the effects of polluted irrigation water on food. A convenient way of monitoring food contaminants with a hand-held X-ray fluorescence analyzer is provided in Chapter 14. The scarcity of water and its impacts are covered in Chapters 15 and 16. The former makes us wonder if we are not already at the brink. Finally, Chapter 17 provides suggestions to address these issues by emphasizing the need to reuse water judiciously.

By now, it should be clear that we need to adopt a nexus approach to dealing with the relationships of food, energy, and water. None of these issues can be effectively resolved in isolation. Development of new technology in any of the sectors should consider the impact on the other sectors as well as on the environment. For example, the impact of a new crop that could be used as an energy crop or a food crop should be evaluated on the basis of water and energy used and the resulting waste products. In the field of energy, fracking (fracturing) delivers abundant levels of natural gas; however, the effect on groundwater has raised concerns. The process uses large amounts of water and generates large volumes of wastewater that cannot be reused for municipal or agriculture purposes. Governmental and international policies need to be developed to assure integrated implementation of technologies after thorough consideration of growing population stresses, the associated energy and food needs, and their impact on the environment, including the potential for global warming. We have to avoid silo approaches and consider the broad effects of new technologies and their impact on food, water, and energy. The nexus approach for providing more food, energy, and water should involve individuals, policy makers, and thought leaders.

I believe that all of the valuable contributions to this book will be found useful by scientists and engineers working in the area of food, energy, and water. Technicians and professionals, including academicians and regulators working in the area of environment will also find this book very useful.

August 1, 2014
Satinder Ahuja

1

Nexus of Food, Energy, and Water

Satinder Ahuja

AHUJA CONSULTING, CALABASH, NC, USA

CHAPTER OUTLINE

Food, Energy, and Water. http://dx.doi.org/10.1016/B978-0-12-800211-7.00001-6

Overview: Nexus of Food, Energy, and Water

If you were to ask a layman which is the most important resource among food, energy, and water, the answer is likely to be "food." As a matter of fact, humans can survive longer without food and energy than without water. However, humans depend on the Sun as the primary source of energy that assures our survival on this planet. Furthermore, the civilized world is highly dependent on energy for comfortable living and mobility. This suggests that energy is a very important requirement. The fact is we need all of them and their interrelationship is discussed below.[1]

Chemistry connects all of these resources. All of them are briefly discussed individually in the order of their relative importance to human beings (assuming the Sun will keep shining on us for a long time), and their relationship with the other resources is shown. We know that water is the most essential requirement for human survival, after air.[2–9] And we need food to assure our survival in the long run. To produce clean water and to raise food crops, we need energy. In turn, energy can be produced from food, e.g., corn is used for producing ethanol. And water is frequently used to produce energy. The nexus of food, energy, and water is depicted in Figure 1; at the nexus, all three are in harmony and our sustainability is assured. This book explores the role that food, energy, and water play in maintaining our lives and lifestyles; and it focuses on the environmental nexus of food, energy, and water. Studies related to this are covered at some length. The book discusses how these three fundamental resources interact with each other for our mutual benefits and how these interactions can be handled to benefit humanity. It explores these areas and discusses ways to improve our utilization of these resources, with minimum damage to our environment.

Water

Water availability and quality must be maintained to assure life on Earth. Sustainability demands that we use water judiciously and reclaim contaminated water, as only 0.06% of freshwater is readily available to us even though Earth is a water planet. A number of

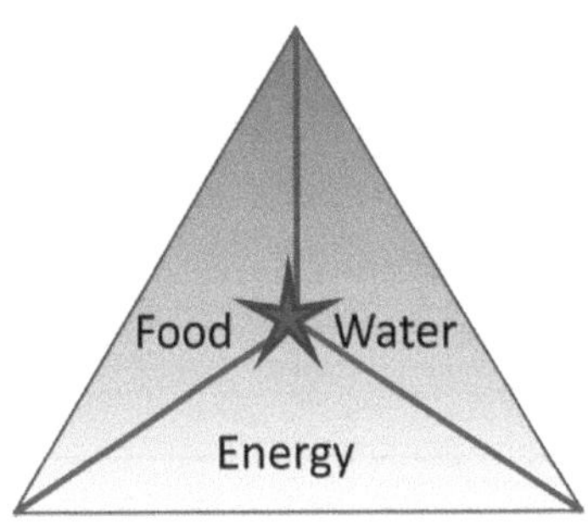

FIGURE 1 Nexus (*) of food, energy, and water.

inorganic and organic compounds, from arsenic to zinc, can pollute our groundwater.[2] For example, even remote regions like the Arctic Sea contain mercury. The ice holds a large amount of methyl mercury that may enter the ecosystem at increasing rates as ice melts because of climate change. Various issues relating to water are discussed below.

Water Availability and Scarcity

Water shortages are being experienced all over the world including the United States (see Water Scarcity: Are We at the Brink?). California is facing severe drought and various states have had disagreements relating to the supply of water. Internationally, water shortages in Africa, Asia, and Latin America are well known. In the Middle East, Iran is facing a water shortage potentially so serious that officials are making contingency plans for rationing in the greater Teheran area, home to 22 million, and also in other major cities around the country. According to figures compiled by the local environmental office, only 5% of the water remains in Lake Urmia. Other lakes and major rivers have also been drying up, leading to disputes over water rights. Water has been identified as a national security issue and we need cooperative efforts locally, nationally, and internationally to address water shortages.

COOPERATION IN THE MIDDLE EAST

In a rare display of regional cooperation, representatives of Israel, Jordan, and the Palestinian Authority signed an agreement to build a Red Sea and Dead Sea water project that is meant to benefit all three parties. The project addresses two problems: the acute shortage of clean freshwater in the region, especially in Jordan, and the rapid contraction of the Dead Sea. A new desalination plant is to be built in Aqaba, Jordan, to convert salt water from the Red Sea into freshwater for use in southern Israel and southern Jordan—each would get 8 billion–13 billion gallons a year. The process produces about the same amount of brine as a waste product; the brine would be piped more than 100 miles to help replenish the already very saline Dead Sea.

Water Reclamation

We need to maintain water quality by monitoring water resources for various known and unknown contaminants rigorously and regularly from point and nonpoint source pollution.[8] However, it is important to note that water reclamation is necessary because of limited water availability. Wastewater can originate from many places: households, industries, commercial developments, road runoff, etc. As diverse as the sources of wastewater are, so too are their potential constituents. The following facts explain how some emerging contaminants have an impact on our water quality: About 12,500 tons of antimicrobials and antibiotics are administered to healthy animals on US farms each year. A 2002 US Geological Survey (USGS) found pharmaceuticals (hormones and other drugs) in 80% of streams sampled in 30 states of the United States. The composition of wastewater affects not only the treatment processes applied but also their source recovery opportunities. Understanding wastewater constituents and their abundance at different stages is a first step in recognizing appropriate opportunities for pretreatment.

Wastewater contains pollutants/contaminants that have to be removed and/or reduced before the water is directed to a surface water source (river, ocean, bay, lake, etc.) or to groundwater (see reference 8 and volume 3 of reference 7). These books cover the entire gamut of natural and engineered processes that are employed to achieve the goal of making the treated wastewater amenable to disposal without any adverse effect on human health and the environment or reuse by humans for drinking.

WASTEWATER AND BIOFUEL PRODUCTION

Biofuels (also see Biodiesel from Plant Oils) can be produced from algae that grow best in polluted waters, such as that contaminated by sewage. The USGS points out that 195 billion gallons of water per day is used for thermoelectric power plant cooling, and about 96% of saline water withdrawals are for thermoelectric power use. Such water presents a disposal problem, which can be easily solved by using wastewater for biofuel production.

Water Quality

The amount of testing necessary for reclaimed water should relate to how it is going to be recycled. For example, if it is recycled into a surface water supply, its quality after purification should match or exceed the requirements of the surface water to which it is being added. Similar rules may be followed for mixing with groundwater. Recycled wastewater for drinking must meet potable water requirements, with the added assurance by ultratrace analysis that no toxic contaminants are present.[9]

Food

Water is necessary to raise various vegetables, fruits, cereals, animals, and their products that we use for food and to make beverages (see Table 1).

Water consumption that is related to water-guzzling crops such as alfalfa deserves special attention, as the water embedded in the alfalfa that the US exported to China in 2012, could supply the annual water needs of roughly 500,000 families in the US. It now costs twice as much (about $45 per ton) to truck alfalfa from a Southern California farm to a dairy in California's Central Valley as it does to ship it from Long Beach, California, to Beijing. Chinese demand has prompted alfalfa prices to double in the past 2 years. Southern California's Imperial Irrigation District gets its water from the Colorado River, 82 miles to the east. Alfalfa farmers in the district use as much as 50% more water than growers in other areas of the state, because of scorching heat, salty soil, and, perhaps most important, their legal rights to an enormous quantity of cheap water. This single irrigation district controls more than 20% of the total annual flow of the Colorado River.

Table 1 Water Usage for Raising Various Vegetables, Fruits, Cereals, Meats, and Beverages[a]

Vegetables	Gallons	Fruits	Gallons
Lettuce	15	Oranges	55
Tomatoes	22	Apples	83
Cabbage	24	Bananas	102
Cucumber	28	Peaches	142
Potatoes	30	Nectarines	142
Corn	107	Mango	190
Olives	22	Avocado	220
Cereals	**Meter³/Ton**	**Animal Products**	**Gallons**
Rice	1903	Eggs	573
Wheat	3419	Chicken	815
Corn	489	Beef	2500–5000
		Pork	1630
		Butter	2044
		Cheese	896
Beverages	**Gallons**		
Tea (8 oz)	7		
Beer, barley (8 oz)	36		
Coffee (8 oz)	29		
Wine (8 oz)	58		

[a]Source (www.waterfootprint.org). These figures represent gallons of water consumed per pound of food (with the exceptions as noted above).

Impact of Contaminated Water on Food

ARSENIC

High concentrations of arsenic (exceeding 1000 $\mu g\ L^{-1}$) were reported in shallow tube wells (STWs) from 17 districts in Bangladesh.[4,10] High levels of arsenic in groundwater occur in many districts. Of the total area of 4 million hectare (ha) under irrigation in Bangladesh, 2.4 million ha is covered via STWs, and 0.6 million ha is covered by 23,000 deep tube wells. In the dry season, 3.5 million ha is used for boro rice, 0.23 million ha for wheat, and 0.27 million ha for other crops. Up to 300% increased accumulation of arsenic has been reported in vegetables grown with arsenic-contaminated water over those grown with uncontaminated water. Total arsenic values found were up to 1.59 $mg\ kg^{-1}$ (dw) in fruits and 0.79 $mg\ kg^{-1}$ (dw) in leafy vegetables. Green leafy vegetables act as arsenic accumulators, with *arum* (kochu), gourd leaf, *Amaranthus*, and *Ipomea* (kalmi) topping the list. Speciation of Bangladeshi rice shows the presence of As(III), dimethyl arsenic acid, and As(V); greater than 80% is in the inorganic form. More than 85% of the arsenic in rice is bioavailable. Several studies[4] have reported a highly increased accumulation of arsenic in vegetables grown with arsenic-contaminated water over that grown with uncontaminated water. It is important to note that arsenic at lower levels is found in all soils, including American farm fields. The fertile soils fanning out across the Mississippi River floodplain are up to five times as high in arsenic as in other parts of Louisiana, Mississippi, and Arkansas, according to studies done by the USGS.

OTHER FOOD CONTAMINANTS

Cadmium rice is a well-known term in China because in 2013, Guangdong Province government reported that 44% of rice samples had excessive levels of cadmium. Other metal-contaminated rices are "arsenic rice," "mercury rice," and "lead rice." It appears that the main sources of cadmium pollution are emissions from smelting plants. For five metals (arsenic, cadmium, lead, manganese, and zinc), strong correlations of concentrations in uncultivated soils indicate a common source, suggesting that emissions from these plants may be a major contributor to elevated concentrations of these five metals in uncultivated soils in this area. The fields are ringed by factories and irrigated with water tainted by industrial waste.

A recent study by USGS found that some of the synthetic compounds found in a wheat field sprayed with biosolids are bisphenol A, hexahydrohexamethyl-cyclopent-2-benzopyran, nonlylphenol ethoxylates, triclosan, and warfarin.[11]

The US produces 251 million tons of municipal waste. Of that, 14.5% is food waste and only 4.8% is recycled.[12] Food waste can be effectively used for resource recovery.[8]

Food and Biofuels

There are two types of liquid biofuels: biodiesel and ethanol (see Coal Use as a Cause of Water Quality Impairment). Biodiesel is produced primarily from triacylglycerol, a lipid. Ethanol is made by using yeast to ferment sugar extracted from sugarcane and sugar beets, or from the starch of grains and cassava. The current production of ethanol requires much less land per unit of biofuel than does biodiesel. In the US in 2011, fermentation of corn-derived sugars yielded about 49 billion liters of ethanol, which was used mostly as a fuel additive. This accounted for 38% of the country's 32.2 million ha used to grow corn for grain. Ethanol makers depend on glucoamylase enzymes to break down starch from corn and other crops, into sugar for fermenting into ethanol. The use of corn for ethanol production is being debated on various grounds.

Producing fuels from a nonedible biomass such as corncobs and cornstalks promises to be a large opportunity for enzyme manufacturers.[13] Scientists are developing enzymes that can degrade cellulose into sugars to produce fuel. One way to increase biofuel production is to gain benefits from the leftover plant materials that are currently being discarded. In addition to taking advantage of these existing sources of biomass, a shift toward cultivation of plants that naturally produce more biomass per ha, using less fertilizer is required.

Some countries are using biomass to provide high levels of fuel. In 2009, 4.6 million ha of Brazilian sugarcane were used to produce about 27 billion liters of ethanol, plus 2 GW of electricity from bagasse, material left over from sugar processing. The adoption of flex-fuel vehicles in Brazil allows the use of a wide range of mixtures of ethanol and gasoline, depending on the relative prices. In fact, pure gasoline is not widely available to consumers in Brazil, but is sold in mixtures containing a minimum of 25% ethanol. In 2008, ethanol replaced about 40% of the volume of gasoline used in Brazil.

Energy

The sources of energy production and their percentages in the US (*WSJ* 4/9/14) are as follows:

Oil 36%
Natural gas 26%
Coal 20%
Hydroelectric 13%
Nuclear 8%
Renewable 6%

In the future, coal usage will drop significantly, while natural gas usage will increase significantly. Here is an interesting comparison of energy sources of other countries: 69% of China's energy comes from coal; 56% of Russia's energy is from natural gas sources; 35% of Brazil's energy is derived from hydroelectric; and 23% of India's energy comes from biomass and waste. Ethanol is by far the most widely produced biofuel in the US, with approximately 13.9 billion gallons produced in 2011. It is present in 90% of the gasoline sold in the US.

The Hoover Dam is an excellent example of hydroelectric energy that has transformed life in nearby states. The dam was built in 1931–35 to control floods, provide irrigation water, and produce electricity. The entire flow of the Colorado River passes through the turbines. The water head is 590 ft high, and the water reaches a speed of about 85 mph. It generates 4.2 billion kWh of electricity, providing power to Nevada, Arizona, and California. Many hydroelectric plants are being built in the world today. For example, China began full operation of Three Gorges Dam recently.

Global hydrocarbon consumption is now about 218 million barrels of oil-equivalent energy per day. It has been suggested that we will have to cover 108,000 square miles with wind turbines every year to keep up with our demand for electricity. It is interesting to note that all energy stored in Earth's reserves of coal, oil, and natural gas can be matched by energy from just 20 days of sunshine.[14] If we were to cover only 4% of the world's desert areas with photovoltaic (PV) cells, we could supply all of the world's electricity. The problem is that the amount of sunlight on Earth is not constant, and solar plants would require transmission lines to deliver power from remote areas. It is encouraging to note that a new solar power system is installed on an American roof every 3 or 4 min.

Energy uses water or water uses energy in two predominant areas:

1. producing steam for energy production
2. pumping water to where it is needed

However, a third important area is now evolving, where significant quantities of energy are required to produce drinking water, because of the water shortages in certain areas of the world. Production of energy is by far the largest water use. Approximately 60% of water withdrawals relate to thermal (steam turbine) power generation. To produce 1 MW of power (enough energy for 1000 homes) requires

- 28–75 M^3 of water for a gas/steam-combined cycle power generation
- 80–200 M^3 of water for a coal- or oil-fired generating station
- 100–240 M^3 for a nuclear energy-generating facility

Based on an average use of 380 L per capita per day of water, the estimated total energy per capita usage in the US is 140 kWh per person per year. This results in an annual expenditure of $5 billion–$8 billion.

It might come as a surprise to some that we need a significant amount of water to travel by car; for example, water requirements for traveling 150 km are[1]:

- 20–60 L for a gasoline-driven vehicle
- 100 L for a hybrid vehicle
- 160 L for a hydrogen fuel cell vehicle
- 500–24,000 L for an ethanol-powered (made from corn) vehicle

It is worth noting that our dependence on fossil fuels is on a collision course with the need of future generations for a habitable environment. With more than 80% of human energy consumption globally, fossil fuel burning adds to the rise of atmospheric greenhouse gases (GHGs) such as CO_2, nitrous oxide, and methane. Various scientific reports say the effects of human-induced climate change are being felt in every corner of the United States, with water growing scarcer in dry regions, torrential rains increasing in wet regions, heat waves becoming more likely and more severe, wildfires growing worse, and forests dying under assault from heat-loving insects. Such sweeping changes have been caused by an average warming of less than 2 °F over most land areas of the country in the past century. If GHGs like carbon dioxide and methane continue to escalate at a rapid pace, the warming could conceivably exceed 10° by the end of this century. Climate change, once considered an issue for a distant future, has moved firmly into the present.

Renewable Sources of Energy

We need to use energy sources that have minimal impact on our environment. Some of the renewable sources of energy are tidal and wave, solar, wind, and biofuels. Wind has enjoyed some success in the areas that are suitable for its use. The United States is the top producer in wind power today. China and Germany are second and third respectively. However, Germany is the largest producer of solar energy. A lot of promise is seen in biofuels as they can be produced by minimal use of corn. However, it appears that the cellulosic biofuels industry will arrive at its own launch party several years late. Optimistic forecasts codified in 2007 called for the US to produce 100 million gallons of the renewable fuels in 2010. It was not until 2012 that even 25,000 gallons were produced. The first group of commercial scale facilities was completed in 2013. Most analysts believe that start-up facilities will validate the technologies, confirm financial returns, and draw additional investment.

Biofuel producers say that they can compete as long as the price of oil remains above $70 a barrel and gasoline stations offer higher blends of ethanol. The South Dakota-based ethanol producer Poet, in partnership with the Dutch company Royal DSM with its start-up Emmetsburg, Iowa, plant, will produce up to 25 million gallons a year, making use of corncobs, husks, and leaves. DuPont's biggest plant will make ethanol from corn waste and will have a capacity of 30 million gallons a year.

Abengoa, S. A. is planning to produce 25 million gallons of biofuel a year and has already started a 21-MW electricity plant at the site powered by biomass. It has developed a proprietary enzyme to mix with cornstalks and wheat straw to produce sugars that will then be fermented and distilled to produce cellulosic ethanol. The more efficient process can increase yields and decrease costs. Over the last 4 years, it has improved yields from 55 gallons of ethanol per ton of biomass to 80 gallons per ton. What is more, because cellulosic ethanol relies on the waste products of corn rather than corn itself, it does not raise demand for corn or raise corn prices. Thanks to advances in molecular genetics and DNA sequencing technology, the San Diego start-up has, in a few years, succeeded in domesticating jatropha, a process that once took decades. SG Biofuels is growing hybrid strains of a plant that produces biofuel in quantities that are competitive with petroleum priced at or above $99 a barrel. The company's jatropha plantations are in Central America. It is one of the few biofuels that has the potential to supply a large fraction of the aviation fuel currently used today.

The reader may want to refer to an interesting book on future sources of energy.[15] Various perspectives on the nexus of food, energy, and water are further addressed in a number of chapters in this book.

Foundations of Water Quality Monitoring and Assessment in the United States

Reliable information from water quality monitoring is needed for all important beneficial water uses (Chapter 2). This was possible because major technical and scientific achievements were made in methods and instruments for analysis and interpretation of data on organic and inorganic chemistry, isotope chemistry, aquatic biology, toxicology, hydrology, and hydrogeology. These advances produced new knowledge about the sources, transport, and fate of human-derived and naturally occurring contaminants. The scientific evidence from water quality monitoring and assessment led to the development of drinking water standards to protect human health, water quality criteria to protect aquatic life, and laws to manage solid and liquid waste to protect groundwater and surface water quality.

Mass Transport and Chemistry at the Air–Water Interface of Atmospheric Dispersoids

Atmospheric aerosols contain a significant quantity of water in the form of both bulk water and thin water films (Chapter 3). These aerosols provide very high surface areas and adsorptive surfaces for organic compounds of low solubility and low vapor pressures. They also provide highly active sites for oxidation reactions with hydroxyl, singlet oxygen, and ozone species in the gaseous atmosphere. Thus, organic compounds are transformed readily by the reaction at the air/water interface of fog/cloud droplets and in the thin water film in atmospheric aerosols. Mathematical models suggest that the reactions in fog and cloud droplets are not limited by gas phase diffusion or mass accommodation at the surface; they primarily depend on the partition constant and reaction rates at the surface. Laboratory data on both of the latter parameters are available in the literature for a variety of organic compounds and they are also supported by field data. Considerable data already exist on the aqueous processing of organic species that are highly water-soluble in atmospheric aerosols. The studies reported in this chapter suggest that aqueous processing of organic chemicals (possessing low solubility and low vapor pressure) at the interface should also be considered in fate models for secondary organic aerosols in the atmosphere.

Complete β-Lactam Antibiotic Activity Removal from Wastewaters

Public concern about the significant concentrations of antibiotics and their metabolites remaining in wastewater after standard treatments is significant because it can affect drinking water or crops if it is reused. This is motivating the study of approaches such as the use of advanced oxidation processes (Chapter 4). However, to ensure complete removal of antibiotics using these processes at large-scale, full-degradation mechanisms, kinetics, and reaction efficiencies must be accurately determined. A study was conducted to quantitatively determine the efficiency of only hydroxyl radical reactions with multiple β-lactam antibiotics, with these radicals generated by radiolysis and activity measured using *Staphylococcus aureus* bacterial growth. It was found that four to five hydroxyl radical oxidations per antibiotic molecule were necessary to completely remove the antibiotic activity in high-quality water solutions. The effects of the real-world water matrix components dissolved organic matter and alkalinity were also separately investigated. The multiple oxidations required to completely remove β-lactam antibiotics and metabolite activity has profound implications for advanced oxidation process treatments of antibiotic-contaminated waters.

Treatment of Mine-Impacted Water with Biochemical Reactors

Chapter 5 reviews the current understanding of the function and design of biochemical reactors (BCRs) for the treatment of water impacted by mining. The term BCR was adopted to describe a treatment unit that uses the biological reduction of sulfate as a basis to treat mine-impacted water. This definition may include systems also termed wetlands, sulfate-reducing bioreactors, or in some cases, permeable reactive barriers designed for the purpose of sulfate reduction to capture metals. The terminology overlaps but does not coincide with deep, organic-rich wetlands in which an extensive anaerobic zone can reduce sulfate. A BCR is most often a basin filled with a permeable substrate including a solid phase organic in which mine-impacted water is treated as a result of sulfate reduction. Water is distributed by pipes into a gravel layer so that flow will be even, then it flows through the solid substrate, and is collected in a manner similar to the distribution. BCRs reduce sulfate to treat mine-impacted water. Sulfate in the metal-containing water is transformed to sulfide, which precipitates many metals in an insoluble form. Adsorption and coprecipitation are other possible mechanisms of metals removal in BCRs. The key design parameter is the rate of sulfate reduction that is determined by the rate at which the organic, usually a woody material, is consumed. A wide range of sulfate reduction rates have been reported, presumably because of differences in degradability of the organic substrate.

Water–Energy–Food: Our Existence Will Require Natural Gas

Chapter 6 provides a compelling argument that natural gas is a key and predominant source of energy for the sustainable future, which could last for the next 400 years. Water and energy have become two limited resources, which are not only affecting our quality of life but are the largest barriers to economic growth and the future of our communities. Today's energy generation infrastructure has caused severe water limitations. Simultaneously, the water used for energy generation has caused severe energy consumption. Today's energy and water infrastructure is a dichotomy of counter interdependence that is doomed for failure. Our focus on energy conservation has been misaligned. However, the good news is that there is a readily available solution. The use of community-based water and energy generation facilities will have significant impact on the sustainability of our communities. In fact the development of such an infrastructure to replace traditional centralized facilities is essential. Success for community-based generation facilities will be dependent on the generation, management, and analysis of reliable data. The technology evolution in smart data systems is an important success story. The ability to manage microcommunity systems is now well underway with the continuous improvement in identifying inefficiencies and reducing operational risk. There are numerous opportunities to use water and energy resources more efficiently through the use of new design methods, materials, and monitoring and control systems. Through the use of simple, already commercialized, community-based energy and water generation technology and the implementation of data utilities, energy consumption can be reduced as much as 80%. In addition the use of community energy management systems, which do not depend on cooling waste heat with freshwater, will reduce the depletion of freshwater inventories in a quantity that would be sufficient volume to satisfy the potable water demands of North America. Finally, the concept of community energy generation will allow an easy reduction in CO_2 emission to satisfy North American 2017 CO_2 targets.

The world's ability to have communication connection between the smallest of instruments to unlimited networking will be changing the installation and operation of all of our community utilities and services. The chapter addresses this game-changing environment. Electrical distribution systems have initiated the change through the use of smart grid-type concepts. The generation of data will become as important as the need for energy, water, and food. Without the data, energy and water services will become unmanageable. The data generation will lead to a plethora of applications for the data that will change the way we analyze and manage operations of all utilities. Data analytics will improve dramatically resulting in better definition of data generation needs and technology. Our energy and water environments will become microenvironments and natural gas as well as data analytics will ensure the sustainability of these essential microutility environments.

The Role of Water in Unconventional In Situ Energy Resource Extraction

Global trends toward developing new energy resources from lower grade, larger tonnage resources not generally accessible using "conventional" extraction methods involve variations of subsurface in situ extraction techniques that could affect groundwater and surface water (Chapter 7). Such methods include hydraulic fracturing for shale gas production, in situ retorting for synthetic oil production, and in situ leach mining for uranium recovery. Although these methods are economically feasible and perhaps result in a smaller aboveground land use footprint, uncertainties remain regarding potential environmental subsurface impacts. Specifically, the role of water in unconventional in situ resource extraction examined includes[1] releases of chemical additives, mining fluids, trace metals, organics, radionuclides, and produced waters outside of the subsurface extraction zones[2]; perturbations to subsurface groundwater equilibrium such that restabilizing the geochemistry following energy resource production is difficult[3]; large volumes of wastewaters disposed of in deep aquifers; and[4] uncertainties in groundwater flow paths and connections to drinking water sources.

Water Management in Unconventional Oil and Gas Development

The US natural gas industry, and by extension the industry globally, has witnessed tremendous change over the past decade (Chapter 8). During this period, the US natural gas production levels have risen from a 20-year low of 18 trillion cubic feet (Tcf) in 2005, to an all-time high of 24 Tcf in 2012. At the same time, natural gas prices have fallen to levels not seen since the period immediately following the US gas market deregulation in the mid-1990s. The underlying driver of these dynamics has been the very rapid growth in the production of unconventional natural gas resources, and in particular shale gas resources that were considered unrecoverable. Technical advances in the areas of drilling and reservoir stimulation have been the key to unlocking shale gas. Today's shale gas is largely produced from wells drilled with horizontal bores that have been subjected to large-scale hydraulic fracture stimulation. The combined efficacy of these technologies in enabling gas production from shale formations (and indeed other low-permeability reservoir settings) is such that their development has led to enormous upward revisions in the assessed scale of the total recoverable natural gas resource in the United States. Today, analysis by organizations including the US Energy Information Administration and the Potential Gas Committee suggest that the likely recoverable US shale gas resource is in the 800–1000 Tcf range. This contrasts with the National Petroleum Council 2003 assessment that estimated the shale resource at 35 Tcf. The prospect of a much larger and indeed lower-cost domestic natural gas resource in the US is having major impacts on both the US and international energy sectors. In the US, unsurprisingly, many are now projecting a more gas-centric future than was envisioned even a few years ago. For example, in the power generation sector, gas-fired generation is now expected to make up a much higher proportion of total US output over the coming 20–30 years than was anticipated before the full extent of the so-called shale gas revolution became apparent.

The widespread use of hydraulic fracturing technology to extract hydrocarbons from shale formations has resulted in a number of potentially harmful environmental and public health consequences. One of the key concerns is the procurement of water for fracturing a well. Wells are often not located adjacent to water resources, thus the process water is either piped or trucked to the well site. The increasing cost of transportation of water to well site has made the local water withdrawals as the first choice for natural gas producers to meet their water demand. Although water withdrawals in major shale formations represent a relatively minor volume of the total water used in the specific area, the problem is that large volumes of water are procured over a relatively short period of time, which could compete with the availability of water for other local users and create a transient stress on the water source. Furthermore, rapid and concentrated procurement of water could result in regional shortages and altered flow regime, impacting the habitat of aquatic biota.

As mentioned above, recent developments in hydraulic fracturing have enabled a dramatic increase in the production of unconventional oil and gas resources. Contemporary hydraulic fracturing treatments are water-intensive and require several million gallons of water per well. Environmental concerns associated with shale gas development, particularly related to flowback water quality and quantities have drawn significant public and regulatory attention. Because of the large flowback water volumes and high concentrations of solids and other pollutants, there are significant concerns about the management of this water and its environmental impacts, with the potential release of inadequately treated or untreated wastewater to the environment. Furthermore, the variability of the chemical profile of flowback and its dependence on the geology, fracture fluid composition, and sampling time exacerbates the problem of managing it. With technological advances in multipad drillings and refracturing of wells, these environmental concerns will increase unless an effective flowback management framework is developed, which encompasses the temporal and spatial dynamics in fluid composition and quantity, economics of technology, the availability of physical infrastructure, and impact of water withdrawal rates on the sustainability of local surface water bodies. Previous attempts to integrate the various dimensions in flowback management provide a consistent approach to decision-making, but complete integration of the diverse interests of multiple stakeholders is challenging. In this chapter, an alternative approach based on system dynamics is presented. Using this approach, flowback management strategy can be tailored to the specific spatial and temporal data. This type of holistic approach is essential for the long-term responsible shale gas development and for secure availability of water and energy at large.

Water Management in the Oil Sands

The global demand for oil is high and, with the projected growth in world population, this demand is expected to grow (Chapter 9). It is important to understand that the production of oil comes at the cost of energy and water. Energy is required to produce oil including the treating and transporting of water; while water is required to extract the oil from the reservoir. Increased production rates continue to drive the consumption of water and energy. Moreover, higher energy consumption yields an increase in GHG emissions released into the atmosphere. It is believed that GHGs contribute to a climate change that negatively impacts the supply of freshwater. Actions continue to be taken in the form of regulations and policies to promote water conservation and mitigate the release of GHGs. There has been a surge in unconventional production over the last decade, which signals a shift in oil production from accessible reserves to those that are less accessible. One such example is the Alberta oil sands, which is a home to the third largest proven oil reserve in the world. The importance and relationship of energy and water is perhaps most apparent in this region. Oil sands in situ methods use 0.5 barrels of water per barrel of bitumen, while conventional oil uses 0.1–0.3 barrels of water per barrel of oil produced. With the current technologies, large amounts of water are required to extract oil from Alberta's hydrocarbon reserves, which creates a significant demand for water in the region and sets up a natural competition between advancing the region's economy and protecting and conserving its limited source of freshwater. Statistics show that over the last three decades, stream flows in the Athabasca River, which is the main supply of freshwater in the region, have declined by 30%. It is possible that the future of the oil sands may ultimately be limited by water supply in the region. The projected growth of the industry coupled with climate changes could potentially reduce the river's flow significantly. Eventually, the rate of growth of the oil sands will be dependent on the availability of water to meet future production demands while preserving the environment.

Coal Use as a Cause of Water Quality Impairment

The use of coal as a major industrial fuel has prompted major public and policy concerns about air quality, including GHG emissions and the storage of large amounts of coal ash in situations where ash spills and groundwater contamination by ash leachate have become problematic (Chapter 10). Coal use has historically had significant impacts on water quality through several pathways, including direct impact of mining activities on surface and groundwater, air pollution effects on surface waters (notably 'acid rain' and mercury emissions), as well as other airborne toxicants and global warming impacts on the hydrological cycle, and through accumulation of truly massive quantities of coal ash at numerous waste storage sites. Coal ash contains many elements and some organic compounds, reflecting coal's biological and geological origins, and many of those materials can be leached from coal ash, leading to contamination of surface and groundwater. Toxic metals commonly found in coal ash and in its leachates include arsenic, cadmium, chromium, lead, and selenium, but many others may also be found, depending on the coal source. Alternative, benign uses of coal ash do not yet have the potential to consume the vast quantities of coal ash now extant, so costly mitigation practices are likely to be required to contain this environmental hazard.

Coal use has supported industrialization and economic growth for centuries, but has also generated a wide variety of environmental impacts. This chapter focuses on how coal use has an impact on aquatic ecosystems, whereas most considerations of the impacts of the use of coal have addressed air quality concerns. Recent attention to coal ash spills, however, has prompted concerns about water pollution effects from coal use, particularly the long-term accumulation and storage of truly immense quantities of solid wastes. Interaction of coal ash with rainfall and groundwater clearly poses threats to the integrity of aquatic ecosystems and, ultimately human health, if these effects are not adequately mitigated. Cleanup costs may ultimately represent one of the largest costs of coal use, an aspect of coal pricing not currently reflected in energy markets.

Biodiesel from Plant Oils

Increasing demand of conventional sources of fuel has made us consider alternate fuels that can satiate the present demand (Chapter 11). Researchers are focusing much more on renewable sources of energy including biofuels, which include ethanol and biodiesel that can be derived from biomass or waste feedstocks. Biofuels are obtained from living organisms, from metabolic by-products, i.e., organic or food waste products, and from plants whose seeds can be used to extract oil. The chemical process involved in the production of biofuel and economics involved in the process are discussed in this chapter. The main focus is on the identification of plants that can be used for the production of biofuels. The current status of biodiesel as a fuel is discussed in this chapter. Biodiesel is basically fuel that can be produced from seeds of edible and nonedible seeds of plants and hence can be classified as a renewable source of energy. The oil from these seeds is extracted by different methods, after which the oil undergoes a chemical process known as transesterification. This process leads to the formation of fatty acid methyl ester and various by-products. Methyl ester with lower viscosity and properties similar to conventional petrodiesel can be used as fuel in internal combustion engines without any modifications. A number of scientists have done experiments carrying out performance analysis of engines using biodiesel, a blend of biodiesel with petrodiesel, and found that engine performance parameters are in line with conventional engines using petrodiesel as fuel. This shows that in the future biodiesel can fulfill the demand of fuel in case of conventional fossil fuel scarcity. Also, the engine emissions can meet the various environmental norms set by different countries.

Solar PV Technology

Solar PV technology is one of the most matured and field-proven technology among different renewable energy technologies. Chapter 12 provides a comprehensive introduction to PV technology, including the necessary technical knowledge for understanding this technology. It encompasses classification of solar PV technology and the manufacturing processes involved in various PV technologies. Each section constitutes specifically an independent learning entity where the technology has been reviewed in detail. Traditional solar cell technology, such as single-crystal silicon technology and newer technologies, has been described. The chapter also covers the comparative assessment of different PV technologies in terms of their efficiency, advantages, and disadvantages.

Contaminated Irrigation Water and the Associated Human Health Risks

The current proportions of freshwater withdrawals that go to agriculture are 86.7%, 81.4%, 68%, 38.7%, and 32.4% for Asia, Africa, Latin America, North America, and Europe respectively (Chapter 13). The use of contaminated water, particularly wastewater, in agriculture is quite widespread and is positively correlated with poverty. The major drivers of the practice include natural scarcity amid increasing demand; contamination of traditional sources of irrigation water and lack of alternatives; affordability, availability, and nutritional value of wastewater; and market incentives. Unfortunately, water, soils, and crops are getting contaminated, and illnesses have been linked to the contaminants through diet, as well as occupational and recreational activities. The major polluters of water are the population-driven activities that include industry, urbanization, and agriculture, with the major contaminants being heavy metal(loid)s, micro- and nanomaterials, pharmaceuticals, dioxins, polychlorinated biphenyls, polycyclic aromatic hydrocarbons, phthalates, and pathogens. The farmers and consumers in most countries where untreated wastewater is routinely used in agriculture have little to no knowledge of the risks involved, which betrays the lack of risk analysis data on the issue. The lack of data complicates the enactment, implementation, and enforcement of regulations because regulating the unknown makes no sense, if at all possible. This could be the major reason some countries that purport to have regulations cannot enforce them. Robust policy that includes investment in water technology can only result from a government-mandated risk analysis that involves scientists, farmers, consumers, and the media. It is risk analysis that should inform public policy, and not the other way around. Continuing the practice is just like dumping hazardous waste at dumpsites, yet in reality, these are agricultural lands, and health impacts will be severe. Something drastic has to happen, and poverty and droughts cannot continue to be the scapegoats.

The benefits of wastewater use in agriculture are truly insignificant when compared to the risks involved. However, in the absence of complete and accessible data and sound national and international policy to the contrary, those involved will remain "benefiting" from the practice. Risks will be manifesting, of course, but the cause may not be fully understood early enough for corrective measures to be implemented with economic viability. The loading of contaminants in the soil over a long period will essentially turn agricultural lands into hazardous waste sites, which will need extensive remediation to make them usable for food production. Continuing the practice will not only affect the current generation but also future generations. It is hardly sustainable and should be stopped. Because food is an international commodity, food safety must be given the global attention that it deserves, but ultimately individual nations must do everything in their power to protect their citizens.

Analysis of Dietary Supplements with a Hand-held X-ray Fluorescence Analyzer

Nine dietary supplements purchased from various dollar stores and two standard reference materials (SRMs) obtained from the National Institute of Standards and Technology were analyzed using a hand-held X-ray fluorescence (XRF) analyzer

(Chapter 14). This method of analysis produced rapid results while still keeping the costs low. The results from the analysis of the SRMs with the hand-held XRF were within 23% of the reported values for potassium, calcium, and zinc; 11% for manganese and lead; and 6% for titanium and iron. The XRF-determined values for titanium, manganese, iron, zinc, and lead were not statistically significantly different from those reported for the SRMs. Each dietary supplement contained less than the reported amounts of the expected element and several of the supplements contained additional elements.

Water Scarcity: Are We at the Brink?

In many parts of the world, water is a luxury and is extremely difficult to obtain. This has profound impacts on humanity (Chapter 15). More than 3,600,000 people die each year from water-related disease and 4000 children die daily because water is polluted or not existent. In addition, 884 million people on the planet—one in eight—live without safe drinking water and basic sanitation services.

Regrettably, the mainstream of the American public, as well as most members of the US Congress do not seem to comprehend the reality of the global water crisis. Public and private utilities provide inexpensive, free-flowing, high-quality, clear, clean drinking water to the US consumer, and it is available whenever desired. As a consequence, water is taken for granted by US citizens, and government leadership is slow in confronting the evolving crisis by not implementing critical international and US public policy and strategies to confront the issue globally and domestically. The planet is in deep trouble and heading for a calamity unless decision-makers begin implementing constructive solutions and recognize the physical evidence. They seem unwilling to accept or even recognize the actuality of significant global population growth, which is projected to exceed 9 billion by 2050, as well as the intense drought conditions and severe weather events that are occurring around the world.

There is energy scarcity, ongoing potential for disease pandemics and epidemics around the world, and there is a profound absence of sanitation, a growing lack of agriculture land and food supply, severe poverty conditions, and degradation of the environment. It is clear that the United States is not exempt from the crisis. Its infrastructure is in decay, and drought and extreme weather events are threatening the water supply, the US economy and national security. The recent West Virginia chemical spill, which activated a tap water ban for an estimated 300,000 citizens as well as the potential for a public health calamity, should more than demonstrate how vulnerable the United States is to an environmental disaster. A list of water strategy principals and a holistic environmental planning process, which will be a key in implementing vital solutions for 2050, are included in this chapter.

Water Scarcity: Global Challenges for Agriculture

Exponential population growth in a world with uneven freshwater distribution will sharply increase water demands (Chapter 16). Rising global temperatures exacerbate violent storms and prolonged drought. Salinization occurs at coastal deltas from rising sea levels and in groundwater from overpumping. The World Bank estimates that 25–30% freshwater is wasted, costing the global economy $14 billion annually. This chapter highlights recent technologies developed to augment finite freshwater systems, and presents potential remediation to relieve agricultural stress. Improving water management in agriculture rests on implementing sustainable irrigation techniques, developing crop modifications that help tolerate water stress, and promoting cooperation among stakeholders. Innovative techniques permit desalination that is less energy-intensive and preserves membranes. Recycling highly purified sewage water is a powerful tool for replenishing groundwater. Although growing crops for biofuels is incompatible with global demands for freshwater and food, recent studies on cellulosic biomass and microalgae cultivation hold promise as economically viable energy sources.

The Need for Water Reuse

By the year 2050, it is anticipated that Earth's population will exceed 9 billion. We need to take steps now to assure we will be able to meet human needs for food, water, and energy, while minimally impacting the environment (Chapter 17). This can be accomplished only if we apply improved technology to deliver secure and safe sources of water for production of food and energy. There is no more land that can be devoted to agriculture. To meet our food needs in the future, we must improve our agricultural production, which traditionally utilizes the largest portion of available freshwater. Energy production also requires considerable amounts of water. The supply of freshwater for food and energy production will increasingly be the limiting factor in our ability to meet human needs. We must establish technology to recover or remediate contaminated water for reuse in energy and food production, and we must become more efficient in its use. All solutions must also deal with the overlay of global climate change.

Meeting our needs for food, water, and energy is extremely complex; it must deliver sufficient and cost-effective energy to meet the needs of all humans, including transportation, water recovery and transport, and food production. Understanding and meeting the three way interactions are critical to our ability to establish a sustainable global food, energy, and water system to meet our needs. Currently, we are contaminating water at an unsustainable rate and we must learn to remediate and reuse water more efficiently. In addition, we must significantly reduce the amount of water required for the production of agricultural commodities while maintaining its availability for ecosystem services.

Conclusions

The chemistry connection between food, energy, and water allows us to deal with this multidimensional problem that links resources, national policies, politics, and quality of life for all of us on the planet now, and for the 9 billion people that will inhabit it in the next 40 years. To address issues relating to food, water, and energy we must look at their nexus. This chapter discusses issues relating to water, food, and energy in terms of their availability, the impact on the environment, and their interdependence. It is important to recognize that our efforts to pursue more energy, food, or water should not degrade any of these valuable resources.

References

1. Ahuja S. Nexus of water, food, and energy. In: *American chemical society meeting*; 2013. New Orleans.
2. Ahuja S. Assuring water purity by monitoring water contaminants from arsenic to zinc. In: *American chemical society meeting*; 2006. Atlanta.
3. Ahuja S. *Handbook of water purity and quality*. Amsterdam: Elsevier; 2009.
4. Ahuja S. *Arsenic contamination of groundwater; mechanism, analysis, and remediation*. New York: Wiley; 2008.
5. Ahuja S. *Monitoring water quality: pollution assessment, analysis and remediation*. Amsterdam: Elsevier; 2013.
6. Ahuja S, Hristovski K. *Novel solutions to water pollution*. Washington DC: American Chemical Society; 2013.
7. Ahuja S. *Comprehensive water quality and purification*, vols 1–4. Amsterdam: Elsevier; 2013.
8. Ahuja S. *Water reclamation and sustainability*. Amsterdam: Elsevier; 2014.
9. Ahuja S. *Ultratrace analysis of pharmaceuticals and other compounds of interest*. New York: Wiley; 1986.
10. Ahuja S. Remediation of arsenic contamination of groundwater in Asia and USA. In: *American chemical society meeting*; 2013. New Orleans.
11. *J Am Water Resour Assoc* 2014;**50**:343. http://dx.doi.org/10.1111/jawr.12163.
12. *Wall Str J* June 6, 2014:A3.
13. *Chem Eng News* Oct. 8, 2012:22.
14. *Time Mag* June 16, 2014:42.
15. Letcher Trevor. *Future energy*. Amsterdam: Elsevier; 2014.

2

Foundations of Water Quality Monitoring and Assessment in the United States

Donna N. Myers

U.S. GEOLOGICAL SURVEY, RESTON, VA, USA

CHAPTER OUTLINE

Introduction

Water quality activities in the United States began around 1800. The period from 1800 to 1975 was characterized by accelerated growth in water use to support economic development and the quality of life. Water resources development and use created a need for water quality monitoring and assessment. Reliable water quality information was needed over time to support the orderly development and use of water for public, industrial, irrigation, and domestic supplies. Water quality information also was needed to support the necessary treatment of drinking water, sewage, and industrial wastewater so that receiving waters could be reused while important beneficial uses for drinking water, recreation, and aquatic life could be protected and sustained.

The goal of this chapter is to describe the important contributions of water quality monitoring and assessment to the development of sound national water policies to guide

Food, Energy, and Water. http://dx.doi.org/10.1016/B978-0-12-800211-7.00002-8
2015 Published by Elsevier Inc.

water resources development, use, and protection in the United States. The objective of this chapter is to answer the question: "What is the history of water quality monitoring and assessment and how has it changed over time?" The approach taken is to provide historical information and a timeline on water quality monitoring and assessment at multidecadal to century scales. The societal relevance of the science, technology, and practice of water quality monitoring and assessment are described in an historical context.

For purposes of this chapter, water pollution is defined as the presence of any foreign substance such as organic, inorganic, radiological, or biological substances in water that degrades its quality so as to constitute a hazard or to impair the usefulness of the water. Sewage means any water containing human waste or animal or vegetable matter in suspension or solution, and may include household wastes from residences and from commercial, institutional, or similar facilities. Industrial waste means any liquid, gaseous, or solid waste substance resulting from any process of industry, manufacture, trade, business, or from the development, processing, or recovery of any natural resource, together with any sewage that is present. Wastewater is defined as used water containing sewage and/or industrial waste.

Prelude to Organized Water Quality Monitoring in the United States

Before the advent of organized water quality monitoring in the United States, methods of water analysis were being developed by chemists in England, France, Germany, and other nations. The first water analysis methods described the mineral content of springwaters for their purported health benefits.[1] Friedrich Hoffmann (1660–1742) invented gravimetric analysis.[2] Hoffman applied gravimetric analysis to describe the quality of mineral and springwaters. Richard Kirwan (1733–1812) published *Essays on the Analysis of Mineral Waters* as a summary of all known methods of mineral water analysis as of 1799.[3,4] Progress in water chemistry in Europe became important to water analysts in the United States by 1800.

All known chemical analyses of water in the United States before 1817 were mineral analyses of popular springwaters made by a handful of medical doctors to promote the purported health benefits of the springs.[5] Test results were interpreted by each analyst as hypothetical combinations to describe common minerals such as epsom salt (calcium sulfate), Glauber's salt (sodium sulfate), and muriated lime (calcium chloride). The limit of detection of most water analysis methods was approximately 1 part per 1000.[4] Analyses of the same springwater by different chemists often resulted in disparate results. A statement frequently made at that time was that mineral water analysis was "the most difficult operation in chemistry."[1]

Some of the first mineral water analyses in the United States by a professional scientist were made by Prof Benjamin Silliman, Jr (1816–1885) of Yale College in New

Haven, Conn. Silliman reported analyses of the observed physical characteristics, mineral content, and organic matter of the Croton River in New York, Schuylkill River in Pennsylvania, and from streams and ponds near Boston that were being used as water supplies.[6] The results described taste, color, odor, specific gravity, effects noticed on boiling, total nonvolatile and volatile matter, and mineral quality.[6] Silliman concluded that "all natural waters may be properly called mineral waters" and that "they each possess a specific and peculiar character dependent upon the nature and amount of solid matters which they contain."[6] Silliman further concluded that "the mineral characteristics depended on the geologic structure of the country where the waters are found." These statements represent some of the first geochemical interpretations of water chemistry data in the United States.[6]

By 1850, over 90 chemical elements had been discovered including all those of immediate interest to water chemists.[2] In 1854, Carl Remigius Fresenius (1818–1897) at the Wiesbaden Agricultural Institution in Germany published a manual on analytical chemistry titled *A System of Instruction in Quantitative Analysis.*[7] The manual was translated into English for use in the United States and Canada by Samuel W. Johnson (1830–1909) of Yale College in New Haven, Conn. and became an important source of methods of mineral analysis of water. Gravimetric analysis methods provided for determinations in freshwater of carbonic acid, chlorine, nitric acid, sulfuric acid, silicic acid, lime, soda, magnesia, total residue, organic matter, specific gravity, and water hardness.[7] Analyses were reported as hypothetical combinations of minerals according to a set of rules for combining acids and bases.[7] After 1865, analysis by titrimetry, or volumetric analysis, joined gravimetric analysis as one of the most widely used approaches for water analysis.[2,4]

A second type of water quality activity associated with water pollution arose due to serious outbreaks of waterborne diseases. Cholera epidemics were recorded in the United States from 1832 to 1834.[8] Other major cholera outbreaks occurred from 1848 to 1849 and from 1865 to 1873 in cities along major rivers such as Baltimore, Maryland; Columbus and Cincinnati, Ohio; Memphis, Tennessee; New York City; Philadelphia, Pennsylvania; and Washington D.C.[8] Typhoid fever, another waterborne and foodborne disease, was responsible for more total deaths each year in the United States than cholera because typhoid fever was more widespread, occurred more frequently, and was more persistent. An average of 30,000 deaths per year was attributed to typhoid fever in 1900.[9,10]

Diseases like cholera and typhoid fever were attributed to "bad drainage, bad sewerage, bad water supply, and poor removal of refuse."[11,12] John Snow's study of the cholera outbreak from 1853 to 1854 in London implicated drinking water contaminated with the excreta of cholera victims as the source of the cholera illness. Snow did not identify the specific agent responsible for cholera but thought that it might be living organized matter.[1,12] Results for mineral content and organic matter in water, the types of water analyses available at the time, proved to be unreliable for indicating the immediate health dangers associated with drinking polluted water.[1] Water chemists actively searched for reliable water analysis methods to answer the question: "Is the water safe to drink?" Chemical analysis methods were developed after 1865 for free ammonia, albuminoid (organic) ammonia, and organic matter to indicate the presence in water of organic wastes.[1,13,14] Organic and inorganic forms of nitrogen and organic matter were useful indicators of past sewage pollution but poor indicators of the immediate health dangers of drinking polluted water.

Arthur H. Hassall (1817–1894), an English physician and chemist who investigated cholera and its transmission, applied the technique of microscopy to examine drinking water sources for the cholera agent. Hassall discovered what he called "animalcules, vegetable productions, and earthy matter" in the Thames River and London's drinking water.[15] None of the living or dead material described by Hassall caused cholera but was a revelation to the public and chemists at the time. Findings from Snow and Hassall led to the relocation of London's most polluted water supply intakes to sections of the Thames River upstream from the worst sewage sources.[1]

The first water quality survey in the United States was undertaken by the State of Massachusetts because of concerns about cholera and typhoid fever. The State Board of Health of Massachusetts, established in 1869, was directed to document water pollution by an act of the state legislature.[16,17] In response to this charge, the seventh annual report of the State Board of Health of Massachusetts in 1876 contained information from the first sanitary survey of pollution sources and water quality by a state. The sanitary survey focused its investigations in the valleys of five rivers: the Blackstone, Charles, Chicopee, Neponset, and Taunton rivers. The Blackstone River was described as "probably the most polluted in the State" due to the presence in its waters of the refuse of 36 woolen mills, 23 cotton mills, 6 iron works, 1 tannery, and 1 slaughterhouse in addition to sewage from the City of Worcester.[16]

The germ theory of disease was proposed by the French chemist Louis Pasteur (1822–1895) in 1878.[12] This discovery provided the understanding to link microorganisms to human illness.[12] In 1883, Robert Koch (1843–1910), a bacteriologist, discovered that cholera was caused by a pathogenic bacterium that he named *Vibrio cholera*. It can inhabit the human gut and be transmitted in drinking water and food by the excreta of infected people through the oral/fecal route.[12] Koch and other microbiologists developed culture methods in the 1880s for live bacteria. While researching *V. cholera* in 1888, Theodor Escherich (1857–1911) described the colon bacterium now known as *Escherichia coli*, which he named *Bacillus coli communis (B. coli)*.[18] *Escherichia coli* is naturally occurring in high numbers in the colons of humans and other warm-blooded animals, is abundant in fresh sewage, and is not typically pathogenic. These characteristics held great potential to make *B. coli* a reliable indicator of drinking water quality if a test could be developed.

Progress in water quality monitoring and assessment through 1889 set the stage for an organized set of activities that began around 1890 in the United States. These activities are described in the following sections for two time periods: 1890–1929 and 1930–1975. The sections describe water quality technology and scientific advances, interpretation of water quality data and information, and water quality investigations for both time periods. Activities described were foundational to the development of water quality monitoring and assessment in the United States.

Water Quality Technology and Scientific Advances, 1890–1929

Water quality monitoring and assessment in the United States as of 1890 was still a relatively new area of investigations and research. A common source of water analysis methods before 1900 was a manual by J. Alfred Wanklyn (1834–1906) and Ernest T. Chapman (1845–1872) titled *Water Analysis: A Practical Treatise on the Examination of Potable Water*, which included methods for water analysis of albuminoid ammonia, free ammonia, nitrate, nitrite, organic matter, and oxygen consumed in water samples.[14] Another commonly cited manual was titled *Sanitary Examinations of Water, Air, and Food* by Cornelius B. Fox (1839–1922), which included methods for analysis of the mineral and sanitary quality of water including bacteriological tests.[19]

Development and standardization of water analysis methods in the United States as of 1890 were made possible by several well-established American academic institutions and public agencies—the Massachusetts Institute of Technology (MIT), Columbia College in New York City, Yale College, the University of Pennsylvania, Johns Hopkins University in Baltimore, Harvard University in Cambridge, Massachusetts, the Louisville Water Company in Louisville, Kentucky, State Boards of Health in Massachusetts and New York, U.S. Army Corps of Engineers (USACE), U.S. Department of Agriculture (USDA), U.S. Geological Survey (USGS), and U.S. Public Health Service (USPHS). The State Board of Health of Massachusetts established the Lawrence Experiment Station in 1888 as a public health and water research laboratory.[17] The mission of the Lawrence Experiment Station, in part, was to develop chemical, bacteriological, and microscopical methods of water analysis.[17]

In 1895, a broad group of researchers and officials from some of the aforementioned organizations formed a laboratory section and a committee on *Standard Methods of Water Analysis* (Committee) under the auspices of the American Public Health Association (APHA).[20] The Committee proposed a set of standardized methods for water temperature, color, turbidity, odor, residue on evaporation free ammonia, albuminoid ammonia, nitrate, oxygen consumed, hardness, alkalinity, iron, sulfuric acid, carbonic acid, dissolved oxygen, microscopic examination of water, a culture method for *B. coli*, and a standard plate count for bacteria in water.[20] The methods represented "the best current practices" of American water analysts.[20]

The first edition of *Standard Methods for the Analysis of Water* (*Standard Methods*) was published in 1905 and contained just over 90 physical, chemical, microbiological, and microscopical sampling and analysis procedures and methods.[21] About every 5 years, *Standard Methods* was revised and published in sequential editions. New and improved methods were added, and methods found to be unreliable or no longer useful were removed. *Standard Methods* became the single most important source for methods of collection and analysis of samples from the water environment throughout most of the twentieth century (Figure 1). The full title of *Standard Methods* changed over time when the APHA was joined in its publication by the American Water Works Association and the Water Pollution Control Federation. For purposes of this chapter, the title *Standard Methods* refers to the manual published by the APHA in 1905 through the 14th edition published in 1975.

Physical and chemical analysis methods: Volumetric and gravimetric methods were the most commonly used approaches for chemical analysis of water samples prior to 1890. In 1891, Walther Nernst (1864–1941) discovered electrochemistry and opened the way to potentiometric titrations, voltammetry, amperometric titrations, coulometry, and polarography.[2] The works of the physiological chemist Søren P.L.Sørensen (1868–1939) led to the concept of pH and methods for measuring acidity in 1909.[2] The analysis of free and albuminoid ammonia could be made to a detection level of about 0.01 mg L^{-1} using the Nessler colorimetric reaction. *B. coli*, albuminoid ammonia, free ammonia, nitrate, oxygen consumed, and organic matter were the methods used most frequently to indicate the sanitary quality of water.[14] Water methods for mineral

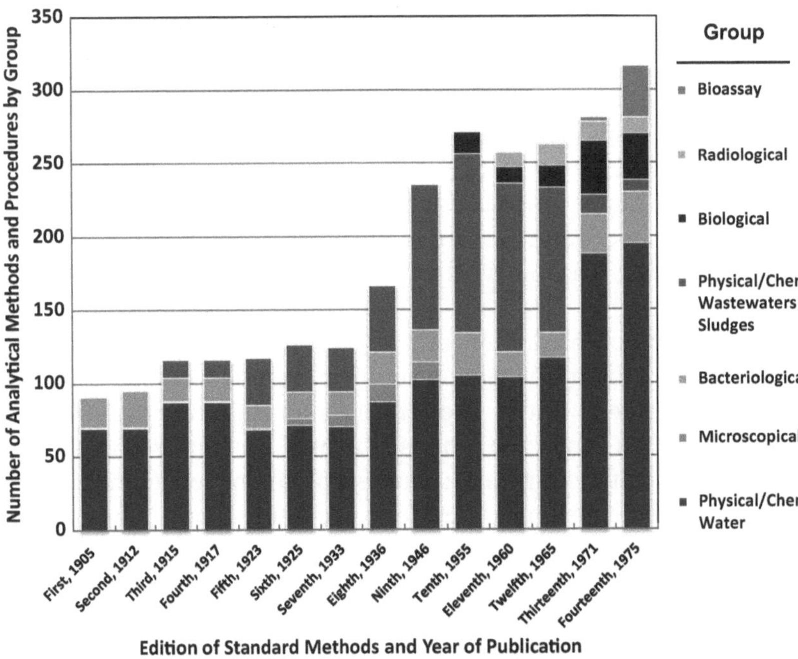

FIGURE 1 Total number of methods and procedures by group as found in *Standard Methods for the Examination of Water and Wastewater* from 1905 to 1975. Each edition added, revised, or eliminated procedures and methods to address the need for changes in physical, chemical, microscopical, bacteriological, biological, radiological, and bioassay methods in water, wastewater, and sludges.[21–34]

analyses included major ions such as calcium, magnesium, potassium, sodium, chloride, nitrate, and sulfate; hardness; turbidity; and total and filtered residue on evaporation. These analyses were commonly used to describe the quality of irrigation water, industrial and public water supplies, and for geochemical characterizations (Table 1).

As of the early 1900s, results of mineral water analyses were no longer reported as hypothetical combinations of acids and bases. Instead, results were reported as the form of the ion radical such as nitrate as (NO_3^{-1}). This method reflected the new ionic theory by Svante Arrhenius (1859–1927) on the dissociation of solutes in water into their ionic

Table 1 The Purpose of a Water Analysis Determined the Specific Constituents to be Tested. Sanitary Purposes Required Analysis of Many Unstable Constituents, whereas Industrial, Agricultural, and Geochemical Purposes Required Analysis of Relatively more Stable Constituents often Called Mineral Analyses. For Example, a Mineral Analysis Calls for Specific Determinations of Dissolved Solids, Alkalinity and Acidity, Iron and Aluminum, and Major Cations and Anions. Constituents Listed are those with Methods Described in the Seventh Edition of *Standard Methods*[26] *of Water Analysis*

	Range in number of constituent determinations by purpose of analysis			
Constituent	**Sanitary**	**Industrial**	**Geochemical**	**Agricultural**
Bacillus coli, streptococci	1–2			
Microscopical examination	1			
Biochemical oxygen demand	1			
Chlorine, hydrogen ion	1–2			
Dissolved oxygen, water temperature	1–2			
Oxygen consumed	1			
Odor	1			
Organic nitrogen and albuminoid nitrogen	1–2			
Inorganic nitrogen as NH_4, NO_3, NO_2	1–3			
Heavy metals (Cu, Pb, Sb, Zn)	1–4			
Color	1	1		
Suspended matter, loss on ignition	1–2	1–2		
Total and dissolved solids, loss on ignition	1–3	1–3	1–3	
Turbidity	1	1	1	
Total hardness	1	1	1	1
Alkalinity, acidity, carbon dioxide	1–3	1–3	1–3	1–3
Cations (Al, Ca, Fe, K, Li, Mg, Mn, Na, Si)		1–9	1–9	1–9
Anions (Br, Cl, CO_3, HCO_3, I, NO_3, SO_4)		1–7	1–7	1–7
Boron				1
Range in number of constituent determinations	16–30	8–27	6–24	5–21

NH_4, ammonium; NO_3, nitrate; NO_2, nitrite; As, arsenic; Cu, copper; Pb, lead; Sb, tin; Zn, Zinc; Al, aluminum; Ca, calcium; Fe, iron; K, potassium; Li, lithium; Mg, magnesium; Mn, manganese; Na, sodium; Si, silica; Br, bromide; Cl, chloride; CO_3, carbonate; HCO_3, bicarbonate; I, iodide; SO_4, sulfate.

forms.[35–37] The units of expression became standardized as part per million or milligram per liter.

Beginning with the first edition, *Standard Methods* included procedures for sample types such as natural water, polluted water, and wastewater with consideration for differences in clarity, turbidity, and color.[21] In this way, the method types reflected the broad range of natural and polluted waters in the United States. For example, albuminoid ammonia was recommended for analysis of relatively unpolluted surface and groundwater but not for highly polluted natural waters or for wastewater. The Kjeldahl method was recommended for all types of water and sediment samples due to its relatively more concentrated digestion reagents.[21]

The Winkler titration for analysis of dissolved oxygen was included in the second edition of *Standard Methods*.[22] Two tests for biochemical oxygen demand (BOD) were introduced in the third edition of *Standard Methods*.[23] A phenol red indicator test for hydrogen ion concentration (pH) was introduced in the fourth edition of *Standard Methods*.[24] New methods for collection and analysis of sewage effluent, industrial wastewater, mud, and sludge were introduced starting in the fifth edition.[25] An improved test for BOD was developed by the USPHS and introduced in the seventh edition of *Standard Methods* replacing the two previously approved BOD methods.[26,38]

A specific set of methods for mineral water analyses were included in the second through the seventh editions of *Standard Methods*.[22,26] The first reliable method for fluoride analysis in water was developed by Margaret D. Foster (1895–1970), the first woman chemist working at the USGS in 1919 (Image 1).[39] Fluoride was an important

IMAGE 1 Margaret D. Foster was the first woman chemist in the US. Geological Survey, Washington, D.C., 1919. Dr Foster developed a test method to analyze water samples for fluoride.[39]

drinking water analysis because elevated levels of fluoride lead to mottled teeth and modest levels prevent dental caries.

Microbiological analysis methods: The science of bacteriology led to the development of the first reliable test to answer the question, "Is the water safe to drink?" The first presumptive test for *B. coli* was developed in 1891 by Theobald Smith (1859–1934) at the State Board of Health of New York.[40] Smith's live culture method for the colon bacterium, *B. coli*, provided an approach to indicate the presence in water of the wastes of humans and other warm-blooded animals and the possible presence of pathogens.[40] With this achievement, American researchers took the lead over their European counterparts in the area of sanitary bacteriology.[41] The *B. coli* test was shown to recover a number of gram-negative, lactose-fermenting bacteria with similar characteristics to *E. coli*. Over time, continuous improvements to the *B. coli* test were made to achieve improved selectivity and specificity for *E. coli*. A laboratory manual titled *Elements of Water Bacteriology with Special Reference to Sanitary Analysis* was authored by Samuel C. Prescott (1872–1962) and Charles-Edward Amory Winslow (1877–1957) while both were at MIT.[41] The manual and its revisions became important sources of microbiological analysis methods for *Standard Methods*.

By 1925, *B. coli* was shown to be a member of the larger coliform bacteria group that were defined as all aerobic and facultative anaerobic gram-negative non-spore-forming bacteria that ferment lactose with gas formation on agar or litmus lactose agar after incubation from 24–48 h at 37 °C. This group, as defined, is equivalent to the organisms recovered by the *B. coli* test in all editions of *Standard Methods* prior to the sixth edition.[42] The *E. coli* (EC) medium was adopted as the primary growth medium in the sixth edition of *Standard Methods*.[27] The EC medium is highly specific for *E. coli*. The change to EC medium also reduced the primary incubation period from 48–24 h.[43]

Certain enterococci bacteria were confirmed in 1900 as being of fecal origin and also were adopted as indicators of the presence in water of sewage and animal waste.[44] *Clostridium perfringens*, an anaerobic spore-forming bacteria common in sewage, became another indicator test organism.[44] As defined by Ashbolt et al.[45] fecal indicator organisms are defined as "a group of organisms that indicate the presence of fecal contamination. Hence, their presence suggests that pathogens may be present."

Aquatic biological field and laboratory analysis methods: For sanitary analysis of drinking water sources, researchers at the Lawrence Experiment Station devised quantitative methods to collect and describe algae and microscopic animals from rivers and lakes and to determine the presence of organisms that were known to cause taste and odor problems in drinking water.[17] A device that is still in use today, the Sedgwick–Rafter cell, was invented at MIT by William T. Sedgwick (1855–1921) and George W. Rafter (1872–1920) for purposes of enumerating the number of microscopic cells in a water sample.[46,47]

For purposes of research, field methods for limnology were developed in the United States by Edward A. Birge (1851–1950) of the University of Wisconsin in Madison, Wisconsin, and Chancey Juday (1871–1944) of the Wisconsin Geological and Natural

History Survey.[48] Limnology and potamology are defined as the study of lakes and rivers, respectively.[48,49] The methods developed by Birge and Juday were used in lake surveys across the country.[48] Collection of microscopic aquatic organisms in water samples employed new types of apparatus such as settling tubes, nets, and various types of filters to concentrate small organisms for counting. In 1887, Stephen Forbes (1844–1930), Director of the Illinois Natural History Survey, published *The Lake as a Microcosm*, a classic treatise in the history of limnology developed from his investigations of the small lakes of the Illinois River system.[50]

The use of aquatic organisms such as fish, insect larvae, aquatic worms, and algae to describe water quality in rivers and lakes was dependent upon the knowledge gained from antecedent biological and natural history surveys. Identification, classification, and inventories of resident native plants and animals inhabiting lakes, rivers, estuaries, coastal areas, and wetlands of North America accelerated after 1871 with the creation of the U.S. Bureau of Fisheries (USBF) within the U.S. Department of Commerce. State Natural History Surveys also contributed knowledge of native aquatic organisms and species. Natural history surveys of native aquatic biota provided a basis of knowledge leading to the use of biological indicators in water quality studies.

Collection methods of aquatic insects, worms, other invertebrates, and algae in streams began to be developed in the late 1880s by the Illinois Natural History Survey and in the early twentieth century by the USPHS. For example, a method for collecting aquatic invertebrates from soft substrates was first described by USPHS in the survey of the Potomac River.[51] Stephen Forbes of the Illinois Natural History Survey first identified a comprehensive approach to stream quality assessment that fully described the stream and its channel, the chemical quality of water, the species of planktonic organisms, physical and chemical condition of the bottom sediments, and types of bottom dwelling aquatic organisms.[52,53]

The book *Fresh Water Biology* by Henry B. Ward (1865–1945) and George C. Whipple (1866–1924) contained 18 taxonomic keys to the identification of organisms from freshwater bacteria to vertebrates in streams and lakes of North America.[54] In 1919, one of the first academic textbooks on freshwater biology, *The Life of Inland Waters*, was published at Cornell University in Ithaca, New York.[55]

The first tests of the toxic effects of pollution on fish are attributed to experiments in Scotland by C. Penny and C. Adams described in the Fourth Report of the Royal Commission on Pollution of Rivers in Scotland and reported in Winslow and Phelps.[56] These studies examined the toxicity to fish of 71 industrial effluents comprising acids, alkalies, metallic oxides, and salts.[56] The State of Massachusetts carried out some of the first toxicity tests in the United States of the effects on fish from exposure to sewage and inorganic chemicals.[57] The tests exposed minnows to effluents and demonstrated that some compounds were persistent, toxic to aquatic life, and could not be decomposed by sewage treatment. The studies also demonstrated the thermal and dissolved oxygen limits necessary to protect fisheries.[57] The USBF developed tests based on exposing fish and other aquatic organisms to various dilutions of industrial wastewater.[58] The test organisms often were chosen for their multiyear life cycles or because of their availability from hatcheries or aquarium suppliers.[59]

Sample collection, preservation, and field measurements: The first edition of *Standard Methods* specified cleaning and sterilization procedures for sample bottles. Hard glass bottles were recommended to prevent alteration of the chemical composition of samples by containers made of other materials. Also specified were methods to collect a representative sample, maximum holding times prior to analysis, conditions of preservation and transport, and expressions of results.[21] Samples for analysis of fecal bacteria such as *B. coli* were collected in sterilized glass bottles, were not to exceed 10 °C prior to analysis, and analysis was required to begin within 6 h for sewage and polluted water and within 24 h for drinking water. Sample analyses for unstable nitrogen compounds such as free and albuminoid ammonia, BOD, and oxygen consumed were to be completed in the laboratory within 24–48 h of collection.[60]

The relatively short holding time requirements for analysis of unstable constituents such as free ammonia were a limitation for laboratories located more than a few hours away from field sites. Analysis at a nearby laboratory was recommended for unstable constituents with short holding times.[61] The USGS developed and used field test kits from 1903 to 1917 to do chemical assays of water samples at the field site.[35] Field assays were simplified versions of laboratory analysis methods. About 1912, motorized vehicles began to replace the horse and wagon for field work. Samples were transported by vehicle to the local post office and shipped to the laboratory by train.

Surface water quality sampling: From 1890 to 1929, water quality surveys and water quality monitoring became the two major approaches for water quality assessments. A monitoring approach was characterized by sample collection and hydrographic measurements at one or more stations over periods of time such as 1 year or more. A survey approach was characterized by sample collection at many related sites with or without hydrographic measurements over shorter periods of time such as several days to several weeks. The USGS developed a surface water quality monitoring approach in which daily samples were collected at a streamflow gaging station for up to 1 year or more.[36] In the laboratory, equal volumes of 10 sequentially collected daily samples were combined into a single composite sample for analysis. Three composite samples were analyzed each month.[36] The composite method was economical and captured semimonthly, monthly, and seasonal variations. Surface water samples were collected by lowering a bottle suspended from a pole near the center of flow.[36] In lakes, samples were collected at weekly to monthly frequencies for 1 year or more and were individually analyzed rather than composited over time prior to analysis.[36] In deep rivers and lakes, samples were collected from bridges or boats. Dip samples were collected from lakes at about 45.7 cm (18 in) below the surface.[36] An improved airtight sampler for collecting water samples at depth was developed by Prof George Kemmerer (1879–1928) of the University of Wisconsin that remains in use today.[62] In shallow Wisconsin lakes, a can was submerged to the desired depth and inverted while tubing and a pump were used to bring water from the can to the surface.[63]

For short-term water pollution surveys in streams, samples were collected at upstream and downstream locations and at wastewater discharges. Similar designs were applied to lakes where samples were collected at varying distances from wastewater discharges. Surveys were made during the summer months when pollutants exerted the greatest stress on receiving waters due to higher seasonal temperatures and lower seasonal streamflows affecting dilution. The USPHS collected water samples from just below the surface of shallow streams during low or stable streamflows. In a survey of the deeper reaches of the Potomac River in 1913, surface water samples were collected from a boat at a minimum of three locations at a river cross-section.[51] Vertically integrated samples for analysis of water chemistry and dissolved oxygen were obtained with a sampler that was lowered and then raised in the water column. The sampler filled two 250-mL glass bottles twice during the transit in the water column. The bottles were stoppered to prevent changes in dissolved oxygen and other unstable constituents prior to analysis.[51]

Groundwater quality sampling: Groundwater samples were collected from domestic, industrial, irrigation, and public supply wells fitted with various types of pumps such as hand pumps, suction pumps, rotary pumps, and deep well pumps (Image 2).[64] A windmill was a common method of pumping water from a large well such as that used for public or irrigation supply. A windlass (a winch-like apparatus) or a well sweep (a bucket on a long poll) was used to obtain water from wells that were open at the surface.[64] Well casings were most often made of cast iron. Dug wells were lined with bricks. If drilled in bedrock, wells were not cased. Poor construction at the surface was a problem described as "all too common."[64] Groundwater samples were collected directly

IMAGE 2 Groundwater sampling in the Mohave Desert northeast of Daggett, San Bernardino County, California, 6 September 1917. *Photograph of W. Mendenhall by D.G. Thompson, U.S. Geological Survey.*

at the surface of flowing wells and springs.[64] Small-diameter pipe wells driven into the ground were used as the first monitoring wells to study shallow groundwater quality.[65]

Bowman[66] prescribed the collection of critical ancillary information during well sampling. Ancillary information included a well log of the depth and character of the strata penetrated by the well bore. Other data included water level in the well and in nearby wells, well yield of water-bearing formations, and a periodic test of the soundness of the well casing. Bowman[66] recommended that interpretation of sample results should be made only after consideration of these factors.

Data management, storage, and retrieval: Publications summarizing the availability of water quality data and the scope of data-collection efforts were periodically assembled to keep track of published and unpublished water quality data collected and analyzed by government, academic, and private organizations.[67,68] These types of reports served as important inventories of water quality data prior to the creation of computerized databases. Keeping track of the published and unpublished data was a large challenge that grew over time as more organizations began producing water quality data and information.

Interpretation of Water Quality Data and Information, 1890–1929

A major purpose of water quality monitoring and assessment was to answer key practical questions such as the following: (1) Is the water safe to drink? (2) Is the water safe for swimming? (3) Is the water suitable for industrial or irrigation purposes without treatment? (4) Is the water suitable for recreation and to sustain healthy populations of fish and aquatic life? Water quality assessment goes beyond monitoring to answer these and other questions through the interpretation and appraisal of water quality.

Characterizing water quality to answer the preceding questions was not simple, because surface and groundwater in a natural state are not pure but contain dissolved and suspended substances.[69] The fact that unpolluted water is not free of naturally occurring substances led to difficulties in interpreting water quality from a chemical, biological, and microbiological perspective. Natural water was known to carry suspended materials such as soil, sediment, natural microbes, such as plankton algae, and that these conditions were not necessarily the same as in polluted and unhealthful water.[69] In an early USGS report titled *Normal and Polluted Waters of the Northeastern United States* normal meant the quality of natural water to which no added pollution was detected.[70] Polluted water was described as water not fit for human use.[70] By one approach, normal and polluted waters were differentiated in terms of *B. coli* levels, by incidence of typhoid fever in a community, and from potential economic damages from the need to treat water that was polluted.[70] By another approach, water was described as polluted if there was a nuisance condition because of its smell, appearance, or if a stream was unsuitable for fish and other aquatic life.[69] Definitions of polluted and normal water varied from one investigator to another.

Table 2 Terms Used to Describe the Quality of Water in the Early Twentieth Century[69]

Terms of Description	Described as Containing
Impregnated	Dissolved solids
Putrid	Decomposing organic matter
Soiled	Eroded soils, clay, and silt
Stained or colored	Extracts from plant leaves or other substances
Tainted	Odorous from algae or decomposing leaves
Wholesome	Not containing any of the above and preferably aerated
Descriptions of Natural Water Artificially Admixed with Substances	
Befouled	Foul, unsightly, ill-smelling but not from sewage
Contaminated	Excrement from humans or animals similar to polluted
Infected	Pathogens capable of producing disease
Polluted	General term to indicate substances artificially admixed with water
Poisoned	Poisonous substance such as lead, tin, other toxic chemicals
Pure	Not containing any of the above artificially admixed and practically unaffected by algae, color, odor, turbidity, and dissolved solids

Negative water quality descriptions used for polluted water included words such as befouled, contaminated, corrupted, impure, infected, putrid, stained, and tainted (Table 2).[69] Conversely, positive water quality descriptions often were limited to words such as wholesome or pure.

Surface water quality data: A criticism of surface water quality monitoring was the reliance on a single sample or a few samples to characterize natural stream waters.[70,71] Leighton[70] wrote "The occasional isolated sanitary analysis of water is positively without value." Collins[71] wrote "...the water in a stream varies so much that an opinion based on the results of an examination of a single sample of water from a given river would nearly always be very different from an opinion based on the results of a series of analyses, where the samples were taken regularly for some time." In monitoring projects, samples were recommended to be collected daily and composited at various time intervals to capture and describe changes in water chemistry over weeks, seasons, and years.[36,70]

Geochemical data: The geochemistry of water held an important role in interpretation of the chemical quality of natural water. Geochemical characterization of surface and groundwater required analysis of hardness, dissolved solids or residue of evaporation, and major anions and cations (Table 1).[72] Palmer[73] examined major ion analyses of water samples collected in rivers throughout the country and concluded that cations and anions are present in water as a chemical system of balanced values. Milligram-per-liter values were converted to milliequivalents per liter by multiplying the milligrams per liter with the reciprocals of the combining weights of the appropriate ions. In an analysis expressed in milliequivalents per liter, unit concentrations of all ions are chemically equivalent. This means that if all ions have been correctly determined, the total milliequivalents per liter of anions should exactly equal the total milliequivalents per liter of cations. This procedure was used for determining the quality of a mineral analysis of a water sample.[73]

The geochemical characteristics of natural water were the basis for the first classification of water into five water types by major inorganic chemical content—three types of freshwater, ocean water, and mine water.[73] An approach to interpreting geochemical results for comparative purposes was to replace tables with graphs. An advantage of a graphical approach was the ability to summarize large amounts of data in a single figure. Graphical methods of presenting major ion data as stacked histograms were first introduced to describe the mineral quality of water.[74] The purpose of graphical analyses was to demonstrate water type. When histograms from sites are plotted on the same graph, differences in water types could be determined.[74,75]

Microbiological data: As of the 1850s, it was known that if polluted water was consumed, it could cause illness. Furthermore, the polluted condition was not always obvious to the individual by unusual color, odor, or taste. Often people became conditioned to drinking polluted water and this raised arguments as to the importance of water pollution, especially as it was deemed difficult to control and required expenditures of money and other resources for treatment.[16] Doctors recognized that populations of differing ages, health, economic status, and education were differentially affected by drinking contaminated water.

Results from the *B. coli* method became the most relied-upon sanitary indicator of the presence in water of human or animal waste and the possible presence of pathogens (Table 1).[40] The value of the test for *B. coli* relies on several traits of the bacterium. *Bacillus coli* cells are abundant in the intestines of warm-blooded animals, are abundant in sewage and animal waste, and are found in waters freshly contaminated with these wastes. Conversely, *B. coli* are not found in large numbers in water not contaminated by wastes. Pathogens causing typhoid fever and cholera also are found in the intestines of sickened people and persist in water and sewage similar to *B. coli*. The ability to judge the health threat of drinking water contaminated with sewage, as indicated by *B. coli* results and epidemiological studies, led to the conclusion that sewage-contaminated drinking water must be treated before distribution as drinking water. As a result of this knowledge, the period from 1890 to 1920 became a "Golden Age of Sanitation" during which filtration and disinfection of drinking water became widely adopted.[76] Waterborne disease outbreaks precipitously declined in major cities of the United States from 1908 to 1920 because of drinking water treatment.[16,77]

Aquatic biological data: The saprobien theory was the earliest interpretive approach to the use of aquatic biological organisms to indicate sewage-polluted and unpolluted water. An aquatic indicator species is one that provides qualitative information about the quality of a stream, lake, or other aquatic habitat. The theory held that certain species of aquatic plants and animals within narrow limits reflect or indicate their chemical environment.[78,79] About 300 plants such as algae, cyanobacteria, and certain fungi, and animals such as protozoans, benthic invertebrates, and fish that occur naturally in

European waters were classified into one of three categories. These categories represented a gradient from sewage-polluted systems (polysaprobic) to systems undergoing recovery from pollution (mesosaprobic 1 and 2), to systems that had recovered from sewage pollution or were unpolluted (oligosaprobic). The approach did not address the toxicity of industrial effluents.[80] However, the indicator organism approach first described by the saprobien theory was practical and informed future interpretive methods in Europe and the United States.[81]

Drinking water standards and water quality criteria: The first water quality criteria were developed by the Second English Rivers Pollution Commission. The following were given as "the chief characteristics of unpolluted water: it (1) is tasteless and inodorous, (2) possesses a neutral or faintly alkaline reaction, (3) rarely contains in 100,000 pounds more than one-half pound of carbon and one-tenth pound of nitrogen in the form of organic matter (5 mg L^{-1} and 1 mg L^{-1}, respectively), and (4) is incapable of putrefaction, even when kept for some time in closed vessels at a summer temperature."[56] No such standards existed in the United States at the time. Geographic and other natural differences in streams in the United States from those in England made numerical and narrative stream criteria from the Second English Rivers Pollution Commission imperfect for use in the United States. For example, naturally occurring concentrations of nitrogen in some streams in the Mississippi River Valley were found to naturally range well beyond those recommended as not to be exceeded in English rivers.[61]

Interpretation of water analyses of polluted drinking water led to the first standards developed in the United States. A standard plate count of bacteria accompanied by results of tests for *B. coli* became the basis for the first drinking water standard in 1914.[82] The standard states that the total number of bacteria developing on standard agar plates, incubated for 24 h at 37 °C, shall not exceed 100 per cubic centimeter (cc) and not more than one out of five 10-cc portions of any sample examined shall show the presence of organisms of the *B. coli* group. The Treasury Department drinking water standards were revised in 1925 with the addition of maximum permissible concentrations for copper, lead, and zinc.[83,84] The addition of trace elements in 1925 signified concerns about the presence of lead in water distribution systems and the presence of other toxic metals from industrial water pollution.

Linkages were made between body-contact recreation in polluted waters and waterborne disease. The informal recommendation for bathing waters was based upon bacterial counts from the *B. coli* presumptive test.[85] The standard read as follows: "No more than 10% of samples covering any considerable period shall contain more than 1000 bacteria per mL." "No single sample may contain more than 5000 bacteria per mL, and not more than two out of five samples collected on the same day, and not more than three out of any ten consecutive samples collected on different dates shall show a positive presumptive test for *B. coli*."

Harold W. Streeter, a senior sanitary engineer with the USPHS, summarized major approaches to develop effluent standards in the United States and compared them with those in Great Britain (England, Scotland, and Wales).[86] The first effluent standards in the United States for sewage and industrial effluents were similar to those in Great Britain and were based on effluent dilution ratios. The first effluent dilution ratios for sewage called for dilution with an in-stream equivalent of 0.056–0.198 $m^3\ s^{-1}$ of streamflow to cubic meter per second of effluent flow per 1000 population.[86] Sewage effluent dilution ratios were issued by Chicago in 1887, Massachusetts in 1890, and Ohio in 1898.[86,87] The State of New York recognized that adequate dilution of industrial wastes by receiving waters was the only method available at the time to reduce effluent toxicity to aquatic organisms.[88] The State of New York recommended the dilution of industrial wastes by factors ranging from 10:1 for less toxic wastes to 1,000,000:1 for highly toxic wastes such as copper sulfate.[88] These approaches began to establish a body of knowledge on effluent limitations for primary and secondary levels of wastewater treatment and for industrial effluent treatment.

Water Quality Investigations, 1890–1929

The earliest water quality investigations can be described as surveys by recent definition. A survey is defined as a set of activities of a finite duration and intensive nature to measure and observe the quality of the aquatic environment for a specific purpose.[89,90] Monitoring on the other hand, is defined as long-term, standardized measurement and observation of the environment to define status and trends.[89,90]

Two major purposes for water quality surveys and monitoring dominated investigations. The first major purpose was to assess the quality of the untreated water resource prior to use. The purpose was described as a sanitary analysis if untreated drinking water was analyzed for unstable chemical compounds and, as of 1900, indicator bacteria.[67,68] Agricultural analyses, geochemical analyses, or industrial analyses reflect the mineral quality and turbidity of natural water quality before water was treated or used for those specific purposes (Table 1). Water uses for public, industrial, ice supply, steam power, and irrigation rely on water of an acceptable quality, and without it, water treatment is needed or water uses can be limited. The outcome of these surveys and monitoring studies contributed to the planning and development of new water treatment systems, dams, water diversions, and water conveyances to deliver water of acceptable quality to water users and developers.

The second major purpose of water quality activities was to document water pollution, its extent, sources, and its relation to the incidence of waterborne disease in what was termed a sanitary survey. Sanitary surveys could include epidemiological studies of waterborne illness in communities, inventories of pollution sources, coupled with water quality sampling surveys of rivers, lakes, groundwater, and water supplies for unstable chemical compounds and, as of 1900, indicator bacteria. These surveys provided the evidence needed by towns and cities to identify sources of drinking water contamination. Sanitary surveys were also used as justification to build a sewage collection and removal system known as a water carriage system.

Water quality surveys and monitoring related to water supply and use: The quality of untreated natural water was measured by a variety of analytical methods for stable and unstable constituents that affected its use and the cost of water treatment across all sectors of supply—public, rural domestic, industrial, and irrigation. Water of appropriate quality is important for public supply and for the manufacture of goods and services that use or incorporate water such as food and beverages. Water filtration and disinfection were the primary water treatment technologies used by public water supplies in the early twentieth century to meet USPHS health standards.[91] A sanitary analysis of water could include a physical examination of color, turbidity, odor, and taste.[92] Laboratory methods for sanitary analyses consisted of a microscopic examination for taste and odor organisms, a bacteriological examination consisting of a standard plate count and analysis for *B. coli*, and a chemical analysis to determine the nature and amount of chemical impurities.[92] Chemical impurities were measured by analyses of dissolved oxygen, oxygen consumed, BOD, chemical oxygen demand, various forms of unstable nitrogen in water—total, dissolved, suspended, and volatile solids—and a list of important trace elements (Table 1). Trace elements include metals such as copper, lead, and zinc and metalloids such as arsenic that occur at levels less than 1 mg L^{-1}.

Water containing a relatively low dissolved solids and hardness content was needed by certain industrial water supplies and for steam electric utilities because of the buildup of scale in boilers from water of relatively higher dissolved solids and hardness. Lime softening came into use in the early twentieth century to remove hardness in water to meet industrial and aesthetic requirements.[93] Ion exchange resins to produce water with low dissolved solids content became available after 1903. Laboratory methods for testing water quality for industrial purposes included mineral water analysis for major anions and cations; hardness; dissolved solids; total, suspended, and volatile solids; and turbidity (Table 1). The geochemistry of water was an important purpose of mineral water analysis because of its relevance to water treatment.

Water treatment can be costly for a given use, and for this reason, some natural waters with objectionable quality for an intended purpose, if not treated, were avoided. For example, irrigation for production of crops can be limited by dissolved solids and toxic trace elements such as arsenic and the element boron. Laboratory analysis for irrigation supply, therefore, involved analysis of major ions, dissolved solids, certain trace elements, and boron (Table 1).

Two reports serve as important inventories of data on chemical analyses of natural waters in the United States from surveys and monitoring completed from 1886 to 1931.[67,68] The inventories are an index to comprehensive publications that contained data from a wide variety of water quality investigations. Water quality reports summarizing only the sanitary quality of water were not included.[67] The original purpose of the two inventories was to provide a link to information of a general character from "a moderate number of samples used so as to have the value of a much greater number."[67] These studies also compared water quality analyses with geologic conditions to allow, for example, "the prediction of the probable quality in a well not yet drilled."[67]

A total of 30 national- and regional-scale summary reports represent data analyzed for approximately 4800 sites nationwide for the period from 1886 to 1931.[67,68] Water analyses at each site ranged from 1 to 30 inorganic chemical, physical, and suspended matter measurements. About 55% of the samples were obtained from streams and lakes, 29% were from springs, 7% were bottled water or from treated public supplies, 5% were from water wells, and about 4% were mine waters or from samples of produced water from oil and gas wells.

A total of 387 state reports were inventoried over the 45-year period from 1886 to 1931.[67,68] Western states such as California reported having more publications containing water quality data compared to eastern states (Figure 2). The western reports were in the form of numerous bulletins from agricultural experiment stations on irrigation quality of streams, lakes, groundwater, and springs, including 10 reports for Arizona; 8 for California; 5 for North Dakota; 4 each for Colorado, New Mexico, Texas, and Utah; 3 each for Oklahoma and Wyoming; 2 for South Dakota; and 1 for Nevada.[67,68]

Examples of reports published on groundwater quality included surveys of "rock water" for deep systems in Illinois, Ohio, and in the lower peninsula of Michigan.[94–96] The Illinois State Water Survey collected and analyzed groundwater quality samples along with information on water-bearing formations and other aspects of geochemistry and hydrogeology.[97–101] Samples from wells were collected quarterly, for example, for purposes of determining average conditions. From 1897 to 1912, more than 24,000 samples were collected and analyzed in Illinois; about 41% of the samples were obtained from wells.[100] From 1913 to 1915, the total number of results had grown by almost 8200, reflecting analyses of over 2700 samples per year.[101]

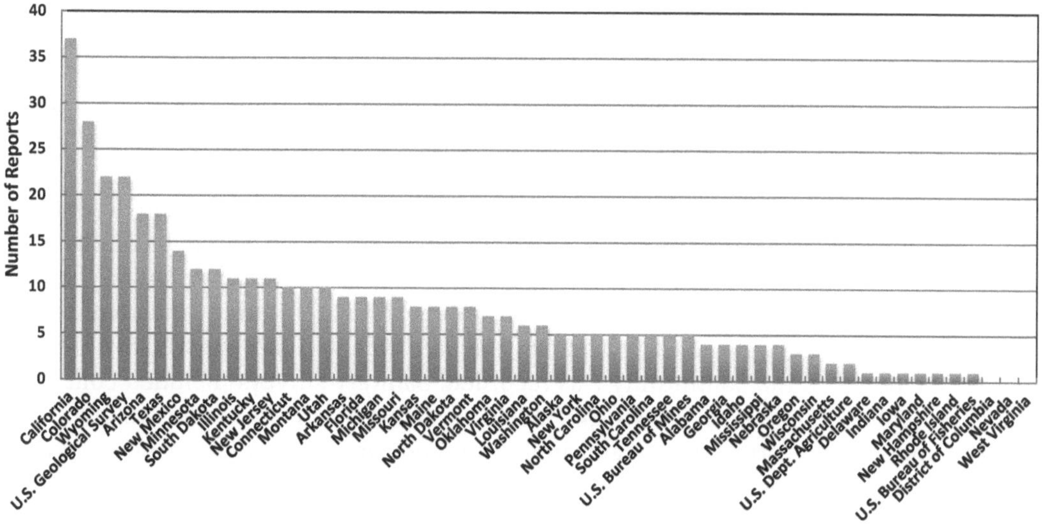

FIGURE 2 Total number of reports summarizing water quality data collection by federal and state agencies from 1886 to 1931. U.S. Geological Survey reports published cooperatively with a state are included with statewide total. U.S. Geological Survey and other federal agency reports of general, regional, or national scope are included with federal agency total.[67,68]

Another important national water quality report was the assessment of the suitability of untreated and treated water for industrial supply in 307 cities, 287 of which were major cities with populations over 25,000.[102] The report provided average conditions for each state, by surface water, groundwater, and all supplies. The analyses were made because of the concern for the effects of elevated levels of major ions, particularly calcium and magnesium, on boilers used for steam power, laundries, domestic uses, and the food and beverage industry. The analyses included results for total dissolved solids, silica, iron, calcium, magnesium, sodium plus potassium, bicarbonate, sulfate, chloride, nitrate, and total hardness. Water hardness maps at the national scale included in the report were the first of this type to be published.[102]

In 1906, USGS began its first national water quality monitoring program in surface water called "An Official Examination of the Surface Waters of the United States" (Official Examination) by Marshall Leighton (1874–1958), Chief Hydrographer of the USGS.[103] The Official Examination called for sampling in 104 rivers, the Great Lakes, and 4 large reservoirs in the conterminous United States that served as the nation's most important sources of water supply. Samples were collected to fully describe annual, seasonal, hydrological, and climatic changes affecting surface water quality. All 5 Great Lakes and 62 rivers were sampled east of the Mississippi River from 1906 to 1908.[36] A similar effort was made for 42 streams and 4 large reservoirs west of the Mississippi River from 1906 to 1907.[104] These surveys, at the time, represented the largest, most consistent, and most comprehensive body of work on chemical water quality of surface waters in the world. Water analyses included those of interest for industrial, irrigation, and geochemical purposes (Table 1).

Campaigns similar to the Official Examination were carried out in cooperation with state governments for 22 rivers and 4 lakes in California,[105] 27 rivers in Illinois,[71] 20 streams in Oregon,[106] 14 streams in Washington,[107] and 22 streams in Kansas.[108] In 1915, the Yukon River and selected major tributaries were sampled to determine the chemical characteristics of polar streams.[109] These and many other results from the USGS monitoring efforts were summarized for the United States, providing an early-twentieth-century baseline data set of inorganic chemical quality.[110]

Sanitary surveys and surface water pollution studies: The following section describes a few of the influential water pollution surveys and their relation to public health and the chemical, microbiological, and biological quality of receiving waters. A sanitary survey included comprehensive information on the location, volume, number, and character of manufacturing wastes and sewage sources. As part of a sanitary survey, waterborne disease incidence in the community was recorded. Sanitary surveys included collection of water quality samples as shown in Table 1, a sanitary survey of the watershed to identify and describe sources of water pollution, and an examination of how water is handled and distributed by the water supply system.[92]

After the completion of the first water pollution survey in 1876, the Massachusetts Legislature passed the Inland Waters Act in 1886.[16,17] The Inland Waters Act provided for a comprehensive program to monitor and protect the State's streams, surface and groundwater water supplies, and public health.[111] In 1887, the State of Massachusetts began what later became known as the Great Sanitary Survey. The Great Sanitary Survey from 1887 to 1894 was the most extensive effort of its time, with more than 40,000 samples collected in rivers, streams, and wells serving as sources for "every water supply in the State existing at the time."[112,113] Sanitary water analyses included observations of turbidity, color, and sediment and chemical analyses of nitrate, nitrite, albuminoid and free ammonia, and chlorine (chloride). All analyses were made under the supervision of Ellen Swallow Richards (1842–1911). Richards was the first woman graduate and instructor in chemistry at MIT and a research chemist for the State of Massachusetts at the Lawrence Experiment Station.[113]

Reports from the Great Sanitary Survey provided advice to towns on the quality of their water resources and water resources to avoid or treat because of water pollution.[111,112] Maps were created that described the river and its valley, locations and types of factories, their discharge points, discharge quantities, and waste characteristics. An example of a map from the 1876 report showing the locations and wastes of factories is shown for the Neponset River (Figure 3).[16]

William T. Sedgwick (1855–1921), Prof of Biology at MIT and advisor to the State of Massachusetts, established for the first time an association between the incidence of typhoid fever in the communities of Lowell and Lawrence, Massachusetts, and sewage contamination in water supplies of the two towns.[114] Soon after the initiation of the Great Sanitary Survey in Massachusetts, state boards of health of New York, Connecticut, and Ohio initiated sanitary surveys and published results in annual reports on the chemical quality of water supplies, pollution sources, and cases of typhoid fever and other communicable diseases.[115–117] Established in 1895, the Illinois State Water Survey collected and analyzed samples and published results for the State's drinking water resources after the typhoid epidemics in 1893.[97–101]

A number of reports dealt directly with sewage pollution such as *Sewage Pollution in the Metropolitan Areas near New York City and Its Effects on Inland Water Resources.*[118] The streamflow, geology, and water quality of the Potomac River were assessed in 1897.[119] A regional report titled *Normal and Polluted Waters in the Northeastern United States* synthesized information from state boards of health in the States of New England, the Mid-Atlantic, and the Midwest.[70] A report was published titled *Laws Forbidding Pollution of Inland Waters.*[120] Another report was titled *Preliminary Report on Pollution in Lake Champlain.*[121] Working with the Illinois State Water Survey, the Illinois Natural History Survey collected samples for chemical analysis and aquatic biological samples to document the pollution of the Illinois River from sewage diverted into it by Chicago starting in 1900.[122–125] To protect the quality of the Chicago water supply source, wastewater that once flowed into the source waters of Lake Michigan was diverted to a canal flowing to the Illinois River and eventually downstream to the Mississippi River.

From 1910 to 1914, studies of oxygen depletion of New York Harbor and the Hudson River were carried out by General William Black (1855–1933) of the USACE and Earle B. Phelps (1876–1953), a consulting engineer and graduate of MIT under the authority of

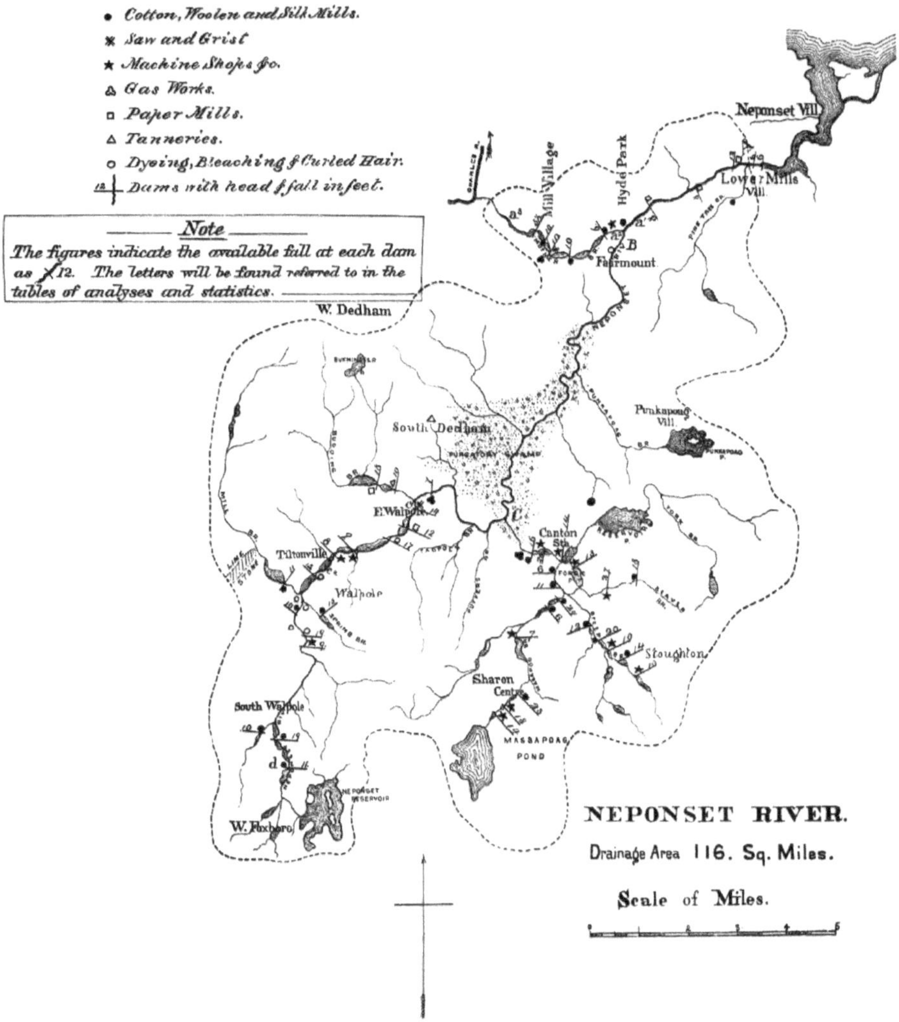

FIGURE 3 Map of the Neponset River showing approximate watershed boundaries, river network, location of towns, and industrial discharges. Maps were important in early sanitary surveys and provided a geographic context for human factors and natural features.[16]

the Rivers and Harbors Act of 1899.[126,127] The New York Harbor and Hudson River studies were the first to determine the amount of oxygen required to decompose sewage to an inert condition.[126]

The federal agency charged by the U.S. Congress with the study of waterborne disease and its prevention was the USPHS. As of 1902, the USPHS (known at the time as the U.S. Public Health Service and Marine Hospital Service) focused on analysis of *B. coli* in natural water, drinking water, food, and the incidence of illnesses like typhoid fever. The USPHS was active in investigating cholera and typhoid fever outbreaks along the Great Lakes, the Potomac River in Washington D.C., and the Missouri River from Sioux City, Nebraska, to Kansas City, Missouri.[128–131]

In 1912, legislation enacted by Congress broadened the authority of the USPHS to investigate the sanitary quality of water supplies; wastewater treatment and disposal; and water pollution in rivers, streams, and groundwater. The USPHS also was directed by the legislation to develop drinking water standards and water quality criteria for interstate waters. Additional responsibilities of the USPHS included development of new chemical and microbiological test methods and sanitary survey designs. The USPHS developed methods to collect, identify, and quantify aquatic biota as part of water pollution surveys.[92] The USPHS did not have regulatory authority to enforce pollution laws. Regulatory authority over pollution control remained vested in the states with whom the USPHS worked in partnership.

The interstate waters of the Potomac River were the subject of the first comprehensive sanitary survey by the USPHS after passage of the 1912 legislation.[51] The survey of the Potomac River watershed examined the waste sources discharging to the main stem and major tributaries in the Blue Ridge Mountains of West Virginia and western Maryland, through Washington D.C., to the mouth at the Chesapeake Bay.[51] The 1912 survey also included estuarine sections of the Potomac River in Virginia and Maryland used as shellfish-growing waters. The survey collected physical, chemical, microbiological, and aquatic biological data in freshwater sections and examined the microbiological quality of shellfish-growing waters in the estuary.[51]

Concepts developed in the survey of the Potomac River in 1912 and studies in New York Harbor led to a new approach to water pollution surveys.[51,127] The purpose of the Ohio River survey in 1913 was to measure and understand the natural purification process of sewage discharged to the Ohio River from cities and towns along its course.[38,132,133] The approach was designed to determine the capacity of a river for sewage dilution, decomposition, assimilation, and recovery from pollution.[134]

The Ohio River survey developed inventories of location, nature, and volume of wastewater discharges, location of public water supplies, and waterborne disease incidence in major cities. Intensive data collection included hydrographic measurements, analyses of water chemistry, and bacteriological analyses with time and distance downstream from sewage and industrial wastewater sources. Results were produced from 15,280 microbiological samples, 15,000 samples for turbidity, 17,588 measurements of dissolved oxygen, 805 results of 5-day BOD, 120 major ion analyses, and 520 and 313 microscopical examinations of water and mud, respectively.[132]

The rate of stream reaeration and the rate of decay of oxygen-demanding wastes were computed for the Ohio River.[135] Measurements were made of the influence of natural hydrographic factors such as the size, depth, morphology, streamflow, and water temperature that affect natural processes of stream purification.[133] Biochemical processes mediated by stream microorganisms affecting deoxygenation of water and physical processes of stream reaeration were found to be much more important than dilution alone in the assimilation of wastes.[133] This new understanding was said to have saved millions of dollars in potential costs in the design and construction of wastewater treatment systems that would have been based on dilution alone to determine the level of treatment necessary to sustain adequate levels of dissolved oxygen in receiving waters. Replication of this study in other locations eventually led to a better understanding of natural stream purification processes in all streams. In addition to the Ohio River survey, the USPHS published information and data from sanitary surveys in 11 other major river basins draining to the Atlantic Ocean in New England, New York, and New Jersey and in the Des Moines and Raccoon Rivers in Iowa.[131,136]

Very few studies were made of industrial water pollution in the United States before 1940.[77,137] A few studies of industrial waste treatment were made by the Lawrence Experiment Station.[17] The treatability of six common types of industrial wastes, their effects on the value of water supplies, and dangers to stream life were described by Leighton.[138] Publications included the treatment of strawboard wastes and oil well wastes,[139] distillery waste,[140] acid-iron wastes,[141] textile and other factory wastes,[142] and mining waste.[108]

The first organized inquiry into the relation between bathing water quality and the incidence of waterborne disease came from a 1922 opinion survey on swimming pools and beaches by the Committee on Bathing Places of the APHA.[85] A total of 2000 questionnaires were distributed to medical doctors in the United States. Of the 571 surveys returned from doctors in 41 states and the territory of Hawaii, over 70% were of the opinion that bathing places were associated with the transmission of disease. The diseases observed in patients that doctors reported as possibly attributed to bathing places included typhoid fever, dysentery, upper respiratory infections, eye and ear infections, skin infections, and two sexually transmitted diseases. About 68% of the doctors thought that the transmission of disease at bathing places was primarily related to the quality of water and related factors.

This survey became the basis for the first publication to suggest safe recreational levels of *B. coli* at bathing beaches and swimming pools. The survey results were published by the Conference of State Sanitary Engineers and the APHA.[85] The levels that were recommended were not enforceable but were adopted in California and several other states.[143] A key finding from the survey was the need to develop improved microbiological methods to assess recreational water quality.[85,144] The report by Simon et al. resulted in two important water quality assessment studies designed to better understand microbiological water quality at public beaches.[85,143,145]

Groundwater quality surveys: Groundwater was known to be vulnerable to various pollution sources yet disposal of wastes in pits and old mines was allowed by many states and towns at the time.[146,147] Groundwater contamination from sources such as oil and gas wells, highly saline and mineralized waters, poorly constructed and sealed well heads, deteriorated well casings, abandoned borings, and other causes were recognized in the early twentieth century as being responsible for pollution of groundwater.[66,148] Phenolic and tarry wastes from manufactured gas plants were frequently cited as groundwater pollutants imparting taste and odor problems.[66] The disposal of sewage in abandoned coal mines, sinkholes, and wells also was cited as a source of pollution of groundwater supplies.[149]

Natural filtration of sewage in sandstone was thought to proceed by two processes: the first was mechanical filtration through sandstone that removed particles and the second was natural oxidation and bacterial decomposition of organic compounds in soils.[150] At the time, natural purification was thought to take place by filtration of liquid wastes through soil. As a result, it was a common practice to allow leachates to infiltrate soils in order to treat the wastes with recommended safe distances for placement of wells from sewage and other waste sources.[150] Matson described the mechanisms of pollutant transport in groundwater in limestone formations.[151] The report dispelled a commonly held misconception that polluted water undergoes natural purification by passing through a short distance of underground rock. Matson[151] stated that polluted water receives virtually no natural purification as it passes through limestone due to sinks, joints, caverns, pores, and connections between surface streams and groundwater. Recommendations were made against placement of refuse and other wastes into sinks to protect groundwater quality in areas underlain by limestone from which drinking water was withdrawn.[151] The disposal of refuse and liquid wastes into sinks continued in some places into the 1950s.[147] Chemical tracers such as salts and fluorescein dye were techniques used to demonstrate the movement of pollutants in groundwater to wells.[152]

One of the first investigations to quantitatively examine the transport and fate of chemical contaminants and *B. coli* from privies to shallow wells was jointly undertaken by the USPHS and USGS at Fort Caswell, N.C. (Image 3).[65] In this survey, chemical contaminants and *B. coli* were found to travel in the direction of groundwater flow from shallow trenches containing sewage effluents toward a network of shallow wells. *B. coli* and chemical contaminants traveled over 100 m during rainy weather toward shallow wells and were detected in samples collected from those wells. This survey provided information relevant to rural sanitary issues associated with sewage contamination from privies and cess pools, which were common disposal methods used during this era.[65]

IMAGE 3 Groundwater was sampled in 400 small pipe wells to measure the movement of *Bacillus coli* and a chemical tracer from privy wastes in trenches to shallow groundwater.[65] *Image shows field work by Nora Dowell Stearns of the U.S. Geological Survey.*

Water Quality Technology and Scientific Advances, 1930–1975

Before 1930, nearly all laboratory methods for inorganic chemical analysis of water samples were gravimetric, colorimetric, or volumetric producing results in the milligram-per-liter range.[153,154] A revolution in analytical chemistry took place beginning around 1930 because of the development of advanced physical instrumentation.[155] The revolution in analytical chemistry affected water analysis in three ways. First, new types of analytical instruments expanded the ability to detect a wide variety of chemical compounds using small amounts of sample.[155] Second, the time needed for analysis was reduced and precise measurements became possible.[155] Third, with each passing decade from 1930 to 1975, the detection capabilities of new analytical instruments improved by about an order of magnitude or more. Detection limits decreased from tens of milligrams per liter in 1890 to tenths of micrograms per liter by 1950. This change allowed for the detection of trace levels of inorganic and organic pollutants. In aggregate, these technological advances provided for the first time the ability to qualitatively and quantitatively characterize the complex characteristics of industrial wastewaters.

From 1936 to 1965, in the 8th through 12th editions of *Standard Methods*, the number of new procedures and methods expanded in sections on physical and chemical analysis of polluted water, sewage, industrial wastewaters, muds, and sludges.[28–32] These new methods and procedures were needed for monitoring the effectiveness of wastewater treatment as required by new state and federal water pollution control laws and policies (Figure 1).[28–32] From 1971 to 1975, in the 13th and 14th editions of *Standard Methods*, the "practice of separating test methods for water from those for wastewaters, sludges, and muds was discontinued. All methods and procedures for a given constituent were provided under a single heading."[33,34] The reorganization of *Standard Methods* was attributed to changes and improvements in wastewater treatment as well as a transition from older methods to new instrumental methods (Figure 1).[33,34]

Microscopical methods for taste and odor organisms were incorporated into a new biological methods section in the 10th edition (Figure 1).[30] The 11th edition of *Standard Methods* for the first time included tests for radiological constituents and toxicity tests for aquatic organisms.[31] The 13th edition of *Standard Methods* more than doubled the number of methods for identification and enumeration of aquatic organisms over previous editions.[33] A bioassay method to assess the toxicity of wastewater effluents to fish was introduced in 1965 in the 12th edition of *Standard Methods*.[32] The number of bioassay methods increased to 35 in the 14th edition of *Standard Methods*.[33,34] Overall, the total number of procedures and methods in *Standard Methods* increased from about 90 in the 1st edition in 1905 to about 316 in the 14th edition in 1975 (Figure 1).

The USGS laboratory manual, *Notes on Practical Water Analysis*, was based on *Standard Methods* and remained in use up to 1950.[72] In 1960, *Methods for Collection and Analysis of Water Samples* replaced an unpublished USGS laboratory methods manual in use from 1950 to 1959.[156] The USGS chemical methods manual by Rainwater and Thatcher included 77 laboratory and field procedures for determining 53 water properties.[156] The Federal Water Pollution Control Administration (FWPCA) was established in 1965 as the successor agency to the USPHS. The FWPCA published the manuals titled *Methods for Chemical Analysis of Surface Waters* in 1969.[157,158] The FWPCA manual included 36 methods for analysis of inorganic pollutants and oxygen-demanding wastes.[158]

In 1970, the FWPCA became the Federal Water Quality Administration, which was incorporated into the newly formed U.S. Environmental Protection Agency (USEPA). In 1971, USEPA published a manual titled *Methods for Chemical Analysis of Water and Wastes*.[159] The methods published by USEPA in 1971 were predominantly based on methods published in the 12th and 13th editions of *Standard Methods*.[32,33] The USEPA accepted certain test methods of other organizations as well, such as those of the American Society for Testing and Materials (ASTM) and USGS.[159]

Use of nonstandard methods of analysis continued despite efforts to coordinate water quality monitoring over time through the use of *Standard Methods*.[160] In 1964, the Office of Management and Budget (OMB) issued Circular A-67, which established common guidelines for federal agencies collecting water data to establish more uniformity.[161] OMB Circular A-67 gave the responsibility to the U.S. Department of Interior, delegated to the USGS, "for exercising leadership in achieving coordination of national networks and standardized water data acquisition activities."[161] In 1973, the USEPA established requirements for the use of specific methods under the 1972 Federal Water Pollution Control Act.[162] A few years later, the USEPA specified methods to be used under the Safe Drinking Water Act.[163]

Physical and chemical analysis methods: Measurement of specific conductance of soils and water–soil mixtures for irrigation quality purposes dates back to 1897.[164] Specific conductance became a common water quality measurement after 1935.[165,166] Specific conductance serves as an inexpensive and direct measurement that can be used to indicate changing water quality conditions as a guide to water sampling, as an indicator of irrigation quality, and as a surrogate for dissolved solids.[165,167]

The Beckman pH meter and glass electrode were major advances that replaced colorimetric indicator chemicals to determine the hydrogen ion activity (pH) of water during the 1930s.[165,168] The electrometric measurement of pH also provided a new way to determine alkalinity by titration. The 10th edition of *Standard Methods* in 1955 recommended use of a pH meter and probe to titrate alkalinity.[30] Precautions were given to begin laboratory measurements of pH and alkalinity immediately after the sample bottle was opened to reduce potential changes upon contact with the atmosphere.[72,156]

Starting in the 1940s, the quantitative analysis of organic and inorganic compounds by more sensitive spectrophotometric methods began to replace less sensitive visual colorimetric methods. The development of spectrophotometers was pioneered at the American Cyanamid Corporation and Perkin Elmer, a new company formed in 1938.[155] These spectrophotometric methods allowed, for example, the analysis of low levels of

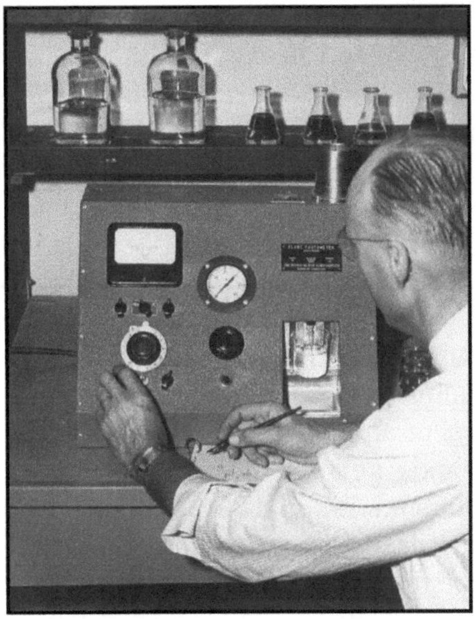

IMAGE 4 Flame photometer used for the analysis of potassium in water samples in the 1950s. *Image from U.S. Geological Survey.*

certain industrial aromatic compounds and various forms of nitrogen and phosphorus.[169] Inorganic phosphorus methods were developed in the 1930s. Several inorganic phosphate methods were published in the ninth edition of *Standard Methods* to test industrial water supplies to which synthetic forms of phosphorus were added as water softeners.[29] Flame photometry was used in the 1950s for analysis of potassium (Image 4).[31,156] Before 1950, the only organic chemical analyses performed were 5-day BOD, chemical oxygen demand, methylene blue active substances, oil and grease, and phenolic compounds.

Prior to the 1960s, mats of asbestos were used to filter particulate matter from samples prior to analysis of major ions and dissolved forms of nutrients. This procedure was replaced in the 1960s by filtration of the sample through cellulose acetate filters with 0.45-μm pore size.[156] Micropore filtration separates the dissolved from the particulate fraction and differentiates these fractions from whole water samples prior to analysis.[159] Dissolved is defined as the solutes passing through the 0.45-μm pore-size filter.[159] However, it was discovered that the filtered fraction passing through the 0.45-μm pore-size filter includes colloidal-sized particles that are retained on filters of 0.1–0.2 μm pore sizes.[170] Therefore, the sample produced by filtration through a 0.45-μm pore-size filter does not exclusively represent the dissolved fraction.[170,171] Filtration also removes bacteria and other organisms to help stabilize the geochemical and microbial changes that take place in water samples during storage prior to analysis.

As of 1960, the USGS and USPHS methods became available for total, hydrolyzable, and dissolved phosphorus in water.[156,172,173] Different fractions of phosphorus and nitrogen were separated for analysis by treatments such as filtration and digestion. As of

1971, the rising interest in phosphorus enrichment in natural water resulted in the inclusion of methods for analysis of hydrolyzable, ortho-, dissolved, and total phosphorus in water in *Standard Methods* and other manuals.[33,159,174] The FWPCA methods in 1969 were among the first to introduce automated procedures for analysis of nutrients and major ions. Compared to manual methods, automated methods increased the capacity for sample analysis per unit of time, shortened the time needed for analysis, increased analytical precision, and lowered the cost.

New analytical instruments made possible the analysis of chemical compounds that were not possible to analyze in the past. Analyses could be made with low detection limits in the microgram per liter level.[155,158] The flame atomic absorption spectrometer was invented in 1955 and introduced in the late 1960s as an analytical approach for elemental analysis of environmental samples.[175] Atomic absorption spectrometry was commonly applied to analysis of aluminum, arsenic, cadmium, calcium, chromium, copper, lead, magnesium, manganese, mercury, nickel, potassium, selenium, sodium, and zinc.[33,34,158,159]

Beginning in the late 1950s, analysis of water samples for low levels of organic contaminants, in the low microgram per liter range or lower, became possible. The organic contaminants that could be analyzed included chlorinated hydrocarbon and organophosphate pesticides, detergents and cleaners, and semivolatile organic compounds. Starting in the 1960s, extraction of samples by solvents such as chloroform and hexane followed by appropriate cleanup steps produced extracts of acid, base, and neutral fractions. Sample types included water, sediment, and fish tissue. Analyses were made by gas chromatography and other chromatographic methods.[176,177] The electron capture detector invented by James Lovelock (1919–) in the 1960s revolutionized the field of organic analysis.[155] The electron capture detector allowed for detection of halogenated hydrocarbons in the nanogram per liter range.[155] Types of pesticides amenable to detection by the electron capture detector included chlorinated hydrocarbon pesticides, organophosphate insecticides, and chlorinated phenoxy acid herbicides.[177]

Mass spectrometry had a long history in the fields of nuclear and particle physics prior to use in environmental analysis.[155] The mass spectrometer was invented in 1919 by Francis William Aston (1877–1945). In 1921, Aston became a fellow of the Royal Society and received the Nobel Prize in Chemistry for his discovery of 212 of the 287 naturally occurring isotopes using the mass spectrometer.[155] Stable isotopes such as deuterium and oxygen-18 began to be analyzed in the 1950s by mass spectrometry and put into use as water tracers.[178]

Beginning in the 1960s, gas chromatography coupled with mass spectrometry served as a gold standard for substance identification in environmental samples because the mass spectrometer can positively identify the actual presence of a particular substance in a sample. The use of a mass spectrometer as the detector coupled with gas chromatography was initially developed during the 1950s.[155] Purge and trap gas chromatography coupled with mass spectroscopy as a confirmation method was applied to the analysis of volatile organic compounds in the early 1970s by the USEPA and became the method of choice in the coming decades for almost all trace organic contaminant analyses.[155]

Microbiological analysis methods: Two improvements in the analysis of coliform bacteria from the 1940s came into standardized use in the 1960s. The first improvement was an increase in the temperature of incubation from 37 °C in the previous *B. coli* methods to 44.5 °C in the thermotolerant fecal coliform test. The elevated temperature improved the selectivity for organisms of fecal origin.[179] The thermotolerant fecal coliform test was approved as an alternate analysis method in the 13th edition of *Standard Methods* and a fully accepted method in the 14th edition of *Standard Methods*.[33,34]

The second improvement in indicator bacteria testing came with the development of membrane filter (MF) technology and methods. A new MF technique using 0.45-μm pore-size cellulose acetate filters effectively separated microorganisms from water after which the MFs were incubated on selective growth media for 24 h. The application of this technology resulted in the development of several MF methods for water monitoring by the USPHS.[180–183] The MF method showed improved precision compared to the multiple tube fermentation methods. The MF method for total coliforms in water was the first to be approved followed by approval of MF methods for fecal coliforms and fecal streptococci in water in the 12th and 13th editions of *Standard Methods*, respectively.[32,33]

Aquatic biological field and laboratory analysis methods: Major advances were made in the taxonomy and enumeration methods of aquatic organisms in the 1930s. An important comprehensive source of taxonomic information on aquatic invertebrates by Robert Pennak (1912–2004) was provided in *Freshwater Invertebrates of the United States*.[184] Pennak's publication also included comparatively more information on methods of collection and life histories of the aquatic insects and other aquatic organisms than similar contemporary books.[184,185] Many freshwater biological sampling and analysis methods for lakes and streams were described in *Limnological Methods* by Paul S. Welch (1882–1959).[186]

The use of aquatic organisms as water quality indicators began to increase in the 1950s. The 10th edition of *Standard Methods* contained 8 general references on aquatic invertebrates, 3 references on annelid worms, 20 on aquatic insects, 6 on crustaceans, 4 on aquatic mites, 3 on mollusks, and 3 on protozoa, whereas the ninth edition contained none.[29] The second edition of Ward and Whipple's *Fresh Water Biology* edited by W.T. Edmondson (1916–2000) of the University of Washington was published in 1960.[187] This volume, at over 1200 pages, was a major source for the taxonomic identification of aquatic organisms. A chapter on identification of diatoms was authored by Ruth Patrick (1907–2013) of the Philadelphia Academy of Natural Sciences. Patrick and the FWPCA found that diatoms were valuable for characterizing water quality and published a guide to the identification of common diatoms found at routine river sampling stations.[188] Other important foundational books published during this period were *The Biology of Polluted Waters* and *The Ecology of Running Waters*.[189,190] The two books by H. B. N. Hynes (1938–1990) focused on the use of aquatic invertebrates in freshwater systems to aid interpretation of the quality of streams.

The FWPCA published the first field manual on the use of aquatic biology in water pollution studies in 1969.[191] The characteristics of aquatic environments, their biota, and the effects of various pollutants upon them are discussed and illustrated by examples of designs of field investigations involving organic wastes, silts, toxic wastes, mine drainages, eutrophication, and radioactive wastes. Separate chapters describe the biology of water supplies and the biological processes of wastewater treatment. The nature and control of biological nuisances such as slimes were described.[191]

The 13th edition of *Standard Methods* contained a greatly expanded section of methods for collection and analysis of algae, macrophyton or large aquatic plants, aquatic invertebrates, fish, aquatic bioassays, and methods for investigations of fish kills.[33] The USGS began to use aquatic biology as an assessment tool to describe water quality in the 1960s and published formal methods for collection and analysis of aquatic biological and microbiological samples in 1973.[192] Methods for spectrophotometric and fluorometric analysis of chlorophyll pigments in water were published in the 14th edition of *Standard Methods* and by the Fisheries Research Board of Canada, respectively.[34,193] These methods provided for analysis of the standing stock of algae in fresh and marine waters.

Toxicity tests were developed by researchers at the USBF to examine the acute toxicity of industrial wastes on freshwater aquatic life.[194,195] The Committee on Research of the Federation of Sewage and Industrial Wastes Association established a group for the purpose of developing a standardized test of pollutant toxicity to aquatic life.[196] The first standard toxicity procedure titled *Methods for the Evaluation of Acute Toxicity of Industrial Wastes and Other Substances to Fish* was published in 1960 in the 11th edition of *Standard Methods* and updated in subsequent editions.[31]

The next major advance was the publication of *Methods for Acute Toxicity Tests with Fish, Macroinvertebrates, and Amphibians* describing a 96-h toxicity test and methods to statistically determine the 96-h median lethal toxicity or LC50 for fish species.[197] The LC50 is defined as the theoretical concentration lethal to 50% of the test organisms.[197] For aquatic invertebrates, the median effective concentration, EC50, was determined from 48-h tests and based on immobilization of 50% of the test organisms.[197] The publication made several recommendations applied to toxicity testing under the 1972 Federal Water Pollution Control Act.

Beginning in 1968, the U.S. Fish and Wildlife Service (USFWS) began testing the toxicity of 271 pesticides and industrial organic chemicals to a wide variety of fish and aquatic invertebrates.[198] The toxicity tests from which LC50 and EC50 values were derived conformed closely to standardized procedures.[33,197] The information in these references became widely disseminated and used in the years that followed.

Sample collection, preservation, and field measurements: Sample stabilization techniques were limited to chilling through the ninth edition of *Standard Methods*.[29] The recommended sample holding times prior to chemical analysis in *Standard Methods* through the 13th edition were not more than 72 h for samples of natural water and not more than 6–12 h for samples of polluted water and wastewater.[33] Some analyses, such as major ions, could be held up to 3 months prior to analysis without chilling but in tightly capped bottles to avoid evaporation of the sample.[72]

Chemical preservation, filtration, and pH adjustments became widely used along with chilling to temporarily stabilize water samples prior to analysis in the 10th and later editions of *Standard Methods*.[30] Holding times became more explicit for selected constituents and depended on the method of chemical preservation and stabilization such as filtration.[157,174,177] Starting with the 10th edition of *Standard Methods* in 1955, inert plastic containers made of polyethylene were recommended for collection and storage of samples for many types of inorganic analyses.[30] Glass or Teflon bottles were recommended for collection of trace organic compounds in the 14th edition of *Standard Methods*.[34]

The trend toward lower detection limits in instrumental analysis methods gave rise to concerns about contamination of samples in the field and laboratory. Special cleaning and testing procedures were recommended to remove low levels of contamination from glassware, containers, sampling and filtration equipment, and preservatives prior to analysis of trace-level organic and inorganic compounds.[156,174,177] Problems with results of trace element analysis due to contamination by containers, preservatives, laboratory practices, reagents, or sampling equipment made of metal were known to experts at the time.[171] Loss of trace (minor) elements on container walls was alleviated by preservation with acids and other types of chemical preservatives. Nutrient samples were preserved with chloroform in the 1960s to stop microbial decomposition.[156] After 1970, mercuric chloride or sulfuric acid replaced other types of preservatives for nutrient samples.

Field measurements: Field studies in the 1960s demonstrated that pH, alkalinity, and specific conductance are more accurately measured as close to the time of sample collection in the field as possible. Differences between field and laboratory measurements of pH ranged from no difference to 0.9 Standard Units (SU) with lower values in laboratory samples compared to field samples.[199,200] As of 1970, analyses of alkalinity, acidity, pH, and specific conductance were recommended to be made in the field.[33,174,201] Another common field measurement, oxidation–reduction potential began to be used in the field in the 1970s as a determinant of groundwater geochemistry.[201] Chemical constituents that are sensitive to oxidation–reduction potential can change quickly once exposed to the atmosphere.[200]

The need to measure water quality constituents and properties quickly for screening purposes led to the development of commercial test kits. The Hach Chemical Company based in Ames, Iowa, introduced the first commercial field test kits for the analysis of hardness in 1949 and introduced a continuous reading turbidity meter in 1957.[202] The DR Colorimeter was introduced in 1954 with 39 different colorimetric tests for water analysis.[202] The first thermistors for measuring water temperature in the field were developed in 1965 at the Yellow Springs Instrument Company (YSI) in Yellow Springs, Ohio.[203] Leland C. Clark (1918–2005) invented the first membrane-covered polarographic oxygen electrode for the measurement of dissolved oxygen in liquids.[204] By 1963, the first field meter for the measurement of dissolved oxygen in water using the Clark electrode was developed by YSI, Inc. Field meters with one to four sensors to measure dissolved oxygen, pH, water temperature, and specific conductance became widely available in the 1960s.[203]

Surface water quality sampling: Two common challenges were encountered in determining the appropriate frequency at which to collect field measurements and water quality samples. The first challenge was to adequately represent the spatial variation in water quality in a river basin at a stream cross-section, in a lake or reservoir, or in an estuary or coastal area. The second challenge was to adequately represent the temporal changes in water quality. Very general guidelines were given in *Standard Methods* up to and including the 13th edition pertaining to methods of collection of representative samples.[33]

The practice of combining equal sample volumes under conditions of changing daily streamflow was shown to produce biased results for computation of a constituent load measured at a stream cross-section.[165] Starting in 1935, daily specific conductance measurements were used as a guide to determine the number of daily samples to composite to capture similar intervals of streamflow and to minimize changes in quality in each set of composite samples.[165] Beginning in 1946, daily streamflow measurements were used to created volume-weighted composite samples for analysis.[165] As of 1970, the use of composite samples as a standard practice in monitoring surveys was replaced by the analysis of individual samples and the use of composite samples was determined by project objectives rather than a uniform approach.[165] The development of the first automatic samplers in the 1970s provided for the collection of discrete samples over daily or shorter intervals.

The BOD sampler provided a way to collect samples for dissolved oxygen, other dissolved gases, and BOD by filling the bottles several times while the sampler was submerged (Image 5). Point samplers were used for collection of surface and groundwater samples in lakes and wells (Image 6). Other types of samplers were developed for

IMAGE 5 Equipment for collection of samples for dissolved oxygen, other dissolved gases, and biochemical oxygen demand. The samplers held one or more glass bottles that were filled directly while submerged in a river or lake. Note glass stoppered bottles for the Winkler titration of dissolved oxygen concentration. This sampler was described in the 10th edition of Standard Methods.[30] *Image from U.S. Geological Survey.*

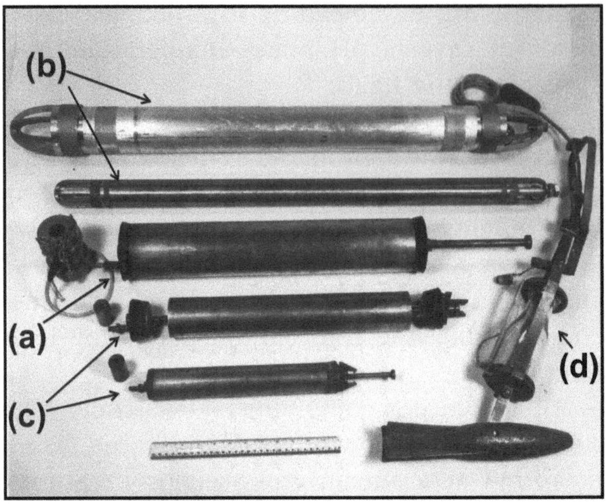

IMAGE 6 Water quality samplers: (a) stream and lake bottom sampler, (b) samplers for collection of dissolved substances and gases in groundwater, (c) large and small point samplers for collection of surface and groundwater, and (d) sampler for collection of surface water.[156]

surface water to collect samples for dissolved oxygen, major ions, and dissolved gases at specific depths (Image 6).

Selection of sites to represent well-mixed conditions was important in water quality studies.[205] Mixing of solutes and suspended materials in the vertical, horizontal, and longitudinal dimensions was discussed in relation to the selection of sampling sites.[206] Surface water grab samples were considered representative of the cross-section of the stream or river when collected at well-mixed sites.[156,206,207] A well-mixed location can be difficult to find and verify under a wide range of hydrologic conditions. Collection of grab samples, integrated samples, and composite samples was given more attention in the 14th edition of *Standard Methods*.[34]

In locations of incomplete mixing, near-surface grab samples are known to provide a poor estimate of the amount of suspended and dissolved chemical constituents passing through a river cross-section. As of the 1950s, surface water sampling apparatus included depth and width integrating samplers for collection of dissolved and suspended materials. Monitoring rivers during all stages of flow requires depth–width integrating samplers, cranes, weights, and reels for raising and lowering samplers from bridges and cableways (Image 7).[156,174] Many water pollution surveys were undertaken during stable low-flow conditions to avoid sampling during changing or higher streamflows.

Most pollution surveys were short term in nature lasting from several hours to a few days at a given study location.[206] The objectives of water pollution surveys dictated the network design.[206] Key measurements of the processes of dilution, decay, and settling of oxygen-demanding wastes in stream bed sediments were made in relation to time and distance from a wastewater discharge. Key variables measured were dissolved oxygen, 5-day BOD, water temperature, streamflow, suspended solids, and indicator bacteria such as fecal coliforms.[208] Sampling frequency was established in water pollution studies to detect changes in dissolved oxygen and 5-day BOD in relation to time of day, solar

IMAGE 7 Collection of a depth–width integrated water quality sample using a bridge crane and cable mounted weight, 1951. *Image from U.S. Geological Survey.*

radiation, tidal cycles, streamflow conditions, incomplete stream mixing, and with time and distance downstream from pollution sources.[207,209] Timing of sampling was also dependent on the schedule of waste discharge releases. A method for estimating travel time of water along a stream course was to dose the stream with a known amount of a salt such as sodium chloride and to measure the time of travel of the solute through the system.[206] As of 1970, Rhodamine WT, a fluorescent dye, injected gases, and isotopes also were used as tracers in time-of-travel studies to measure the transport characteristics of dissolved constituents from release upstream to downstream sampling sites.[208,210]

Haney and Schmidt[205] discussed collecting samples during emergencies such as chemical spills and how to collect a sample as evidence for legal purposes. The growing complexity of surface water quality studies created the need for a comprehensive guide on conducting water quality field studies.[211] The first field manual to recommend a comprehensive set of procedures and practices for water pollution studies was published by the FWPCA.[206]

Groundwater sampling: Groundwater was sampled predominantly from supply wells that were equipped with a pump. Groundwater samples were collected directly from a motor-pumped well, windmill, hand-pumped well, and flowing wells. Springs and seeps also were directly sampled. If a well without an in-place pump was sampled or if the in-place pump could not be used to collect samples, point samplers with a ball-check value were used to collect samples at specific depths for the analysis of dissolved gases and other water quality constituents (Image 6).[156]

For groundwater, sampling the same well at various frequencies demonstrated that water quality constituents can change with time, although at a slower rate than in surface water. The appropriate frequency of collection of groundwater samples was first investigated in the 1950s. Special studies were made to determine the variability of groundwater—in some systems the changes over a day were found to be as large as changes over a season or a year.[212] Groundwater quality was found to be dependent on factors such as season, well pumping, geochemistry of the aquifer, recharge and discharge of groundwater, groundwater and surface water interactions, and climate. As a result, groundwater sampling frequency for routine descriptions of water quality conditions ranged from one to four times per year, but was dependent on the hydrogeology of the subsurface and other factors.[156,174,201] A procedure for sampling groundwater for the analysis of carbon-14 was published in 1963.[213]

To collect a representative groundwater sample from a monitoring well, several volumes of well water were displaced, including the water in the gravel pack, well casing, and distribution system between the aquifer and the point of collection.[156] For sampling domestic water wells, it was recommended that the pump be turned on and sample collected at a spigot or pump after the water was left to run for at least 5–15 min before samples were collected. Advice was given not to collect samples from spigots or pumps located in the distribution systems of treated drinking water.

More detailed instructions for well sampling became available as the number and type of groundwater quality investigations increased in the 1970s.[201] Information to collect with groundwater samples included geographical and legal location, depth of well, diameter of well, length of casing and position of screens, and method of collection. Other data to be collected included point of collection, water-bearing formation(s), water level, yield of well in normal operation, water temperature, principal use of the water, name of the collector, date of collection, water appearance at time of collection, weather, and other well-related, natural, and land-use factors that may assist in interpreting chemical quality.[156]

The spatial design of groundwater monitoring networks within an area was challenging for several reasons. At issue is the fact that the volume, movement, and quality of water in aquifers usually have no visible expression at the land surface. Consequently, many surveys of groundwater quality were subject to disappointing results because of poor knowledge of subsurface conditions.[214] The spatial design encompassed the hydrogeologic setting of the well or well network. The hydrogeologic framework of groundwater systems was found to be critical for designing the number, location, and depth of wells for groundwater monitoring networks.[214] The necessary hydrogeologic framework, for example, includes knowledge of porosity and permeability, hydraulics, and direction and rate of natural movement and discharge of groundwater.[214,215]

For pollution studies, the development of the environmental framework included consideration of the possible sources and types of contaminants, either human derived or naturally occurring. Pathways by which pollutants enter the subsurface, such as direct injection of wastes, indirect contamination from septic tanks or landfills, infiltration of pollutants applied at the land surface, or pumping-induced migration, were demonstrated to be important to the design of a sampling network.[216] The anticipated fate of contaminants in the subsurface was found to be an important factor to understand before installation of a monitoring well network.[217] The factors affecting the fate of contaminants included processes that may remove or degrade the contaminants in the unsaturated and saturated zones, such as filtration, sorption, ion exchange, dilution, dispersion, oxidation or reduction, and biogeochemical processes of transformation.[216]

Data management, storage, and retrieval: A report issued in 1936 estimated that about 90% of all water quality data were collected by hundreds of state and local agencies or private businesses for their own purposes.[160] State and local data, while valuable to the data owner, were largely unavailable for use by outside organizations. One way of keeping track of and disseminating water quality data had been through the publication of bibliographies and indexes of reports summarizing the data-collection efforts by various federal agencies.[218–221]

Keeping track of the organizations involved in water quality monitoring and their published and unpublished data reports was a large challenge. As a solution to the challenge, the USPHS began to devise a computerized data storage and retrieval system in 1961 called the Storage and Retrieval system (STORET) that was completed in 1964 by the FWPCA.[222,223] STORET was adopted for use by the USEPA; other federal agencies; interstate, state, tribal, and local agencies; and universities. The system was designed for water quality data collected in streams, lakes, wetlands, coastal zones, estuaries, groundwater, and point sources collected by public agencies. STORET was not intended to serve as a database for effluents analyzed by private or public entities as required by the Federal Water Pollution Control Act of 1972.

The location of each data-collection site was defined by latitude and longitude coordinates. Each data-collection site received a unique station name, station identification number, and state code. Data-collection agencies contributing data to STORET were given unique agency identification codes. The STORET system accepted field measurements such as pH, dissolved oxygen, and discrete water, sediment, and tissue results from physical, chemical, and microbiological analyses and toxicity tests. Each result included units of measurement and result-level remark codes such as a less than sign (<) to indicate that a result was below detection. Results were stored by year, month, day, hour, and minute. Composite sample results could be stored by beginning and ending times. Samples also could be stored by depth. Data could be retrieved in many ways—primarily by geographic polygon or station, parameter code, state code, and agency code. STORET was an important step in implementing some of the reporting requirements of the Federal Water Pollution Control Act of 1972.

In 1968, the USGS developed a computerized system for input, management, storage, and retrieval of data on groundwater, surface water, and water quality. The WATSTORE system accepted water quality results by parameter codes. WATSTORE was used for data management within the USGS as well as for delivery of water data to the public. Differences between the WATSTORE and STORET systems included the ability of the former to accept continuous data of high temporal resolution for streamflow, groundwater level, and water quality measurements. Because water quality data from WATSTORE were uploaded annually to STORET through 1999, the systems were developed to be somewhat compatible.

The amount of data entered into STORET was inventoried by the USEPA Water Quality Office.[224] In 1968, just over 67,000 water quality monitoring sites had been provided to STORET by state and federal agencies (Figure 4, Table 3). About 33% of these water quality monitoring sites were sampled at a monthly frequency. The remaining two-thirds of sites were sampled at daily, weekly, quarterly, annual or other frequencies; each of these frequency categories comprised 20% or less of the total number of sites. In the 1968 STORET inventory, no single water quality measurement or constituent was reported at more than 12% of the sites. The most commonly measured constituents in 1968 were field measurements of pH and water temperature reported at about 11–12% of sites, respectively.[224] From 7.2% to 8.2% of the total number of sites reported major inorganic constituents, hardness, and total dissolved solids. Nitrogen and phosphorus compounds were reported at 5.2% and 5.6% of sites, respectively. Coliform bacteria were reported at 4.6% of sites.[224]

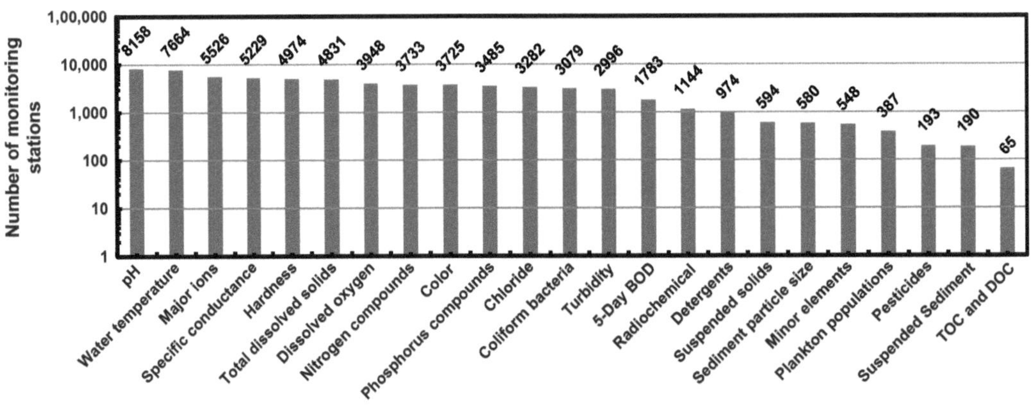

FIGURE 4 Number of active water quality monitoring stations nationwide in 1968. Water quality monitoring results were retrieved from the STORET database inventory by constituent group.[224] (BOD, Biochemical Oxygen Demand; TOC, Total Organic Carbon; DOC, Dissolved Organic Carbon).

Table 3 Total Number and Percent of Total Number of Water Quality Monitoring Stations in the United States, by Constituent and Sample Collection Frequency, 1968[224]

Number of Stations Measuring Given Constituent at Indicated Sample Collection Frequency									
Constituent	**Continuous**	**Daily**	**Weekly**	**Monthly**	**Quarterly**	**Annually**	**Other**	**Total**	**Percent of Total**
Physical Measurements									
Dissolved oxygen	85	45	345	1379	500	913	681	3948	5.9
Water temperature	871	822	529	2270	832	1064	1276	7664	11.4
pH	93	506	2070	2659	1121	1520	1832	8158	12.2
Color	14	87	205	939	611	1122	747	3725	5.6
Turbidity	31	285	332	1053	615	57	623	2996	4.5
Specific conductance	241	432	97	1636	491	1287	1045	5229	7.8
Major Ions and Hardness									
Hardness	6	362	225	1490	674	1310	907	4974	7.4
Major ions	20	412	119	1775	555	1357	1288	5526	8.2
Chloride	10	121	118	1196	354	552	931	3282	4.9
Oxygen-Demanding Wastes and Nutrients									
BOD	0	12	239	866	460	30	176	1783	2.7
Nitrogen	14	32	243	1266	507	896	775	3733	5.6
Phosphorus	23	58	135	853	609	889	918	3485	5.2
TOC and DOC	0	4	4	22	0	31	4	65	0.1
Solids and Sediment									
Dissolved solids	42	188	232	1495	624	1187	1063	4831	7.2
Suspended solids	0	2	48	328	213	2	45	594	0.9
Suspended sediment	1	8	5	97	2	6	71	190	0.3
Particle size	5	56	7	134	12	51	315	580	0.9
Other									
Minor elements	0	12	3	164	45	149	175	548	0.8
Radiochemical	21	3	37	339	448	31	265	1144	1.7
Pesticides	1	1	7	69	11	24	80	193	0.3
Coliform bacteria	13	240	423	1261	637	160	345	3079	4.6
Detergents	0	1	85	363	463	33	114	974	1.5
Plankton	6	3	102	127	60	13	76	387	0.6
Total number	1497	3692	5610	21,781	9844	12,684	13,752	67,088	
Percent of total	2.2	5.5	8.4	32.5	14.7	18.9	20.5	100.0	

BOD, biochemical oxygen demand; TOC, total organic carbon; DOC, dissolved organic carbon.

Interpretation of Water Quality Data and Information, 1930–1975

The breadth and scope of water quality monitoring and assessment in terms of the numbers and types of samples collected and analyzed and measurements made in surface and groundwater greatly increased from 1930 to 1975. The purposes of data collection also expanded with the laws governing water quality and pollution, and water and wastewater treatment. Water quality monitoring integrated the subjects of physics, chemistry, microbiology, aquatic biology, toxicology, and statistical hydrology. Interpretation of these data was greatly enhanced by the advent of computers and databases after 1960.

Surface water quality data: Beginning in the 1940s, the study of the transport characteristics of suspended sediment and dissolved substances revealed statistically significant relations between streamflow and the concentration or mass of a constituent passing through a river cross-section. These correlations were shown to be more complicated than the simple dilution model relied upon earlier in the century to explain the effect of streamflow on the concentrations of major ions.[110,171] Consequently, discharge-weighted mean concentration came into common use to represent the average water quality condition in a stream.[171] Graphical procedures to statistically summarize constituent concentrations included frequency distributions and cumulative frequency distributions. Scatterplots came into use to describe the relation of water discharge to concentration, to mass per unit of time or load, and to mass per unit area or yield. Dissolved solids were plotted against chloride or other major ions to examine geochemical relations. Dissolved constituent concentrations were shown to be correlated to specific conductance at selected sites.[171]

The development of simple descriptive statistics led to statistical analysis of surface water quality in relation to hydrology. Rating curves are one example of graphical analysis and were developed in the 1940s and 1950s. A rating curve can be a statistical or empirical relation between two variables: streamflow and observed concentration or load of a water quality constituent. Rating curves were developed so that predictions of a constituent concentration or load could be computed from the statistical equation or simply interpolated from a line fitted to measured quantities.[225–227] These predictions were used to estimate constituent concentrations or loads at times when no water quality sample result was available. Rating curves for sediment samples and daily discharge were statistically analyzed to provide estimates of sediment discharge per unit time and per unit area of a watershed.[226]

The concepts of how oxygen-demanding wastes react in receiving waters were described in the book titled *Stream Sanitation,* which was published in 1944.[228] The objective of the book was to describe water pollution studies and how to determine the assimilative capacity of the water body for oxygen-demanding wastes under varying hydrologic and temperature conditions. Prof Clarence Velz (1901–1992), a student of

Earle B. Phelps and a well-known environmental engineer in his own right, published the book *Applied Stream Sanitation*.[207] A chapter in this book described various approaches for statistical analysis of stream quality data and another chapter described computer programs for common graphical measures such as dissolved oxygen profiles.[207]

The development and use of surface water quality models was driven by requirements of the Federal Water Pollution Control Acts of 1956 and 1972 that required treatment of sewage and industrial waste discharges to protect the quality of receiving waters.[229] The first water quality models for rivers were rooted in the graphical and analytical equations for estimating the temperature-dependent decay rate of oxygen-demanding wastes in a stream as first described by Streeter and Phelps.[133] Over the next 50 years, the approach was refined by introducing several additional processes affecting dissolved oxygen in rivers, including sediment oxygen demand, surface reaeration rate, nitrification of organic wastes, and photosynthesis of algae.

In the 1960s with the advent of mainframe computers, the model QUAL1 was the first numerical simulation model for streams that used field data from water quality studies to model the effects of point source effluents on stream quality.[229] QUAL1 and subsequent models of the QUAL family became the main tools for calculating the decay of oxygen-demanding wastes and bacteria as a function of temperature, travel time, and distance downstream from a wastewater source. Simulation models were used to estimate these processes under conditions such as the 7-day, 10-year (7Q10) low flow, when dilution of wastes in a stream is at a minimum.

Watershed models were developed to address nonpoint sources of various contaminants on water quality in streams and lakes. Two watershed models emerged in the 1960s. The rainfall-runoff model known as Hydrologic Simulation FORTRAN Program was originally developed in the early 1960s as the Stanford Watershed Model. In the 1970s, water quality processes were added.[229] The Storm Water Management Model has been under continuous development since 1969 to simulate urban runoff.

Geochemical data: With more data being produced by new types of instrumental methods, there was a need to summarize and present large amounts of water quality results using meaningful and concise methods. Special graphs were used to show water type based on chemical composition, relations among ions and relations among groups of ions in individual waters, and to compare the quality characteristics of many waters. The study of geochemistry of water was critical to the study and interpretation of water quality data.

The interpretation of inorganic chemical data for assessing the quality of water supplies gained an important tool through the use of diagrams for display of major ion data.[171] Piper diagrams are plots of cation and anion percentages for each sample calculated from the ratios of milliequivalents per liter concentrations of major ions to the respective cation or anion total.[230] The relative percentages of cations and anions are plotted in two triangles and each triangle is then projected into a quadrilateral polygon that describes the water type.[171,230,231]

Stiff diagrams are used to compare the ionic composition of water samples between different locations, depths, or aquifers.[232] The Stiff diagram is a polygon created from three horizontal axes extended on both sides of a vertical axis. Cations are plotted on the left side of the axis and anions are plotted on the right side, both in milliequivalents per liter. A greater distance from the vertical axis represents a larger ionic concentration. The cation and anion concentrations are connected to form an asymmetric polygon.[232] Durov plots were developed for characterizing groundwater into 18 water types within 5 classes ranging from relatively dilute calcium bicarbonate water types to transition water types, and finally to relatively complex mineralized water types.[75,165] Other graphical and numerical approaches for determination of water type included circular and star pattern diagrams.[75]

Hydrochemical facies were first described in the early 1960s as distinct zones of water that have cation and anion concentrations that can be categorized into diagnostic compositional groups.[231] Examples of geochemical groundwater studies in the 1960s include mineralogical factors in natural water composition and carbonate equilibria in groundwater.[233,234] The redox potential, or oxidation–reduction potential (Eh), is a numerical index of the intensity of oxidizing or reducing conditions within a system.[165] The Eh of solutions was found to indicate the form of the multivalent elements that dominate aquatic systems, such as ferrous ion (Fe^{2+}) in reducing conditions versus ferric ion (Fe^{3+}) in oxidizing conditions.[235] Graphical *x–y* plots of pH versus Eh called stability field diagrams were developed to predict the equilibrium state of redox-dependent multivalent elements such as iron, manganese, sulfur, and nitrate in water.[235]

The publication *Study and Interpretation of the Chemical Characteristics of Natural Water* was the first treatise on the aqueous chemistry of fresh and mineralized waters.[236] The publication described the occurrence of 70 common inorganic and organic chemical constituents and physical properties of rivers, lakes, and groundwater.[165,171,236] The textbook titled *Aquatic Chemistry: An Introduction Emphasizing Chemical Equilibria in Natural Waters* was a basic text for water chemists.[237] The authors, Werner Stumm (1924–1999), a professor at the Swiss Federal Institute for Water Resources and Water Pollution Control in Zurich, Switzerland, and James P. Morgan, a professor at the California Institute of Technology in Pasadena, California, are widely credited with "fundamental discoveries about the physical and chemical processes that take place in the water environment."[238] In 1999, Stumm and Morgan won the Stockholm Prize for Water, which is awarded by the Stockholm Water Foundation and is equivalent to the Nobel Prize.

The computer program WATEQF was one of the first programs used to calculate the equilibrium distribution of inorganic aqueous species of major and important trace elements in natural waters using chemical analysis and in situ measurements of temperature, pH, and redox potential.[239] Solute transport models were developed as early as the 1950s and became a common technique after 1970 to describe chemical reactions during solute transport in groundwater systems.[240] Solute transport models are necessary to understand and predict contaminant fate and transport in the subsurface.[240]

Isotopes are atoms of the same element that differ in the number of neutrons in the nucleus giving them different atomic masses. Mass spectrometry enabled the precise measurement of ratios of stable isotopes. Of particular interest were ratios of deuterium to hydrogen and the oxygen-18 stable isotope to oxygen-16 of water. Small mass differences in these isotopic ratios caused by fractionation during evaporation and condensation of water can be used to trace the sources of surface and groundwater.[241] "In the process of evaporation, the heavier molecules tend to become enriched in residual water, and the lighter species are more abundant in water vapor, rain, snow, and most freshwater; the heavier forms are more abundant in the ocean."[165]

The unstable (radioactive) hydrogen isotope tritium or hydrogen-3 with a half-life of 12.26 years and its decay product helium-3 were first used in the 1950s to estimate the apparent age of groundwater up to 50 years since recharge.[242] Except for radioactive decay, water molecules containing tritium behave identically in most respects to water molecules containing stable hydrogen and deuterium. Although tritium is naturally occurring, it was released in amounts well above background levels as a result of aboveground nuclear weapons testing in the 1940s through 1960s.[242]

Carbon-14 with a half-life of about 5730 years is an isotope used in age dating older groundwater and was first used to determine the ages of groundwater in the range of hundreds to thousands of years.[243–245] Measurement of groundwater age became important in the refinement of numerical groundwater models and to determine the rates of geochemical reactions in aqueous solutions.[246]

Aquatic biological data: The use of biological measures of water quality became widely used for interpretive studies from 1930 to 1975. Biological organisms are sensitive measures of changes in the environment.[81,189] Many groups of organisms were suggested over time as suitable indicators, including bacteria, fungi, algae, diatoms, freshwater clams and snails, aquatic insects, and fish.[247] The indicator approach evolved into assessments of community composition of larger groups of aquatic organisms such as aquatic invertebrates.[189] However, a shortcoming of the indicator approach was that one or a few key indicator organisms or groups of organisms often were found to be insensitive to water quality change. Another shortcoming was that a key indicator group might not be representative of organisms not measured that are affected by changing water quality conditions. Biological indices using the full community composition of taxa, however useful, lacked information beyond the structure of the resident community of organisms. Further, biological indices of community composition did not indicate the preferences, functions, or tolerances with which to explain the disappearance of taxa from some areas and abundance in other areas.[81]

In a landmark survey of the Conestoga River Basin in Pennsylvania, Ruth M. Patrick proposed a relative abundance index for organisms that could be displayed as a histogram.[248] Sorting organisms into pollution-tolerant or pollution-intolerant forms added power to improve interpretation. As taxonomic identification and methods of enumeration improved, biological indices evolved to become semiquantitative and quantitative measures of the change in aquatic community composition in response to changes in the water environment.[81]

As a next step, diversity indices were proposed in the 1950s and 1960s.[81] These approaches were based on the theory that a large change in the environment that does not provide sufficient time for the evolution of a new aquatic biological community to form will reduce the diversity of the resident community.[81,249] Community diversity measures relied on the number of species or taxa in a sample, with a greater number of species or taxa being rated superior to a lesser number of species or taxa. Diversity indices are dimensionless values based on information theory.[250] In this theory, the more diverse a community, the more uncertain is the chance of finding a single organism of a given species in a random draw from the community.[250]

Community and diversity indices are not indicative of the causative agent(s) of change in an aquatic system. To improve assessment techniques, toxicity tests were added to the list of tools for aquatic system assessment. M.M. Ellis (1887–1953), a senior aquatic physiologist at the USBF and at the USFWS, brought together findings from the widely scattered scientific literature and supplemented those studies with national-scale experimental field work covering (1) the conditions that should be maintained for good communities of fish fauna and (2) the specific effects of various types and chemical characteristics of effluents.[194] For example, a minimum dissolved oxygen concentration of 5.0 mg L^{-1}, pH range from 6.5 to 8.5, and temperature ranges targeted to cold-water, cool-water, and warm-water fish fauna were established and remained in use for many years.[194,195]

The studies by Ellis[194] also had a laboratory component. Ellis[194] established the first acute toxicity thresholds for 114 pollutants using goldfish and a microscopic aquatic invertebrate, *Daphnia magna*. Ellis also described the action of various pollutants on external organ systems and tissues such as gills, lining of the mouth, and scales and the internal gastrointestinal tract of fish.[194] The pollutant categories tested included acids, metallic salts, and various other organic and inorganic chemicals found in industrial wastes.

The diagnostic approach used multiple measures in the field and laboratory for assessment of receiving water quality and aquatic health.[251] The diagnostic approach included field and laboratory studies of physical and chemical quality of stream water and bottom sediments; a description of the resident aquatic biological populations of algae, fish, and aquatic invertebrates; and toxicity tests on one or more types of organisms. This combination of approaches provided multiple lines of evidence to the interpretation and assessment of water quality conditions in the coming decades.

Drinking water standards and water quality criteria: With the exception of drinking water standards, the phrase *water quality standard* had little regulatory meaning in the 1930s. In 1962, drinking water standards included 11 inorganic compounds and the organic compounds alkyl benzene sulfonate, compounds detected in a carbon chloroform extract, and phenol. Radionuclides were added for the first time in 1962 with maximum allowable limits for concentrations of gross alpha and gross beta radiation, radium-226, and strontium-90. These limits were recommended by the National Radiation Council and approved by President John F. Kennedy.[83,252] President Gerald Ford signed the Safe Drinking Water Act in 1974 (42 U.S.C. §300f et seq.). Under the Safe Drinking Water Act, the USEPA developed many more drinking water standards than were previously available. These standards, called the National Interim Primary Drinking Water Regulations, specify maximum levels of drinking water contaminants and monitoring requirements for public water supply systems. The contaminant limits are set at levels to protect public health to the extent feasible, using available techniques.[163]

The concept of beneficial use embodied the idea that water resources in their natural state should meet a certain predetermined quality that supports all important human uses and that sustains fish and populations of other aquatic life. The beneficial use criteria were divided into two types. The first type was based on those uses of natural water that were important to public health such as the quality of untreated drinking water, quality of shellfish-growing waters, and quality at bathing beaches.[86] Standards for phenols and cresols were established for source waters to control objectionable taste and odor problems that these compounds imparted in treated drinking water.[86] Problems associated with naturally occurring salinity, dissolved solids, boron, *B. coli*, and toxic compounds were a concern in irrigation water used on fresh vegetable crops to be consumed without cooking. Standards to support recreational uses were based almost exclusively on maximum levels of *B. coli* and other fecal coliform bacteria not to be exceeded at various temporal scales such as daily, monthly, or as an annual average.

As part of the concept of beneficial uses of water, a group of water quality criteria was developed to maintain fish and other aquatic life, to prevent nuisance conditions, and for water recreation such as fishing. Criteria were established for minimum dissolved oxygen levels to protect fish. Ellis[194] reported that healthy populations of most fish species require a minimum concentration of at least 5 mg L^{-1} of dissolved oxygen and that only some tolerant species could withstand lower levels temporarily. Some states in tropical and subtropical climates adopted lower minimum levels of dissolved oxygen.[87] Narrative criteria for odor control and control of floating debris, sludges, and floating solids were established to protect recreation and to prevent nuisance conditions.[86]

Some state and interstate agencies classified receiving waters into two to four water use types or zones. Classes were identified as A–D or 1–4, with sequentially lower quality being accepted for waters with fewer declared uses.[87] Standards initially were recommended for only four pollutants: coliform bacteria, 5-day BOD, pH, and phenols.[87] Criteria were applied individually by states and often did not protect high-quality waters that had not yet been developed. Criteria were limited to a few compounds or elements that were known to be associated with degradation of water quality for human uses. By 1949, 15 states had officially and 10 states had unofficially adopted stream standards.

California was the first state to formally publish narrative and numerical water quality criteria in 1952 and to revise them in 1963.[83] The criteria were used as a model by many other states and public health officials. The California water quality criteria were extensively researched, scientifically based, comprehensive in terms of pollutant types, and addressed all important water uses.[83] By 1962, almost all states had some type of pollution control law with water quality standards and criteria adapted to suit the designated use of the water resources.[83] Some states had only narrative criteria, while others established narrative and numerical criteria.

The Water Quality Act of 1965 required all states to establish water quality standards based on minimum federal water quality criteria as of 1968. The FWPQA defined standard as "a plan that is established by a governmental authority for water pollution prevention and abatement."[253] A criterion was defined as "a scientific requirement on which a decision or judgment may be based concerning the suitability of water quality to support a designated use."[253] As required by the Water Quality Act of 1965, the first water quality criteria were published in 1968.[253] The FWPCA applied these water quality criteria only to interstate and coastal waters.[253] Numerical and narrative criteria were developed for uses, including fresh and marine recreational waters, public water supplies, fish and wildlife propagation, and agricultural and industrial supply.

After its formation in 1970, the USEPA convened a National Technical Advisory Committee (NTAC) under the auspices of the National Academies of Science and Engineering. The NTAC published a revised and expanded set of water quality criteria in 1972.[254] The 1972 revisions greatly expanded the scope of the 1968 criteria to include intrastate as well as interstate and coastal waters, and included new types of pollutants such as viruses, polychlorinated biphenyls, and other industrial chemicals and byproducts. The 1972 criteria included for the first time requirements for marine and freshwaters used for fish and wildlife propagation, limits on toxicants, and restrictions on ocean disposal of solid wastes. The 1972 publication also redefined criteria to mean "the scientific data evaluated to derive recommendations for characteristics of water for specific uses." The 1972 criteria provided minimum narrative and numerical water quality requirements for surface waters and called for review and revision of criteria to track progress of scientific knowledge of various pollutants.

The need for effluent limits as part of a comprehensive approach to control the point sources of water pollution and to protect the beneficial uses of receiving waters became well recognized in the 1940s.[87] Yet, only three states had adopted effluent limits by 1948.[87] As part of a comprehensive water quality policy, effluent limits on sewage and industrial water pollutants were required under the Federal Water Pollution Control Act of 1972. The provisions of the 1972 Act were revolutionary for several reasons including the establishment of a pollutant discharge permit program for all wastewater dischargers and a timetable of 1985 after which the nation's waters were to be fishable and swimmable. The fishable and swimmable goals of the 1972 Act were responsible for changing a century or more of public policy that allowed the nation's waters to be relied upon for waste assimilation in lieu of uniform technology-based standards for wastewater treatment that applied to all sources.

Water Quality Investigations, 1930–1975

Water quality monitoring and assessment for purposes of characterizing sanitary, agricultural, public and industrial supplies, and geochemistry of water continued through the early 1950s. The purpose was described as sanitary if untreated water was analyzed prior to use as drinking water. Agricultural analyses, geochemical analyses, or industrial analyses reflect analyses of natural water quality before water was treated or used for those specific purposes.[218,219] The concept of beneficial uses of water for public, industrial, and agricultural uses resulted in adoption of federal water quality criteria in 1968 and 1972. Federal standards were reissued for drinking water in 1962 and the Safe Drinking Water Act was passed in 1974, establishing maximum contaminant levels. Assessment of beneficial uses became a major purpose for water quality monitoring after 1970.

Objectives of water quality monitoring describe how activities are carried out to achieve a stated purpose. Meybeck et al.[89] identified three major objectives of water quality activities: (1) monitoring, (2) surveys, and (3) surveillance. As mentioned above, monitoring is defined as "the long-term, standardized measurement and observation of the aquatic environment in order to define status and trends." Surveys are defined as "finite duration, intensive programs to measure and observe the quality of the aquatic environment for a specific purpose." Surveillance is defined as "continuous, specific measurement and observation for the purpose of water quality management and operational activities."

The following sections describe important water quality surveys and monitoring activities associated with beneficial uses of water and water pollution. The first regional and national water quality monitoring networks are described.

Water quality surveys and monitoring related to water supply and use: The quality of untreated water resources was of great interest to water users and water developers because of the cost and requirements of water treatment prior to use. Quality requirements included new health-based standards for drinking water, salinity limits for irrigation water, and industrial water guidelines.[149,252] Regulatory requirements and an increase in water use of 819% nationwide from 1900 to 1970 created a greater need than ever for the collection and reporting of water quality data and information.[91,255]

Surface water was the dominant source for public, industrial, irrigation, and thermoelectric power supplies through 1970.[255] To support the data needs of these water uses, published and unpublished chemical analyses of surface waters in the United States were indexed and disseminated in four published bulletins covering all known and available data sources held by federal agencies through 1961. Bulletins 2 and 9 have the title *Inventory of Published and Unpublished Chemical Analyses of Surface Waters in the Western United States*, and together cover water quality monitoring for the periods prior to 1 October 1955.[218,219] Bulletin 6, *Inventory of Published and Unpublished Chemical Analyses of Surface Waters in the Eastern United States*, covers water quality monitoring prior to 1 October 1951.[220] A fourth publication covers the entire United States and fills the time gap from the end of the period covered in Bulletins 2, 6, and 9 to 1 October 1961.[221] Together, these four publications summarize more than 80 years of water quality data-collection efforts in the United States.

Water quality data are categorized in Bulletins 6 and 9 according to four purposes of collection: sanitary, industrial, geochemical, and miscellaneous. Bulletin 2 was limited to waters that were likely to be used for irrigation. The fourth publication made no distinction as to purpose of data collection.[221] The monitoring sites inventoried in these reports were organized by drainage basin and stream name, location, drainage area when known, period of record, number of analyses, and indication of the colocation of the water quality data with streamflow records. Also reported were frequency of sampling, types of analyses, numbers of analytical values reported, and references to the data reports. The second part of each report references the published and unpublished data reports. Water suppliers relied on their own tests but also on water quality information from government sources such as the reports identified in these bulletins, particularly before selecting a geographic location to establish a new supply. The availability of water quality data in electronic databases beginning in the 1960s eliminated the need for publications like these bulletins.

A total of 42 federal, interstate, and state agencies produced 124 reports containing water quality data over the 10-year period from 1951 to 1960 (Figure 5).[221] Twelve reports were published by the USACE; 8 each by the State of Ohio and the USGS; 7 each by California, Kentucky, and Louisiana; 6 by New York; and 5 each by Pennsylvania and Wyoming (Figure 5).[221] Notably, one or more reports were published by the Tennessee Valley Authority (TVA), OSANCO, the Interstate Commission on the Potomac River, and the Sabine River Compact Commission.

Bulletin 6 referenced a report published in 1934 that summarized analyses of the inorganic chemical composition of the public water supplies of large cities in the United States.[256] Data were given for 670 places representing 46% of the total population of the United States. The report included water quality results for every system serving a population of 20,000 or more. The report included tables of constituent values by city and a generalized map showing hardness.[256] Data were reported for alkalinity and major

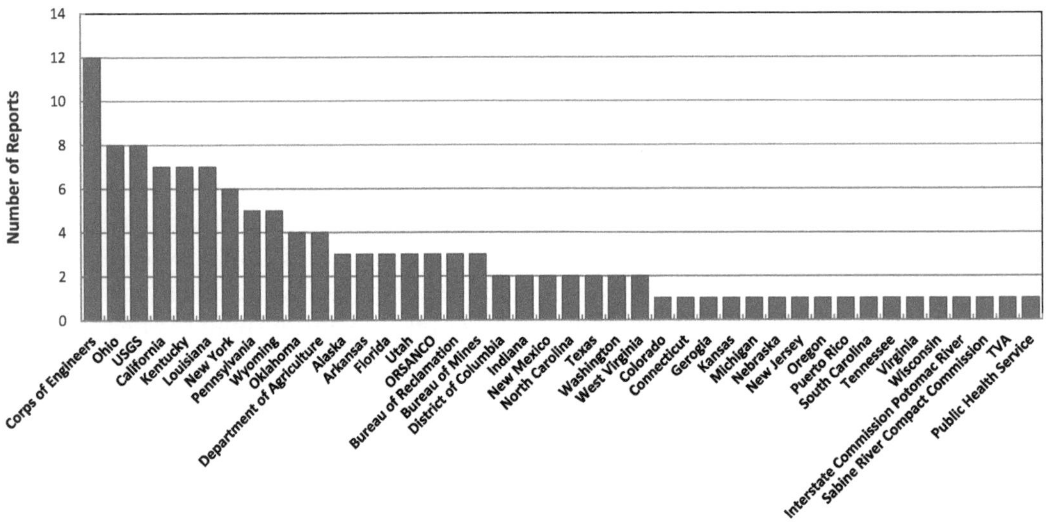

FIGURE 5 Total number of published reports containing water quality data from federal, interstate, state, and local agencies, 1951–1960.[221]

ions, trace elements, boron, fluoride, and iron. Physical analyses included turbidity, dissolved solids, and suspended sediment.

Other examples of data reports include 134 Water Supply Papers published from 1941 to 1976 by the USGS in an annual series under the general title, *Quality of Surface Waters in the United States*.[153] Two summary reports provided water quality data for 1315 of the larger cities with populations of 15,000 or more throughout the United States.[257,258] The population of these cities represented 58.3% of the total population of the United States from the 1950 census and more than 90% of the total urban population of the United States. Of the total of 1315 places reported, 711 relied on surface water supplies, 472 relied on groundwater supplies, and 132 relied on blended supplies from surface and groundwater sources. The hardness of the large public supplies ranged from less than 5 to 700 mg L^{-1}.[257,258] Over 2500 chemical analyses were provided. Results of these studies indicated that surface water supplies generally were more variable in chemical composition than groundwater supplies, but contained lower dissolved solids on average.[257,258] A similar publication was issued for the public water supplies of the 100 largest cities in the United States in 1962.[259]

Streeter[260] reported on the efficiency of drinking water treatment systems for removing *B. coli* from raw river water at 31 municipal water treatment plants in New England, New York, and the Midwest. This survey provided insights into the levels of indicator bacteria and pathogens in source waters at drinking water intakes that could be effectively removed by drinking water treatment. The report indicated that source (untreated) water containing a *B. coli* level at or above 20,000 per 100 mL was unsuitable for a drinking water supply. The survey also found that waters with *B. coli* levels from 2000 to about 20,000 per 100 mL required additional treatment steps such as prechlorination of raw water to meet Department of Treasury standards.[260]

Many natural waters in the 1950s and 1960s, without treatment, did not meet the quality requirements for intended uses.[149] Treating water to make it acceptable for intended uses, or importing water if treatment of the local supply was not feasible, were the greatest water resource challenges in 1960, particularly in regions with water shortages.[149] Many of the public supplies required additional water treatment to produce a satisfactory product for certain industrial uses that require dissolved solids less than 0.5 mg L^{-1}.[258,259] After 1960, drinking water treatment with aeration, coagulation, flocculation, and activated carbon filtration began to be employed by public water supplies to further remove turbidity, color, and organic and inorganic pollutants.[91,261] Awareness of the occurrence in water of trace levels of organic and inorganic pollutants from industrial wastewater discharges led to requirements for water treatment prior to water reuse for public supply.[262]

In 1970, the USPHS published *Analysis of National Survey Findings of the National Community Water Supply Study*.[263] In this study, 969 representative water supply systems located in the United States in 1969 were examined. Systems included 22 large metropolitan systems serving in aggregate over 13 million people, 947 systems serving 5 million people in communities with less than 100,000 population, and 760 systems serving populations less than 5000 people.[263] USPHS drinking water standards[252] were used to evaluate the quality of these systems. Findings on the quality of surface water, groundwater, and blended supplies indicated that a total of 36% of 2600 tap water samples contained one or more bacteriological or chemical constituents exceeding the limits of one or more drinking water standards.[263] The study noted that expanding water use came at a time of greatly increased pollution of groundwater aquifers and streams, lakes, and rivers. Historically, groundwater had been considered of good sanitary quality and safe to drink.

As of 1970, chemicals in the environment had been on the increase for 25 or more years. The number of chemicals used in industry, agriculture, and domestic purposes amounted to about 12,000 toxic chemical compounds in use in 1970, and 500 new compounds were developed each year with little known about their possible environmental and human health effects.[263] A survey in 1972 found 36 chemicals in treated water taken from the Mississippi River near New Orleans, Louisiana.[264] The trace organic compounds in the drinking water supplies were deemed a "potential threat to the 1.5 million people who consume this water."[264] The increased awareness of chemical contamination of water supplies led to the passage of the Safe Drinking Water Act of 1974.[91]

Sanitary surveys and surface water pollution studies: New federal legislation to control water pollution began with the passage of the first temporary Federal Water Pollution Control Act in 1948. A permanent act was passed in 1956 and amended in 1965, 1966, and 1972. The most comprehensive of these laws was the Federal Water Pollution Control Act of 1972. New laws created new regulatory purposes for water pollution surveys and monitoring such as compliance monitoring with end-of-pipe effluent limitations and emergency response to spills and releases of oil, sewage, and chemicals.

Before the passage of the 1948 Federal Water Pollution Control Act, the USACE under the authority of the Rivers and Harbors Act of 1899 was directed by Congress in 1938 to undertake a comprehensive investigation of the Ohio River and its tributaries in cooperation with the USPHS.[265] In 1937, the surface water resources of the Ohio River Basin provided drinking water for over 7 million people and received the combined wastes of over 8.5 million people each day.[265] More than two-thirds of these wastewaters received no treatment before discharge. Industrial wastes with the equivalent oxygen demand of an additional 10 million people were discharged to the Ohio River and its tributaries each day.[265] Adding to that amount, over 1.63 million metric tons of acidic wastewater flowed or was released to the Ohio River and its tributaries on an annual basis from active and abandoned coal mines in the Appalachian coal fields in Kentucky, Ohio, Pennsylvania, and West Virginia.[265] The combined impacts of acidic mine discharges caused damage to water and wastewater treatment processes and infrastructure. Pollution abatement prior to reuse of water was an important issue in the Ohio River Valley.[266]

The objectives of the 1938 comprehensive investigation of the Ohio River and its major tributaries were to identify all important sources of pollution and measure the effects of the sources on water quality, drinking water, and aquatic life. The goal was to prescribe levels of pollution abatement and wastewater treatment needed to improve the quality of the Ohio River and its tributaries to meet all important water uses and potential costs of water pollution abatement. The two lead federal agencies were joined in the investigation by 8 of 14 Ohio River Basin states and the TVA. A number of other federal agencies assisted in the investigation including the USFWS, the successor agency to the USBF, Bureau of Mines, and the USGS.

The survey design included

1. an epidemiological and microbiological assessment of the suitability of the Ohio River as a source of drinking water;
2. investigations of taste and odor problems in 25 drinking water systems resulting from the discharge of phenolic wastes from coking facilities in the upper basin;
3. investigations of the sources and loadings of suspended and deposited oxygen-demanding wastes and fecal bacteria;
4. investigations of the effects of water pollution on fish and other aquatic organisms;
5. an inventory of municipal and industrial wastewater treatment plants and their locations, levels of treatment, discharge volumes, and waste characteristics; and
6. an assessment of acid mine drainage inputs and effectiveness of underground mine seals.[267]

About 131,000 laboratory analyses were made from 71,000 samples collected at 2000 stream locations through 1939. Measurements of 5-day BOD, dissolved, suspended, and volatile solids and fecal coliform bacteria were made for all sampling locations. Dissolved solids, pH, acidity, and alkalinity were analyzed in studies to assess acid drainage conditions in selected locations in the Ohio River and its tributaries. Extensive data tables were published.[267] The outcome of this survey was the development of wastewater treatment goals for sewage and industrial wastes to protect the beneficial uses for water supply, recreation, fisheries, and aquatic life.[268]

The first prospective epidemiological studies of the incidence of illness in swimmers of various age-groups were carried out in 1951 at three beaches where water quality was rated from good to poor: Lake Michigan near Chicago, Illinois; the Ohio River near Dayton, Kentucky; and Long Island Sound, New York.[269] Results showed that swimmers had a higher incidence of illnesses ranging from infections of the ear, eye, nose, skin, throat, and intestinal tract compared to nonswimmers. In all studies, the youngest age-groups experienced 100% more illness than older age-groups. In two studies, swimmers became ill more frequently on days when water quality was rated poor than on days when water quality was rated good. Although not considered conclusive, the studies showed an association between swimming-related illness and bacterial contamination of water.[269] These studies led to the first proposed water quality criteria for recreational waters.[87]

Nutrient enrichment became an important water quality issue in the late 1940s in North America, Europe, and Japan. Excessive phosphorus enrichment caused the accelerated eutrophication of freshwater lakes.[270–272] Eutrophication is a natural aging process in lakes that is accelerated by excessive nutrient inputs. In Lake Washington near Seattle, Washington, accelerated eutrophication was caused by phosphorus in wastewater discharges released directly to the lake.[271] In Lake Washington, diversion of wastewater discharges away from the lake resulted in a reversal of many symptoms of eutrophication.[271]

The eutrophication of Lake Erie became widely known after articles in the press reported massive algal blooms, fish kills, and dissolved oxygen deficits in the lake's bottom waters.[273,274] The causative agent of the algal blooms and associated problems in Lake Erie was identified as phosphorus that derived primarily from discharges of sewage and industrial wastewater and phosphorus fertilizers in agricultural runoff.[273,274]

The Great Lakes Water Quality Agreement of 1972 between the United States and Canada placed a monthly average limit of 1.0 mg L^{-1} for phosphorus in effluents from all municipal wastewater treatment plants discharging at or above 3785 cubic meters per day.[275] Annual limits were placed on Lake Erie tributary loads to control nonpoint sources of phosphorus fertilizer runoff from agriculture. Phosphate detergent bans were enacted in the Lake Erie and other Great Lakes basin states and Canadian provinces to help reduce phosphorus levels in wastewater. Actions taken in the 1970s and 1980s were successful in reducing phosphorus levels and excessive algal growth in Lake Erie. Dissolved oxygen levels increased in many parts of the Lake that were suboxic prior to phosphorus controls.[275] The encouraging results from Lake Erie and Lake Washington demonstrated that accelerated eutrophication is reversible if phosphorus levels are adequately reduced.

In 1972, the USEPA initiated the first National Eutrophication Survey (NES) as a nationwide assessment of the nutrient enrichment of inland public freshwater lakes and reservoirs.[276] The NES was designed to assess the trophic status of freshwater lakes and reservoirs in the United States to determine the major point and nonpoint sources of nutrients to the lake and to prescribe nutrient reduction management plans such as detergent phosphorus bans and phosphorus removal during wastewater treatment. Over 800 lakes and reservoirs, 4200 tributaries and lake outlets, and 1000 wastewater treatment plants were sampled and assessed.[276] Measurements included concentrations of total and dissolved phosphorus, water temperature, water transparency, pH, specific conductance, and dissolved oxygen levels at various depths in the lakes. Algal identifications, chlorophyll a, and laboratory assays of the lake water to determine algal growth helped establish the degree of phosphorus limitation. Trophic state assessments placed each of the 800 lakes into one of four categories: oligotrophic—nutrient poor, mesotrophic—moderately nutrient enriched, eutrophic—enriched, and hypereutrophic—highly enriched. Data, information, and a nutrient reduction plan were provided to the lake owner in a report.[276]

Surface water quality monitoring networks: Water quality monitoring networks are sets of related stations that are sampled over time in an organized and perennial manner.[90] This type of monitoring involves repeated collection of samples at fixed or varying time intervals and at fixed locations to describe water quality status and trends and surveillance activities at various spatial scales. These water quality monitoring activities are collocated with hydrologic measurements of streamflow, lake level, or groundwater level as appropriate. These types of monitoring stations or networks of stations, often with no projected ending point, are known as ambient, fixed station, water quality monitoring networks.[277]

In 1936, a report titled *Deficiencies in Basic Hydrologic Data* called for the development of a nationally extensive water monitoring network.[160] The report found that only 23 federally operated stations in 10 states were monitored regularly over time for water quality. Most of the analyses were for major ions, alkalinity, hardness, dissolved solids, and suspended materials.[160] The report recommended that a federal surface water quality monitoring network be established at streamflow gaging stations.[160] A 200-station network was proposed to be operated by federal agencies and 400 additional stations to be operated cooperatively by federal, state, and local agencies. Data collection would include analysis of major ions and suspended silt load and measurements of pH, acidity, turbidity, and oxygen content. While not implemented at the time, the design served as a model for future surface water quality monitoring networks.

The need for water quality networks signified that a yardstick to assess water quality improvements was of policy interest because of the growth in investments in wastewater treatment and pollution control after 1948. Surveillance networks are used for monitoring water quality for management purposes such as compliance with water quality criteria, pollution control regulations, or goals such as limitations on the transport of mass (loads) delivered to sensitive receiving waters.[90] There are several other types of networks designed to meet specific objectives such as spill detection and warning, to describe baseline water quality, and for multipurpose monitoring to cover a broad range of human water uses and to support tracking of water quality criteria to protect aquatic life and fisheries.[90]

The Ohio River Sanitation Commission (ORSANCO) surveillance and spill notification network was the first interstate surface water quality monitoring network. The ORSANCO network started in 1951 with 11 stations that expanded to 35 stations at or near drinking water intakes on the Ohio River and its 19 major tributaries.[278] The network was sampled for a total of 21 chemical and microbiological constituents and physical properties on a weekly to monthly basis along with daily streamflow. An annual report summarized results for each station and compared values with water quality standards established for the Ohio River.[278] The purpose of the network was to "judge the performance of pollution abatement efforts and to determine what additional efforts are needed."[278]

With the passage of the Federal Water Pollution Control Act in 1956 as amended by Public Law 87–88 in 1961, the USPHS was authorized to establish a surface water surveillance and multipurpose monitoring network. The objectives of the network were to provide information (1) on changes in water quality affected by changes in water use and development, (2) on the nature and extent of pollutants affecting water quality, (3) for development of comprehensive water resources programs, and (4) to assist state, interstate, and other agencies in their water pollution control programs and in the selection of stations for water uses.[279]

Each sampling station in the USPHS network met one or more of the following criteria: stations were either located on major waterways used for public water supply, agriculture or industrial uses, propagation of fish and wildlife, and recreation; or on interstate, coastal, and international boundary waters.[279] Sampling stations were selected after consultation with local, state, federal, and other agencies. Network activities began in October 1957 at 50 fixed-location fixed-frequency sampling stations in 15 major river basins in the lower 48 states and in Alaska.[279] In 1963, the network was named the National Water Quality Surveillance System (NWQSS) and the number of stations increased to 128 (Figure 6).[279] In 1966, the responsibility for network operations was transferred from the USPHS to the FWPCA. In 1971, NWQSS was transferred to the newly created USEPA. The network continued to be operated for more than a decade afterward by the states and the USEPA.

Samples were collected at NWQSS stations on a weekly basis for 5-day BOD, dissolved oxygen, chemical oxygen demand, chlorine demand, coliform bacteria, ammonia nitrogen, pH, color turbidity, temperature, alkalinity, hardness, phosphates, and total dissolved solids. Major ions, trace elements, and radioactivity of gross alpha, gross beta, and strontium-90 were determined on 3-month composites of weekly samples. Plankton populations were collected semimonthly, and chlorinated hydrocarbon pesticides were collected monthly in some years and in short-term surveys in other years. The USPHS published an annual compilation of data as well as journal articles describing results.[279]

In 1958, the USGS established the Hydrologic Benchmark Network (HBN) at 57 stream monitoring stations in 37 states as a baseline network.[280] The HBN consisted of small streams with an average watershed size of 142 km^2 within watersheds that were

FIGURE 6 Map of the conterminous United States showing water quality monitoring stations in the National Water Quality Surveillance System network; 1962–1963.[279]

expected to remain minimally affected by water management and other human activities for the foreseeable future.[280] Baseline networks are located in relatively undisturbed areas to serve for comparison to stations affected by human disturbance.[90] According to Luna Leopold (1915–2006), Chief Hydrologist of the USGS at the time, the HBN was established to answer "A recurring question of our times, and one that we anticipate will be increasingly vexing to posterity, to know how much of the change in our environment is caused by man and how much is natural."[281]

Concentrations of chlorinated hydrocarbon pesticides were reported for a subset of stations in the NWQSS network from 1957 to 1965 and in the HBN network from 1968 to 1970.[282,283] Analyses were made because of concerns about pesticide occurrence in water, sediments, and fish; about immediate toxicity to aquatic life; and about bioaccumulation in organisms such as birds and people.[284] DDT, DDD, and DDE were the most frequently detected pesticides at HBN stations. DDT compounds were detected in about 20% of the water samples and in about 85% of bed sediment samples.[283] At NWQSS stations, dieldrin, DDT, DDD, and endrin were the most frequently detected compounds. Dieldrin was detected at half or more of the stations in 12 major river basins.[282] These results demonstrated a widespread pattern of contamination of rivers and streams by chlorinated hydrocarbon pesticides.

Three stations in the NWQSS network were routinely monitored in the lower Mississippi River; one station each was located at West Memphis, Tennessee; at Vicksburg, Mississippi; and at New Orleans, Louisiana. Detections of endrin, a chlorinated hydrocarbon pesticide, in water and bed sediments at the three lower Mississippi River stations were among the highest nationwide.[282] About 5.4 million fish were killed in a series of incidents in the fall and winter months from 1960 to 1964 in the lower Mississippi River.[285] Endrin contamination in the lower Mississippi River was detected at the three NWQSS sites. The State of Louisiana and the USPHS investigated the lower Mississippi River fish kills and implicated endrin as the causative agent.[285] Endrin was one of the most toxic of the chlorinated hydrocarbon pesticides and was used on cotton and sugarcane grown in the lower Mississippi River valley.[286] Major sources of endrin were applications to cotton and sugarcane and releases from waste ponds and drains at an endrin manufacturing plant in West Memphis, Tennessee.[285,287]

Public concerns were raised over the immediate and long-term effects of pesticides after the publication of *Silent Spring* in 1962.[284] Pesticides were the third leading cause of fish kills nationwide in the 1960s and 1970s.[288] Highly visible incidents of water bird poisonings and reproductive failures from eating fish contaminated with DDT, DDD, and other chlorinated hydrocarbon pesticides were documented at Clear Lake, California and at Klamath Lake, Oregon.[285] Reproductive failures in birds led to the discovery of bioaccumulation of chlorinated hydrocarbon compounds in the food web.[285] DDT was restricted from use in 1972.[289] Endrin was restricted from use in 1980 by the USEPA.[290] Many other chlorinated hydrocarbon pesticides were cancelled by the USEPA in 1987.[291] On 21 October 1972, the Federal Environmental Pesticides Control Act, a far-reaching amendment to the Federal Insecticide, Fungicide, and Rodenticide Act (FIFRA) was enacted. These amendments provide the USEPA with more effective pesticide regulation mechanisms than were previously available under FIFRA.[291]

Beginning in the 1970s, there was a nationwide interest in trend detection to answer the question of whether the nation's water quality was getting better or worse. The suitability of existing data for detecting trends in water quality became an important topic in the early 1970s.[292] A study by Enviro Control published in the Council for Environmental Quality's *Third State of the Environment Report* evaluated data in STORET and concluded that there were severe limitations in available water quality data for trend detection.[293,294] The reported shortcomings of data in STORET for trend analysis included (1) the relatively short length of water quality records, (2) changes in location and frequency of observations, (3) lack of hydrologic data to accompany water quality data, and (4) lack of data on the temporal variability of constituent concentrations. The Enviro Control study noted that of about 70,000 stations in the STORET system, only 142 stations had 8 or more years of samples taken at quarterly or more frequent intervals.[293] A follow-up analysis found that there were only 80 sites nationwide with sufficient long-term data to detect trends.[295]

In response to these challenges, the USGS established the National Stream Quality Accounting Network (NASQAN) to address the need to detect long-term trends.[296] NASQAN was initiated with 50 stations in 1973, increasing to 100 stations in 1974, and eventually to 345 stations in 1975 with more than 500 stations by 1978. The original NASQAN stations were located on the nation's largest rivers.[296] Trend networks such as NASQAN are designed for long-term monitoring of water quality to describe temporal and spatial changes over multidecadal periods of time to track the influences of human and natural factors, and to track changes in water policy implementation.

Groundwater quality surveys: Use and reliance on groundwater as a safe source for public and domestic supply increased as surface water resources became more polluted.[297] Slowly and without much notice at first, incidents of groundwater contamination began to reach serious proportions before there was a national awareness of the problem.[297] As stated by LeGrand, "Many problems of groundwater contamination seemed to have developed casually, and to the geologist and hydrologist, busy in other work, these problems appear isolated and easy to evaluate—at least in hindsight. Consideration of groundwater contamination as a multitude of independent problems, separately solvable as each arises, is outmoded; wise policies, relating water supply to contamination, are needed to alleviate and to forestall problems."[214]

One of the largest problems threatening the quality of groundwater was disposal of solid waste.[217,298] In 1970, one estimate was that about 635,000 kg of solid wastes were generated each day in the United States—enough to fill 1.62 km^2 to a depth of over 3 m.[217] Solid waste included food waste or garbage, solid wastes from manufactured materials known as rubbish, ashes from industrial and home incinerators and wood burning, street refuse and litter, dead animals, and abandoned vehicles and appliances.[217,299] Sometimes, industrial liquid wastes contained in barrels were disposed of with solid waste or were stored at industrial sites.

Disposal methods for solid waste included open dumps, sanitary landfills, on-site solid waste disposal, and solid waste composting.[217,299] Pathways of groundwater contamination were either through direct disposal of waste in ponds, mines, sinkholes, and pits, from fugitive dust emissions, from wastes and agricultural chemicals spread on the land surface, and from wastes disposed of in improperly sited or poorly constructed waste injection wells. A waste injection well is one in which liquid waste is disposed of at depths below freshwater formations.[217]

Open dumps were the most common method of solid waste disposal prior to the 1960s. Open dumps were located in almost any area having the capacity to accept fill material including ponds, lakes, and pits, even abandoned mines and quarries. These types of disposal sites provided a direct connection to groundwater.[217] Landfills were managed by alternating layers of soil over compacted solid waste and were covered each day by soil to impede the penetration of rainfall. However, designs might not have required a liner of clay or other impervious material at the bottom to impede leachates from escaping.[299] Land disposal sites were selected without regard to geologic controls such as soil texture and depth or rock type underlying the disposal site.[217] When precipitation fell on open dumps and uncovered landfills, infiltration of rain into the waste created leachates that leaked from the landfill to pollute surface waters or infiltrate to contaminate groundwater. Leachates were known to contain elevated levels of 5-day BOD, chemical oxygen demand, total organic carbon, nutrients, chloride, sulfate, and dissolved solids.[298] Many industries disposed of their solid wastes in open dumps and sanitary landfills where lack of regulation permitted the practice. Alternative disposal methods included on-site storage and disposal at industrial and manufacturing sites.

By the 1950s, the nature, extent, and severity of groundwater contamination from improper disposal of solid municipal wastes, agricultural chemicals, and various classes of liquid and solid wastes were well known. National policies to control these sources required an additional 20 years of scientific evidence before action was taken.[299] The first major evidence of the deterioration of groundwater quality came from a summary of the contaminants that were found nationwide in groundwater based on 700 published studies.[147] This summary produced recommendations to better understand groundwater contamination for 19 classes of priority contaminants.[147] The top 10 contaminants or contaminant classes in priority order were bacteria, detergents, chlorides, nitrates, pesticides, petroleum oil wastes, radiological waste, contaminants regulated by the drinking water standards of the USPHS, viral contaminants, and trace elements.[147]

California and Illinois, respectively, were the first states to develop regulations to protect groundwater quality from disposal of solid and industrial wastes.[299] Passage of the USPHS Drinking Water Standards in 1962 placed requirements on public water supplies obtained from groundwater sources. The Safe Drinking Water Act of 1974 established maximum contaminant limits on pollutants in treated drinking water. However, these two laws did not fully protect groundwater from unregulated sources of contamination such as disposal of solid and hazardous wastes. One of the most notable environmental incidents reported in 1974 was the contamination of surface and groundwater by industrial wastes at Love Canal, New York. Incidents such as Love Canal led to the passage in 1976 of the Resource Conservation and Recovery Act (RCRA). RCRA became the principal federal law in the United States governing the safe management and disposal of solid and hazardous wastes to protect human health, safety, and the environment.

Conclusions

Water quality monitoring and assessment faced major scientific and technical challenges throughout its history to address important public health, environmental, and water resource issues. Reliable water quality data and information initially were needed to answer fundamental questions such as the following: “Is the water safe to drink?” “Are water resources of appropriate quality for industrial and irrigation uses?” “Is the water safe for recreation and conducive to the growth and reproduction of fish and other aquatic life?” “Is water quality getting better or worse over time?”

The stage was set by 1890 for the accelerated and organized development of water quality activities in the United States. Advances in water chemistry led to reliable methods to indicate water pollution from sewage and other oxygen-demanding wastes. Methods to describe the mineral content of water were applied to the analysis of waters needing treatment to remove hardness and to understand the geochemistry of naturally occurring elements and minerals in water. The discoveries of microbiologists led to a new branch of water analysis of importance to human health—water bacteriology.

High-quality data have proved effective in providing objective information on many societally relevant water quality issues. Answers to the first three of the four questions stated above were possible as a result of development and standardization of physical, chemical, microbiological, and aquatic biological monitoring and assessment methods. Objective and reliable information provided by water quality monitoring and assessment was essential to improve the sanitary quality of the nation’s drinking water, to determine the capacity of streams for natural purification, to determine the availability and quality of water resources for beneficial uses, and to restore the water environment from point and nonpoint sources of water pollution.

Acknowledgments

The USGS Publications Warehouse and the USEPA National Service Center for Environmental Publications were major sources of digital reports. Digital reports were also accessed through the WebArchive, a source of digital publications in the public domain. Digitized public domain books not available from the previous sources were accessed from other institutions and through Google. This chapter serves as a tribute to the individuals and institutions whose accomplishments provided the foundations of water quality monitoring and assessment in the United States. Any use of trade, firm, or product names is for descriptive purposes only and does not imply endorsement by the U.S. Government.

References

1. Hamlin C. *A science of impurity, water analysis in nineteenth century Britain.* Berkeley and Los Angeles, Ca: University of California Press; 1990. pp. 18–46.
2. Karayannis MI, Efstathiou CE. Significant steps in the evolution of analytical chemistry—is the today's analytical chemistry only chemistry? *Talanta* 2012;**102**:7–15.
3. Kirwan R. *An essay on the analysis of mineral waters.* London: J.W. Myers Printing; 1799. 279 pp.
4. Beck II CM. Classical analysis: a look at the past, present, and future. *Anal Chem* 1994;**66**:224A–39A.
5. Davis SN, Davis AG. Saratoga springs and early hydrogeochemistry in the United States. *Ground Water* 1997;**35**(2):347–56.
6. Silliman B. Chemical examination of several natural waters. *Am J Sci Arts* 1846;**52**:218–24.
7. Fresenius CR. English translation. In: Johnson SW, editor. *A system of instruction in quantitative chemical analysis.* New York: John Wiley & Sons; 1871. p. 483–7.
8. Smoot F. *King cholera; the cholera epidemic years in the United States.* 2002. http://www.tngenweb.org/darkside/cholera.html [accessed 11.06.12].
9. Smart C, Abbott SW, Ashmun GC, Daniels WW, Playter E. Report of the commission on the pollution of water supplies. *Am Public Health Assoc, Public Health Pap Rep* 1888;**14**:62–78.
10. Centers for Disease Control and Prevention. Achievements in public Health–1900 to 1999. *Morb Mortal Wkly Rep* 1999;**48**:621–3.
11. Melosi M. *The sanitary city, environmental services in urban America from colonial times to the present.* University of Pittsburg Press; 2008. 354 pp.
12. Bynum R. On the mode of communication of cholera. *Nature* 2013;**495**:169–70.
13. Frankland E. *Water analysis for sanitary purposes.* London: John Van Voorst; 1880. 139 pp.
14. Wanklyn JA, Chapman ET. *Water analysis: a practical treatise on the examination of potable water.* 8th ed. London: Kegan, Paul, Trench, Trubner, & Co., Ltd; 1891. 213 pp.
15. Hassall AH. *A microscopic examination of the water supplied to the inhabitants of London.* London: Samuel Highly; 1850. 67 pp.
16. Kirkwood JP. *A special report on the pollution of river waters, annual report, State Board of Health of Massachusetts.* New York: Arno Press Reprint; 1876.
17. Massachusetts State Board of Health. *A brief history of its organization and its work; 1869–1912.* Boston: Mass; 1912. 1–38.
18. Shulman ST, Friedmann HC, Sims RH. Theodor Escherich: the first pediatric infectious diseases physician. *Clin Infect Dis* 2007;**45**:1025–9.
19. Fox CB. *Sanitary examinations of water, air, and food.* 2nd ed. London: J.A. Churchill; 1886. pp. 9–216.
20. Fuller GW, Whipple GC, Clark HW, Gehrmann A, Wyatt J, Jordan EO, et al. Standard methods of water analysis. *Science* 1900;**12**(311):906–15.
21. American Public Health Association. *Report of the committee on standard methods of water analysis to the laboratory section of the American public health assoc.* Boston: Mass.; 1905. 141 pp.
22. American Public Health Association. *Standard methods for the examination of water and sewage.* 2nd ed. Boston, Mass: American Public Health Assoc; 1915. 144 pp.
23. American Public Health Association. *Standard methods for the examination of water and sewage.* 3rd ed. Boston, Mass: American Public Health Assoc; 1917. 115 pp.
24. American Public Health Association. *Standard methods for the examination of water and sewage.* 4th ed. Boston, Mass: American Public Health Assoc; 1920. 115 pp.

25. American Public Health Association. *Standard methods for the examination of water and sewage.* 5th ed. New York, N.Y: American Public Health Assoc; 1923. 111 pp.
26. American Public Health Association. *Standard methods for the examination of water and sewage.* 7th ed. New York, N.Y: American Public Health Assoc; 1933. 119 pp.
27. American Public Health Association. *Standard methods for the examination of water and sewage.* 6th ed. New York, N.Y: American Public Health Assoc; 1925. 119 pp.
28. American Public Health Association and American Water Works Association. *Standard methods for the examination of water and sewage.* 8th ed. New York, N.Y: American Public Health Assoc., American Water Works Assoc; 1936. 309 pp.
29. American Public Health Association and American Water Works Association. *Standard methods for the examination of water and sewage.* 9th ed. New York, N.Y: American Public Health Assoc., American Water Works Assoc; 1946. 286 pp.
30. American Public Health Association, American Water Works Association, Federation of Sewage and Industrial Wastes Association. *Standard methods for the examination of water and sewage.* 10th ed. New York, N.Y: American Public Health Assoc., American Water Works Assoc., Federation of Sewage and Industrial Wastes Assoc; 1955. 522 pp.
31. American Public Health Association, American Water Works Association. *Water pollution control federation, standard methods for the examination of water and wastewater.* 11th ed. New York, N.Y: American Public Health Assoc., American Water Works Assoc., and Water Pollution Control Fed; 1960. 626 pp.
32. American Public Health Association, American Water Works Association. *Water pollution control federation, standard methods for the examination of water and wastewater.* 12th ed. New York, N.Y: American Public Health Assoc., American Water Works Assoc., Water Pollution Control Fed; 1965. 769 pp.
33. American Public Health Association, American Water Works Association, and Water Pollution Control Federation. *Standard methods for the examination of water and wastewater.* 13th ed. New York, N.Y: American Public Health Assoc., American Water Works Assoc., Water Pollution Control Fed; 1971. 874 pp.
34. American Public Health Association, American Water Works Association, Water Pollution Control Federation. *Standard methods for the examination of water and wastewater.* 14th ed. Washington, D.C: American Public Health Assoc., American Water Works Assoc., and Water Pollution Control Fed; 1975. 1193 pp.
35. Leighton MO. *Field assay of water.* U.S. Geological Survey; 1905. Water Supply and Irrigation Paper 151, 77 pp.
36. Dole RB. *The quality of surface waters in the United States, Part 1. Analyses of waters east of the one-hundredth meridian.* U.S. Geological Survey; 1909. Water Supply Paper 236, 118 pp.
37. Clarke FW. The data of geochemistry, U.S. Geological Survey; *Bulletin* 330, 1908: 716 pp.
38. Theriault EJ, Hommon HB. The determination of biochemical oxygen demand of industrial waste and sewage. *U.S. Public Health Serv Public Health Bull* 1916;**97**:71–86.
39. Foster MD. Colorimetric determination of fluoride in water using ferric chloride. *Industrial Eng Chem Anal Ed* 1933;**5**(4):234–6.
40. Smith T. A new method for determining quantitatively the pollution of water by fecal bacteria. In: *Thirteenth annual report of the State Board of Health of New York for 1892*; 1893. p. 712–22. Albany, N.Y.
41. Prescott SC, Winslow C-EA. *Elements of water bacteriology with special reference to sanitary analysis.* 1st ed. New York: John Wiley and Sons; 1904. 162 pp.
42. Kehr RW, Butterfield CT. Notes on the relation between coliforms and enteric pathogens. *Public Health Rep* 1943;**58**(15):589–607.

43. Perry CA, Hajna AA. Further evaluation of EC medium for the isolation of coliform bacteria and *Escherichia coli. Am J Public Health* 1944;**34**(7):735–8.
44. National Research Council. *Indicators for waterborne pathogens, committee on indicators for waterborne pathogens, National Academy of Science, National Research Council.* Washington, D.C: National Academies Press; 2004. pp. 30–37.
45. Ashbolt NJ, Grabow WOK, Snozzi M. Indicators of microbial water quality. Chap. 13. In: Fewtrell L, Bartram J, editors. *Water quality: guidelines, standards and health.* Geneva, Swiss: World Health Organisation; 2001. p. 289–316.
46. Rafter GW. *The microscopical examination of potable water.* New York: Van Rostrand Co.; 1900. pp. 65–83.
47. Whipple GC. *The microscopy of drinking water.* Revised. 2nd ed. New York: John Wiley and Sons; 1910. 323 pp., 19 Pl.
48. National Research Council. *Freshwater ecosystems: revitalizing educational programs in limnology, committee on inland aquatic ecosystems.* Washington D.C: National Academies Press; 1996. 384 pp.
49. MacDonald RME. *An analytical subject bibliography of the publications of the Bureau of Fisheries, 1871–1920.* Bureau of Fisheries Document 899, Washington, D.C., 306 pp. 1921.
50. Forbes SA. *The lake as a microcosm.* Peoria, Ill: Bulletin Scientific Assoc; 1887. pp. 77–87.
51. Cumming HS, Purdy WC, Ritter HP. Investigation of pollution and sanitary conditions of the Potomac River watershed. In: *U. S. Public Health Service, Hygienic Laboratory Bulletin No. 104*; 1916. p. 239.
52. Forbes SA, Richardson RE. *Studies on the biology of the upper Illinois River.* Urbana, Ill: Bulletin Illinois State Laboratory of Natural History; 1913; 9(10): pp. 478–574.
53. Kehr RW, Purdy WC, Lackey JB, Placek OR, Burns WE. A study of the pollution and natural purification of the Scioto River. *Public Health Bulletin* 1941;**276**:153. U.S. Public Health Service.
54. Ward HB, Whipple GC. *Fresh water biology.* 1st ed. New York: John Wiley and Sons; 1918. 1111 pp.
55. Needham JG, Lloyd JT. *The life of inland waters.* Ithaca, N. Y: The Comstock Publishing Co; 1919. 438 pp.
56. Winslow C-EA, Phelps EB. Investigations of the purification of Boston sewage made at the Sewage Experiment Station of the Massachusetts Institute of Technology with a history of the sewage disposal problem. In: *Water Supply and Irrigation Paper No. 185.* U.S. Geological Survey; 1906. p. 164.
57. Clark HW, Adams GO. *Studies of fish life and water pollution, State Board of Health of Massachusetts.* Forty-fourth Annual Report. 1913. Boston, Mass., pp. 336–345.
58. Tarzwell CM. A brief history of water pollution research in the United States, Section 2. In: *Proceedings of the first and second U.S.A.-U.S.S.R. symposia on the effects of pollutants upon aquatic ecosystems.* Cincinnati, Ohio: EPA 600/3-76-078; 1978. p. 11–31.
59. Ankley GT, Villaneuve DL. The fathead minnow in aquatic toxicology: past, present and future. *Aquat Toxicol* 2006;**78**:91–102.
60. U.S. Geological Survey. *Chemical and bacteriological examination of water.* U.S. Geological Survey; 1903. Hydrographic Branch Circular n. 9, 3 pp.
61. Dole RB. Sanitary inspection versus sanitary analysis. *Public Health Rep* 1905;**31**:59–67.
62. Kemmerer G, Bovard JF, Boorman WR. Northwestern lakes of the United States: biological and chemical studies with reference to possibilities in production of fish. *Bull Bureau Fish* 1923;**944**: 56–7.
63. Birge EA, Juday C. *The inland lakes of Wisconsin, the dissolved gases of the water and their biological significance.* Madison, Wis: Wisconsin Geological and Natural History Survey; 1911. Bulletin 22, n. 7, pp. 5–7.

64. Clapp FG, Bayley WS. *Underground waters of Southern Maine with records of deep wells.* U.S. Geological Survey; 1909. Water Paper 223, pp. 49–53.

65. Stiles CW, Crohurst HR. The principals underlying the movement of *Bacillus coli* in groundwater, with resulting pollution of wells. *Public Health Rep* 1923;**38**(24):1350–3.

66. Bowman I. *Problems of water contamination.* U.S. Geological Survey; 1906. Water Supply and Irrigation Paper 160, pp. 92–95.

67. Collins WD, Howard CS. *Index of analyses of natural waters in the United States.* U.S. Geological Survey; 1925. Water Supply Paper 560-C, pp. 53–85.

68. Collins WD, Howard CS. *Index of analyses of natural waters in the United States, 1926 to 1931.* U.S. Geological Survey; 1932. Water Supply Paper 569-C, pp. 191–209.

69. Whipple GC. *The value of pure water.* New York: John Wiley and Sons; 1907. 61 pp.

70. Leighton MO. *Normal and polluted waters in the Northeastern United States.* U.S. Geological Survey; 1903. Water Supply and Irrigation Paper 79, 102 pp.

71. Collins WD. *Quality of the surface waters of Illinois.* U.S. Geological Survey; 1910. Water Supply Paper 239, 102 pp.

72. Collins WD. *Notes on practical water analysis.* Water Supply Paper 596-H. Washington, D.C: U.S. Geological Survey; 1928. pp. 235–258.

73. Palmer C. *Geochemical interpretation of water analyses.* U.S. Geological Survey; 1911. Water Supply Bulletin 479, 31 pp.

74. Collins WD. *The industrial utility of public water supplies in the United States.* U.S. Geological Survey; 1923. Water Supply Paper 496, 59 pp.

75. Zaporozec A. Graphical interpretation of water quality data. *J Ground Water* 1972;**10**(2):32–43.

76. Winslow C-EA. They were giants in those days. *Am J Public Health* 1953;**43**:15–9.

77. Andreen WL. The evolution of water pollution control in the United States—State, local, and federal efforts, 1789–1972: Part II. *Stanf Environ Law J* 2003;**22**:215–94.

78. Kolkwitz R, Marrson M. Ecology of animal saprobia, Reports of the German Botanical Society. translation by United States Joint Publication Research Service, Washington, D.C., 6 pp. In: *Biology of water pollution, a collection of papers on water pollution.* Federal Water Pollution Control Admin; 1908. p. 47–52 (1968).

79. Kolkwitz R, Marrson M. Ecology of plant saprobia, International Review of Hydrobiology and Hydrogeography. Translation by United States Joint Publication Research Service, Washington, D. C., 10 pp. In: *Biology of water pollution, a collection of papers on water pollution.* Federal Water Pollution Control Admin; 1909. p. 85–95 (1968).

80. Cairns Jr J. Indicator species vs. the concept of community structure as an index of pollution. *Bull Am Water Resour Assoc* 1974;**10**(2):338–47.

81. Warren CE, Doudoroff P. Biological indices. Chap. 20. In: *Biology and water pollution.* Philadelphia, Pa: W.B. Saunders Company; 1971. p. 322–48.

82. U.S. Department of Treasury. Bacteriological standard for drinking water. *Public Health Rep* 1914; **29**(45):2959–66.

83. McKee JE, Wolf HD. *Water quality criteria.* Publication No. 3-A, California. 2nd ed. Water Resources Control Board; 1963. p. 89.

84. Taylor F. Experience with the coliform standard under the interstate carrier program. In: Hendricks CW, editor. *Evaluation of the microbiological standards for drinking water.* Washington, D.C: U.S. Environmental Protection Agency; 1978. p. 153–7. EPA 570/9-78-OOC.

85. Simons Jr GW, Furgeson HF, Gage SDe M. Report of the committee on bathing places. *Am J. Public Health* 1922;**12**(2):121–3.

86. Streeter HW. Tendencies in standards of river and lake cleanliness. *Sew Works J* 1934;**6**(4):721–33.
87. Streeter HW. Standards of stream sanitation. *Sew Works J* 1949;**21**(1):115–26.
88. Suter R, Moore E. *Stream pollution studies.* Albany, N. Y: New York State Conservation Comm; 1922. 34 pp.
89. Meybeck M, Kimstach V, Helmer R. Strategies for water quality assessments. Chap. 2. In: Chapman D, editor. *Water quality assessments.* 2nd ed. London: Chapman and Hall; 1996. not paginated.
90. Helmer R. Water quality monitoring: national and international approaches. In: *Proceedings of the Rostov-on-Don Symposium*, vol. 219. International Assoc. of Hydrological Sciences; 1994. p. 3–17.
91. U.S. Environmental Protection Agency. *25 years of the safe Drinking water act: history and trends.* U.S. Environmental Protection Agency; 1999. EPA 816-R-00–007, 53 pp.
92. Rosenau MJ. Sanitary analysis of water. Part VI. In: *Preventive Medicine and Hygiene.* New York: Appleton and Co.; 1913. p. 691–763.
93. U.S. Bureau of Reclamation. *Lime softening fact sheet.* U.S. Department of Interior; 2010. 4 pp.
94. Leverett F. *The water resources of Illinois, Part II. Economic geology and hydrography.* U.S. Geological Survey; 1896. Seventeenth Annual Report, pp. 116–134.
95. Orton E. *Rock waters of Ohio.* Nineteenth Annual Report of the U.S. Geological Survey. 1898. pp. 633–717.
96. Lane AC. *Lower Michigan mineral waters, a study into the connection between their chemical composition and mode of occurrence.* U.S. Geological Survey; 1899. Water Supply Paper 31, 97 pp.
97. Palmer AW. *Chemical survey of the waters of Illinois.* preliminary report. Urbana, Ill: University of Illinois; 1897. 53 pp.
98. Palmer AW. *Chemical survey of the waters of Illinois: report for the years 1897–1902.* Urbana, Ill: University of Illinois; 1903. 254 pp.
99. Bartow E. *Municipal water supplies of Illinois. Water survey series n. 5,* **v. 5**. University of Illinois Bulletin; 1907. n. 7, Urbana, Ill., 123 pp.
100. Bartow E. *Chemical and biological survey of the waters of Illinois. Water survey series n. 10,* vol. 10. Urbana, Ill: University of Illinois Bulletin; 1913. n. 36, pp. 10.
101. Bartow E. *Chemical and biological survey of the waters of Illinois. Water survey series n. 13,* **v. 14**. Urbana, Ill: University of Illinois Bulletin; 1916. n. 5, p. 12.
102. Collins WD. *Relations between quality of water and industrial development in the United States.* U.S. Geological Survey; 1926. Water Supply Paper 559, 42 pp.
103. Leighton MO. An official examination of the waters of the United States. *Public Health Pap* 1902;**28**: 272–7.
104. Stabler H. *Some stream waters of the western United States with chapters on sediment carried by the Rio Grande and the industrial application of water analyses.* U.S. Geological Survey; 1911. Water Supply Paper 274, p. 188.
105. Van Winkle W, Eaton FM. *The quality of surface waters of California.* U.S. Geological Survey; 1910. Water Supply Paper 237, 134 pp.
106. Van Winkle W. *The quality of surface waters of Oregon.* U.S. Geological Survey; 1910. Water Supply Paper 363, 137 pp.
107. Van Winkle W. *The quality of surface waters of Washington.* U.S. Geological Survey; 1914. Water Supply Paper 339, 105 pp.
108. Parker HN. *Quality of the water supplies of Kansas, with a preliminary report on stream pollution by mine waters in southeastern Kansas.* U.S. Geological Survey; 1911. Water Supply Paper 273, 375 pp.

109. Waring GA, Dole RB, Chambers AA. *Mineral springs of Alaska, with a chapter on the chemical character of some surface waters of Alaska.* U.S. Geological Survey; 1917. Water Supply Paper 418, 114 pp.

110. Clarke FW. *The composition of waters of the United States.* U.S. Geological Survey; 1924. Professional Paper 135, 198 pp.

111. Massachusetts State Board of Health. *Twenty-sixth annual report of the State Board of Health of Massachusetts, 2. Water supply and sewerage.* 1895. Boston, Mass., pp. 3–712.

112. Massachusetts State Board of Health. *Water supplies and inland waters of Massachusetts; 1887–1890.* Report on water supply and sewerage, Part 1, Boston, Mass., 857 pp. 1890.

113. Durant E. *Ellencyclopedia.* Massachusetts Institute of Technology Review; 2007. accessed at http://www.technologyreview.com/article/408456/ellencyclopedia/page/3/.

114. Sedgwick WT. On recent epidemics of typhoid fever in the cities of Lowell and Lawrence due to infected water supply. In: *State board of health of Massachusetts*; 1893. p. 667–704. Twenty-Fourth Annual Report, Boston, Mass.

115. New York State Board of Health. *Tenth annual report of the State Board of Health for the State of New York.* 1890. Albany, N.Y., p. 112.

116. Connecticut State Board of Health. *Fourteenth annual report of the state board of health of the state of Connecticut.* 1892. New Haven, Conn. pp. 231–425.

117. Ohio State Board of Health. *Thirteenth annual report of the State Board of Health of the State of Ohio.* 1898. Columbus, Ohio, pp. 345–593.

118. Leighton MO. *Sewage pollution in the metropolitan area near New York City and its effects on inland water resources.* U.S. Geological Survey; 1902. Water Supply and Irrigation Paper 72, 75 pp.

119. Parker HN, Bailey W, Bolder RH, Ashe WW, Marsh MC. *The Potomac river Basin.* U.S. Geological Survey; 1907. Water Supply and Irrigation Paper 192, 364 pp.

120. Goodell EB. *A review of the laws forbidding pollution of inland waters of the United States.* U.S. Geological Survey; 1905. Water Supply and Irrigation Paper 103, 120 pp.

121. Leighton MO. *Preliminary report on the pollution of Lake Champlain.* U.S. Geological Survey; 1905. Water Supply Paper 121, 117 pp.

122. Kofoid CA. *The plankton of the Illinois River, 1894–1899*, vol. 8. Urbana, Ill: Bulletin Illinois State Laboratory of Natural History; 1908. n. 1, 360 pp.

123. Forbes SA, Richardson RE. *The fishes of Illinois, natural history survey of Illinois.* Danville, Ill: State Laboratory of Natural History; 1908. 357 pp.

124. Forbes SA, Richardson RE. Some recent changes in Illinois River biology. *Bull Ill Nat Hist Surv* 1919; **13**:139–56. Springfield, Ill.

125. Richardson RE. *The bottom fauna of the Illinois River, 1913–1925, State of Illinois.* Urbana, Ill: Division of the Natural History Survey; 1928. Bulletin 17, n. 12, pp. 387–475.

126. Black WM, Phelps EB. *Report of Col. William M. Black, U.S.A. and Prof. E.B. Phelps concerning the location of sewer outlets and the discharge of sewage to New York Harbor, City of New York Board of Estimate and Apportionment.* 1911. New York, N. Y., 193 pp.

127. Phelps EB. Studies of the pollution of New York Harbor and the Hudson River. *Sew Works J* 1934; **6**(5):998–1008.

128. McLaughlin AJ. Sewage pollution of interstate and international waters with special reference to the spread of typhoid fever: I, Lake Erie and the Niagara River, Public Health and Marine Hospital Service. *Hyg Lab Bull N* 1911;**77**:169.

129. Rosenau MJ, Lumsden LL, Kastle JH. *The origin and prevalence of typhoid fever in the District of Columbia*, vol. 52. U.S. Public Health and Marine Hospital Service of the United States, Hygienic Laboratory, Bulletin; 1909. Report no. 3, pp. 1–157.
130. McLaughlin AJ. *Sewage pollution of interstate and international waters with special reference to the spread of typhoid fever: the Missouri River from Sioux City to its mouth*. U.S. Public Health Service, Hygienic Laboratory Bulletin; 1913. n. 89, 84 pp.
131. Frost WH. A review of the work of the U.S. Public Health Service in investigations of stream pollution. *Public Health Rep* 1926;**41**(3):75–85.
132. Hoskins TK. Planning the organization and conduct of stream pollution surveys. *Public Health Rep* 1938;**53**:729–35.
133. U.S. Public Health Service. *Annual report of the Surgeon General of the Public Health Service of the United States for fiscal year 1915*. 1915. Washington, D.C., pp. 99–112.
134. Theriault EJ. The rate of deoxygenation of polluted waters. *Public Health Rep* 1926;**41**(6):207–17.
135. Streeter HW, Phelps EB. A study of the pollution and natural purification of the Ohio River: III Factors concerned in the phenomena of oxidation and reaeration. *US Public Health Serv Public Health Bull* 1925;**146**:75.
136. U.S. Public Health Service. *Annual report of the surgeon general of the public health service of the United States*. Washington, D.C: U.S. Public Health Service; 1918. 373 pp.
137. Tarr JA. Industrial wastes and public health; some historical notes, part 1, 1876–1932. *Am J Public Health* 1985;**75**(9):1059–67.
138. Leighton MO. Industrial wastes and their sanitary significance. *Public Health Pap Rep* 1905;**31**(1): 29–40.
139. Sackett RL, Bowman I. *The disposal of strawboard and oil-well wastes*. U.S. Geological Survey; 1905. Water Supply Paper 113, 51 pp.
140. Stabler H. *Prevention of stream pollution by distillery refuse, based on investigations at Lynchburg, Ohio 1906*. U.S. Geological Survey; 1906a. Water Supply Paper: 179. 34 pp.
141. Stabler H. *Stream pollution by acid-iron wastes, a report based on investigations made at Shelby, Ohio 1906*. U.S. Geological Survey; 1906. Water Supply Paper: 186, 37 pp.
142. Stabler H, Pratt GH. *The purification of some textile and other factory wastes*. U.S. Geological Survey; 1909. Water Supply Paper: 235, 75 pp.
143. Winslow C-EA, Moxon D. Bacterial pollution of bathing beach waters in New Haven Harbor. *Public Health Rep* 1929;**44**(3):130–4.
144. Mallman WL, Sypien A. Pollution indices of natural bathing places. *Am J Public Health* 1934;**24**(7): 681–8.
145. Scott WJ. Survey of Connecticut's shore bathing waters. *Am J Public Health* 1932;**22**(3):316–21.
146. McCallie SW. *Experiment relating to problems of well contamination in Quitman, Georgia*. U.S. Geological Survey; 1905. Water Supply Paper 110, pp. 45–54.
147. Stanley WE, Eliassen R. *Status of knowledge of groundwater contaminants*. Federal Housing Admin., Technical Studies Program; 1961. 465 pp.
148. Fetter Jr CW. Hydrogeology: a short history, Part 2. *J Ground Water* 2004;**42**(6):949–52.
149. McGuiness CL. *The role of groundwater in the national water situation*. U.S. Geological Survey; 1963. Water Supply Paper 1800, 1121 pp.
150. Fuller GW, Clapp FG, Matson GC, Sanford S, Wolff HC. *Underground water papers*. U.S. Geological Survey; 1911. Water Supply Paper 258, pp. 57–65.
151. Matson GC. *Pollution of underground waters in limestone*. U.S. Geological Survey; 1911. Water Supply Paper 258.

152. Dole RB. *Use of fluorescein in the study of underground waters.* U.S. Geological Survey; 1906. Water Supply and Irrigation Paper 160, pp. 73–85.

153. Durum WH. *Historical profiles of quality of water laboratories and activities, 1879–1973.* U.S. Geological Survey; 1978. Open-File Report 78-432, 235 pp.

154. Baird D. Analytical chemistry and the "big scientific instrumentation revolution". In: Morris PJT, editor. *From classical to modern chemistry, the instrumental revolution.* London, UK: Royal Society for Chemistry and the Chemistry Museum; 2002. p. 29–56.

155. Travis AS. Instrumentation in environmental analysis, 1935–1975. In: Morris PJT, editor. *From classical to modern chemistry, the instrumental revolution.* London, UK: Royal Society for Chemistry and the Chemistry Museum; 2002. p. 285–308.

156. Rainwater FH, Thatcher LL. *Methods for collection and analysis of water samples.* U.S. Geological Survey; 1960. Water Supply Paper 1454, 301 pp.

157. Federal Water Pollution Control Administration. *Interim methods for chemical analysis of surface waters.* Cincinnati, Ohio: U.S. Department of Interior, Federal Water Pollution Control Admin; 1968. not paginated.

158. Federal Water Pollution Control Administration. *Methods for chemical analysis of surface waters.* Cincinnati, Ohio: U.S. Department of Interior, Federal Water Pollution Control Admin; 1969. not paginated.

159. U.S. Environmental Protection Agency. *Methods for chemical analysis of water and wastes, 1971.* Cincinnati, Ohio: U.S. Environmental Protection Agency; 1971. 312 pp.

160. National Resources Committee. *Report of the Special Advisory Committee on standards and specifications for hydrologic data, deficiencies in basic hydrologic data.* Washington, D.C: Water Resources Committee of the National Resources Committee; 1936. 72 pp.

161. U.S. Geological Survey. *Goals of the U.S.* Geological Survey, U.S. Geological Survey; 1986. Circular 1010, 17 pp.

162. U.S. Environmental Protection Agency. *Guidelines establishing procedures for analysis of pollutants.* 1973. Federal Register, Title 40 Part 136, n. 199, 28758–60.

163. U.S. Environmental Protection Agency. *Summary of the safe drinking water act.* 2014. http://www2.epa.gov/laws-regulations/summary-safe-drinking-water-act [accessed 07.12.14].

164. Scofield CS. *Measuring the salinity of irrigation waters and of soil solutions with the Wheatstone bridge.* U.S. Department of Agriculture; 1932. Circular 232, 20 pp.

165. Hem JD. *Study and interpretation of chemical characteristics of natural water.* Water Supply Paper 2254. 3rd ed. U.S. Geological Survey; 1985. 264 pp.

166. Collins WD, Howard CS, Love SK. *Quality of surface waters of the United States, 1941.* U.S. Geological Survey; 1943. Water Supply Paper 942, 73 pp.

167. Miller RL, Bradford WL, Peters NE. *Specific conductance: theoretical considerations and application to analytical quality control.* U.S. Geological Survey; 1988. Water Supply Paper 2311, 16 pp.

168. Gallwas J, Beckman AO. 1900–2004. *Anal Chem* 2004:264

169. U.S. Public Health Service. Laboratory research, field investigation, and training program of the Robert A. Taft Sanitary Engineering Facility at Cincinnati, Ohio. *Public Health Rep* 1954;**69**(5): 508–12.

170. Kennedy VC, Zellweger GW, Jones BF. Filter pore-size effects in the analysis of Al, Fe, Mn, and Ti in water. *Water Resour Res* 1974;**10**(4):785–90.

171. Hem JD. *Study and interpretation of chemical characteristics of natural water.* Water Supply Paper 1473. 2nd ed. U.S. Geological Survey; 1970. 363 pp.

172. Murphy J, Riley J. A modified single solution for the determination of phosphate in natural waters. *Anal Chim Acta* 1962;**27**(31).

173. Gales Jr M, Julian E, Kroner R. Method for quantitative determination of total phosphorus in water. *J Am Water Works Assoc* 1966;**58**(10):1363.
174. Brown E, Skougstad MW, Fishman MJ. *Methods for collection and analysis of water samples for dissolved minerals and gases.* U.S. Geological Survey; 1970. Techniques of Water Resources Investigations, Book 5, Chapter A-1, 160 pp.
175. Walsh A. Atomic absorption spectroscopy and its applications—old and new. *Pure Appl Chem* 1977; **49**:1621–8.
176. Breidenbach W, Lichtenberg JJ, Henke CF, Smith DJ, Eichelberger Jr JW, Stierli H. The identification and measurement of chlorinated hydrocarbon pesticides in surface waters. *Fed Water Pollut Control Admin* 1964:63.
177. Goerlitz DF, Brown E. *Methods for analysis of organic substances in water.* U.S. Geological Survey; 1972. Techniques of Water Resources Investigations, Book 5, Chap. A3, 40 pp.
178. Friedman A. Deuterium content of natural waters and other substances. *Geochim Cosmochim Acta* 1953;**4**:89–103.
179. Kabler PW, Clark HF, Geldreich EE. Sanitary significance of coliform and fecal coliform organisms in surface water. *Public Health Rep* 1964;**79**(1):58–60.
180. Clark HF, Geldreich EE, Jeter HL, Kabler PW. The membrane filter in sanitary bacteriology. *Public Health Rep* 1951;**66**(30).
181. Geldreich EE, Kabler PW, Jeter HL, Clark HF. A delayed incubation membrane filter test for coliform bacteria in water. *Am J Public Health* 1955;**45**:1462–74.
182. Kenner A, Clark HF, Kabler PF. Fecal streptococci II, quantification of streptococci in feces. *Am J Public Health* 1960;**50**(10):1553–9.
183. Geldreich EE, Clark HF, Huff CB, Best LC. Fecal coliform organism medium for the membrane filter technique. *J. Am Water Works Assoc* 1965;**57**:208–13.
184. Pennak RW. *Fresh-water invertebrates of the United States.* New York: Ronald Press; 1953. 769 pp.
185. Macy RW. Book review of Fresh-water Invertebrates of the United States. *Science* 1953;**118**:34.
186. Welch PS. *Limnological methods.* Blakiston, Philadelphia, Pa., 381 pp. 1948.
187. Ward HB, Whipple GC. Fresh water biology. In: Edmonson WT, editor. 2nd ed. New York: John Wiley and Sons; 1960. p. 1248.
188. Federal Water Pollution Control Administration. *A guide to the common diatoms at water pollution surveillance system stations.* Cincinnati, Ohio: Federal Water Pollution Control Admin.; 1966. 101 pp.
189. Hynes HBN. *The Biology of polluted waters.* Liverpool: Liverpool Univ. Press; 1960. 555 pp.
190. Hynes HBN. *The biology of Running waters.* Liverpool: Liverpool Univ. Press; 1970. 202 pp.
191. Mackenthun KM. *The practice of water pollution biology.* Washington, D.C: Federal Water Pollution Control Admin; 1969. 293 pp.
192. Slack KV, Averett RC, Greeson PE, Lipscomb RG. *Methods for collection and analysis of aquatic biological and microbiological samples.* U.S. Geological Survey; 1973. Techniques of Water-Resource Investigation: 05–A4, 165 pp.
193. Strickland JDH, Parsons TR. A practical Handbook of Seawater analysis. *Fish Res Board Can Bull* 1972;**167**:310. Ottawa, Canada.
194. Ellis MM. Detection and measurement of stream pollution. *Bull US Bureau Fish* 1937;**48**:365–437.
195. Hunn JB. *History of acute toxicity tests with fish, 1863–1987, in Investigations in Fish Control.* La Crosse, Wisconsin: U.S. Fish and Wildlife Service; 1989. 10 pp.
196. Doudoroff P, Anderson BG, Burdick GE, Galtsoff PS, Hart WB, Patrick R, et al. Bioassay methods for the evaluation of acute toxicity of industrial wastes to fish. *J Sew Ind Wastes* 1951;**23**:1380–97.

197. U.S. Environmental Protection Agency. *Methods for acute toxicity tests with fish, macroinvertebrates, and amphibians.* Corvallis, Ore: U.S. Environmental Protection Agency; April 1975. 79 pp.
198. Johnson WW, Finley MT. *Handbook of acute toxicity of chemicals to fish and aquatic invertebrates,* vol. 137. U.S. Fish and Wildlife Service, Resource Publication; 1980. 98 pp.
199. Roberson E, Feth JH, Seaber PR, Anderson P. *Differences between field and laboratory determinations of pH, alkalinity, and specific conductance of natural water, Article 115, in Short Papers in Geology and Hydrology.* U.S. Geological Survey; 1963. Professional Paper 475-C, pp. 212–215.
200. Barnes A. *Field measurement of alkalinity and pH.* U.S. Geological Survey; 1964. Water Supply Paper 1535-H, 17 pp.
201. Wood WW. *Guidelines for collection and field analysis of groundwater samples for selected unstable constituents.* U.S. Geological Survey; 1976. Techniques of Water Resources Investigations, Book 1, Chap. D-2, 24 pp.
202. Hach Chemical Company, History and heritage, http://www.hach.com/history. [accessed 30.03.14].
203. Yellow Springs Instruments, A brief history of YSI, http://www.ysi.com/history.php. [accessed 07.07.14].
204. Heineman WR, Jensen WB, Leland C. Clark, Jr. (1918–2005) *Biosens Bioelectron* 2006;**21**:1403–4.
205. Haney PD, Schmidt J. Representative sampling and analytical methods in stream studies. *J Sew Ind Wastes* 1958;**30**(6):812–20.
206. Kittrell FW. *A practical guide to water quality study of streams, Report CWR-5.* Federal Water Pollution Control Admin; 1969. 135 pp.
207. Velz CJ. *Applied stream sanitation.* New York, New York: Wiley-Interscience; 1970. 619 pp.
208. Schroepfer GT. An analysis of stream pollution and stream standards. *Sew Works J* 1942;**14**(5): 1030–63.
209. Velz J, Clark RN. 1950, Sampling for effective evaluation of stream pollution. *J Sew Ind Wastes* 1950; **22**(5):666–84.
210. Tsivoglou C, Cohen JB, Shearer SD, Godsil PJ. Tracer measurement of stream reaeration II: field studies. *J Water Pollut Control Fed* 1968;**40**(2):285–305.
211. Kittrell FW, West AW. Stream survey procedures. *J Water Pollut Control Fed* 1967;**39**(4):627–41.
212. Morris DA, Hackett OM, Vanlier KE, Moulder EA, Durum WH. *Groundwater resources of Riverton irrigation project area.* Wyoming, U.S. Geological Survey; 1959. Water Supply Paper 1375, pp. 74–75.
213. Feltz HR, Hanshaw BB. *Preparation of water sample for carbon-14 dating.* U.S. Geological Survey; 1963. Circular 480, 3 pp.
214. LeGrand HE. Monitoring changes in quality of groundwater. *J Groundw* 1968;**6**(3):14–8.
215. LeGrand HE. Environmental framework of groundwater contamination. In: Pettyjohn WA, editor. *Water quality in a stressed environment.* Minneapolis, Minn: Burgess Publishing Co; 1972. p. 90–8.
216. Deutsch M. Natural controls involved in shallow aquifer contamination. *Groundwater* 1965;**3**(3): 37–40.
217. Schneider WJ. *Hydrologic implications of solid-waste disposal.* U.S. Geological Survey; 1970. Circular 601-F, 10 pp.
218. Federal Interagency River Basin Committee. *Inventory of published and unpublished chemical analyses of surface waters in the western United States.* Bulletin 2, Washington, D.C., 114 pp. 1948.
219. Federal Interagency Committee on Water Resources. *Inventory of published and unpublished chemical analyses of surface waters in the western United States.* 1956. 1947–1955, Bulletin 9, Washington, D.C., 65 pp.
220. Federal Interagency River Basin Committee. *Inventory of published and unpublished chemical analyses of surface waters in the eastern United States.* Bulletin 6, Washington, D.C., 97 pp. 1954.

221. Woodward TH, Heidel SG. *Inventory of published and unpublished chemical analyses of surface waters in the continental United States and Puerto Rico, 1961*. U.S. Geological Survey; 1964. Water Supply Paper 1786, 490 pp.

222. Green RS. *The storage and retrieval of data for water quality control.* U.S. Public Health Service; 1964. 36 pp.

223. Dubois DP. *Storet II, Storage and retrieval of data for open water and land areas.* Federal Water Pollution Control Admin; 1966. 25 pp.

224. Sayers WT. Water quality surveillance: the federal-state network. *Environ Sci Technol* 1971;**5**(2): 114–9.

225. Campbell FB, Bauder HA. A rating-curve method for determining silt-discharge of streams. *Trans Am Geophys Union* 1940;**21**(2):603–7.

226. Colby BR. *The Relationship of sediment discharge to streamflow.* U. S. Geological Survey; 1956. Open File Report 56-27, 170 pp.

227. Colby BR. *Fluvial sediments—a summary of source, transportation, deposition, and measurement of sediment discharge.* U. S. Geological Survey; 1963. Bulletin 1181-A, 47 pp.

228. Phelps EB. *Stream sanitation.* London: John Wiley and Sons, Inc; 1944. 275 pp.

229. Ambrose RB, Wool TA, Barnwell Jr TO. Development of water quality modeling in the United States. *Environ Eng Res* 2009;**4**:200–10.

230. Piper AM. A graphic procedure in the geochemical interpretation of water analyses. *Am Geophys Union Trans* 1944;**25**:914–23.

231. Back W. *Origin of hydrochemical facies of groundwater in the Atlantic Coastal Plain.* U.S. Geological Survey; 1961. Professional Paper 434-D, pp. 380–2.

232. Stiff Jr HA. The interpretation of chemical water analysis by means of patterns. *J. Petroleum Technol* 1951;**3**(10):15–7.

233. Bricker OP, Garrels RM. Mineralogic factors in natural water equilibria. In: Back W, Freeze RA, editors. *Chemical hydrogeology*; 1967. p. 275–6 (1983).

234. Hanshaw BB, Back W, Rubin M. Carbonate equilibria and radiocarbon distribution related to groundwater flow in the Floridan Aquifer, U.S.A. In: Back W, Freeze RA, editors. *Chemical hydrogeology*; 1965. p. 93–206 (1983).

235. Hem JD, Cropper WH. *Survey of ferrous-ferric chemical equilibria and redox potentials.* U.S. Geological Survey; 1959. Water Supply Paper 1459, pp. 1–32.

236. Hem JD. *Study and interpretation of chemical characteristics of natural water.* Water Supply Paper 1473. 1st ed. U.S. Geological Survey; 1959. 269 pp.

237. Stumm W, Morgan JJ. *Aquatic chemistry: an introduction emphasizing chemical equilibria in natural waters.* John Wiley and Sons Inc; 1970. 583 pp.

238. Betz K. Stumm and Morgan win Stockholm Water Prize. *Environ Sci Technol* 1999;**33**(11):233a.

239. Plummer LN, Jones BF, Truesdell AH. *WATEQF a Fortran IV version of WATEQ, a computer program for calculating chemical equilibrium of natural waters.* U.S. Geological Survey; 1976. Water-Resources Investigations Report 76-13, 59 pp.

240. Grove DB, Stollenwerk KG. Chemical reactions simulated by groundwater quality models. *Water Resour Bull* 1987;**23**(4):601–15.

241. Craig H. Isotopic variations in meteoric waters. *Science* 1961;**133**:1702–3.

242. Rhodehamel EC, Kron VB, Dougherty VM. *Bibliography of tritium studies related to hydrology, through 1966.* U.S. Geological Survey; 1971. Water-Supply Paper 1900, 174 pp.

243. Hanshaw BB, Back W, Rubin M. Radiocarbon determinations for estimating groundwater flow velocities in central Florida. *Science* 1965;**148**(3669):494–5.

244. Hanshaw BB, Back W, Rubin M, Meyer RL, Wait RL. Relation of carbon-14 concentration to saline water contamination of coastal aquifers. *Water Resour Res* 1965;**1**:109–14.

245. Hoffman CM, Stewart GL. *Quantitative determination of tritium in natural waters.* U.S. Geological Survey; 1966. Water Supply Paper 1696-D, 18 pp.

246. Rosenshein JS, Moore JE. *History of the development of hydrogeology in the United States.* National Ground Water Assoc. Press; 2012. pp. 143–148.

247. Keup LE, Ingram WM, Mackenthun KM. *Biology of water pollution, a collection of selected papers on stream pollution, waste water, and water treatment.* Federal Water Pollution Control Admin., Department of Interior; 1967. 127 pp.

248. Patrick R. A proposed biological measure of stream conditions based on a survey of Conestoga Basin, Lancaster County, Pennsylvania. *Proc Acad Nat Sci Phila* 1949;**101**:277–341.

249. Patrick R. Biological measure of stream conditions. *Sew Ind Wastes J* 1950;**22**(7):926–38.

250. Margalef DR. Information theory in ecology. *General Syst* 1958;**3**:36–71.

251. Patrick R, Gaufin AR, Tarzwell CM. Aquatic organisms as an aid in solving waste disposal problems. *J Sew Ind Wastes* 1953;**25**(2):210–4.

252. U.S. Public Health Service. *Public health service drinking water standards.* revised, 1962. Washington, D.C: U.S. Public Health Service; 1962. 61 pp.

253. Federal Water Pollution Control Administration. *Water quality criteria, a report of the National Technical Advisory Committee to the Secretary of Interior.* 1968. Washington, D.C., 234 pp.

254. U.S. Environmental Protection Agency. *Water quality criteria, 1972–A report of the committee on water quality criteria.* Washington, D.C: National Academy of Sciences and National Academy of Engineering; 1972. p. 594.

255. U.S. Department of Commerce. *Historical statistics of the United States, Colonial times to 1970, volume 1, chapter J., land, water, and climate.* Census Bureau, Washington, D.C: U.S. Department of Commerce; 1975. p. 434.

256. Collins WD, Lamar WL, Lohr EW. *The industrial utility of public water supplies of the United States, 1932.* U.S. Geological Survey; 1934. Water Supply Paper 658, 135 pp.

257. Lohr EW, Love SK. *The industrial utility of public water supplies in the United States, 1952, Part 1: states east of the Mississippi River.* U.S. Geological Survey; 1952. Water Supply Paper 1299, 639 pp.

258. Lohr EW, Love SK. *The industrial utility of public water supplies in the United States, 1952, Part 2: states west of the Mississippi River.* U.S. Geological Survey; 1952. Water Supply Paper 1300, 462 pp.

259. Durfor CJ, Becker E. *Public water supplies of the 100 largest cities in the United States, 1962.* U.S. Geological Survey; 1968. Water Supply Paper 1812, 364 pp.

260. Streeter HW. Experimental studies of water purification. *Public Health Rep* 1933;**48**(15):377–400.

261. Brezonik PL, Arnold WA. Water chemistry: fifty years of change and progress. *Environ Sci Technol* 2012;**46**:5650–7.

262. Shuval HI. The public health significance of trace chemicals in waste water utilization. *Bull World Health Organ* 1962;**27**:791–800.

263. McDermott JE. Significance of the national community water supply study. In: Pettyjohn WA, editor. *Water quality in a stressed environment.* Minneapolis, Minn: Burgess Publishing Co; 1972. p. 1–11.

264. U.S. Environmental Protection Agency. *Industrial pollution of the lower Mississippi River in Louisiana.* 1972. Dallas, Tex., 155 pp.

265. U.S. Public Health Service. *Ohio River pollution study, final report to the Ohio River Committee,* vol. 1. U.S. Public Health Service; 1942. 373 pp.

266. Tarzwell CM. Augmentation of water resources through pollution abatement. *Ohio J Sci* 1963;**63**(2): 75–82.
267. U.S. Public Health Service. *Ohio River pollution study, supplements A-F, collection of data on sources of pollution.* U.S. Public Health Service; 1942. 618 pp.
268. Smith Jr JE, Kreissl JF, Logsdon GS, Schomaker N. Environmental research in Cincinnati: a century of federal partnership. *Environ Eng* 2011;**13**:11.
269. Stevenson AH. Studies of bathing water quality and health. *Am J Public Health* 1953;**43**(5 Pt 1):529–38.
270. Hasler AD. Eutrophication of lakes by domestic drainage. *Ecology* 1947;**28**:383–95.
271. Edmondson WT, Anderson GC, Peterson DR. Artificial eutrophication of Lake Washington. *Limnol Oceanogr* 1956;**1**:47–53.
272. Dillon PJ, Rigler FH. The phosphorus-chlorophyll relationship in lakes. *Limnol Oceanogr* 1974;**19**: 767–73.
273. Beeton AM. Eutrophication of the St. Lawrence Great Lakes. *Limnol Oceanogr* 1965;**10**:240–55.
274. Federal Water Pollution Control Administration. *The Lake Erie report, a plan for water pollution control.* Great Lakes Region: Federal Water Pollution Control Admin; 1968. pp. 53–65.
275. Environment Canada, history of the Great Lakes water quality Agreement, https://www.ec.gc.ca/grandslacs-greatlakes/default.asp?lang=En&n=647DC488-1. [accessed 30.06.14].
276. U.S. Environmental Protection Agency. *National eutrophication survey methods, 1973–1976, working Paper No. 175.* U.S. Environmental Protection Agency; 1975. 91 pp.
277. Ward RC, Loftis JC, McBride GB. *Design of water quality monitoring systems.* New York, N. Y: Van Nostrand Reinhold; 1990. p. 10.
278. Cleary EJ. Perspective on river-quality diagnosis. *J Water Pollut Control Fed* 1978;**50**:825–32.
279. U.S. Public Health Service. *Water pollution surveillance system annual compilation of data.* 1963. October 1, 1962 to September 30, 1963, v. 9, pp. 1–17.
280. Cobb ED, Beisecker JE. *The national hydrologic benchmark network.* U.S. Geological Survey; 1971. Circular 460-D, 38 pp.
281. Leopold LB. *A national network of hydrologic bench marks.* U.S. Geological Survey; 1962. Circular 460-B, 4 pp.
282. Breidenbach W, Gunnerson CG, Kawahara FK, Lichtenberg JJ, Green RS. Chlorinated hydrocarbon pesticides in major river basins, 1957-65. *Public Health Rep* 1967;**82**(2):139–56.
283. Biesecker JE, Leifeste DK. *Water quality of hydrologic benchmarks; an indicator of water quality in the natural environment.* U.S. Geological Survey; 1975. Circular 460-E, 21 pp.
284. Carson R. *Silent Spring.* New York, N.Y: Houghton Mifflin Co.; 1962. 378 pp.
285. Cottam C. The ecologist's role in the problems of pesticide pollution. *BioScience* 1965;**15**(7):457–63.
286. U.S. Environmental Protection Agency. *Aspects of pesticidal use of endrin on man and the environment.* U.S. Environmental Protection Agency; 1973. 392 pp.
287. U.S. Public Health Service. Conference in the matter of pollution of the interstate waters of the lower Mississippi River; 1964, v. II. In: *Proceedings of a conference held May 15–16, 1964, New Orleans, La.* Washington, D.C: U.S. Department of Health, Education, and Welfare; 1964.
288. U.S. Environmental Protection Agency. *Fish kills caused by pollution: fifteen year summary, 1961-1975, EPA 440/4-78-011.* Washington, D.C: U.S. Environmental Protection Agency; 1978. 78 pp.
289. U.S. Environmental Protection Agency. *Persistent bioaccumulative and toxic (PBT) program.* 2014. http://www.epa.gov/pbt/pubs/ddt.htm. [accessed 20.08.14].

290. U.S. Environmental Protection Agency, Technical fact sheet on: endrin, U.S. Environmental Protection Agency, (undated), http://www.epa.gov/safewater/pdfs/factsheets/soc/tech/endrin.pdf. [accessed 20.08.14].

291. U.S. Environmental Protection Agency. *Federal insecticide, fungicide, and rodenticide act (FIFRA)*. 2014. http://www.epa.gov/agriculture/lfra.html. [accessed 20.08.14].

292. Wolman MG. The Nation's rivers: problems are encountered when appraising trends in water and river quality. *Science* 1971;**174**:905–18.

293. Enviro Control. *National assessment of trends in water quality*. U.S. Department of Commerce, National Technical Information Service; 1972. PB-210, p. 669.

294. Council on Environmental Quality. *Third annual report of the Council on Environmental Quality*. Washington, D.C: Council on Environmental Quality; 1972. pp. 13–15.

295. Steele TD, Gilroy EJ, Hawkinson RO. *An assessment of areal and temporal variations in streamflow quality using selected data from the National Stream Quality Accounting Network*. U.S. Geological Survey; 1974. Open-File Report 74-217, 210 pp.

296. Ficke JF, Hawkinson RO. *The national stream quality Accounting network (NASQAN) – some questions and answers*. U.S. Geological Survey; 1975. Circular 719, 23 pp.

297. Colten CE. A historical perspective on industrial wastes and groundwater contamination. *Geogr Rev* 1991;**81**(2):215–28.

298. Weaver L. Refuse disposal and its significance. *J Groundw* 1964;**2**(1):26–30.

299. Zanoni AE. Groundwater pollution and sanitary landfills—A critical review. *J Ground Water* 1972; **10**(1):3–16.

3

Mass Transport and Chemistry at the Air–Water Interface of Atmospheric Dispersoids

Kalliat T. Valsaraj, Franz S. Ehrenhauser, Aubrey A. Heath, Mickael Vaitilingom
CAIN DEPARTMENT OF CHEMICAL ENGINEERING, LOUISIANA STATE UNIVERSITY, BATON ROUGE, LA, USA

CHAPTER OUTLINE

Introduction

Aerosols play a dominant role in the chemistry of the atmosphere and affect the climate.[1,2] Atmospheric aerosols are composed of solids and liquids in sizes ranging from submicron to hundreds of microns; they are typically called "dispersoids." The water content of atmospheric aerosols can vary greatly such that fog, mist, rain, and cloud droplets (collectively called "hydrometeors") are mostly water, whereas dry aerosol particles have only traces of water. Small aerosol particles of diameters in the 0.1- to 1-μm range are inhalable and pose health concerns.[3] Typical aerosol particles have long residence times and can act as cloud condensation nuclei (CCN).[4] Aerosols that are incorporated into high atmospheric clouds or lower atmospheric fog have lower lifetimes. Besides the atmospheric lifetimes, aerosols also have high specific areas and facilitate the adsorption, absorption, and reactions (oxidations) of gaseous species

Food, Energy, and Water. http://dx.doi.org/10.1016/B978-0-12-800211-7.00003-X

Table 1 Typical Properties of Atmospheric Aerosol Systems[7–11]

Property	Aerosol Particle	Fog	Cloud	Rain	Snow
Particle (droplet) size/μm	0.1–10	1–10	$10–10^2$	$10^2–10^3$	$10^3–10^5$
Liquid water content/g m^{-3}	10^{-5}	0.05	1–10	1–100	3–10 (Vol %)
Surface area/m^2 m^{-3}	$10^{-2}–1.2 \times 10^{-6}$	0.3–0.03	6–0.06	6–0.006	1.5×10^3
Terminal settling velocity/cm s^{-1}	$1.7 \times 10^{-4}–0.60$	$3.4 \times 10^{-3}–0.3$	0.3–29.5	29.5–2946	100–200

Note: Surface area is per cubic meter air; assumed spherical droplet size for snow is based on measured specific surface area and the liquid water content; aerosol particle assumed between 1 nm and 20 μm film thickness. For aerosol particles, the value 10^{-2} "square meters of water surface per cubic meter air" would be achieved for a 1-nm layer on only 0.1-μm particles and for a 1-μm film on a 10-μm particle a value of 1.2×10^{-6} at the given liquid water content.

present in air. A listing of typical deposition rates and surface areas of atmospheric dispersoids is given in Table 1.[5,6]

Although much is known about atmospheric aerosols, in general, the specific composition is a matter of much debate. This is especially so because of its dependence on the source type, and its chemical reactivity and age.[12] The typical composition of an aerosol particle is shown in Figure 1, as is the composition of a hydrometeor (fog). The predominant fractions are inorganic, composed of ions such as SO_4^{2-}, NH_4^+, NO_3^-, NO_2^-, Cl^-, Na^+, K^+, Ca^{2+}, and Fe^{2+}. There is a significant portion of the aerosol that is carbonaceous in nature, both elemental (carbon) and organic. The inorganic fraction (mostly sulfates and nitrates) gives rise to the acidity in the aerosols, whereas the carbonaceous matter provides the hydrophobic/hydrophilic balance for the aerosols. Early work on atmospheric aerosols was to study the acidity and hygroscopic nature of aerosols. Both organic and inorganic components of aerosols can condense water from

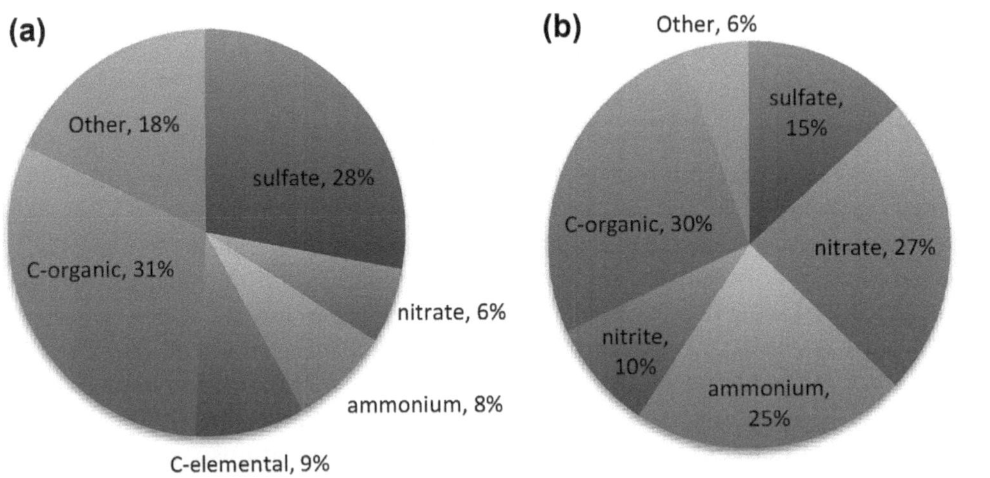

FIGURE 1 Composition of a typical (a) particle aerosol (Heitzenberg et al., 1989) and (b) fog droplet (Collett et al., 2008).

the atmosphere and act as CCN. In most cases, the mixing state of the atmospheric aerosol determines the water uptake capacity.[13,14] Temperature and relative humidity determine the water content of particles in the atmosphere.[15] The water present in aerosol particles can be both "bound" and "unbound" depending on whether water is in the coordination sphere of the ion; in these cases, water exists as a thin film and the surface area to volume ratio will be large. As we transition to hydrometeors such as fog, and mist, the water is the main ingredient in which particles and ions are dispersed; these are only a few tens of microns in diameter and hence the specific surface area will be large. As we move to large hydrometeors such as rain and cloud droplets, the specific surface area will be comparable to the bulk. Thus, atmospheric aerosols facilitate both surface and bulk processes.

It is important to note the origins of the aerosol particles and their transformations to hydrometeors. The sources are both anthropogenic (industrial, automobiles, and fuel burning) and natural (dust, sea spray, plants, and animals). Both sources are responsible for direct emissions of particles (called primary aerosols) and gases. Particles in the atmosphere can adsorb gases and react to form secondary aerosols. Organic material in the aerosol can be subdivided into primary organic aerosol derived directly from surface sources and secondary organic aerosols (SOA) formed in the atmosphere from chemical reaction and partitioning of organic compounds. SOA can be formed from three pathways: (1) semivolatile organic compounds (SVOCs) can partition onto preexisting particles, (2) SVOCs can react on particles, nucleate, and grow into new particles, and (3) when incorporated into hydrometeors, both gases and water-soluble organic carbon (WSOC) from particles can chemically react in the aqueous medium to convert to SOA via sunlight-induced photochemistry or reaction with reactive species such as hydroxyl radicals; this pathway is termed "aqueous processing" of aerosols and gases and therefore arises from heterogeneous, multiphase reactions. Particles, gases, and hydrometeors can be removed from the atmosphere via deposition to soil, water, and biota; this can occur via direct (dry deposition of gases or particles) or facilitated (wet deposition of hydrometeors) deposition. Thus, significant chemical processing of aerosols and gases occur during their atmospheric life cycle. A schematic of the various processes is shown in Figure 2.

The aqueous (cloud/fog) processing of particles and gases in atmospheric aerosols occurs at the air–water interface and would necessarily involve multiphase processes. For example, a gas species will have to diffuse to the air–water interface, adsorb and react at the interface, or absorb and react in the bulk phase. Following the reaction, volatile product species will diffuse out of the aqueous phase to the gas phase, whereas nonvolatile product species will undergo further transformations within the aqueous phase. The very high specific surface area of fine aerosol particles and hydrometeors provide efficient routs of heterogeneous reaction routes for conversion of reactive trace gases with prevalent atmospheric oxidants. These interactions can change not only the chemical composition of the aerosols, but also their physical characteristics such as hygroscopicity and cloud condensation properties. Figure 3 represents the various multiphase processes that occur at the surface of an atmospheric hydrometeor.[16]

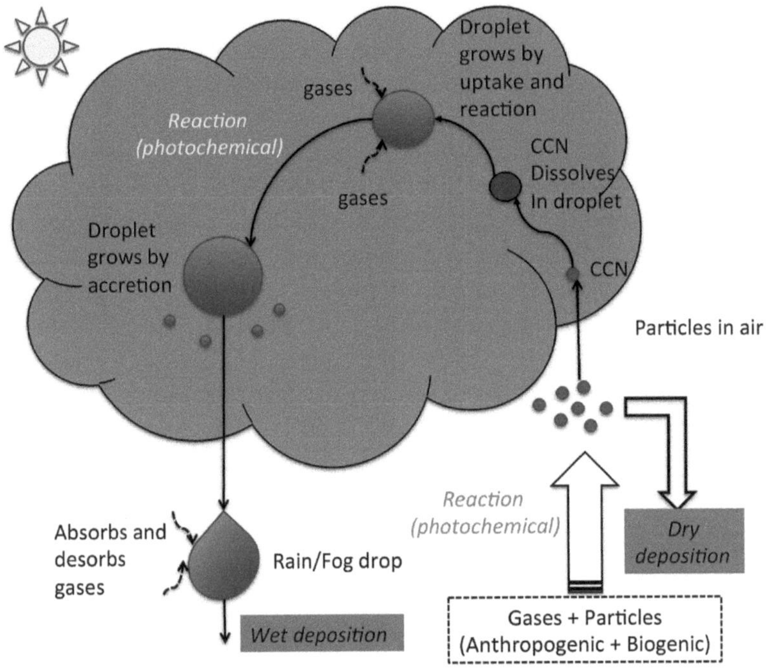

FIGURE 2 Atmospheric fate and transport of gases and aerosols.

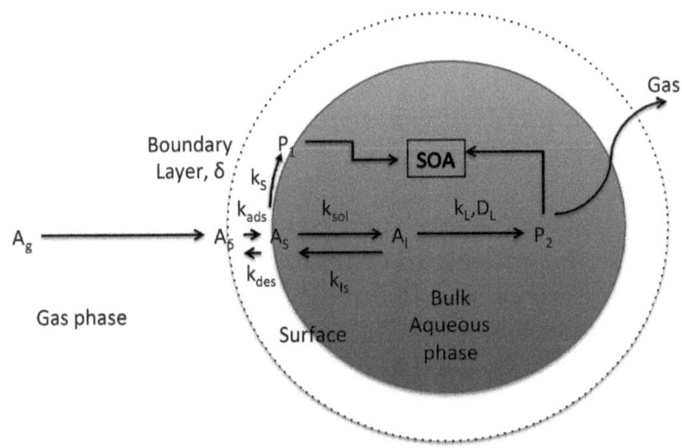

FIGURE 3 Various mass transport and reaction processes during the uptake and processing of organic vapors from the gas phase in a fog or cloud droplet.

A Mass Transport, Kinetic Model Framework for Aqueous Processing

The various processes shown in Figure 3 can make a general mass balance conservation equation for a compound that is subject to the aqueous processing in a fog or cloud system.[6,17] Atmospheric particles and droplets are different from laboratory bulk systems since there is facile exchange of material with the large reservoir of gases and oxidants present in the atmosphere. Thus, not only do we have to know about the equilibria, but we also need to consider the kinetic rates of exchange between phases. Generally, the fog or cloud system consists of various droplet sizes. However, since gas transfer is generally much faster than the chemical reactions involved, the calculations are not sensitive to droplet sizes and one typically uses an effective droplet size. Let us consider a fog/cloud system with liquid water content of L_w dispersed in the form of droplets of diameter d_D.

The change in the gas phase concentration is given by the following mass balance equation[6,18]:

$$\frac{dC_G}{dt} = k_{mt}L_w\left(\frac{C_D}{K^*_{WA}} - C_G\right) + \langle R_G \rangle \tag{1}$$

The following equation represents the change in concentrations in the liquid (droplet) phase[19]:

$$\frac{dC_D}{dt} = k_{mt}\left(C_G - \frac{C_D}{K^*_{WA}}\right) - Q\langle R_D \rangle \tag{2}$$

where C_G is the gas phase concentration of the species, C_D is the aqueous phase concentration at the surface of the droplet, k_{mt} is the mass transfer coefficient for the species across a spherical boundary, K^*_{WA} is the effective partition constant for the species between the water and gas phases, $\langle R_G \rangle$ is the gas phase reaction rate, and $\langle R_D \rangle$ is the overall rate of conversion on the droplet (aqueous) surface, given by kC_D. The above equations represent the changes in the interstitial air phase concentration of a compound resulting from mass exchange to the droplets determined primarily by the gas phase and interfacial mass transport limitations. Changes in the droplet phase concentration are determined by reactions in the droplet and are dominated by aqueous phase diffusion given by Q.

The term Q is obtained from

$$Q = 3\left(\frac{\coth q}{q} - \frac{1}{q^2}\right); \quad q = \frac{d_D}{2}\sqrt{\frac{k}{D_L}} \tag{3}$$

where d_D is the radius of the droplet, k is the first-order reaction rate constant, and D_L is the diffusivity of the reactant in the liquid phase. Note that Q is a correction factor that becomes important when liquid (droplet) phase mass transfer becomes rate limiting.

Let us now consider a fog/cloud system with liquid water content of L_w in which the gas phase concentration is large such that it remains constant during scavenging by the fog droplet. For the above case one can solve the ordinary differential equation (Eqn (2)) to obtain the following:

$$C_D(t) = C_G\left(\frac{K^*_{WA}}{1+\left(\frac{kQ}{k_{mt}}\right)K^*_{WA}}\right)\left(1-e^{-\left(\frac{k_{mt}}{K^*_{WA}}+kQ\right)t}\right) \quad (4)$$

Now if we consider the case where a given fixed mass of material in a fixed volume of cloud-filled air is being transferred to the droplets, the amount of material in the gas phase will significantly change. We now have to consider the total mass of material to be distributed among the gas and liquid phases such that $C_0 = L_w C_D + (1 - L_w)C_G$. The mass balance for the droplet phase will now be given by the following:

$$\frac{dC_D}{dt} = k_{mt}\left(\frac{L_w}{1-L_w}\right)\left[\frac{C_0}{L_w} - \frac{C_D}{\varepsilon}\right] - Q\langle R_D\rangle \quad (5)$$

The solution to the above equation gives

$$C_D(t) = \frac{C_0}{L_w\left[\frac{1}{\varepsilon} + \left(\frac{1-L_w}{L_w}\right)\frac{kQ}{k_{mt}}\right]}\cdot\left[1-e^{-\left(k_{mt}\left(\frac{L_w}{1-L_w}\right)\frac{1}{\varepsilon}+kQ\right)t}\right] \quad (6)$$

where $\frac{1}{\varepsilon} = 1 + \left(\frac{1-L_w}{L_w}\right)\frac{1}{K^*_{WA}}$ and ε gives the fraction residing in solution. Note that in the limit as $\varepsilon \ll 1$ we regain the earlier equation for constant gas phase concentration. If, further, mass transfer is fast, then the above equation provides limiting values of the droplet concentrations. For the above case of fixed volume of air, in the limit $\varepsilon \ll 1$, $C_D(t) \sim K^*_{WA} C_0$, whereas for large values of ε, we have $C_D(t) \approx \frac{C_0}{L_w}$.

The effective partition constant is given by the bulk phase partition constant (Henry's law coefficient), K_{WA} (dimensionless), and the air–water interface partition constant, K_{IA} (micrometers)[20] in the following equation.

$$K^*_{WA} = K_{WA} + \frac{6}{d_D}K_{IA} \quad (7)$$

The mass transfer coefficient for the species is given by the following equation, which connects the resistances for transfer between the gas and the droplet:

$$k_{mt} = \left[\frac{d_D^2}{12D_g} + \frac{2d_D}{3\bar{c}\alpha}\right]^{-1} \quad (8)$$

In the above equation, the first term on the right-hand side is the diffusional resistance in the gas phase and the second term is the interfacial limitation to mass transfer based on the average thermal speed, $\bar{c}$, and the mass accommodation coefficient, α. Thus, the characteristic time for mass transfer delineated above is a combination of diffusional and interfacial resistance time constants, i.e., $\tau_{mt} = \tau_{diff} + \tau_{int}$. The value of the mass accommodation coefficient, α, at the surface of the droplet determines the interfacial resistance. The value of α is not directly obtained from the experiment. The experimental value[21] that is accessible is an analogous term called the steady state reaction probability, γ, which is the fraction of molecular collisions with the droplet surface from the gas phase that lead to reactive loss to the droplet. The reaction probability is connected to various resistances to mass transfer as follows[16,22,23]:

$$\frac{1}{\gamma} = \frac{1}{\Gamma_g} + \frac{1}{S} + \frac{1}{\Gamma_s + \frac{1}{\frac{1}{\Gamma_b f\left(\frac{d_D}{2l}\right)} + \frac{1}{S\left(\frac{k_{sol}}{k_{des}}\right)}}} \tag{9}$$

where $\Gamma_b = \frac{4K_{WA}\sqrt{k_L D_L}}{\bar{C}}$ relates to the bulk phase reaction, $f\left(\frac{d_D}{l}\right) = \coth\left(\frac{d_D}{2l}\right) - \frac{2l}{d_D}$ is a multiplicative factor and $\Gamma_s = \frac{4k_s K_{IA}}{\bar{C}}$ relates to the surface reaction. Both Γ_b and Γ_s are restricted to values ≥ 0, but not limited to values ≤ 1. Γ_g is the gas phase resistance given by $\frac{8D_g}{d_D \bar{C}}$. S is called the sticking coefficient and is related to the mass accommodation coefficient, α, through the following equation:

$$\alpha = S\left(\frac{k_{sol}}{k_{sol} + k_{des}}\right) \tag{10}$$

Thus, Eqn (9) has three resistances in series, the gas phase diffusion, mass accommodation at the interface, and the reaction (bulk and surface) in the aqueous phase of the droplet. Ideally, the mass accommodation coefficient $\alpha = S$ only if $k_{sol} >> k_{des}$. Further, if all other resistances are negligible (gas and liquid phase in Eqn (9)), then the mass accommodation coefficient, α, is identical to the experimental reaction probability, γ. Depending on the nature of the experimental set up for measuring γ, different terms in Eqn (9) are retained for analyzing the experimental data.[24] Table 2 lists the values of α obtained for organic compounds that are subject to aqueous processing by fog.[25,26]

The values of α for soluble, nonvolatile organic compounds are generally higher than less soluble, volatile organic compounds. Even for hydrophobic compounds, if the volatility is small, then the probability of them sticking to the water surface and getting incorporated into the droplet is higher. As α augments the mass transfer, it becomes less dominated by the aqueous phase.

Let us consider the time evolution of droplet concentrations for two gaseous compounds of environmental significance for which physicochemical parameters are currently available (Table 3).[20,26] Benzene is more water soluble (less hydrophobic) and volatile, whereas phenanthrene is less water soluble (more hydrophobic) and less volatile. The bulk air–water partition constant and the air–water interface partition constant are both larger for phenanthrene indicating that it prefers to stay at the droplet interface, and hence its uptake rate is dependent on surface area.

Figure 4 shows the uptake rates of benzene and phenanthrene from the vapor phase into droplets of various sizes for a typical fog water content, L_w, of 1×10^{-7}. In both cases, we see that the droplets reach equilibrium with the gas phase rapidly. For benzene the effects of the droplet size on the equilibrium values of the uptake are minimal since the value of the effective air–water partition constant is affected minimally by the droplet

Table 2 Values of Mass Accommodation Coefficients for Selected Organic Compounds at the Air–Water Interface Reported in the Literature

Compound	T Range	α/(-)
Aliphatic Alcohols		
Methanol	260–291	0.12–0.02
Ethanol	260–292	≥0.02
n-propanol	260–291	0.06–0.02
i-propanol	260–291	0.10–0.02
Aliphatic Aldehydes		
Formaldehyde	260–270	0.04
Acetaldehyde	267	>0.03
Aliphatic acids		
Formic acid	260–191	0.10–0.02
Acetic acid	258–292	≥0.02
Aromatic hydrocarbons		
Benzene	298	$(9.7 \pm 1.8) \times 10^{-5}$
Naphthalene	298	$(2.2 \pm 0.6) \times 10^{-4}$
Phenanthrene	298	$(1.4 \pm 0.4) \times 10^{-2}$

Table 3 Physicochemical Parameters for Calculating Uptake Rates

Physicochemical Parameter	Benzene	Phenanthrene
Air–water bulk phase partition constant, K_{WA}/(-)	4.5	705
Air–water interface partition constant, K_{IA}/m	4.5×10^{-7}	0.1
Mean molecular speed, $\overline{C}$/m s^{-1}	284	188
Gas phase diffusivity, D_G/m^2 s^{-1}	8.8×10^{-6}	4.8×10^{-6}
Aqueous phase diffusivity, D_L/m^2 s^{-1}	9.0×10^{-10}	5.8×10^{-10}
Mass accommodation coefficient, α/(-)	9.7×10^{-5}	0.014
Aqueous phase OH reaction rate constant, k_b/s^{-1}	4.47×10^{-6}	2×10^{-5}
Surface reaction rate constant, k_s/s^{-1}	–	2×10^{-5}

diameter, whereas that for phenanthrene shows a substantial effect as its K_{IA} value predominates. A comparison of terms in τ_{uptake} in both cases shows that the reaction rate is orders of magnitude smaller than the mass transfer to the droplet surface.

Further, the dependence of the droplet phase concentration on the liquid water content for a typical 10-μm fog droplet is shown in Figure 5. For benzene the cloud liquid water content minimally effects the droplet phase concentration, whereas that for phenanthrene shows a sharp change at water contents of 10^{-5} and more. Thus, the partitioning to the droplet is inversely proportional to the liquid water content at values

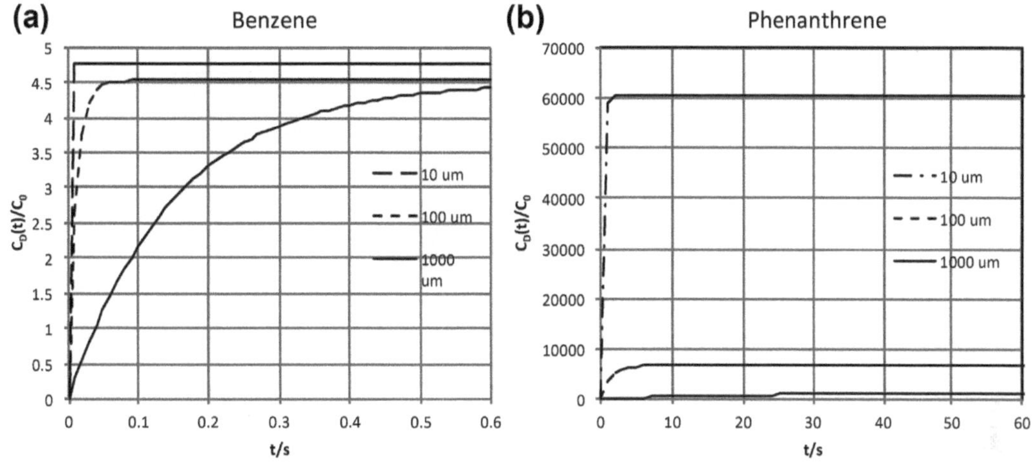

FIGURE 4 Approach to equilibrium uptake for benzene and phenanthrene vapor toward different sizes of a water droplet.

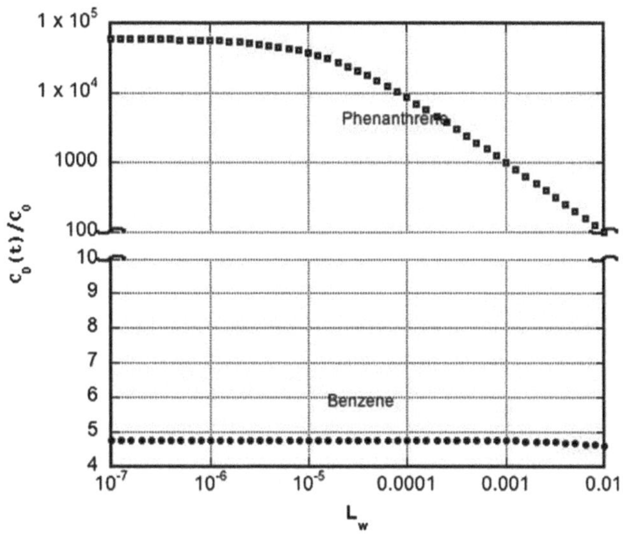

FIGURE 5 Dependence of liquid water content in the fog cloud on equilibrium uptake for benzene and phenanthrene on a 10-μm droplet.

greater than 10^{-5}, whereas for values less than 10^{-5} it is independent of the liquid water content. As we discussed earlier, this is in keeping with the two limiting values given by Eqn (6).

It is useful in atmospheric systems to estimate the characteristic time constants associated with mass transport to the droplet, reaction on/in the droplet and diffusion within the droplet. Table 4 gives the expressions for each of these characteristic times and typical values for benzene and phenanthrene.

Table 4 Characteristic Times for Mass Transport, Reaction, and Diffusion in the Droplet (see Figure 3)

Description of Characteristic Time	Equation	Compound	
		Benzene	Phenanthrene
Mass transfer to the droplet from the gas phase	$\tau_{mt} = \frac{k_{mt}}{K^*_{WA}}$	863 s	3.9 s
Reaction on/in the droplet	$\tau_{rxn} = \frac{1}{k_b + k_s\left(\frac{3}{R_p}\right)\frac{K_{IA}}{K_{WA}}}$	2.2×10^5 s	588 s
Diffusion within the droplet	$\tau_{diff} = \frac{d_D^2}{16D_L}$	6.9×10^{-3} s	1.0×10^{-2} s

The discussion above has shown that the two primary factors determining the uptake into the fog droplets are the droplet size and the liquid water content of the fog. The mass accommodation coefficient and gas phase diffusion are unimportant in fog processing of aromatic hydrocarbons. The effective partition constant, K^*_{WA} of the compound between the air and the droplet determines the uptake capacity of the droplet. K^*_{WA} is composed of bulk air–water partitioning, K_{WA}, and the interface partitioning, K_{IA}. Since we have established that mass transfer is not limiting, the only factor that determines the change in the droplet concentration of aromatic compounds is the reaction rate during the droplet residence time within a fog cloud. Thus, the following factors are discussed further: the equilibrium partition constants and the reaction rate constant for aromatic compounds in fog.

Equilibrium Partition Constants (K_{WA} and K_{IA})

There is evidence in the literature (experimental and theoretical) that indicates that organic compounds possess an eminent propensity for the air–water interface. Specifically, most organic compounds of low vapor pressure (P*) and low aqueous solubility (C^*_w) have free energy minima at the air–water interface compared to both the bulk air and water phases. Molecular dynamic simulations of organic compounds moving between the two phases show a clear minimum in potential of mean force (PMF). For example, Figure 6(a) shows the PMF for three aromatic hydrocarbons (benzene, naphthalene, and phenanthrene).[27–29] The PMF progressively increases as the molecular size increases while both vapor pressure and aqueous solubility decreases. Figure 6(b) shows the PMF profile for OH radicals transferring between the gas and the aqueous phase. The interface is also the region that receives the maximum light intensity (photon flux), is an energetically unique place where molecules can be oriented and mass transfer limitations from the gas phase are minimal. The conclusion drawn from these simulations is that the reaction probability for organic compounds with radicals such as OH is high at the interface than either of the bulk phases.

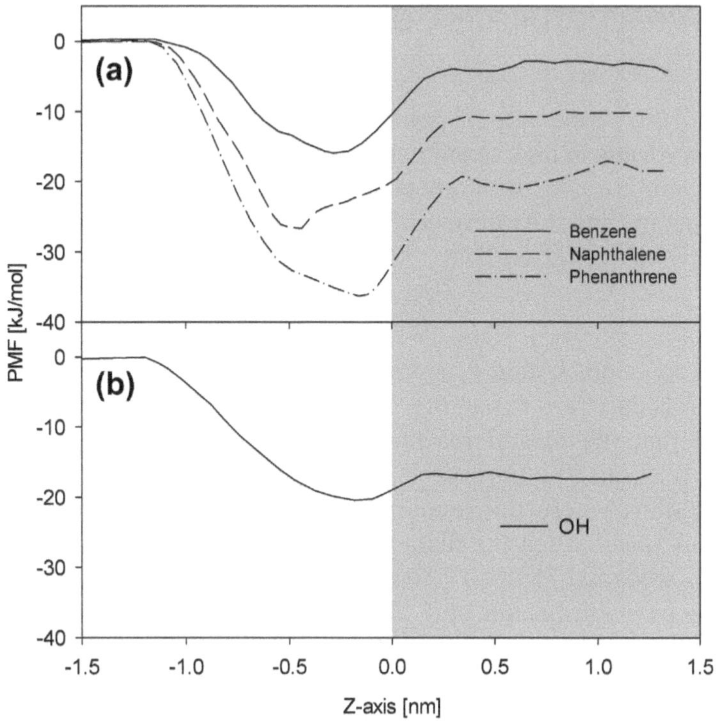

FIGURE 6 PMF for the transfer of (a) benzene, naphthalene, and phenanthrene and (b) OH radical between the gas and the aqueous phases. The shaded region is the aqueous phase.

There is experimental data available in the literature that supports the above theoretical simulations. These include measurements of surface concentrations of organic compounds at the air–water interface using a variety of techniques such as direct surface tension measurements[30–38] and indirect surface concentration measurements (e.g., gas chromatography techniques[20,39–42] and thin film flow reactor methodology[43]). These measurements provide the partition constant, K_{IA} (m), between the interface and air at infinite dilution conditions.

$$K_{IA} = \lim_{C_G \to 0} \left(\frac{\Gamma_I}{C_G} \right) \tag{11}$$

where Γ_I is the surface concentration (moles per square meter) and C_G is the equilibrium gas phase concentration (moles per cubic meter). Values of the partition constant for various classes of compounds are listed in the literature. Based on the experimental values, theoretical correlations have been developed to derive partition constants for those compounds for which experimental values are unavailable.[44–47]

The partition constant between the two bulk phases, viz., K_{WA} is available from several sources, specifically the extensive compilation by Sander[48] and the Jet Propulsion Laboratory data series.[25]

Laboratory Reaction Rate Constants (Bulk and Surface Reactions)

As noted earlier, the propensity of organic molecules and oxidants to remain at the air–water interface leads to high reaction probability at the interface. In any atmospheric water system, there are two simultaneous reaction sites possible, viz., the bulk liquid phase (homogeneous) and the interface (heterogeneous). The overall rate of a reaction is the sum of the two and is given by[43]

$$k = k_L + \frac{6}{\delta} k_s \tag{12}$$

In the above equation, k_L and k_s represent the general first-order homogeneous and heterogeneous reaction rate constants, respectively, and δ represents the water film thickness. Note that the overall reaction rate will increase with decreasing droplet diameter. Such relationships have been observed in laboratory experiments involving very thin water films where the predominant reaction will be on the water surface compared to bulk phase when the films are exposed to a gas phase containing organic compounds and oxidants. Figure 7 show the results of ultraviolet-photochemical oxidations of gaseous naphthalene and fluorene on water surfaces with varying film thicknesses. The experiments were conducted in laboratory flow reactors where precise control of water thickness was possible.[43,49] The reaction rates of the gases were

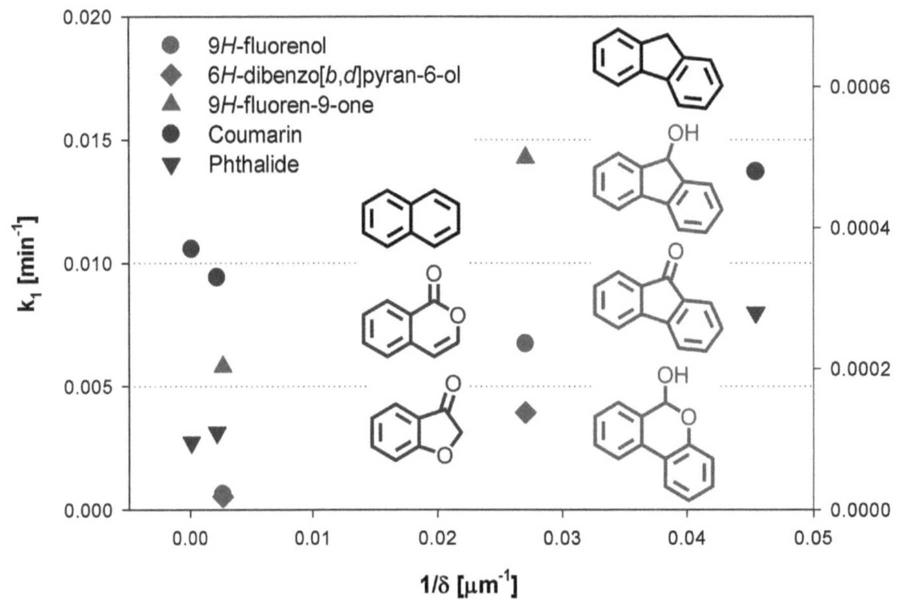

FIGURE 7 First-order rates of formation of oxidation products of naphthalene (in blue) and 9*H*-fluorene (in red) vapors on thin water films in a laboratory flow reactor. The left *y*-axis is for naphthalene and the right *y*-axis is for 9*H*-fluorene.

obtained from the disappearance of the compounds from the liquid phase via high-performance liquid chromatography/fluorescence detection.

Clearly as δ decreased (or δ^{-1} increased), the rate of product formation increased. Thus, as the film thickness decreased, the surface reaction became more important. It is also clear that the heterogeneous reaction is faster and the reaction rate therefore became larger. The increased heterogeneous reaction rate was suggested to be a result of the partial solvation of the reactants at the interface and subsequent increased collisions between the reactants at the surface.[30] More recently, there have been several reports of direct confirmation of aqueous phase reactions in droplets.[50–54] Recent studies have also shown that the interfacial water in aerosols supports a variety of reactions but depends on the material properties and internal mixing states of aerosols.[14,55–57] General rate constants, for example, those of OH radicals with the dissolved organic carbon (DOC) in cloud and fog waters, have been provided by Arakaki et al.[58]

Field Data in Support of Air–Water Interface Reactions

Two field studies were conducted on aqueous processing of specific organic compounds during fog in the environment. Of special interest in these studies were many polycyclic aromatic compounds (PAH) and their oxidized polycyclic aromatic compounds (OPAC) products. The field campaigns were conducted in the Central Valley of California during 2010–2011 in Fresno and Davis.[59] Along with fog water samples, both particulate matter and interstitial air were sampled continuously during several days of fog episodes. Thus, samples before and during the fog provided concentration changes for various compounds in the three phases. Although several compounds were tracked, we focus here only on two, one which is highly volatile (naphthalene) and another that is nonvolatile (9*H*-fluorene). A number of oxidation products for both compounds were also tracked. Figure 8 shows the molar ratio of the oxidation products to the parent compound for the case of naphthalene and Figure 9 shows the same for 9*H*-fluorene.

The median as well as the 10- and 90 percentiles of the ratios is given. For naphthalene, seven specific known oxidation products were tracked, whereas for 9*H*-fluorene, two known oxidation products were measured. For the case of naphthalene, the $PM_{2.5}$ is rich in naphthalene oxidation products, whereas the air contains most of the naphthalene. It is clear that during fog there is very little change in the ratio indicating similar gas phase and dry particulate matter (PM) phase chemistry. However, there is an increase in OPAC/PAH in fog water of about two orders of magnitude, which was attributed to dissolution from particles and reaction in the fog water. The total fog concentration (composed of fog water and the insoluble fraction) had a high OPAC/PAH ratio suggesting a high OPAC content and therefore reactive transformation. The postfog ratio for fine particles (i.e. less than 2.5 μm in diameter: $PM_{2.5}$) showed only marginal change, which is expected as naphthalene is predominantly in the vapor phase, and the reaction in the fog droplet does not affect its atmospheric composition. The case of 9*H*-fluorene, on the other hand, is different since it is much less volatile and exists predominantly on the particles.

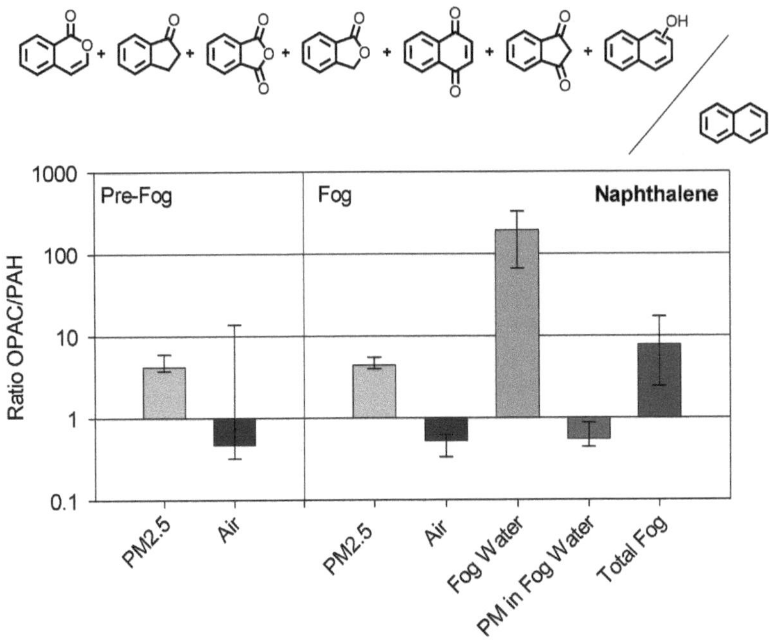

FIGURE 8 The ratio of oxidation products to parent compound (naphthalene) before and during fog events in Davis, CA (USA).

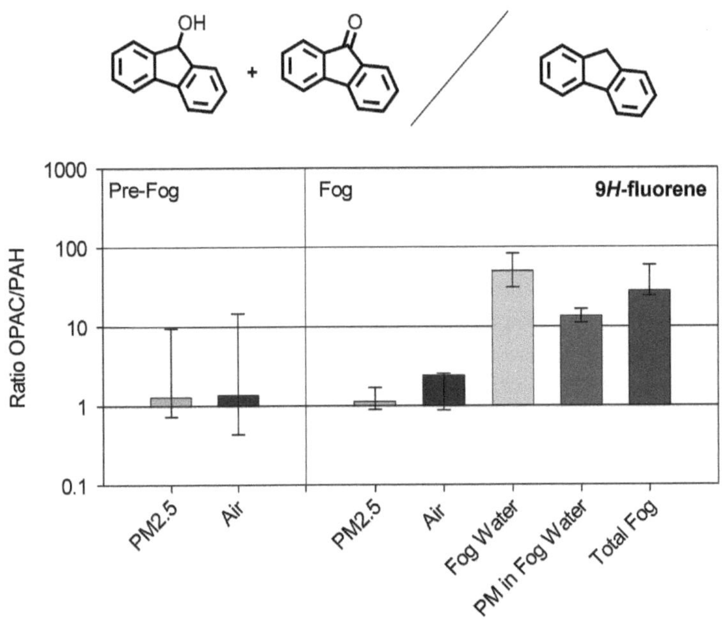

FIGURE 9 The ratio of product to parent compound (9*H*-fluorene) before and during fog events in Davis, CA (USA).

Therefore, it has a small OPAC/PAH ratio before the fog. As the fog evolves, the OPAC/PAH ratio seemed to increase above 1 indicating the possibility of oxidation in the gas phase. The fog water analysis indicated that the oxidation products were in high concentrations of the insoluble particulate matter in fog, in contrast to the observation for naphthalene. More importantly, the OPAC/PAH ratio for fluorene was at least an order of magnitude higher for both insoluble matter and fog water and therefore for the total fog water concentration in the droplet as well. This clearly indicated enhanced reactivity in the presence of fog droplets. The postfog ratio was closer to that of prefog, indicating that the oxidized products were removed via fog droplet settling, viz., wet deposition.

Although settling and washout of inorganic species was observed during fog, the concentration of both OPAC and PAH in postfog particulates was substantially higher than the prefog values. Unlike other forms of wet deposition (e.g., rain) in which the particulate matter concentration can be lowered substantially, organic matter scavenged by fog tends to remain in the atmosphere as water evaporates. Upon evaporation, the particles with high concentration, as described previously, will be left behind, which increases the OPAC concentration in the particulate matter in air. Additionally, water-soluble OPAC from particles will be stripped from particles resulting in insoluble organic matter, which is high in PAH and low in OPAC. Thus, the particulates not associated with the fog water will remain substantially unaffected in PAH or OPAC content. However, within the fog water, enhanced reactivity and conversion of PAH to OPAC was distinctly observed. This is more evident for very-slow-reacting chemicals such as naphthalene and 9*H*-fluorene and not for fast-reacting compounds, e.g., acenaphthene.[59]

A Practical Evaluation of Interface Reaction Rates versus Bulk Phase Processes

A practical means of delineating the importance of reactions versus diffusion is the use of the Thiele modulus, ϕ_P, which is defined as the ratio of the reaction on a droplet versus diffusion within the bulk phase.[5,60]

$$\phi_P = \frac{\textit{Surface reaction rate}}{\textit{Diffusion rate}} = \frac{V_D}{A_D} \cdot \sqrt{\frac{k}{D_{aq}}} \tag{13}$$

where $\frac{V_D}{A_D} = \frac{d_D}{6}$ for a spherical droplet. The necessary condition for reaction limitation is that $\phi_P << 1$. Let us consider naphthalene vapor reacting with hydroxyl radical in a fog droplet. For a spherical droplet in air, the fraction of the naphthalene adsorbed on the surface is given by $f_{ads} = \frac{K_{IA}}{\left(\frac{d_D}{6}\right) K_{WA} + K_{IA}}$. Buxton et al.[61] give the reaction rate constant for naphthalene with OH radical in water. The values of partition constants for naphthalene are given in Raja and Valsaraj.[62] With these estimates, the Thiele modulus for various droplet radii can be obtained from Eqn (13) above. The calculations can be repeated for different rate constant estimates. Figure 10 shows the plot of Thiele modulus versus

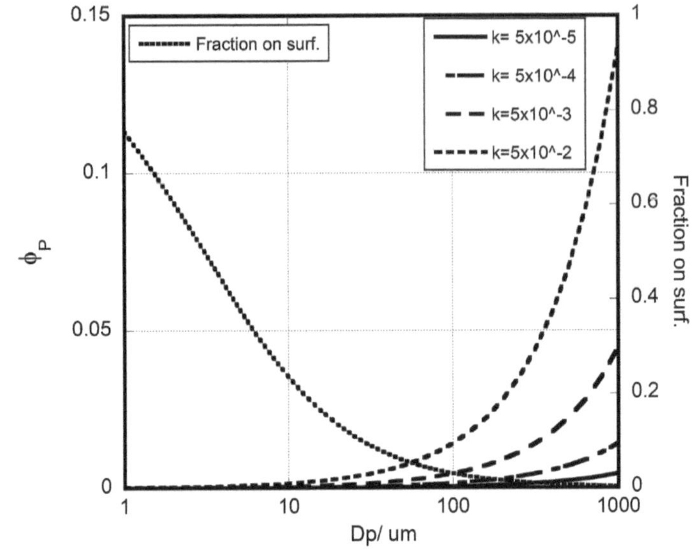

FIGURE 10 A plot of Thiele modulus versus droplet diameter for the oxidation and diffusion of gaseous naphthalene in droplets. Also shown is the fraction of naphthalene adsorbed on the surface of the droplet. Liquid phase reaction rate with OH radicals obtained from Buxton et al. (1988) and diffusion parameters from Raja and Valsaraj (2004).

droplet diameter for the case of the gaseous naphthalene reaction on fog droplets. Also shown is the fraction of surface-adsorbed naphthalene versus droplet diameter. The first point to note is that for all values of rate constants, the Thiele moduli are much less than 1 for droplet sizes $\leq$10 μm indicating that reaction rate and not diffusion is rate limiting. Secondly, we see that for the same-size droplets, the fraction adsorbed is greater than 0.2 with values approaching 0.75 for a 1-μm droplet. Thus, clearly between the interface and the bulk droplet, there is a ratio of 3:1 in the organic reactant distribution for a 1-μm droplet and the reaction is rate limited. The reactants (organic and OH) are present freely on the surface, enclosed in partial solvent cages, and facilitate the reaction between them. Molecular collisions on the surface easily replace lost reactants. In the bulk phase, not only is the reactant availability reduced (due to bulk phase diffusion limitations), but also they are present in fully developed solvent cages. Thus, the solvent cages have to be broken and the reactants have to diffuse within the aqueous phase to react. These limitations clearly point to a high degree of reactivity on the surfaces of small droplets for most organic compounds that have affinity for the air–water interface.[19,30,50,63–65] For fog droplets, particularly, the sizes are of the order of 10 μm and therefore the predominance of surface reactions is evident.

Conclusions

The air–water interface is the one entity that is admittedly the most significant part of the air–water system in terms of high concentrations and lifetimes for most organic compounds of low aqueous solubility and low vapor pressure. It is also the site for large lifetimes for oxidizing agents such as hydroxyl, peroxide, and singlet oxygen. Thus, the interface enables accelerated degradation of organic compounds. It can be the predominant reaction venue for chemical reactions in atmospheric water that exists as fog droplets, cloud droplets, and thin films on aerosols. Unlike the two bulk phases, this interface is a good venue for the aqueous degradation of most nonvolatile and nonsoluble materials. The reasons are not only the increased concentrations at the interface, but also the augmented reactive collision probabilities, and the reduced solvation at the interface resulting in decreased diffusion limitations for reactants at the interface. There is much evidence both in the laboratory and in the field for enhanced reactions at the air–water interface, which contributes to efficient aqueous processing of organic compounds in fog, clouds, and atmospheric water in aerosols.[66]

To adapt a famous statement by the Nobel Prize winning physicist, Richard Feynman,[67] one might restate: "There is plenty of ~~room~~ *interface* at the bottom" (replaced room with interface). In other words, the air–water interface is a rich area for further research in atmospheric systems.

Acknowledgments

The authors thank the National Science Foundation (Grant No: AGS 1106569, AGS 0907261, ATM 0355291, ATM 0082836) for financial support. Parts of this paper were presented at the ACS meeting in New Orleans in April 2013.

References

1. Dunstone NJ, Smith DM, Booth BBB, Hermanson L, Eade R. Anthropogenic aerosol forcing of Atlantic tropical storms. *Nat Geosci* 2013;**6**(7):534–9.
2. Ghan SJ, Schwartz SE. Aerosol properties and processes – a path from field and laboratory measurements to global climate models. *Bull Am Meteorol Soc* 2007;**88**(7):1059.
3. Monks PS, et al. Atmospheric composition change – global and regional air quality. *Atmos Environ* 2009;**43**(33):5268–350.
4. Poschl U. Atmospheric aerosols: composition, transformation, climate and health effects. *Angew Chem Int Ed* 2005;**44**:7520–40.
5. Valsaraj KT. *Elements of environmental engineering: thermodynamics and kinetics*. 2nd ed. Boca Raton: Lewis Publishers; 2000. p. 679.
6. Seinfeld JH, Pandis SN. *Atmospheric chemistry and physics: from air pollution to climate change*. 2nd ed. New York: John Wiley & Sons, Inc.; 2006.
7. Valsaraj KT. *Elements of environmental engineering: thermodynamics and kinetics*. 3rd ed. Boca Raton: CRC Press; 2009. pp xv, 468 pp.
8. Valsaraj KT. A review of aqueous aerosol surface chemistry in the atmospheric context. *Open J Phys Chem* 2012;**2**(1):58–66.
9. Gill PS, Graedel TE, Weschler CJ. Organic films on atmospheric aerosol particles, fog droplets, cloud droplets, raindrops, and snowflakes. *Rev Geophys.* 1983;**21**(4):903–20.

10. Domine F, Taillandier AS, Simpson WR. A parameterization of the specific surface area of seasonal snow for field use and for models of snowpack evolution. *J Geophys Res Earth Surf* 2007;**112**(F2): F02031.
11. Liyana-Arachchi T, Valsaraj K, Hung F. Ice growth from supercooled aqueous solutions of reactive oxygen species. *Theor Chem Acc* 2012;**132**(1):1–13.
12. Deguillaume L, et al. Classification of clouds sampled at the puy de Dôme (France) from 10 yr monitoring: mean features of their physico-chemical properties. *Atmos Chem Phys Discuss* 2013; **13**(8):22795–846.
13. Gelenscer A. *Carbonaceous aerosols*. Dordrecht (The Netherlands): Springer; 2004.
14. Shiraiwa M, Zuend A, Bertram AK, Seinfeld JH. Gas-particle partitioning of atmospheric aerosols: interplay of physical state, non-ideal mixing and morphology. *Phys Chem Chem Phys* 2013;**15**(27): 11441–53.
15. Pilinis C, Seinfeld JH, Grosjean D. Water content of atmospheric aerosols. *Atmos Environ (1967)* 1989;**23**(7):1601–6.
16. Hanson DR. Surface-specific reactions on liquids. *J Phys Chem B* 1997;**101**:4998–5001.
17. Berkemeier T, et al. Kinetic regimes and limiting cases of gas uptake and heterogeneous reactions in atmospheric aerosols and clouds: a general classification scheme. *Atmos Chem Phys* 2013;**13**(14): 6663–86.
18. Warneck P. *Chemistry of the natural atmosphere*. 2nd ed. San Diego (CA): Academic Press; 2000.
19. Knipping EM, et al. Experiments and simulations of ion-enhanced interfacial chemistry on aqueous NaCl aerosols. *Science* 2000;**288**(5464):301–6.
20. Raja S, Yaccone FS, Ravikrishna R, Valsaraj KT. Thermodynamic parameters for the adsorption of aromatic hydrocarbon vapors at the Gas-water interface. *J Chem Eng Data* 2002;**47**(5):1213–9.
21. Kolb CE, Davidovits P, Jayne JT, Shi Q, Worsnop DR. Kinetics of trace gas uptake by liquid surfaces. *Prog React Kinet Mech* 2002;**27**:1–46.
22. Moller D. *Chemistry of the climate system*. Berlin (New York): Walter de Gruyter GmbH & Co. KG; 2010.
23. Sander R. Modeling atmospheric chemistry: Interactions between gas-phase species and liquid cloud/aerosol particles. *Surv Geophys* 1999;**20**(1):1–31.
24. Kolb CE, et al. An overview of current issues in the uptake of atmospheric trace gases by aerosols and clouds. *Atmos Chem Phys* 2010;**10**:10561–605.
25. Sander SP, et al. *Chemical kinetics and photochemical data for use in atmospheric studies, evaluation no. 17*. 2011.
26. Raja S, Valsaraj KT. Uptake of aromatic hydrocarbon vapors (benzene and phenanthrene) at the air-water interface of micron-size water droplets. *J Air Waste Manag Assoc* 2004;**54**(12):1550–9.
27. Chen J, et al. Adsorption of Gas-Phase phenanthrene on atmospheric water and ice films. *Polycycl Aromat Compd* 2011;**31**(4):201–26.
28. Liyana-Arachchi TP, Valsaraj KT, Hung FR. Molecular simulation study of the adsorption of naphthalene and ozone on atmospheric air/ice interfaces. *J Phys Chem A* 2011;**115**(33):9226–36.
29. Liyana-Arachchi TP, Valsaraj KT, Hung FR. Adsorption of naphthalene and ozone on atmospheric air/ice interfaces coated with surfactants: a molecular simulation study. *J Phys Chem A* 2012;**116**(10): 2519–28.
30. Valsaraj KT. Trace gas adsorption thermodynamics at the air-water interface: Implications in atmospheric chemistry. *Pure Appl Chem* 2009;**81**(10):1889–901.
31. Blank M, Ottewill RH. Adsorption of aromatic vapors on water surfaces. *J Phys Chem* 1964;**68**(8): 2206.

32. Hauxwell F, Ottewill RH. Adsorption of toluene vapor on water surfaces. *J Colloid Interface Sci* 1968; **28**(3–4):514.

33. Aveyard R, Haydon DA. Thermodynamic properties of aliphatic hydrocarbon-water interfaces. *Trans Faraday Soc* 1965;**61**(514P):2255.

34. Baumer D, Findenegg GH. Adsorption of hydrophobic vapors on the surface of water. *J Colloid Interface Sci* 1982;**85**(1):118–27.

35. Massoudi R, King AD. Effect of pressure on surface-tension of water – adsorption of low-molecular weight gases on water at 25 degrees. *J Phys Chem* 1974;**78**(22):2262–6.

36. Bruant RG, Conklin MH. Dynamic determination of vapor/water interface adsorption of volatile hydrocarbon organic compounds (VHOCs) using axysymmetric drop shape analysis: procedure and analysis of benzene adsorption. *J Phys Chem B* 2000;**104**:11146–52.

37. Donaldson DJ. Adsorption of atmospheric gases at the air-water interface. I. NH3. *J Phys Chem A* 1999;**103**(1):62–70.

38. Donaldson DJ, Anderson D. Adsorption of atmospheric gases at the air-water interface. 2. C-1-C-4 alcohols, acids, and acetone. *J Phys Chem A* 1999;**103**(7):871–6.

39. Dorris GM, Gray DG. Adsorption of hydrocarbons on silica-supported water surfaces. *J Phys Chem* 1981;**85**(24):3628–35.

40. Hoff JT, Mackay D, Gillham R, Shiu WY. Partitioning of organic-chemicals at the air-water-interface in environmental systems. *Environ Sci Technol* 1993;**27**(10):2174–80.

41. Hartkopf A, Karger BL. Study of interfacial properties of water by Gas Chromatography. *Accounts Chem Res* 1973;**6**(6):209–16.

42. Roth CM, Goss K-U, Schwarzenbach RP. Adsorption of a diverse set of organic vapors on the bulk water surface. *J Colloid Interface Sci* 2002;**252**:21–30.

43. Chen J, Ehrenhauser FS, Valsaraj KT, Wornat MJ. Uptake and UV-photooxidation of gas-phase PAHs on the surface of atmospheric water films. 1. Naphthalene. *J Phys Chem A* 2006;**110**(29): 9161–8.

44. Goss K-U. Predicting adsorption of organic chemicals at the air–water interface. *J Phys Chem A* 2009; **113**(44):12256–9.

45. Kelly CP, Cramer CJ, Truhlar DG. Predicting adsorption coefficients at air-water interfaces using universal solvation and surface area models. *J Phys Chem B* 2004;**108**(34):12882–97.

46. Valsaraj KT. On the physicochemical aspects of partitioning of non-polar hydrophobic organics at the air–water interface. *Chemosphere* 1988;**17**(5):875–87.

47. Valsaraj KT. Binding constants for non-polar hydrophobic organics at the air-water interface – comparison of experimental and predicted values. *Chemosphere* 1988;**17**(10):2049–53.

48. Sander R. Henry's law data. In: Lindstrom PJ, Mallard WG, editors. *Vol NIST chemistry webBook, NIST standard reference database 69.* Gaithersburg (MD): U.S. Department of Commerce; 2005.

49. Ehrenhauser FS. Photochemical reaction products of polycyclic aromatic hydrocarbons adsorbed at an air-water interface [Ph.D. dissertation]. Baton Rouge: Louisiana State University; 2012.

50. Kameel FR, Hoffmann MR, Colussi AJ. OH radical-initiated chemistry of isoprene in aqueous media. Atmospheric implications. *J Phys Chem A* 2013;**117**(24):5117–23.

51. Wang X, et al. The secondary formation of inorganic aerosols in the droplet mode through heterogeneous aqueous reactions under haze conditions. *Atmospheric Environ* 2012;**63**(0):68–76.

52. Kaul DS, Gupta T, Tripathi SN, Tare V, Collett JL. Secondary organic aerosol: a comparison between Foggy and Nonfoggy Days. *Environ Sci Technol* 2011;**45**(17):7307–13.

53. Martins-Costa MTC, Anglada JM, Francisco JS, Ruiz-Lopez MF. Reactivity of volatile organic compounds at the surface of a water droplet. *J Am Chem Soc* 2012;**134**(28):11821–7.

54. Ammann M, et al. Evaluated kinetic and photochemical data for atmospheric chemistry: Volume VI – heterogeneous reactions with liquid substrates. *Atmos Chem Phys* 2013;**13**(16):8045–228.

55. Trainic M, Abo Riziq A, Lavi A, Rudich Y. Role of interfacial water in the heterogeneous uptake of glyoxal by mixed glycine and ammonium sulfate aerosols. *J Phys Chem A* 2012;**116**(24):5948–57.

56. Buszek RJ, Francisco JS, Anglada JM. Water effects on atmospheric reactions. *Int Rev Phys Chem* 2011;**30**(3):335–69.

57. Nguyen TB, Coggon MM, Flagan RC, Seinfeld JH. Reactive uptake and photo-fenton oxidation of glycolaldehyde in aerosol liquid water. *Environ Sci Technol* 2013;**47**(9):4307–16.

58. Arakaki T, et al. A general scavenging rate constant for reaction of hydroxyl radical with organic carbon in atmospheric waters. *Environ Sci Technol* 2013;**47**(15):8196–203.

59. Ehrenhauser FS, et al. Processing of atmospheric polycyclic aromatic hydrocarbons by fog in an urban environment. *J Environ Monit* 2012;**14**(10):2566–79.

60. Froment GF, Bischoff KB, Wilde JD. *Chemical reactor analysis and design*. 3rd ed. Hoboken (NJ): John Wiley & Sons, Inc.; 2011.

61. Buxton GV, Greenstock CL, Helman WP, Ross AB. Critical review of rate constants for reactions of hydrated electrons, hydrogen atoms and hydroxyl radicals ([center-dot]OH/[center-dot]O[sup -] in aqueous solution). *J Phys Chem Ref Data* 1988;**17**(2):513–886.

62. Raja S, Valsaraj KT. Heterogeneous oxidation by ozone of naphthalene adsorbed at the air–water interface of micron-size water droplets. *J Air Waste Manage Assoc* 2005;**55**:1345–55.

63. Nissenson P, Knox CJH, Finlayson-Pitts BJ, Phillips LF, Dabdub D. Enhanced photolysis in aerosols: evidence for important surface effects. *Phys Chem Chem Phys* 2006;**8**(40):4700–10.

64. Nissenson P, et al. Evidence of the water-cage effect on the photolysis of and $FeOH^{2+}$. Implications of this effect and of H_2O_2 surface accumulation on photochemistry at the air–water interface of atmospheric droplets. *Atmos Environ* 2010;**44**(38):4859–66.

65. Blando JD, Turpin BJ. Secondary organic aerosol formation in cloud and fog droplets: a literature evaluation of plausibility. *Atmos Environ* 2000;**34**(10):1623–32.

66. Carlton AG, Turpin BJ. Particle partitioning potential of organic compounds is highest in the Eastern US and driven by anthropogenic water. *Atmos Chem Phys Discuss* 2013;**13**(5):12743–70.

67. Feynman RP. There's plenty of room at the bottom: an invitation to enter a new field of physics. *Caltech Eng Sci* 1960;**23**(5):22–36.

4

Complete β-Lactam Antiobiotic Activity Removal from Wastewaters: Hydroxyl Radical-Mediated Oxidation Efficiencies

Shauna C. Otto, Stephen P. Mezyk, Kathrin D. Zimmerman
DEPARTMENT OF CHEMISTRY AND BIOCHEMISTRY, CALIFORNIA STATE UNIVERSITY AT LONG BEACH, LONG BEACH, CA, USA

CHAPTER OUTLINE

Introduction

Alexander Flemming's serendipitous discovery of Penicillin in 1928, and Howard Florey and Ernst Chain's subsequent work to purify the compound in the early 1940s, ushered in an era of rapid progress toward controlling communicable diseases. Tuberculosis, which in the nineteenth century accounted for almost one-quarter of all deaths in

Food, Energy, and Water. http://dx.doi.org/10.1016/B978-0-12-800211-7.00004-1

Europe[1] was soon tamed by Streptomycin, a compound isolated from a soil bacterium. Such inspiration from nature drove the development of even more powerful synthetic, and semisynthetic, chemotherapeutics over the next 30 years.

In response to the sudden selective pressure from these drugs, bacteria have quickly developed multiple survival methods. Direct degradation, decreased uptake, increased efflux of an antibiotic, alteration of binding site topography, overexpression of the antibiotic target molecule, or even complete metabolic rerouting are all common methods of antibiotic resistance.[2] Over the past 40 years, the number of unique β-lactamase enzymes has increased by three orders of magnitude.[3] There are some strains that can be highly—sometimes even completely—resistant to entire classes of antibiotics, such as those harboring the gene for New Delhi metalo-β-lactamase 1[4] or the gene that confers resistance in methicillin-resistant *Staphylococcus aureus*, *mecA*, which codes for a target protein that has an extremely low affinity for β-lactam antibiotics.[5]

Approximately 80 million courses of β-lactam antibiotics were prescribed in 2010.[6] Up to 70% of these antibiotics pass directly into wastewater streams untransformed, with the remainder secreted as metabolites. In addition to their therapeutic use in humans, antibiotics are also used extensively in commercial agriculture[7] and are commonly given at subtherapeutic levels to enhance livestock growth.[8] In 2011 alone over 800,000 kilograms of β-lactam antibiotics were used in animals.[9] The antibiotics and their residues present in manure and other livestock waste inevitably make their way into our water systems. Solution concentrations of β-lactam antibiotics have been found at nanogram per liter levels in both sewage effluent and in surface waters.[10] While this is not a therapeutically relevant concentration, there is evidence that even subinhibitory levels of antibiotics are sufficient to help selection of antibiotic resistance genes,[11] and resistant strains have been isolated at high frequency in wastewater streams.[12,13]

Concern about the significant concentrations of antibiotics remaining even after traditional primary and secondary treatment methods[14] utilized by the majority of municipal treatment systems[15] is driving the study of additional treatment steps for these contaminated waters. Advanced oxidation process (AOP) technologies are an attractive option to destroy these small molecule contaminants, as their high water solubility makes adsorption technologies less efficient. AOPs typically employ highly oxidizing hydroxyl radicals ($E^o = 2.8$ V[16]) that react with, and can ultimately mineralize, a wide variety of contaminants.[17–19] This reactive species can be created in situ using a variety of methods—Ozone/H_2O_2, O_3/UV-C, and H_2O_2/UV-C are common systems used for large-scale treatment,[20] while Fe(II) and H_2O_2 can produce $^{\bullet}OH$ radicals though Fenton's reaction chemistry,[21] and TiO_2/UV creates $^{\bullet}OH$ through charge separation.[22] Hydroxyl radicals are also created in aqueous solutions without added chemicals by techniques that deposit large amounts of energy directly into the water, such as with sonolysis,[23] supercritical water oxidation,[24] cativation,[25] and radiolysis.[20]

Oxidation by hydroxyl radicals has demonstrated nonselective reactivity with a wide variety of dissolved organic compounds in water, including antibiotics.[26] However, an important caveat to consider, regardless of the method of •OH radical generation, is that the oxidized products of these reactions may still retain biological activity. For the β-lactam antibiotics the potency of their putative oxidation products has been under scrutiny for several years now. Previous kinetic studies[27] suggested that the •OH preferentially reacts with the peripheral phenyl ring of the antibiotic, leaving the core β-lactam moiety intact. One of these possible hydroxylation products, *p*-hydroxybenzylpenicillin, has been demonstrated to be biologically active.[28,29]

In addition, there have been three detailed studies investigating the •OH-based destruction of the β-lactam antibiotic activity, particularly penicillin-G. The first[26] study utilized •OH radical generation from O_3/H_2O_2 and found that a single oxidation of a penicillin-G solution by either the hydroxyl radical or ozone did *not* remove all the antibiotic behavior, as measured by changes in reference *Escherichia coli* or *Bacillus subtilis* bacterial strain growth. The oxidized penicillin-G product species retained at least partial activity. This study also demonstrated that low levels of *o*-, *m*-, and *p*-hydroxybenzylpenicillin were formed, as expected. However, these findings were contradicted by a more recent investigation[30] that generated •OH radicals using UV/H_2O_2, and based on the same *B. subtilis* bacterial assay study. This work concluded that only a single oxidation by this radical was required to completely remove all antibiotic activity. The same results were obtained in the latter study in both ultrapure and secondary effluent wastewater from two different treatment plants, and imply that the antibiotic activity of an •OH-treated solution of penicillin-G depends only upon the parent molecule concentration.

The inconsistency in the amount of hydroxyl radical-based oxidation to completely remove penicillin-G antibiotic activity of these two investigations suggests that additional chemistry from the hydrogen peroxide or ozone present obscures the true •OH radical chemistry with these antibiotics. This view is further supported by another recent study[31] that utilized gamma radiation techniques to generate oxidizing hydroxyl radicals in antibiotic solutions in water, and found that four to nine radical oxidations were necessary to completely remove all antibiotic activity for three structurally similar β-lactams (penicillin-G, amoxicillin, and ampicillin). However, this large variation in this extent of oxidation necessary makes quantitative comparison to the previous studies difficult.

To ensure complete deactivation of biologically active compounds using AOPs at large scale, a quantitative understanding of the oxidative degradation mechanisms occurring, •OH-mediated oxidation kinetics, reaction efficiencies, and biological activity of the oxidized products must be accurately determined.[32] Therefore, in this study we have quantitatively investigated the role of *only* hydroxyl radical reactions with multiple members of the β-lactam antibiotic family using radiolysis. The major advantage of this approach is that no additional chemicals need to be added to the water matrix, as the radiolytic energy is directly deposited in the water solvent to quantitatively create hydroxyl radicals. Additionally, as real-world wastewaters are complex, multicomponent systems, the confounding effects of other major components, such as dissolved organic matter (DOM) and alkalinity must also be considered—these effects are explored preliminarily herein. All of these parameters need to be ultimately combined to calculate effective rate constants for the total removal of antibiotic activity via •OH oxidation under AOP process conditions.

Experimental

Chemicals

All chemicals were purchased from the Sigma–Aldrich Chemical Co. or Indofine Chemical Company, with purities of ≥95%. Suwannee River Fulvic Acid Standard I (SRFA I) was purchased from the International Humic Substances Society. Solutions were prepared in Millipore MilliQ® charcoal-filtered and deionized (>18.2 MΩ) water, which was continuously illuminated by a Xe arc lamp (172 nm) to keep organic contaminants below 13 $\mu g\ L^{-1}$.

Solutions were pH adjusted to 7.00 ± 0.05 using perchloric acid or sodium hydroxide as necessary. All solutions were aerated and stored at 4 °C until irradiation. Antibiotic activity of irradiated samples was measured within 48 h.

Bacterial Growth Inhibition Experiments

Staphylococcus aureus cultures were maintained on a 20% glycerol stock in Tryptic Soy Broth (TSB) stored at −20 °C. TSB flasks inoculated from a streaked plate were incubated overnight for 12–18 h. OD_{600} was measured to confirm log phase growth. A 1 mL aliquot taken from this overnight culture was then centrifuged for 5 min at 5000 rpm. The supernatant was removed carefully and the pellet washed with MilliQ water. The sample was washed again as above and diluted to a final volume of 10 mL, which was used for inoculation.

The colorimetric reagent was prepared as previously described[33] immediately before use. In brief, triple strength LB-Miller broth (3XLB) was mixed with Cell Titer 96© Aqueous One Solution Reagent (MTS) to give a 4:1 3XLB to mL MTS ratio.

The minimum inhibitory concentration (MIC) for each antibiotic was found using the microtiter broth dilution method.[34] Working antibiotic solutions were prepared with a concentration threefold higher than their measured MICs. The specific concentrations for the compounds investigated were 4.11 μM amoxicillin, 4.05 μM ampicillin, 3.18 μM cloxacillin, 4.02 μM penicillin-G, 11.1 μM piperacillin, and 28.1 μM ticarcillin.

Standard 96 well-flat bottomed microtiter plates were used to perform the assay. Each reaction volume contained 50 μL of antibiotic solution, 50 μL of colorimetric reagent, and 50 μL of diluted inoculum, added in that order. In the positive controls, MilliQ® water was substituted for the antibiotic solution, in the negative controls MilliQ® water was substituted for both the antibiotic and inoculum solutions. Outer wells were filled with 200 μL MilliQ® water and plates wrapped loosely with parafilm to prevent evaporation losses while being incubated at 37 °C. All samples were measured in triplicate. A Thermo Varioskan Plate reader was used to measure the sample absorbencies at 490 nm. Plates were read every 30 min until the absorbance of the positive control reached an absorbance of 1.75.

Steady State Radiolysis Experiments

Steady state gamma ^{137}Cs (University of California, Irvine) or ^{60}Co (University of Notre Dame) radiolysis was used to quantitatively generate specific quantities of •OH radicals in situ in the antibiotic solutions. These high-energy gamma rays deposit energy directly into the water solvent, which causes water bond breakage and ionization followed by radical combinations according to the general scheme,[16]

$$H_2O \xrightarrow{radiolysis} [0.28]^{\bullet}OH + [0.06]H^{\bullet} + [0.27]e^-_{aq} + [0.05]H_2 + [0.07]H_2O_2 + [0.27]H^+ \qquad (1)$$

The numbers in brackets preceding each species are their ultimate escape yield (G-value) in μmol Gy^{-1} of deposited energy. Under aqueous, aerated, conditions at neutral pH, the reducing radicals preferentially react with the dissolved oxygen present ($[O_2]_{dissolved} = 250$ μM) to produce the relatively inert superoxide radical anion, according to[16] the following:

$$e^-_{aq} + O_2(aq) \rightarrow O_2^-(aq) \quad k_2 = 1.9 \times 10^{10}\ M^{-1}s^{-1} \qquad (2)$$

$$H^{\bullet} + O_2(aq) \rightarrow HO_2^{\bullet}(aq) \leftrightharpoons H^+(aq) + O_2^-(aq) \quad k_3 = 1.2 \times 10^{10}\ M^{-1}s^{-1},\ pK_{a3} = 4.8 \qquad (3)$$

Of the remaining oxidizing species, the H_2O_2 reaction with the β-lactam antibiotics is very slow,[35] and not significant on the timescale of these experiments. Therefore, the only remaining reaction of note is that of the hydroxyl radical, which is as follows:

$$^{\bullet}OH + \beta\text{-lactam} \rightarrow \text{product(s)} \quad k_4 \qquad (4)$$

The exact amount of •OH radicals delivered per minute to each sample was calculated using the standard yield of $G_{OH} = 0.28$ μmol Gy^{-1}[16] and the measured irradiator dose rate (Gy min^{-1}). By simply varying the overall irradiation time, exact [•OH]:[Antibiotic] stoichiometric ratios were achieved.

Pulse Radiolysis Experiments

Hydroxyl radical kinetics were measured using electron pulse radiolysis, as done previously.[27] The linear accelerator at the Radiation Research Laboratory at the University of Notre Dame was used to generate pulses of 2–4 ns of 8 MeV electrons for these experiments. Details of this system have been published previously.[36,37] The high-energy electron radiolysis of water produces the same distribution of radical and molecular species as gamma irradiation, shown in Eqn (1).

Samples for pulsed radiolysis experiments were prepared using 10 mM phosphate buffer buffered to pH 7.0–7.4 in MilliQ water that was presaturated with N_2O gas

($[N_2O] \sim 25$ mM at 298 K). Under these conditions, the two reducing water radiolysis radicals react with the N_2O to produce more •OH radicals, according to[16] the following:

$$e^-_{aq} + N_2O + H_2O \rightarrow N_2 + OH^- + {}^{\bullet}OH \quad k_5 = 9.1 \times 10^9\ M^{-1}s^{-1} \tag{5}$$

$$^{\bullet}H + N_2O \rightarrow {}^{\bullet}OH + N_2 \quad k_6 = 2.1 \times 10^6\ M^{-1}s^{-1} \tag{6}$$

The total absorbed dose for these rate constant determinations was measured using standard KSCN dosimetry[38] using 1.00×10^{-2} M KSCN. The absorption of the transient $(SCN)_2^{\bullet-}$ species created was measured at 475 nm, from which the original •OH concentration was determined, assuming a $G\varepsilon$ value of $5.1 \times 10^{-4}\ m^2\ Gy^{-1}$.

•OH reaction kinetic data was gathered by direct measurement of the transient absorptions of their oxidation products. For reactions that did not yield a detectable product, the method of KSCN competition kinetic analysis was employed. Addition of a standard amount of KSCN with different concentrations of an analyte (X) gives rise to the following hydroxyl radical competition[16]:

$$^{\bullet}OH + X \rightarrow \text{non-absorbing products} \quad k_7 \tag{7}$$

$$^{\bullet}OH + SCN^- (+ SCN^-) \rightarrow OH^- + (SCN)_2^{\bullet-} \quad k_8 = 1.1 \times 10^{10}\ M^{-1}s^{-1} \tag{8}$$

This kinetic competition can be readily integrated to give the following expression.

$$\frac{Abs^o(SCN)_2^{\bullet-}}{Abs(SCN)_2^{\bullet-}} = 1 + \frac{k_7[X]}{k_8[SCN^-]} \tag{9}$$

In the absence of any added analyte, the $(SCN)_2^{\bullet-}$ formed will give a maximum absorbance, $Abs^o(SCN)_2^{\bullet-}$, at 475 nm. As the added concentration of the analyte is increased, less $(SCN)_2^{\bullet-}$ will be formed, and so a decreased absorbance, $(Abs(SCN)_2^{\bullet-})$, is seen. When the ratio of these two absorbances is plotted against the concentration ratio $[X]/[SCN^-]$, a straight line of unity intercept and slope k_7/k_8 is obtained, from which k_7 is readily calculated.

Results and Discussion

Mode of Action for β-Lactam Antibiotics

The bactericidal activity of the β-lactams comes from their disruption of bacterial cell wall synthesis, specifically the peptidoglycan strand cross-linking step performed by D-alanyl-D-alanine carboxypeptidase/transpeptidases (DD-transpeptidases). The bacterial cell wall is necessary to maintain cell shape and viability, and the peptidoglycan component of this framework is one of the more ubiquitous constituents of this complex structure in bacteria. Without sufficient peptidoglycan, cross-linking cell wall synthesis halts, and the growing bacteria lyse, and die.[39]

All species in the penam subclass of β-lactams contain the core β-lactam ring fused to a five-membered heterocyclic ring, and inhibit DD-transpeptidase activity by binding strongly to the enzyme. The carbonyl group on the strained β-lactam ring reacts with a

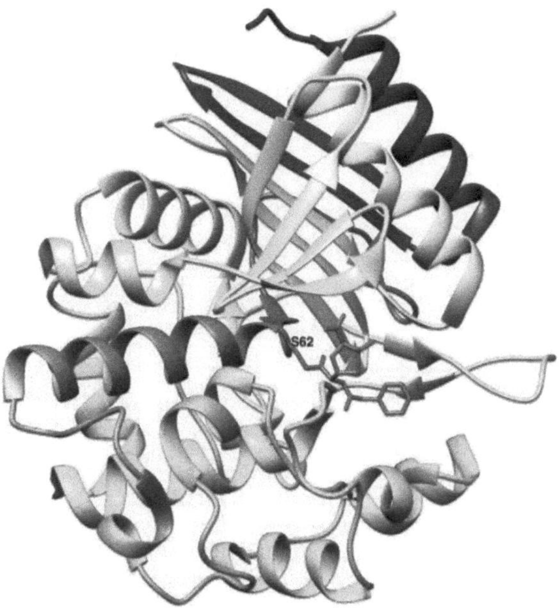

FIGURE 1 Penicillin-G, shown in orange, covalently bound to serine 62, labeled in black, in the active site of the *Streptomyces* sp. R61 DD-transpeptidase. Image of 1PWC[40] created using UCSF Chimera.[41] (For interpretation of the references to color in this figure legend, the reader is referred to the online version of this book.)

serine hydroxyl group in the DD-transpeptidase active site, forming a stable acyl-enzyme intermediate.[40] This covalent complex is seen in Figure 1. It is readily seen that the presence of this core structure is critical for the molecules' antibiotic activity.

The variation of the side chain, denoted as R in Figure 2 below gives rise to the various commercial penicillin antibiotics, which differ in their cell wall permeability, binding affinity to the target enzyme, and solubility.[39]

•OH Reaction Kinetics

Rate constants for the •OH-mediated oxidation of ampicillin, cloxacillin, Penicillin-G, and ticarcillin have been determined previously using direct measurement techniques.[27] However, the second-order rate constants for the •OH reactions for amoxicillin and piperacillin were determined using SCN^- competition kinetics. Figure 3 shows both the kinetic traces and transformed straight-line plot analysis for the reaction of •OH with piperacillin.

The resulting experimental •OH rate constants at ambient temperature for amoxicillin and piperacillin were $k = (6.94 \pm 0.44) \times 10^9\ M^{-1}\ s^{-1}$ and $k = (6.98 \pm 0.18) \times 10^9\ M^{-1}\ s^{-1}$, respectively. These data are summarized in Table 1 along with the measured rate constants for the other β-lactam antibiotics of this study.

Insight into the •OH radical reactivity can also be obtained from a simple structure-reactivity analysis of the different β-lactams. Assuming that the rate constant for the

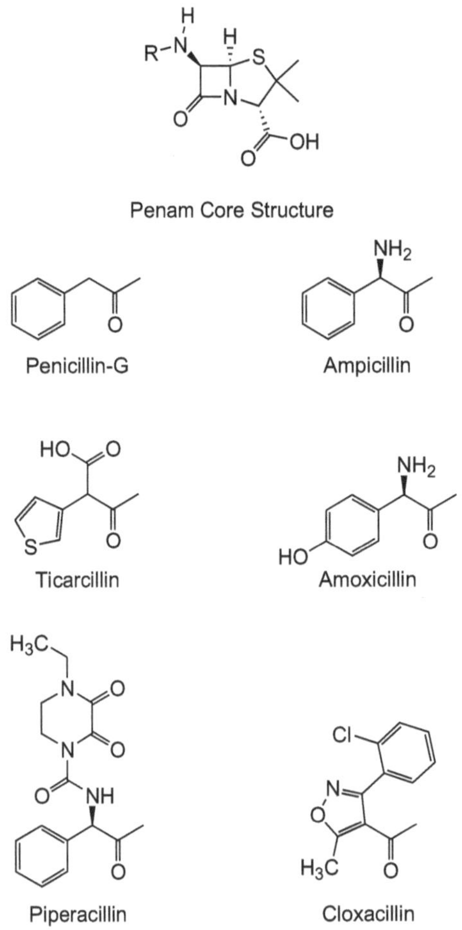

FIGURE 2 The core penam structure and specific substituents of the β-lactam antibiotics investigated.

•OH radical reaction with the entire molecule is the sum of the rate constants for the reactions of the •OH radical with individual structural components, and based on the measured rate constant for (+)-6-aminopenicillanic acid, the core moiety for all these species, of $k = (2.40 \pm 0.05) \times 10^9\ M^{-1}\ s^{-1}$[42] the significantly faster reactivity for all the other species (see Table 1) suggests that most of the reactivity occurs at the side-chain substituents. For example, for penicillin-G whose rate constant is $k = (8.70 \pm 0.32) \times 10^9\ M^{-1}\ s^{-1}$,[27] the factor of 4× faster reactivity suggests that the initial •OH reactivity occurs at the substituent ~75% of the time. Based on analogous rate constant data from the literature for similar compounds[16] for the core moiety, most of the •OH reactivity would be expected to occur at the sulfur atom, while for the substituent it would be expected to occur at the aromatic phenyl ring, ultimately resulting in phenolic species. However, these ultimate products would still have an intact β-lactam ring, and *thus still*

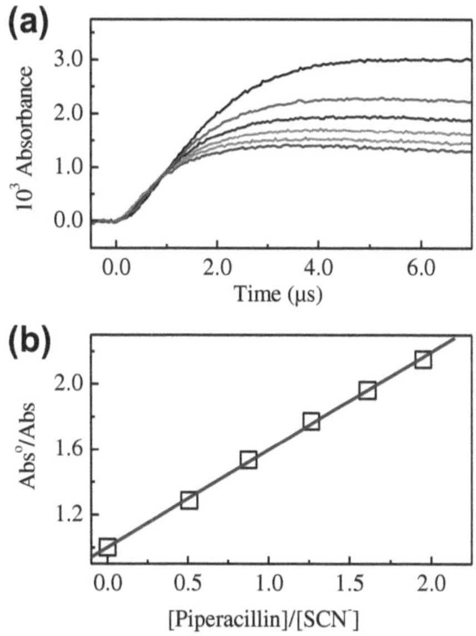

FIGURE 3 (a) Absorbance of $(SCN)_2^{\bullet-}$ decreases with higher concentrations of Piperacillin. (b) The ratio of the $(SCN)_2^{\bullet-}$ absorbance intensities plotted against the ratio of piperacillin and thiocyanate concentrations to give a straight-line plot where the slope represents $k_{piperacillin}/k_8$.

Table 1 Measured Rate Constants and Deactivation Efficiences for Representative β-Lactam Antibiotics

Name	$k_{\bullet OH}$ (10^9 M^{-1} s^{-1})	# •OH Oxidations to Completely Remove Activity
Amoxicillin	6.94 ± 0.44	5
Ampicillin[27]	8.21 ± 0.29	4
Cloxacillin[27]	6.27 ± 0.13	4
Penicillin-G[27]	8.70 ± 0.32	5
Piperacillin	6.98 ± 0.18	5
Ticarcillin[27]	8.18 ± 0.99	4

be functional antibiotics. This overall reactivity is depicted for penicillin-G in Figure 4. Similar inferences can be made for all the other β-lactams of this study.

Bactericidal Activity of Oxidation Products

To quantitatively establish the extent of only •OH radical based oxidation required to fully remove antibiotic activity from waters, specific experiments were conducted using steady state ^{137}Cs or ^{60}Co irradiation to selectively oxidize aqueous solutions of the β-lactams.

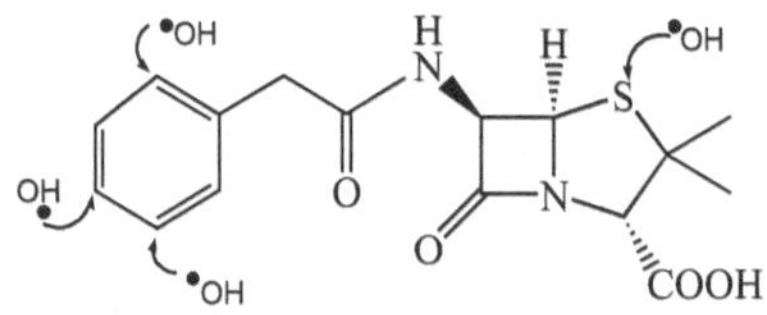

FIGURE 4 Schematic depicting possible •OH radical reaction locations for Penicillin-G. •OH preferentially reacts with the electron-rich phenyl ring. As a result, more than one oxidation may be required to deactivate the molecule.

The major advantage to this technique is that the the photon energy deposited into the water (55.45 M) as compared to the added solutes ($\leq$10 mM). This allows a known concentration of •OH radicals to be consistently produced in all solutions, independent of the β-lactam present, based on the known dose rate of the irradiator. In our experiments, specific amounts of •OH radicals were generated in antibiotic solutions prepared at a concentration threefold higher than the experimentally determined MICs. For each antibiotic the irradiation time points were chosen so that exact [•OH]:[Antibiotic] ratios were achieved.

These irradiated solutions were then used in bacterial growth measurements, measured relative to unirradiated samples. Relative growth percentages were calculated by subtracting the average minimum sample absorbance from the average absorbance of the sample and dividing by the difference difference between the average minimum and maximum absorbances as in Eqn (10).

$$\text{Relative Growth\%} = 100 \times \frac{\text{Abs} - \text{Abs}_{\text{min}}}{\text{Abs}_{\text{max}} - \text{Abs}_{\text{min}}} \tag{10}$$

These relative growth percentages were plotted against the calculated [•OH]:[Antibiotic] ratios. A sigmoidal curve was fit to the data to determine the integer ratio at the onset of the limiting (100% relative growth) value. A representative plot showing cloxacillin, penicillin-G, and amoxicillin is seen in Figure 5. No change in antibiotic activity is observed (zero bacterial growth) until after at least three oxidation events occurs. The point where 100% growth of the bacteria occurred was taken as where the average number of oxidations per molecule was sufficient so that all antibiotic activity was destroyed. For penicillin-G, this occurred at a total of 5 •OH oxidations, similar to the value predicted from the kinetic data.

An equivalent value was obtained for amoxicillin, which is much lower than the original measured value of 9 oxidation equivalents.[31] For cloxacillin, only 4 •OH-based oxidations were required. The values obtained for all the β-lactams of this study are given in Table 1. It is seen that with increasing side-chain complexity, the number of •OH radical oxidations to remove all antibiotic activity increases, as expected.

The finding that multiple (4-5) oxidations are required to completely remove *all* activity for β-lactam antibiotics in water has profound implications for AOP-based treatments, and the reduced efficiency of oxidative degradation definitely needs to be incorporated into kinetic models of these systems at large scale. If one assumes a priori that the •OH radical reactivity with each of the individually oxidized species remains the

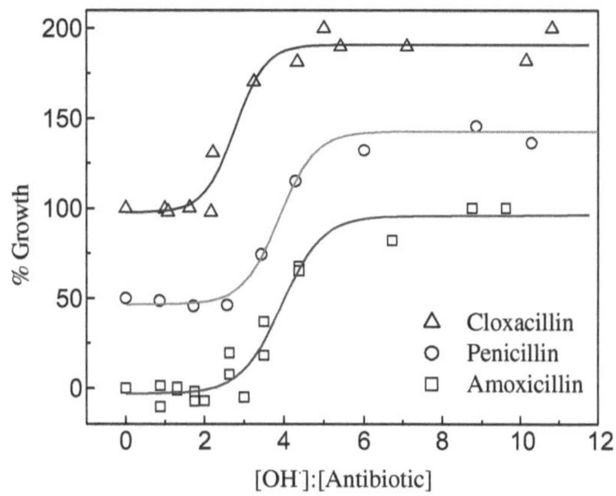

FIGURE 5 Bacterial growth inhibition curves for cloxacillin, penicillin-G, and amoxicillin. Data are offset for clarity. A sigmoidal curve was fit to the data to find the onset of 100% growth recovery.

same, then one can simply divide the overall measured rate constant by the measured oxidation number, to give an effective rate constant for the process. However, for an exact treatment of this sequential mechanistic pathway, specific rate constants for •OH reaction with each of the oxidized intermediates would have to be individually determined.

These calculated oxidative degradation data for these β-lactam antibiotics are quantitative measurements of *only* •OH oxidative-based degradation of β-lactam antibiotic activity. In comparison to the previous investigations[26,30,31] on penicillin-G, our data are in closer agreement to the original finding[26] that there is residual antibiotic activity after initial •OH oxidation of penicillin-G. We also performed some preliminary LCMS measurements, to find that only 1.5 equivalents of •OH are required to almost completely convert penicillin-G to its mono- and dihydroxylated derivatives.

In contrast, our experimental findings directly contradicts the findings of Keen and Linden,[30] who determined that only a single oxidation was required to remove all antibiotic activity. At this time we cannot fully reconcile reconcile these efficiency data, nor how their calculated rate constant for hydroxyl radical reaction with penicillin-G of $(12.6 \pm 2.5) \times 10^9\ M^{-1}\ s^{-1}$ was much faster than the absolute measured values of $(8.70 \pm 0.32) \times 10^9\ M^{-1}\ s^{-1}$[27] and $(7.97 \pm 0.11) \times 10^9\ M^{-1}\ s^{-1}$.[42] It is possible that the presence of excess H_2O_2 used in their method somehow impacts the •OH partitioning across the molecule, or that the intermediate formed in the first oxidation reacts with the peroxide to break the lactam ring.

Impacts of Wastewater Matrix Components

To obtain further insight into the hydroxyl radical reactions occurring for the β-lactams, additional preliminary experiments were performed in this work to specifically investigate the impacts of added DOM and carbonate. In these studies SRFA I was used as a

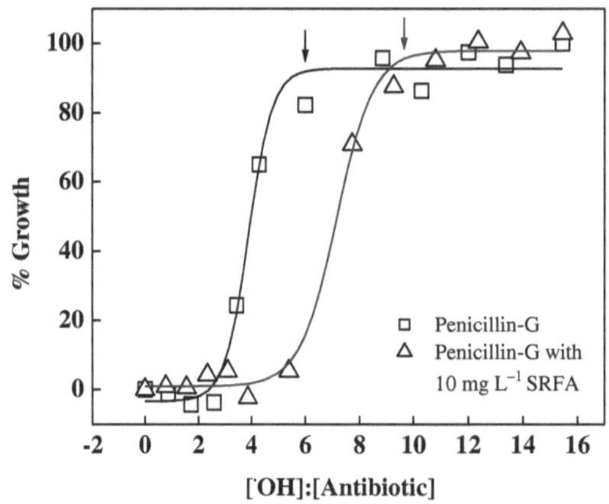

FIGURE 6 Bacterial growth inhibition curves for penicillin-G alone in solution, and in the presence of 10 mg L^{-1} Suwannee River Fulvic Acid.

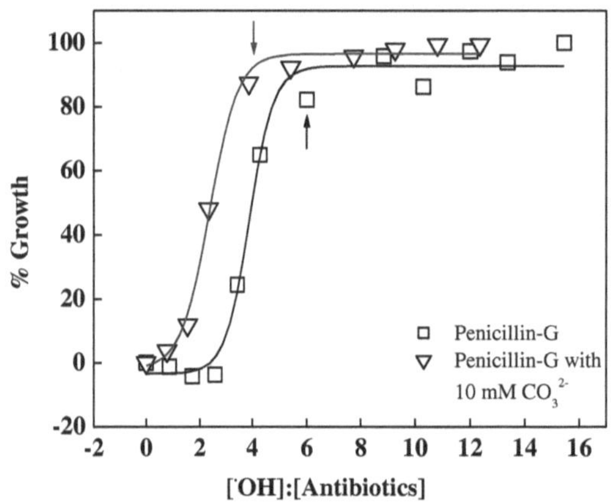

FIGURE 7 Bacterial growth inhibition curves for penicillin-G alone in solution, and in the presence of 10 mM CO_3^{2-}. Integer [OH]:[Antibiotic] values were determined as in Figure 5.

model DOM. Penicillin-G solutions were prepared as before and then separately spiked with 10 mg L^{-1} SRFA I prior to irradiation. Relative bacterial growth curves for these samples were determined as before, and are plotted as a function of [•OH]:[Penicillin-G] in Figure 6. For comparative purposes, the original data for penicillin-G in highquality water is superimposed in this plot (Figure 7).

As expected, the deactivation of penicillin-G in the presence of SRFA required more •OH oxidations than penicillin-G alone (ca, nine •OH radical reactions as compared to five). This increase can be attributed to the loss of hydroxyl radicals by reaction with DOM, but also infers that any DOM radical produced, do not react to deactivate penicillin-G to the same extent.

Analogous experiments were also performed for the separate addition of 10 mM L^{-1} carbonate to penicillin-G solutions, with the solution pH adjusted back to the original value after addition. In this case, the overall deactivation of penicillin-G was seen (see Figure 7) to be to be *more* efficient, with loss of total activity observed after only 3 •OH oxidation events. For this carbonate concentration, and at this pH, a significant fraction of the hydroxyl radical reaction would occur to produce carbonate radicals ($CO_3^{\bullet-}$), and so we can postulate that the subsequent reaction of the carbonate radical with penicillin-G is more efficient at breaking the β-lactam ring than hydroxyl radicals. We are presently performing additional experiments to further investigate this finding. However, even under these most favorable circumstances, we cannot reproduce the single hydroxyl radical oxidation event removing all penicillin-G antibiotic activity finding reported previously.[30]

Conclusions

Hydroxyl radical reaction efficiencies with β-lactam antibiotics in pure water solutions have been quantitatively measured using radiolysis and bacterial growth inhibition techniques. It was found that multiple (4-5) radical oxidations per antibiotic molecule are required to completely remove all antibacterial activity. These data support previous measurements obtained using O_3/H_2O_2 hydroxyl radical generation, but contradict those based on UV/H_2O_2 radical generation. Additional investigations using penicillin-G aqueous solutions containing DOM showed that the deactivation efficiency was decreased, as expected. In contrast, the presence of bicarbonate/carbonate improved the deactivation efficiency, which implied that carbonate radicals are more efficient at breaking the β-lactam ring than hydroxyl radicals. This complex chemistry that encompasses radical kinetics and overall deactivation efficiencies under real-world conditions has to be fully elucidated in order for AOP treatments of antibiotic-contaminated waters at large scale to be fully understood.

Acknowledgments

Molecular graphics and analyses were performed with the UCSF Chimera package. Chimera is developed by the Resource for Biocomputing, Visualization, and Informatics at the University of California, San Francisco (supported by NIGMS P41-GM103311). Kinetic measurements were performed at the Radiation Research Laboratory, University of Notre Dame, which is funded by the United States Department of Energy. We gratefully acknowledge the access to the ^{137}Cs irradiator at the University of California, Irvine and Dr. Julie R. Peller for her help in our preliminary LCMS measurements.

References

1. Bloom BR. *Tuberculosis: pathogenesis, protection, and control.* Washington (D.C.): ASM Press; 1994.
2. Lewis K. Platforms for antibiotic discovery. *Nat Rev Drug Discov* 2013;**12**(5):371–87.
3. Davies J, Davies D. Origins and evolution of antibiotic resistance. *Microbiol Mol Biol Rev* 2010;**74**(3): 417–33.

4. Kumarasamy KK, Toleman MA, Walsh TR, Bagaria J, Butt F, Balakrishnan R, et al. Emergence of a new antibiotic resistance mechanism in India, Pakistan, and the UK: a molecular, biological, and epidemiological study. *Lancet Infect Dis* 2010;**10**(9):597–602.
5. Lowy FD. Antimicrobial resistance: the example of *Staphylococcus aureus. J Clin Invest* 2003;**111**(9): 1265–73.
6. Hicks LA, Taylor TH, Hunkler RJ. U.S. outpatient antibiotic prescribing, 2010. *N Engl J Med* 2013; **368**(15):1461–2.
7. Hollis A, Ahmed Z. Preserving antibiotics, rationally. *N Engl J Med* 2013;**369**(26):2474–6.
8. Gaskins HR, Collier CT, Anderson DB. Antibiotics as growth promotants: mode of action. *Anim Biotechnol* 2002;**13**(1):29–42.
9. FDA Department of Health and Human Services. Summary report on antimicrobials sold or distributed for use in food-producing animals. http://www.fda.gov/downloads/ForIndustry/UserFees/AnimalDrugUserFeeActADUFA/UCM338170.pdf [accessed 04.01.14].
10. Kummerer K. Antibiotics in the aquatic environment – a review – part I. *Chemosphere* 2009;**75**(4): 417–34.
11. Knapp CW, Engemann CA, Hanson ML, Keen PL, Hall KJ, Graham DW. Indirect evidence of transposon-mediated selection of antibiotic resistance genes in aquatic systems at low-level oxytetracycline exposures. *Environ Sci Technol* 2008;**42**(14):5348–53.
12. Szczepanowski R, Linke B, Krahn I, Gartemann KH, Gutzkow T, Eichler W, et al. Detection of 140 clinically relevant antibiotic-resistance genes in the plasmid metagenome of wastewater treatment plant bacteria showing reduced susceptibility to selected antibiotics. *Microbiology* 2009;**155**(Pt 7): 2306–19.
13. Schluter A, Krause L, Szczepanowski R, Goesmann A, Puhler A. Genetic diversity and composition of a plasmid metagenome from a wastewater treatment plant. *J Biotechnol* 2008;**136**(1–2):65–76.
14. *Clean watershed needs survey 2004: Report to Congress.* Washington, DC: U.S. Environmental Protection Agency; 2004.
15. Kolpin DW, Furlong ET, Meyer MT, Thurman EM, Zaugg SD, Barber LB, et al. Pharmaceuticals, hormones, and other organic wastewater contaminants in U.S. Streams, 1999–2000: a national reconnaissance. *Environ Sci Technol* 2002;**36**(6):1202–11.
16. Buxton GV, Greenstock CL, Helman WP, Ross AB. Critical review of rate constants for reactions of hydrated electrons, hydrogen atom and hydroxyl radicals ($\cdot OH/\cdot O^-$) in aqueous solutions. *J Phys Chem Ref Data* 1988;**17**:513–886.
17. von Sonntag C. Advanced oxidation processes: mechanistic aspects. *Water Sci Technol* 2008;**58**(5): 1015–21.
18. Ikehata K, El-Din MG. Aqueous pesticide degradation by hydrogen peroxide/ultraviolet irradiation and fenton-type advanced oxidation processes: a review. *J Environ Eng Sci* 2006;**5**(2):81–135.
19. Huber MM, Ternes TA, von Gunten U. Removal of estrogenic activity and formation of oxidation products during ozonation of 17α-ethinylestradiol. *Environ Sci Technol* 2004;**38**(19):5177–86.
20. Mezyk SP, Rickman KA, Hirsch CM, Dail MK, Scheeler J, Faust T. Advanced oxidation and reduction process (AO/RP) radical generation in the laboratory and at large scale: an overview. In: Ahuja S, editor. *Monitoring water quality – Pollution assesment, analysis, and remediation.* Elsevier; 2013. p. 228–45.
21. Walling C. Fenton's reagent revisited. *Acc Chem Res* 1975;**8**(4):125–31.
22. O'Shea KE, Beightol S, Garcia I, Aguilar M, Kalen DV, Cooper WJ. Photocatalytic decomposition of organophosphonates in irradiated TiO2 suspensions. *J Photochem Photobiol A Chem* 1997;**107**(1–3): 221–6.

23. Mason TJ. *Sonochemistry and sonoluminescence.* Dordrecht: Kluwer Academic; 1999.
24. Kritzer P, Dinjus E. An assessment of supercritical water oxidation (SCWO): existing problems, possible solutions and new reactor concepts. *Chem Eng J* 2001;**83**(3):207–14.
25. Namihira T, Wang D, Akiyama H. Pulsed power technology for pollution control. *Acta Phys Pol A* 2009;**115**:953–5.
26. Dodd MC, Kohler HE, von Gunten U. Oxidation of antibacterial compounds by ozone and hydroxyl radical: elimination of biological activity during aqueous ozonation processes. *Environ Sci Technol* 2009;**43**(7):2498–504.
27. Dail MK, Mezyk SP. Hydroxyl-radical-induced degradative oxidation of β-lactam antibiotics in water: absolute rate constant measurements. *J Phys Chem A* 2010;**114**(32):8391–5.
28. Housewright RD, Berkman S, Henry RJ. The relative effectiveness of pure penicillins against *Bacillus anthracis* in vitro and in vivo. *J Immunol* 1947;**57**(4):343–9.
29. Kavanagh F. Activities of twenty-two antibacterial substances against nine species of bacteria. *J Bacteriol* 1947;**54**(6):761–6.
30. Keen O, Lindin KG. Degradation of antibiotic activity during UV/H_2O_2 advanced oxidation and photolysis in wastewater effluent. *Environ Sci Technol* 2013;**47**:13020–30.
31. Mezyk SP, Otto SC. Quantitative removal of β-lactam antibiotic activity by hydroxyl radical reaction in water: how much oxidation is enough? *J Adv Oxid Technol* 2013;**16**(1):117–22.
32. Crittenden JC, Hu S, Hand DW, Green SA. A kinetic model for H_2O_2/UV process in a completely mixed batch reactor. *Water Res* 1999;**33**(10):2315–28.
33. Kuda T, Yano T. Colorimetric alamar blue assay as a bacterial concentration and spoilage index of marine foods. *Food Control* 2003;**14**(7):455–61.
34. Weigand I, Hilpert K, Hancock REW. Agar and broth dilution methods to determine minimal inhibitory concentration (mic) of antimicrobial substances. *Nat Protoc* 2008;**3**(2):163–75.
35. Asamoto DK, Mezyk SP. Reactivity of H_2O_2 with β-lactam antibiotics, in preparation.
36. Hug GL, Wang Y, Schöneich C, Jiang PY, Fessenden RW. Multiple time scales in pulse radiolysis. Application to bromide solutions and dipeptides. *Radiat Phys Chem* 1999;**54**(6):559.
37. Whitham K, Lyons S, Miller R, Nett D, Treas P, Zante A, et al. In linear accelerator for radiation chemistry research at Notre Dame. In: *Proceedings of the 1995 Particle Accelerator Conference*, vol. 1; 1995. p. 131–3.
38. Buxton GV, Stuart CR. Re-evaluation of the thiocyanate dosimeter for pulse radiolysis. *J Chem Soc Faraday Trans* 1995;**91**(2):279–81.
39. Blumberg PM, Strominger JL. Interaction of penicillin with the bacterial cell: penicillin-binding proteins and penicillin-sensitive enzymes. *Bacteriol Rev* 1974;**38**(3):291–335.
40. Silvaggi NR, Josephine HR, Kuzin AP, Nagarajan R, Pratt RF, Kelly JA. Crystal structures of complexes between the r61 dd-peptidase and peptidoglycan-mimetic β-lactams: a non-covalent complex with a "perfect penicillin". *J Mol Biol* 2005;**345**(3):521–33.
41. Pettersen EF, Goddard TD, Huang CC, Couch GS, Greenblatt DM, Meng EC, et al. UCSF chimera—a visualization system for exploratory research and analysis. *J Comput Chem* 2004;**25**(13):1605–12.
42. Song W, Chen W, Cooper WJ, Greaves J, Miller GE. Free-radical destruction of β-lactam antibiotics in aqueous solution. *J Phys Chem A* 2008;**112**(32):7411–7.

5

Mine-Impacted Water and Biochemical Reactors

Mark Fitch

CIVIL, ARCHITECTURAL AND ENVIRONMENTAL ENGINEERING, MISSOURI UNIVERSITY OF SCIENCE AND TECHNOLOGY

CHAPTER OUTLINE

Food, Energy, and Water. http://dx.doi.org/10.1016/B978-0-12-800211-7.00005-3

Introduction

This chapter reviews the current understanding of the function and design of biochemical reactors (BCRs) for the treatment of water impacted by mining. The term BCR was adopted to describe a treatment unit that uses the biological reduction of sulfate as a basis to treat mine-impacted water (MIW).[1] This definition may include systems also termed wetlands, sulfate-reducing bioreactors, or in some cases, permeable reactive barriers designed for the purpose of sulfate reduction to capture metals. The terminology overlaps but does not coincide with deep, organic-rich wetlands in which an extensive anaerobic zone can reduce sulfate.

A BCR is most often a basin filled with a permeable substrate including a solid-phase organic in which MIW is treated as a result of sulfate reduction. Water is distributed by pipes into a gravel layer so that flow will be even, then continues through the solid substrate and is collected in a manner similar to the distribution, as shown in Figure 1.

There are configurations of BCRs in which the organic is fed to a reactor that also receives the MIW, such as reviewed by Sanchez-Andrea et al.[2] Such liquid-fed systems are at the time of this writing rare in the field; lab-scale studies show a liquid feed provides the potential for much smaller treatment volumes. However, the active feeding of organic requires more complexity than the passive system depicted in Figure 1, in which water percolates through a pile of slowly degrading solid-phase organic. This chapter focuses primarily on such solid-phase BCRs, and is organized to address the major questions when addressing the treatment of MIW using a BCR. These are

1. Is the water suitable for treatment by BCR? A decision tree should be followed when considering the treatment of MIWs, and the water often has a low pH, which must be dealt with prior to treatment by a solid-substrate BCR.
2. What reactions occur in a BCR? Metal removal can occur due to several mechanisms, with sulfide precipitation preferred. The source of that sulfide is the biological processing of cellulose to organic acids, which are in turn used by sulfate-reducing bacteria (SRB) that use sulfate as an electron acceptor. Sulfide is the waste product, analogous to humans breathing out carbon dioxide.
3. Why are systems designed as they are? The initial observations of BCR function were in natural wetlands, which led to the concept of constructing wetlands for

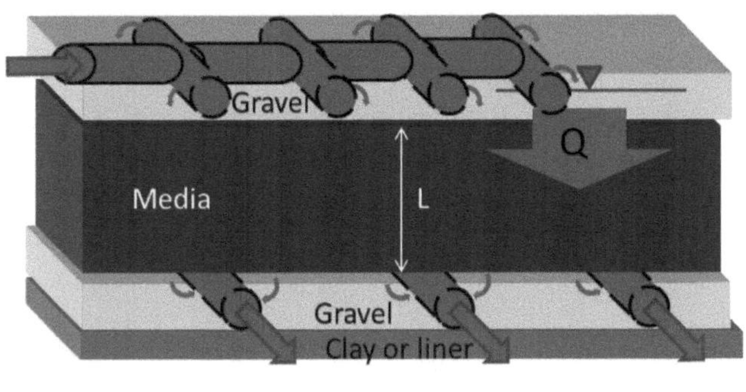

FIGURE 1 A typical vertical-flow biochemical reactor. Dashed lines on pipes indicate perforated pipe.

the purpose. Study of the mechanisms controlling metal removal resulted in realization that the fundamental reactor was not a wetland, but a slow degradation of woody materials resulting in low rates of sulfide production, which designers currently estimate as 300 mmol/m^3 reactor/d.

4. What is the treatment goal? This is governed by local, regional, and national laws.
5. What should be in the solid substrate? The substrate should be permeable to water and contain long-term electron donor such as wood chips, a source of organisms such as manure, and a source of immediate electron donor such as manure (which thus serves two roles). The electron donor degradation rate determines the sulfate reduction rate, which in turn determines the efficiency of operation.
6. What is the sulfate reduction rate? A range of values have been reported and accurate prediction of a rate is not yet possible for woody substrate. Liquid substrates may give a hundred-fold greater rate.
7. How much water can be pushed through a BCR? The hydraulic properties of BCRs depend upon their composition and change with age.
8. How does one design a BCR? The foregoing answers may be combined to give a simple design process.
9. Is the water exiting a BCR ready for release? A settling and aeration step should follow a BCR to reduce both solid concentrations and also the toxicity of sulfide in the effluent.
10. How does one start up a BCR? The system is flooded and allowed to sit until sulfate reduction occurs. Flow then is started at a low rate and ramped up gradually to full flow.
11. What are operating issues? Food may run out. Short-circuiting will occur to some degree. Plugging or gas lock can occur. Pipes may clog or burst. The source water may have low pH and toxins, which may affect operation if upstream controls fail. Erosion and vandalism can occur. Finally, sulfide smells bad.

Applicability of BCRs for Mine-Impacted Water

Although acid mine drainage is the most common impact of mining on water quality, the variety of mineral geology is such that there is no single-characteristic mine water. The term “mine-impacted water” is used because it is more inclusive than terms such as acid mine drainage (AMD) or acid rock drainage (ARD). A heap leaching pad might not be considered a mine, but produces an acidic fluid with high metal concentrations. On the other hand, some metal mines are in limestone or dolomite, producing water that is neutral or even alkaline but contains toxic levels of metals. Mine-influenced water pH values span the range -1 to 9, and combined Zn, Cu, Pb, Co, and Ni concentrations can be more than 60 g/L.[3,4]

Although the term “mine-impacted water” is highly inclusive, in the majority of cases that water is an AMD. This is because iron is a large fraction of the earth’s crust, and most mine sites, regardless of being coal or hard rock mines, include pyrite or a similar form of iron. A set of reactions of that iron lead to AMD. Such water has low pH and also iron present in large amounts due to pyrite oxidation, which may be summarized as follows:

$$4FeS_2(s) + 14O_2 + 4H_2O \rightarrow 4Fe^{2+} + 8SO_4^{2-} + 8H^+$$
$$4Fe^{2+} + O_2 + 4H^+ \rightarrow 4Fe^{3+} + 2H_2O$$
$$4Fe^{2+} + O_2 + 2H_2O \rightarrow 4Fe^{3+} + 4OH^-$$
$$2FeS_2(s) + 14Fe^{3+} + 8H_2O \rightarrow 15Fe^{2+} + 2SO_4^{2-} + 16H^+$$

The overall result of the reactions is the production of a fluid high in sulfuric acid and carrying large amounts of ferrous iron (Fe^{2+}). The water is likely anaerobic, because the little oxygen flux into the subsurface is used in the reactions. When this fluid surfaces and meets plentiful oxygen and pH begins to rise, the ferrous iron reacts to form ferric oxy(hydrox)ides, which precipitate as yellow boy and other generally red solids.

Such seepages, with orange, yellow, or red tails running downstream from the site, are common in mining country. Mining activities have left a legacy of such sites that are often remote and dispersed, and many are abandoned. A conservative estimate of more than 280,000 abandoned mines in the United States may be arrived at by combining the 134,400 sites recorded by the Office of Surface Mining Reclamation and Enforcement[5] with the 161,000 hard rock sites in 12 western states estimated by the Government Accountability Office (GAO).[6] There were, in 2012, an additional 718 operating mines in the Untied States.[7]

Not all mines leave a lasting environmental impact. However, mining has been estimated to have degraded water quality in the headwaters of more than 40% of the watersheds in the Western United States.[8] The GAO[6] estimated more than 33,000 sites as having degraded the environment, largely through impacts on water. Resulting US Federal cleanup costs between 1998 and 2007 were slightly more than $2,400 million.

Extrapolating data from the United States to the whole world, one must conclude there are hundreds of thousands of mine sites that impact water quality.

The treatment of MIW varies according to both the chemistry of the water to be treated and the quantity of that water[9–13] as online guidance[1,14] and EPA studies.[15,16] For large quantities of water, active processes in which chemicals are added, mixed, and precipitation occurs, mixing and clarifiers may be the only practical approach to treatment; passive systems for large quantities of water would require unfeasibly large land footprints. However, a large fraction of sites with MIW have sufficiently small amounts of water flow that more passive methods can be used. These are considered passive because little or no operating energy is used. In some cases, particularly where the topography allows for downhill flow into and out of reactors, the only energy use is for monitoring the site.

The choice of treatment of MIW is illustrated in Figure 2, which the author contributed to the Interstate Technology and Regulatory Council (ITRC) guidance document for regulators.[18] The major considerations are as follows:

1. Neutralization. The pH of water must be raised, assuming it is acidic, to a circum-neutral value. The acidity is most often lowered through the addition of alkalinity in the form of limestone. For AMD/ARD, such neutralization results in large

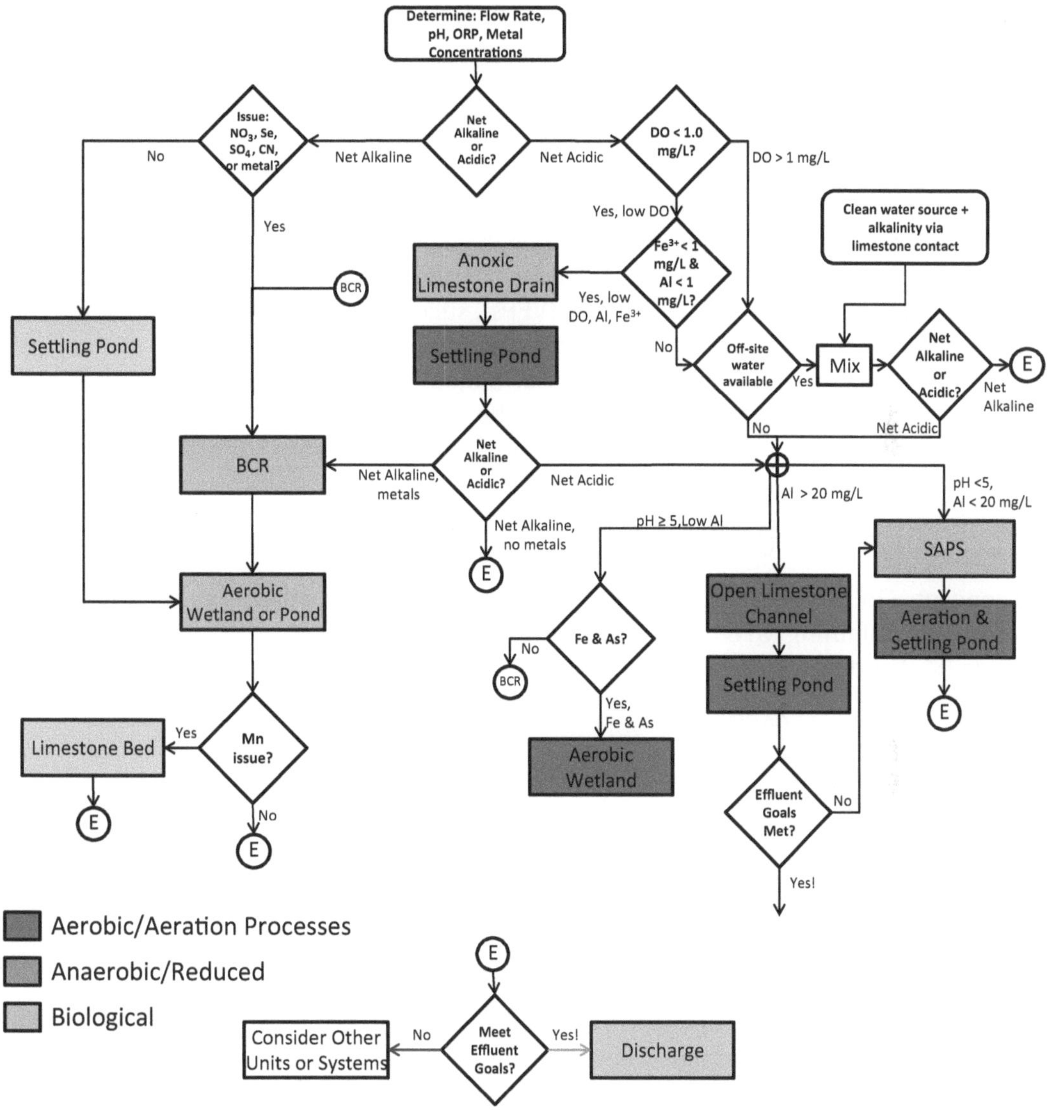

FIGURE 2 Decision tree for treatment of mine-impacted water. BCR, biochemical reactor; DO, dissolved oxygen; ORP, redox potential; SAPS, successive alkalinity producing systems. *Adapted from Ref. 17.*

amounts of ferric iron and/or aluminum, which when exposed to oxygen form bulky precipitates.

2. Iron precipitation. Iron oxy(hydrox)ides are high in specific gravity and will readily settle in settling ponds. As these oxides form and settle, they act as an adsorptive medium to coprecipitate other metals.
3. Remaining metals. After neutralization and precipitation of oxides, if necessary, the neutral or alkaline water still may contain metals in toxic amounts. Additionally,

sulfate is often present at levels that may be adverse to the receiving stream's biota. This is the situation in which BCRs are applied.

The above is not intended as guidance on the treatment of MIW, but rather to describe the setting in which BCRs are appropriate for the treatment of mineral-containing water. Water entering a BCR should be low in iron and aluminum oxides simply because the precipitation of these solids will block pores. A pH below 5 is advisable if the electron donor (biological energy source) is cellulosic.

Although similar in appearance, the definition of a BCR does not include those units that have been called vertical flow ponds (VFPs), successive alkalinity producing systems (SAPS), and reducing alkalinity producing systems (RAPS). These systems are not primarily based on sulfate reduction. Instead, biological electron acceptance is used to create reducing conditions, allowing acid to be treated by an alkalinity source such as limestone without extensive armoring by iron and/or aluminum oxyhydroxides.[19]

Mechanisms in BCRs

In concept, a BCR brings together sulfate-reducing organisms, sulfate, an electron donor, and metal-containing MIW. The SRB use the sulfate as an electron acceptor, producing sulfide (S^{2-}) and water while consuming an electron donor such as hydrogen, acetate, or a simple organic compound. The sulfide reacts with the metal, forming a metal sulfide with low solubility.

Metal Sulfide

Metals that may or may not be precipitated as sulfides are shown in Table 1. The solubility product, K_{SP}, is the product of the concentration of the soluble ions that react to form the solid, such as

$$Pb^{2+} + S^{2-} \leftrightarrow PbS(s)$$

which has a K_{SP} of roughly 7×10^{-29}, describing the equilibrium of the activities of each component at saturation, roughly equal to the molar concentrations:

$$[Pb^{2+}][S^{2-}] = 7 \times 10^{-29} (mol/L)^2$$

It should be noted that the sulfide is present as HS^- at most pH values, and the true reaction is between solid-phase metal sulfide and H^+, yielding the dissolved metal and HS^-. This solubility product is shown in Table 1 for a number of metals of interest.

As indicated in Table 1, some metals are not easily removed by BCRs, with manganese of particular note; this relatively common contaminant is effectively soluble in water containing sulfide; instead, carbonate or other precipitation chemistry should be considered to remove such metals. But for many metals, the waste from the microbial metabolism, sulfide, is a good agent for precipitating the metal. Assuming that precipitation occurs rapidly and ample sulfide is provided to precipitate metal, the observation

Table 1 Minimum Metal Sulfide Solubility and Calculated Metal Concentration if 0.1 mg/L Sulfide (0.003 mM) is Measured in Solution

Metal/Metal Sulfide	−$\log_{10}$ Solubility Product of Metal Sulfide, pK_{SP}	Metal Concentration if 0.1 mg/L HS^-, mg/L
Ag/Ag_2S	36.3	4×10^{-11}
Al/Al_2S_3	−76	Large[a]
Bi/Bi_2S_3	63.1	1×10^{-18}
Cd/CdS	15.9	5×10^{-6}
Co/CoS	13.0	2×10^{-3}
Cu/Cu_2S	22.9	1×10^{-4}
Fe/FeS_2	16.3	0.3
Hg/HgS	39.1	5×10^{-29}
Mn/MnS	0.4	Large[a]
Ni/Ni_3S_4	35.1	2.7
Pb/PbS	15.2	4×10^{-5}
Sn/SnS_2	36.2	8×10^{-21}
Zn/ZnS	11.5	7×10^{-2}

[a]Some metals essentially do not precipitate with sulfides.
Data from Ref. 20.

of very little sulfide in solution corresponds to negligible amounts of many metals. It is worth noting, however, that real water chemistry is more complex than describing the metal species by a single reaction, and wisdom dictates consideration of all of the potential geochemistry such as may be provided by some equilibrium software packages (e.g., PHREEQC).

Less Favorable Precipitates

If oxygen is present in the MIW, which is quite likely as a result of a preceding iron precipitation step, it is possible some additional formation of metal oxides could occur, or that nonsettled, dissolved metal oxides enter the BCR. The concern for having such oxides enter a BCR is that the aerobic metabolism depletes oxygen, causing the precipitate to dissolve back into solution as oxygen concentrations fall to zero.

As pH is raised, such as in limestone treatment of AMD or ARD, the solubility of both metal hydroxides and metal carbonates decrease. Increasing pH corresponds to increasing $[OH^-]$ in solution, and through carbonate chemistry, results in increased $[CO_3^{2-}]$. Hydroxide and carbonate precipitation often are major determinants of metal solubility as a function of pH, with an extreme example being the softening of water by lime addition; the raised pH causes bicarbonate (HCO_3^-) to release a proton and form carbonate, which then reacts with calcium to form calcium carbonate as a precipitate, and in the case of magnesium, the direct formation of $Mg(OH)_2$. These precipitates are generally moderately soluble, and if pH were to fall, the precipitates would dissolve and the metal would be released.

Sulfate Reduction

SRB have been extensively studied for their electron acceptance from sulfate, role in sulfur cycling, reactions involving electron transfers with metals, and for their use of hydrogen as an electron donor. However, the organic electron donors used by these organisms have not been as thoroughly examined.[21] SRB are able to use a variety of organic compounds, including fatty acids, aromatic compounds, amino acids, sugars, and some alkanes and alkenes. However, research indicates complex polymeric organics such as cellulose, starch, and protein are not used directly by these organisms. Instead, different microbes hydrolyze the complex organics and the fermentative products of such metabolism are the "food" for sulfate reducers, as illustrated with wood as the substrate in Figure 3.

Most waters produced from mines have high concentrations of sulfate, hundreds of milligrams per liter, providing ample electron acceptor for sulfate-reducing organisms as water flows through the BCR when compared to the rate at which electron donor is provided. The source of electron donor may be to provide a liquid substrate such as methanol or acetate, but most often is an inexpensive woody material provided as a solid substrate.

The use of a slowly degrading, complex organic such as wood limits the rate at which sulfide is produced; to produce sulfide, sulfate reducers must have small carbon molecules and/or hydrogen. That electron donor is provided by fermenters, and the rate at

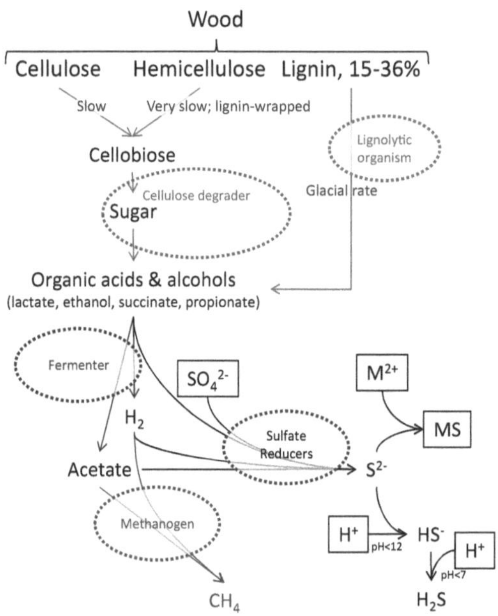

FIGURE 3 Sulfate reduction, the bottom of the typical anaerobic food chain, resulting in metal (M) sulfide formation. CO_2 and H_2O are omitted. *Adapted from Ref. 22.*

which fermentation occurs is in turn limited by the slow degradation of the components of wood. In contrast, mixing a liquid organic into MIW, as done in the type of BCR that has been called a sulfate-reducing bioreactor, not only results in control of the rate but also requires the active addition of the organic.

Adsorption to Ligands

Although thought of as "food for bugs," solid organic substrates such as wood chips also have a capacity to remove metals by adsorption. The organics contain chemical groups such as carboxylic acids, which have the ability to adsorb cationic metals. If the material is a solid, like the surface of a wood chip, then the organic is a solid adsorption media. However, as solid woody material is degraded, some of the organic compounds are dissolved (as components of dissolved organic carbon (DOC)). For example, plant-derived organic material tends to degrade to compounds such as humic and fulvic acids, and adsorption of metal to these colloids means the metal is complexed with an aqueous ligand. The organic-bound metal remains present in the aqueous phase, but is unavailable for other removal mechanisms.

In this way, organic matter both can capture metal (adsorption to solids) and prevent metal removal (DOC-bound metals). Adsorption is not a preferred mechanism for metal removal because the adsorption sites eventually will be exhausted and the reactor thus would appear to fail prematurely. Another issue related to such adsorption is that the decay of the organic material acting as an electron donor to the BCR also results in loss of adsorption sites and therefore release of previously adsorbed metals.

The Evolution of BCR Design

BCRs were originally inspired by natural wetlands[23,24] and therefore were considered a form of constructed wetland. Treatment wetlands are designed largely on areal loading and residence time, and a similar approach was initially used for the treatment of MIW. Kleinman[25] suggested 5 m^2 of wetland per L/min of mine drainage, a loading rate of 0.29 m^3/m^2-d. This value was subsequently revised down to 15 m^2 per L/min (0.096 m^3/m^2-d).[26]

In contrast to an areal hydraulic loading rate, Brodie et al.[27] considered the amount of metal to be removed and determined an area required for treatment of 1.3 mg Fe/m^2-min. Dietz et al.[28] suggested a similar areal design on acidity, based on 6 g of acidity/m^2-d. However, the failure of a single areal rate to describe the performance or design of these BCRs had become apparent. For example, Hedin and Nairn[29] found no clear relationship between iron areal load and removal for three different wetlands.

The idea of a volumetric rate of sulfide production (or sulfate reduction) was eventually put forward to describe BCR design. The idea in this case is that the total rate of sulfide production in moles per day should be greater than the expected total metal load in moles per day. McIntire and Edenborn[30] reported rates comparable to coastal marine sediments, with 2.0–600 mmol S/m^3-d measured at the Friendship Hill (PA) system treating water with pH of around 3. Wildeman's group at Colorado School of Mines offered 300 mmol S/m^3-d as a "conservative estimate" based on the Big Five wetland.[31] It should be noted, however, that lower values have been reported; Mueller et al.[32] determined sulfate removal rates of 128–547 mmol S/m^3-d. The studies at the Big Five tunnel also showed that permeability of the substrate was a critical design factor for successful "constructed wetland" operation.

The use of a single loading factor or rate for design of a BCR will not fully describe the geo- and biochemistry involved. The first published attempt to model geochemical and biogeochemical processes in a BCR seems to have been by Klusman around 1990 (e.g., Ref. 33). MINTEQA2 code was modified to calculate rates, including an assumed sulfide generation rate, and plug flow was simulated by doing time steps on a batch case. Although capable of matching effluent concentrations from several operating BCRs, the authors noted "inadequate experimental data" existed to confirm fully model results against BCR realities.

In addition to the question of design values, specifics of design have been the subject of doubts. Is the direction of flow better as vertical down, vertical up, or horizontal? What redox condition, aerobic or anaerobic, results in better treatment? Are plants important? Dietz and Stidinger[34] found subsurface flow performance to be significantly better for alkalinity impact, acidity reduction, and iron and aluminum removal. They speculated that subsurface flow had improved contact between substrates and the impacted water. Although subsurface flow is now recognized as desirable for removal by sulfide precipitation, the direction of flow has not been shown to affect removal. That noted, there only a few horizontal-flow BCR systems.

The role of plants also came into question in the late 1980s.[35] If removal was by sulfide precipitation and wetland plants aerate the subsurface, perhaps plants were not desirable. Not having plants would also decrease concerns over bioaccumulation. On the other hand, plants increase microbial counts in the subsurface and as the senescent mass decays, replenish the bed organic matter. As noted by Beining and Otte,[36] a natural wetland in Wicklow, Ireland, has treated MIW for more than a century and is calculated, based on plant growth rates, to continue treatment for at least another century. A powerful counterexample of the potential role of plants was the shallow Brewer Pad 5 system operated in 1993–1995. When vegetation was present, the system was aerobic and copper removal efficiency was poor; presumably sulfate reduction was inhibited. Although there is no clear answer to whether plants are a net benefit, current BCR designs often exclude plants through placement of a cap, a thick gravel layer, or a free water surface with a depth of 1 m. Ignoring designer's wishes, plants tend to nonetheless colonize these systems.

The flow of water through BCRs has not been incorporated into designs. The idea of an areal or volumetric load treats the BCR as a batch or plug-flow reactor, and assumes the water will have sufficient contact time to allow produced sulfide to mix into the water. The reality, measured so far only at lab scale by tracer tests, is that water flow in solid media is rarely ideal. Short-circuiting of some portion of the flow is to be expected, and at an extreme will result in lower removal than expected because the part of the BCR suffering this short-circuiting is much more heavily loaded than allowed for in the design.

The preceding are highlights of dozens of studies, which, along with determination that plants remove very little metal, were the departure point from "wetlands," implying plants and surface water, to the current concept of the BCR, in which anaerobic conditions result in sulfide formation. The current rule of thumb is to measure a sulfate reduction rate at pilot scale on site, or to use a value such as 300 mmol S/m^3-d. Looking forward, on-site measurement remains the standard for design, but the prediction of full-scale sizing and operation based on such measurements is likely to evolve to include the concept that sulfide reduction is caused by organic decomposition, and that hydraulics are far from ideal. Given appropriate tools, designers will factor in the loss of reducing power (slowing of electron donation) from solid substrates and will allow for short-circuiting of flow within the BCR.

Treatment Goals

Current court opinion in the United States is that MIW is a point discharge requiring an National Pollution Discharge Elimination System (NPDES) permit regardless of the agent attempting to implement treatment (*West Virginia Highlands Conservancy v. Huffman*, 4th US Circuit Court of Appeals, No. 09-1474, 8/11/10). The broad remedial goal is to make negligible the impact of the effluent on a receiving water body. That impact often is toxicity due to low pH, the metal load carried by the MIW, and/or high sulfate concentrations. Therefore, the treatment goals are usually expressed as meeting some minimum pH and/or maximum concentrations of the metals of concern. BCRs may be designed to provide alkalinity and raise pH and to remove metals from solution, with demonstrated reduction of toxicity.

Media Composition

The media in a typical BCR serves two major roles. First, it must be permeable enough to allow water to flow through the media, allowing contact between the water and the organisms that presumably grow on the surface of the media. Second, the media is the source of electron donor. Table 2 describes the functions served by the components of the media. If an exogenous feed source is supplied, this supply fulfills the first two roles noted in the table.

Most often, the provision of alkalinity and bulk are accomplished through including 10–30% by volume limestone gravel. The long-term organic of choice is often wood, and the size of the wood particle is a currently unresolved source of question: smaller particles increase the rate of organic evolution for SRBs, but larger particles last longer. The rate of organic supply for SRBs is thus at some balance point between longevity (large pieces) and supply rate (small pieces). Characterization of several BCR organic

Table 2 Biochemical Reactor (BCR) Substrate Functions and Common Components

Role	Examples of Components
Source of long-term electron donor (food)	Wood chips, sawdust, walnut shells, mushroom compost, other composts, chitin, rice hulls
Source of short-term electron donor for start-up (fast food)	Manure, hay, straw, chitin, yard waste, brewery waste, beet pulp, corncobs
Source of microbial inoculum	Ruminant manure, other BCRs
Source of alkalinity[a]	Limestone, chitin, seashells, fly ash, cement kiln dust
Bulk to allow for good flow through the bed	Sand, gravel; also above components in centimeter-scale sizes such as wood chips, gravel, fractured nut shells
Solid surface sites for microbial attachment and retention	Provided by all above components; decreases as organic solids degrade

Overall, then, the media placed in a BCR is a mix, from most to least by volume, of wood chips or sawdust, hay, or straw, limestone gravel, sand, and cow manure.
[a]Generally not needed for neutral or alkaline mine-impacted water.

substrates was offered by Seyler et al.[22]; it is notable that most wood has very little rapidly degradable material to be used in consuming influent oxygen and creating initial anaerobic conditions. One interesting possible media component is material made from crustacean shells, which contain chitin, an aminated glucose polymer that is combined with small particles of calcium carbonate when making the crustacean shell; with a chemical composition of approximately 20% chitin, 40% chitin-bound calcium carbonate, 30% protein, and less than 9% nitrogen,[37] the material thus is a source of organic and provides alkalinity. Manure both supplies the organisms and also is a short-term organic source, as the organisms in the manure continue to eat the organics in that matrix.

Electron Donor Sources and Rates

Researchers examining woody degradation consider plant materials to have a biodegradable and a nonbiodegradable fraction. Measuring biological oxygen demand (BOD) of wood directly is probably impossible, so instead volatile suspended solids (VSS) or total organic carbon (TOC) is used. Wood is composed of three major portions, cellulose, hemicellulose, and lignin. Cellulose is a chain of sugars and relatively readily biodegraded. Hard woods have higher cellulose content. Softwoods have more lignin content. Hemicellulose is a structure of cellulose wrapped in the complex aromatic compound that is lignin; the polysaccharide must be extracted from that lignin. Lignin is a highly variable structure including ring structures, and is nearly nondegradable. Rot fungi, as an example, are thought to degrade lignin by extracellular, somewhat nonspecific oxidative attack.

Because lignin is so slowly degraded, woody material can be thought of as having a degradable and a nondegradable fraction, so, for example, $VSS = bVSS + nbVSS$. The term bVSS is the biodegradable VSS, and nbVSS represents the nonbiodegradable VSS. The nbVSS is the VSS remaining after decades of degradation. If the biodegradable portion of the VSS is indeed cellulose and hemicellulose, the composition is approximated as sugar. Given that the molecular weight of the cellulose unit is 162 and that of O_2 is 32, each gram of cellulose requires 1.18 g of oxygen to fully oxidize. Assuming all bVSS is cellulose or hemicellulose, the ratio of theoretical BOD to bVSS is approximately 1.2 g BOD/g bVSS.

If the biodegradable portion of VSS is cellulose and hemicelluloses, then the nonbiodegradable portion of the organic matter, nbVSS, is lignin. The lignin content of plants varies not only among genus and species, but also within a given tree by height. Lignin comprises between 15% and 36% of woody plant mass.[38] One might thus estimate the bVSS of the plant matter as 65–85% of the total VSS.

Taken together, these estimates of content and BOD may be used to predict the BOD realized after some years of degradation. For example, 1 kg of wood with a 12% water content (a typical value from Ref. 39) has 880 g of VSS. If the wood has a 25% lignin content, the cellulose + hemicellulose mass (bVSS) is approximately 660 g. The ultimate BOD is 779 g (=660 g bVSS × 1.18 g O_2/g bVSS). In a BCR, a fraction of that BOD realized would occur as a result of anaerobic sulfate reduction rather than oxygen respiration. Unfortunately, the rate at which woody BOD is used and the fraction of BOD realized as sulfate reduction remains broadly unknown.

Estimating the first step of the process, the rate of woody material degradation in a BCR, is challenging. Limited information is available on rates of degradation in BCRs or BCR-like environments. Also, it is important to remember that each BCR has its own chemistry and environmental setting, which will affect the observed rate. For example, low pH increases cellulose solubility, thus potentially increasing the rate of degradation, but pH below 5 may severely inhibit cellulose degraders and thereby prevent degradation. Temperature can be a large factor as well, with a common rule of thumb in biology that rates approximately double for every 10 °C increase in temperature within the range of an organism's tolerance. Finally, the availability of cellulose for degradation varies tremendously among plant species and rate is a function of the amount of area available to cellulose degraders. With those caveats noted, the literature has described wood degradation using various kinetics.

Fitch et al.[40] applied this Monod model to wetland substrate used in lab-scale BCRs over 7 years, and reported values for kX of 0.098–1.8 mg VSS/mg VSS-d, and for K_S, 6.8–210 mg VSS/mg VSS. The values were all normalized to 1 g of BCR substrate, with an observed nbVSS/VSS in the range of 10–13% for a primarily oak chip bark mixture. A more realistic approach to wood degradation includes that only the surface of wood is available to be eaten. Contois kinetics is such an approach, a Monod-like treatment that can describe the case when there is limited surface area of the food (e.g., wood particles). Hemsi et al.[41] reported rate constants of k_c, specific rate of cellulose degradation, 0.625–1.25 g/g-d; X_c, biomass density of cellulose degraders, 0.0005–0.01 times bVSS, which gave values of 9–223 mg/L; and K_c, the Contois saturation constant, 7.5–37.5 g/g.

The combination of these values resulted in estimated cellulose decomposition on the order of tens of milligrams VSS per liter-day.

A simpler first-order approach aimed at showing the rate of VFA production in landfill-like situations for office paper and cardboard was obtained at 35 °C,[42] with a reported stoichiometric coefficient for cellulose conversion to volatile fatty acids (VFA) of 0.66 g/g and a first-order hydrolysis rate constant of 0.012 d^{-1}. The same study characterized inhibition due to pH as none if pH ≥ 7, else described as $\exp\left\{-3\left[\frac{pH-7}{7-5}\right]^2\right\}$, which predicts no hydrolysis at pH 5, and 47% inhibition at pH 6.

Comparing these various rate models in terms of predicted VSS use rate, as shown in Table 3, does not build confidence that a single value may be used for design.

In addition to the kinetic studies noted above, a number of relevant observations have been made without fitting kinetic values. For example, barley straw resulted in more sulfide produced in pore water after 8 weeks of flow than did Timothy hay or compost.[46] Similarly, sawdust was observed to be less degradable than straw when examined as a bulking agent for compost[47] with an oxygen uptake rate (OUR) for straw reported as 2.14 mg O_2/g VS/h at 35 °C, and for sawdust, 0.29 mg O_2/g VS/h. Those rates are similar to values determined from anaerobic landfill simulations; for example, Wang et al.[43] reported first-order decay rates based on methane production: inoculum pine, 21.5 $year^{-1}$; spruce, 1.8 $year^{-1}$; red oak, 2.3 $year^{-1}$; and eucalyptus decay was effectively zero. The eucalyptus was found to inhibit conversion of organics to methane.

Decommissioning of a pilot-scale BCR by Golder[44] reported a decrease in TOC from 23.2% to 14.4% over 27 months. This corresponds to a first-order rate constant of 0.21 $year^{-1}$. A similar order of magnitude rate constant, 0.41 $year^{-1}$, was observed for corn stover over 13 months in situ, with 36% loss of organic carbon.[45] Over 9 months, a mixture of 70% walnut shell and 25% corn stover was fitted by a first-order rate constant,

Table 3 Comparison of Predicted Rates of Anaerobic Woody Matter Degradation

Approach and Reference	Predicted Initial bVSS Use Rate[a] (mg/L-d)	Observed Rate (mg/L-d)
Contois[41]	14	nd
Monod-like for down flow[40]	192	13–201[b]
First-order[42]	4392	nd
First-order[43]	1.8–21.5	nd
First-order, based on Ref. 44	nd	40
First-order, based on Ref. 45, corn stover	305	330
First-order, based on Ref. 45, 70% walnut shell, 25% corn stover	56	32

nd, no data; BCR, biochemical reactor; VSS, volatile suspended solids.
[a]Assuming BCR substrate is 50 vol% woody mass (estimated density 30 lb/ft^3) with 30% lignin giving a bVSS density of 336 g/L and using middle of reported value ranges.
[b]Normalized to above case, 336 g bVSS/L.

0.061 $year^{-1}$, with only 5.4% loss of organic carbon over 9 months, which may have been entirely attributable to the corn stover.

If coastal sediment is considered to be somewhat similar to BCR substrate, then studies related to biogeochemical cycles may be consulted. Canfield et al.[48] showed carbon oxidation rates to be in the range of 20–105 μmol/cm^3/year; assuming a formula of $CH_2O + O_2 \rightarrow CO_2 + H_2O$ as the measured carbon oxidation, this range would be 1.6–8.6 mg/L-d. Corresponding rates of sulfate reduction were 20–55 μmol/cm^3/year (55–150 mmol S/m^3-d).

Sulfate Reduction Rate

Using sulfate as an electron acceptor requires a substantial energy investment rather than yielding energy. It is thermodynamically favorable only at Eh below about −200 mV, which is why oxygen, nitrate, ferric iron, and some fermentations occur before sulfate reduction is observed in a system. Sulfate reduction competes with methanogenesis to come in last as a useful compound to which organisms transfer electrons. However, in the absence of these higher energy electron acceptors and with electron donor ("food," generally organic) present, SRB produce sulfide at some rate.

Most BCRs receive ample sulfate. Therefore, the SRB are limited for electron donor. That electron donor is generated by bVSS degradation, so the rate of bVSS degradation will determine the rate at which sulfide is produced. When considering woody substrates, the stoichiometry of that conversion is not completely clear because anaerobic cellulose degraders may produce succinate, propionate, ethanol, or other compounds.

Considering the range of potential sulfide yields and assuming the analysis of Hemsi et al.[41] broadly applies to cellulose degradation, that is, about half the carbon goes to cell mass, then the amount of evolved sulfide may be calculated to be equivalent to 6–14 mmol S per g VSS consumed. Given a rate of VSS consumption of 100 g/m^3-d, a midrange value in Tables 1–3, the rate of sulfide generation predicted from stoichiometry is 600–1400 mmol S/m^3-d. This value is surprisingly close to observed values considering the string of approximations required to arrive at the value. A summary of reported sulfate reduction rates is provided in Table 4.

As previously noted, a stoichiometric approach that considers the source of electrons might allow the designer to estimate the sulfide generation rate and to determine how that rate will decline as the organic is depleted. More data would be required for the specific plant matter used and to estimate the change in degradation with time, as no rigorous long-term study has shown values for such degradation rates.

One attraction of BCRs in which electron donor is added in a liquid phase is that the rate of sulfate reduction may be controlled based on the addition of that BOD, and the achievable rate is much greater than relying on slowly degrading cellulosic material. A recent review of such rates was provided by Sanchez-Andrea et al.[2] For simple organics such as methanol, formate, and lactate, reported rates span a range of 0.6–151 mol/m^3/d.

Table 4 Sulfate Reduction Rates in Biochemical Reactors (BCRs)

Reference	Substrate[a]	SRR ($mol/m^3/d$)	Influent pH	Flow Condition	Influent (mg/L)	Notes
49	100% Spent mushroom compost from 11-year-old VFP	−0.2 to 0.05	5.3–6.0	Semibatch, water volume and rate not reported	Ca: 181 ± 14 Cd: 0.09 ± 0.03 Fe: 67.8 ± 8.3 Mg: 65.3 ± 4.7 Mn: 7.3 ± 1.0 Na: 44.8 ± 5.5 Pb: 0.26 ± 0.08 Zn: 3.9 ± 0.6 SO_4: 817 ± 80	6-Month semibatch experiment in 1-L cubitainers
	67 vol% horse manure/33% sawdust mix "with high-quality limestone" from 8-year-old VFP	0.2–0.6				
	100% Spent mushroom compost from 3-year-old VFP	−0.6[b] to 0.3				
	45% Spent mushroom compost, 45% hardwood chips, 10% limestone sand from 1-year-old VFP	0.7–3.5[c]				
	Cattail marsh sediment from natural wetland	−0.8[b] to 0.2				
66	71.5% Aged chipped wood, 3.5% manure, 5.5% ag limestone, 19.5% alfalfa hay[d]	0.02–0.14, median ~0.05 $mol/m^3/d$	3.4	Downflow, 0.21 $m^3m^{-2}d^{-1}$ EBCT: 8 d	Fe: 96 Cu: 0.22 Zn: 1.8 As: 0.09 Ni: ND Co: ND Al: 24.7 Mn: 2.8 SO_4: 554	Dual cells in parallel; values during 4 years of operation
41	45% Sand, 5% limestone, 10% fresh dairy manure, 40% alfalfa hay	0.55 ± 0.04 (SD)	6.0	Upflow, 0.015 $m^3m^{-2}d^{-1}$ EBCT: 12 d	SO_4: 1000 Zn: 50 Mn: 50	0.38-L column; values after 100 days operation

	45% Sand, 5% limestone, 10% fresh dairy manure, 40% corncob	0.59 ± 0.05 (SD)				
	45% Sand, 5% limestone, 10% fresh dairy manure, 40% oak	0.23 ± 0.04 (SD)				
	45% sand, 5% limestone, 10% fresh dairy manure, 40% pine	0.15 ± 0.04 (SD)				
50	10% Hay, 50% wood chips, 30% limestone, 10% cow manure	Not reported	3.8–7.1	Downflow 0.8 $m^3m^{-2}d^{-1}$ EBCT: 1.3 d	Fe: 5.2 Cu: 0.26 Zn: 26 Cd: 0.14 Pb: 0.5 Mn: 11 SO_4: 281	No flow ~15 weeks during reported year
51	100% Spent mushroom compost	0.42	4.1	Downflow 0.4 $m^3m^{-2}d^{-1}$ EBCT: 1.25 d (median values; EBCT for compost layer)	Fe: 202 Mn: 11.4 Cd: 0.01 Pb: 0.12 Zn: 0.13 Na: 1378 SO_4: 5688	Full-scale, 1 year of operational data
	100% Spent mushroom compost	0.25	7.2		Fe: 16 Mn: 8.8 Cd: 0.01 Pb: 0.06 Zn: 0.02 Na: 1277 SO_4: 4902	
52	92 wt% walnut shell, 8 wt% corn stover	0.13–0.24	5.2–5.8	Upflow 0.015–0.03 $m^3m^{-2}d^{-1}$ EBCT: 10–20 d	Fe: LT 0.1 Zn: 50 SO_4: 700	0.59-L lab-scale columns
		0.28	6.3–6.5	0.015 $m^3m^{-2}d^{-1}$ EBCT: 10 d	Fe: LT 0.1 Zn: 1.5 SO_4: 300	

Continued

Table 4 Sulfate Reduction Rates in Biochemical Reactors (BCRs)—cont'd

Reference	Substrate[a]	SRR ($mol/m^3/d$)	Influent pH	Flow Condition	Influent (mg/L)	Notes
53	75% Sugar beet waste, 25% +10 mesh taconite ore	0.65 0.93 w/ethanol	7.9–8.6	Downflow 0.2 $m^3m^{-2}d^{-1}$ EBCT: 3.3 d HRT: 0.7–2 d	Iron ore tailings basin water. Fe: GT 100 SO_4: 810–660	Lab-scale columns ~12 L. Columns with ZVI not included here
	75% Peat, 25% +10 mesh taconite ore	0.27 2.2 w/ethanol				
	10% Manure, 40% sawdust, 25% feed corn, 25% +10 mesh taconite ore	2.5–0.6				
	25% Manure, 40% sawdust, 10% hay, 25% +10 mesh taconite ore	1.5–0.6				
	25% Biosolids, 40% sawdust, 10% hay, 25% +10 mesh taconite ore	2.2–3.1				
54	~20% Spent mushroom substrate, 80% fine inert grit	0.09–0.15	7.8	Surface flow 0.07–0.13 $m^3m^{-2}d^{-1}$ EBCT: 30–55 d	Pb: 0.1 Zn: 5.0 SO_4: 900	Pair of 16 m^2, 2 m deep wetlands, substrate in 0.5 and 1 m beds
55	30% Wood chip and sawdust, 30% leaf compost and poultry manure, 40% sand, creek sediment, urea, and calcium carbonate	0.2–1.6 (7.3 d HRT) 0.09–1.3 (10 d HRT)	5.7–2.9	Downflow HRT: 7.3 and 10 d	Ca: 372 ± 75 Cd: 9.8 ± 1.8 Fe: 504 ± 83 Mg: 85.8 ± 10.5 Mn: 10.1 ± 2.6 Na: 625 ± 231 Ni: 13.7 ± 1.0 Zn: 14.5 ± 2.1 SO_4: 4022 ± 583	3.5-L columns

56	Aquifer material: 25% mica sand, 67% gravel	0.05–0.4 (planted) 0.009–0.18 (unplanted)	6.72 ± 0.25	Horizontal flow 0.23 $m^3m^{-2}d^{-1}$ EBCT: 14 d	Ca: 394 ± 10 Fe: 1 ± 0.6 SO_4: 957 ± 85	3 m^3 wetlands, one with reeds, other unplanted, 4–5 years operation
60	Wetland: 20% clay, 70% sand	0.6 (Winter) 0.03 (Spring) 0.07 (Fall)	9–10	Horizontal, free water surface EBCT: 9 d	Fe: 0.8–34 Cr: 0.05–0.4 Ni: 0.01–0.036 Zn: LT 0.01–0.09 BOD: 11–94 SO_4: 1650 (Winter), 270 (spring), 750 (fall)	2000 m^3 sediment in ~3000 m^3 total FWS wetland w/*Typha domingensis* (southern cattail), treating tool facility wastewater
31	100% Fresh mushroom compost	0.2	3.0	Horizontal flow 0.2 $m^3m^{-2}d^{-1}$ EBCT: 15 d	Al: 4 Ca: 375 Cd: 0.03	Four 9–19 m^3 BCRs studied over 11 months
	0.67 m mushroom compost under 0.3 m hay	0.04		Upflow 1.3 $m^3m^{-2}d^{-1}$ EBCT: 11 d	Cu: 0.57 Fe: 43.2 Mn: 31	
		0.01		Upflow 0.01 $m^3m^{-2}d^{-1}$ EBCT: 36 d	Ni: 0.2 Pb: 0.02 Zn: 9.1	
		0.05		Downflow 0.04 $m^3m^{-2}d^{-1}$ EBCT: 11 d	SO_4: 1700	
	33% Peat, 33% aged steer manure, 33% composted wood shavings and sawdust	0.1		Downflow 0.1 $m^3m^{-2}d^{-1}$ EBCT: 10 d		
57		−1.2 to 1.8 Median: 0.16 Average: 0.09	3.0–6.8	Vertical flow 0.08–0.65 $m^3m^{-2}d^{-1}$ EBCT: 0.3–100 d	Al: 0–48 Ca: 86–320 Fe: 0.2–193 Mn: 2.4–57	23 Vertical flow SAPS, field-scale. Values calculated on compost layer volume

Continued

Table 4 Sulfate Reduction Rates in Biochemical Reactors (BCRs)—cont'd

Reference	Substrate[a]	SRR ($mol/m^3/d$)	Influent pH	Flow Condition	Influent (mg/L)	Notes
					SO_4: 455–1372	
58	180-day-old MSW compost, limestone	−0.25 to 0.37 Median: ~0.12	4.5–5.0	Upflow 0.6–2.0 $m^3m^{-2}d^{-1}$ EBCT: 0.9–2.7 d	Co: 3–4 Zn: 3–4 Ni: 80–100	0.78-L columns operated 5 years after 2–4 years use of substrate in barrel scale
	180-day-old MSW compost	−0.27 to 0.3 Median: ~0.1		1.3, then 0.3 $m^3m^{-2}d^{-1}$ EBCT: 1.3, 6.5 d	Cu: 40–50 SO_4: 1500–2000	
	Yard waste compost	−0.2 to 0.25 Median: ~0.04		1.3 $m^3m^{-2}d^{-1}$ EBCT: 1.3 d		
	45-day-old MSW compost	−0.25 to 0.7 Median: ~0.04		0.6–3.1 $m^3m^{-2}d^{-1}$ EBCT: 0.9–6.5 d		
Liquid substrate						
61	None (upflow anaerobic sludge blanket reactor)	20	6–7	Upflow EBCT: 1 d	SO_4: 2000	Lab-scale 3-L UASB fed lactate
59	100% Ceramic rings	5 (Acetate), 7–8 (propionate)	7–8	Upflow with recycle ratio of 4 EBCT: 1.3 d	SO_4: 1200	Lab-scale 1 1/3-L reactor fed propionate or acetate
53	100% +10 Mesh taconite ore	2.4–3.1 (Ethanol) 2.0 (Molasses) 1.3 (Methanol)	7.9–8.6	Downflow 0.2 $m^3m^{-2}d^{-1}$ EBCT: 3.3 d	Iron ore tailings basin	Lab-scale columns ~12 L
	90% Oxidized ore, 10% coarse tailing	2.8 (Ethanol) 3.0 (Molasses) 1.6 (Methanol)		HRT: 0.7–2 d	water. Fe: GT 100 SO_4: 810–660	

SAPS, successive alkalinity producing systems; VFP, vertical flow ponds; SRR, sulfate reduction rate; EBCT, empty bed contact time; HRT, hydraulic residence time; ZVI, zero valent iron; FWS, free water surface; MSW, municipal solid waste.

[a]Percentages are volume unless otherwise noted.

[b]Substrate dried before experiments, may have oxidized sulfide and thus released sulfate.

[c]Steady decline from 3.5 to 0.7 ±0.2 over 6-month batch experiment.

[d]Estimated assuming 130 kg/m^3 hay.

Hydraulics

BCRs are, in terms of flow, essentially porous media. There have been few published reports of values for the hydraulic conductivity, K, in BCR materials (Table 5). Bolis et al.[63] reported that hydraulic conductivity of initially dry substrate fluctuates significantly, and that soaked material may have values lower by an order of magnitude than seen after a few months. A similar increase in K was seen in column studies.[62]

Unfortunately, real materials like soil or BCR substrate are nonhomogenous and have widely differing properties. In the case of hydraulic conductivity, values may be different by more than an order of magnitude depending on the direction of measurement, and compaction (during construction) may decrease hydraulic conductivity by two or more orders of magnitude.

If one portion of the substrate has a higher conductivity than others, more water will flow through this location. The higher conductivity is probably due to larger pores related to local structure and mixing of the original substrate. The flow of a greater volume of water at this location increases the shear and also delivers more mass of chemicals, which in turn increases the observable rates of reaction. In BCR substrate, this means more particle scour, opening pores larger, and more rapid depletion of organic matter, resulting in some further increase in porosity. Thus, the porosity, already larger than average, is increased, and because pores are larger, hydraulic conductivity increases and even more water flows at the location. At an extreme, all substrate would be cleared and the location has become a pipe, transporting water through the substrate without treatment.

Table 5 Hydraulic Conductivity of Biochemical Reactor (BCR) Substrates

Substrate	Range of K (cm/s)	Reference
77% Limestone (1/8–1/2 in), 14% alfalfa, 9% inoculum from another BCR	0.001–0.012	63
75% Cow manure, 25% potting soil	1×10^{-4}–6×10^{-3}	63
75% Cow manure, 25% top soil	$> 5 \times 10^{-5}$–0.005	63
5% Peat moss, 5% alfalfa hay, 20% chert gravel, 3% aged manure, 2% biosolids, 15% sand, 50% chip bark	1×10^{-4}–7×10^{-4}	62
Ibid, limestone instead of chert gravel, compacted to half initial dry volume	3×10^{-6}–4×10^{-4}	62
Mushroom compost	3×10^{-4}–0.014	67, as reported in Ref. 63
100% Organic compost (biosolids and wood chips)	7.8×10^{-4}	32
70% Organic compost, 30% gravel	7.4×10^{-4}	32
40% Organic compost, 60% gravel	0.032	32
100% Gravel	1	32
100% Gravel	>0.6, >1.02	68
20% Organic compost, 80% gravel	0.8	68
50% Organic compost, 50% gravel	$<5 \times 10^{-4}$	68

Short-circuiting in this way is more likely near edges, and has been repeatedly observed in lab-scale BCRs. In one case, 68% of the flow was found to be short-circuited through a lab-scale vertical BCR after slightly more than 1 year of operation.[40] In larger BCRs, such edge effects will be less likely to have a significant impact.

The best method to determine short-circuiting is to perform a tracer test. A conservative tracer such as sodium bromide is mixed into the influent as a pulse or step-change and the effluent concentration is compared to the influent over time. The resulting curve shows the time distribution for the tracer, and thus fluid element carrying the tracer. Assuming some flow model such as a pair of parallel plug-flow systems, this residence time distribution (RTD) curve may be deconvoluted to find the apparent short-circuiting behavior.

Design Process

The current status of the design process was provided in great detail in the ITRC guidance document.[18] Briefly stated, the designer should evaluate the potential substrate at lab scale, using a mixture that gives sulfate reduction in batch operation with real mining-impacted water. The best performing substrate mix is then used in a pilot-scale unit operated on site. The performance of this pilot-scale unit is evaluated to determine the initial sulfate reduction rate.

For a solid substrate, the rate of sulfate reduction will decline over time as that substrate is consumed. The designer therefore should consider sizing the BCR on the basis of a safety factor. Unfortunately, there are no reported measurements of this decline, so the amount of loss of reducing capacity over years is uncertain. Field-scale operating results have not been shared, either for proprietary reasons or because sulfate data were not collected at these generally remote locations. The few multiyear pilot- or lab-scale results indicate that sulfate reduction may continue at the initial rate, with "fresh" substrate operated on as a moving aerobic front sweeps slowly through the bed.

The concept of aerobic metabolism is surprising, but the water carries oxygen, so the influent portion of the media is exposed to oxygen. Aerobic cellulose degradation is much faster than anaerobic, so the organic will be depleted more quickly at this aerobic front. When the organic is sufficiently depleted, there is insufficient BOD to be used to consume oxygen, and the oxygenated water continues further into the bed. If aerobic metabolism dominates as the source of organics for sulfate reduction, then the observed initial rate might be the sustained rate until the bed is almost completely exhausted.

With a sulfate reduction rate determined as a number of moles per volume per day, such as 300 $mmol/m^3$-d, the amount of metal needed to be removed is summed as moles per liter, and translated to load by multiplying with the expected flow rate, giving moles per day. The minimum volume of the BCR is thus the metal load divided by the sulfate reduction rate.

Hydraulics

The distribution of water into a BCR is not trivial. To limit short-circuiting, water should be applied evenly at the influent and collected evenly at the effluent from the bed of media. This is accomplished by (see Figure 1)

- delivering and collecting water using sufficiently large pipes that head loss along the pipe is small,
- spreading the water into a highly porous bed, generally gravel, using a perforated piping manifold to keep the distance of flow from pipe to media surface roughly even,
- placing the media as evenly as possible, to prevent thin spots and areas that are either less or more compacted than the whole; the media is the largest resistance to flow and most likely place for short-circuiting to develop,
- collecting the water in a manner similar to the delivery, from a porous gravel bed into perforated piping manifold that feeds a pipe, and
- setting the outlet level by weir such that water is above the surface of the media. Water in a gravel bed, or above that bed, will be at constant level and thus assure the same head across the bed of media.

A delivery and collection system following the above guidance will have the head loss dominated by the media, and thus water will enter and exit the bed evenly. As noted above, the media cannot be placed completely homogenously, so there will be differences in hydraulic conductivity that will result in nonideal flow. In at least two cases, BCRs have been designed with an intermediate gravel layer in the media to redistribute flow between the media layers, such that a channel developing in one layer of media will not inevitably lead to failure. The benefit of such design is not proven, and complicates construction. On the other hand, short-circuiting and resulting poor performance or even failure of a BCR has been proven.

Cap

Given the consensus that plants are a net negative for a BCR and that many mine sites are in cold climes, BCRs generally are designed with a low permeability cap. Placing a layer of clay, for example, as a cap over the top of the system prevents excess precipitation from impacting the reactor, decreases the impact of plant roots either adding oxygen or creating channels in the media, and insulates the bed from what can be quite cold conditions.

Deep open water would serve some of these purposes, as a layer of ice insulates, and 1 m depth of water will generally prevent the growth of emergent wetland plants. Another alternative cap is the media itself or the woody component of the media, as a dry layer above the BCR surface. The material acts as insulation and may, as the organic degrades and precipitation runs down into the BCR, add simple organics to the BCR.

Settling and Aeration

Effluent from an operating BCR should be low in metal concentration and also have a lowered concentration of sulfate. It will have gained sulfide and potentially settleable solids. Sulfide is toxic[64] and therefore, a BCR should be followed by aeration to remove that sulfide. If sufficient elevation head is available, an aeration cascade or short waterfall will release the sulfide.

A settling pool is advisable after the BCR to let any solids carried out of the BCR to settle. This is not a required element; during operation, the suspended solids observed from BCRs are quite low. A settling pool may, if less than 1 m deep, become a growth site for wetland plants. Plant growth is aesthetically pleasing, indicates lowered toxicity, should contribute oxygen to the water, and will not impact settling in a noticeable fashion.

Start-up

Sulfate reduction requires anaerobic conditions, so the immediate priority of start-up is to achieve sulfate reduction. The freshly laid media is filled with water, displacing much of the air in the pore space, and thus decreasing the mass of oxygen in the bed. Water contains little oxygen when compared to air, and the diffusion of oxygen through water is much lower than through air. Because the influent water may be saturated for oxygen, more water flow is not good after the bed is filled. As the organic in the media is used, oxygen is depleted and the redox potential (ORP) falls. After the BCR becomes anaerobic, further electron acceptance must occur in order to drive that potential low enough for sulfate reduction to be thermodynamically possible. This process takes at least days if not weeks, as aerobes use oxygen, then fermenters begin to use simple organics, and finally the sulfate reducers begin to chew on fermentation products.

From a practical point of view, the BCR is filled, water flow is stopped, and the Eh/ORP is monitored. One may also monitor dissolved oxygen (DO) to show progress, but an absence of oxygen does not assure sulfate is being reduced. When the redox drops to −200 mV, sulfate reduction should have started; additional powerful evidence may be obtained by monitoring the concentration of sulfate, or determining the presence of sulfide.

Once sulfide generation is assured, flow may begin. Because of the possible fragility of maintaining sulfate reduction and also due to concerns about media consolidation or loss, flow is gradually increased over weeks to the full design value. During this period, and periodically during operation at full flow, the generation of sulfide should be checked. A decrease in sulfide generation or sulfate removal is an indicator that flow should be decreased or even stopped to reestablish the low ORP.

As flow starts to pass out of the BCR, the water will carry suspended and dissolved solids. For some time period, often a few weeks, the water will be brown with humic and fulvic acids being carried from the media. As these dissolved organics are swept out, the water becomes clear.

Operation

It is somewhat too simple to state that solid-phase BCRs work until they fail. However, because they are passive and the solid substrate provides electron donor for sulfate reduction for many years, the operation of these systems has generally followed a pattern of benign neglect.

Because BCRs treat water, regular monitoring of flow rate and effluent contaminants of concern (e.g., metal concentration, pH, and/or toxicity) is a regulatory requirement. In addition to monitoring of the effluent, the status of the system should be assessed visually for damage or changes, especially in water levels. Such monitoring is in, some cases, semiannual, so damage to a system might go unobserved for months.

If the design is completely passive, using gravity flow, then likely the BCR will work until it does fail. Water will flow through the media and organisms will consume sulfate until such time as the water flow stops, the organic is exhausted, or a toxic element enters the system. The modes of failure of a BCR are as follows:

- Organic depletion. This is the "best" mode of failure; the food has been used up. If the BCR has piezometers placed in it at varying locations for monitoring, changes in the ORP should show the moving front of depletion of electron donor. This failure requires, simply, replacement of the media.
- Short-circuiting. Short-circuiting is challenging to observe as it develops unless monitoring includes periodic tracer tests or fine observation of head difference across the BCR. This type of failure requires analysis of the cause of the hydraulic failure, such as a failed joint in a pipe causing scour on the bed surface, and correction of that flaw. The flaw is quite likely to simply be evolution of flow within the media, with pores having opened and allowed more flow, causing more scour and enlargement of those pores. Generally, the failure is in the media and replacement of the media is warranted.
- Plugging. Within the BCR, precipitates may form and block pores, causing plugging. This issue has not been reported for sulfide formation, but more often is a result of iron oxide precipitation. If the water to be treated is high in iron, pretreatment for iron is a preventative measure. Evidence of plugging is a decrease in hydraulic conductivity, observable as an increase in the water level above the bed, or in constant-head systems, a decrease in flow. Since finding plugging within the media is unlikely, replacement is the most likely solution.
- Gas lock. It is possible for evolved gases, such as CO_2, to form bubbles in the media. These bubbles plug the media much as would solids. In piping, gas lock is a well-understood phenomenon, preventable by the installation of air relief valves at high spots. Gas lock in BCR piping has been a concern, and gas vents should be considered when designing piping placed within the BCR for distribution and collection. Gas lock within media, rather than within pipes or impermeable components of a bed, has not been reported as a significant problem.

- Pipe clogging or bursting. Pipes exposed to the sun will age rapidly. Changing temperatures cause contraction and expansion at joints. Excessive cold may cause ice blockages. Systems with exposed pipes are vulnerable to such failures. The formation of precipitates in the pipes can result in plugging; iron and aluminum oxides in particular may be present in such amount that pipe fouling is excessive unless iron and aluminum pretreatment (e.g., Figure 2) is adequate. Fortunately, pipe failures are often easy to find and to fix, and partial clogs may be addressed by treatment with chelating agents such as citrate.[65] Prevention of further failure likely involves relaying and burying pipes.
- Acidity. If an upstream component of the system fails, then low pH may impact the BCR. Solid-substrate BCRs rely on cellulose degraders to supply food to sulfate-reducing organisms. Cellulose degraders are generally sensitive to pH below 6, with no apparent activity below pH 5. Thus, if an alkalinity system is exhausted, the influent to the BCR would fall below pH 5 and the system would become inactive. In the case of pH, recovery of activity may be possible through a repeat of the start-up procedure, as the organisms likely are still present, as is the food.
- Toxins. It is conceivable that a change in the source water could release greatly increased metal or metalloid concentrations with resulting toxicity. In such a case, the BCR failure is likely, and a new treatment system must be provided to cope with the high concentrations.
- Erosion. If storm water flow is high and berms, spillways, and cover are not resistant to scour, erosive failure may occur. If erosion is noted when still small in impact, rebuilding with more appropriate materials may be a simple fix. If erosion is due to storm water, berms to direct water around the BCR are recommended.
- Vandalism. Most BCRs are at remote sites. The design should make the system unattractive to vandalism by people and wildlife. Beavers are particularly notable as vandals in pondlike systems, and antiwildlife measures such as screened outfalls and avoiding exposed falls of water (such as over a weir) helps limit the impact of these and other wildlife.

If a BCR is more active, requiring power and pumps, the mode of failure is likely to be loss of power, electronic malfunction, or motor failure. In such cases, more frequent maintenance is recommended, and automated failure reporting may be wise.

An operating BCR should produce sulfide, and at neutral pH a small fraction of that sulfide will form H_2S and evaporate. Thus, BCRs may smell of rotten eggs, and if extensive sulfate reduction is occurring, the odor may be offensive. That odor is a positive indicator of proper operation, but odor control may be required to mitigate complaints.

Conclusions

BCRs fill a niche, the treatment of MIW, which is not a small need in the industrial world. Passive BCRs require little to no operating energy, and thus are a low carbon footprint treatment technology compared to others. The rate of decay of organic materials is key to the rate of treatment, but this rate remains poorly understood. Instead, a range of values from operating BCRs has been reported and applied for design. The lack of knowledge of sulfate reduction rates and the difficulty of assuring consistent and even permeability in the bed are the two major design uncertainties in BCRs.

References

1. ITRC The Interstate Technology & Regulatory Council Wetlands Team. *Mining waste treatment technology selection.* 2010. http://www.itrcweb.org/miningwaste-guidance/ [accessed 30.04.12].
2. Sanchez-Andrea I, Sanz JL, Bijmans MFM, Stams AJM. Sulfate reduction at low pH to remediate acid mine drainage. *J Hazard Mater* 2014;**269**:98–109.
3. Rose AW, Cravotta III CA. Geochemistry of coal mine drainage. In: Brady KBC, Smith MW, Schueck J, editors. *Coal mine drainage prediction and pollution prevention in Pennsylvania.* Harrisburg: The Pennsylvania Department of Environmental Protection; 1998. P. 1-1–1-22. http://www.dep.state.pa.us/dep/deputate/minres/districts/cmdp/main.htm. [accessed 02.04.12].
4. Plumlee GS, Smith KS, Montour MR, Ficklin WH, Mosier EL. Geologic controls on the composition of natural waters and mine waters draining diverse mineral-deposit types. In: Filipek LH, Plumlee GS, editors. *The environmental geochemistry of mineral deposits, part B: case studies and research topics. Reviews in economic geology*, vol. 6B. Society of Economic Geologists, Inc; 1999. p. 373–432.
5. OSM Office of Surface Mining Reclamation and Enforcement. *National mine map repository.* 2012. http://mmr.osmre.gov/ [accessed 16.04.12].
6. GAO United States Government Accountability Office. *Hardrock mining: information on abandoned mines and value and coverage of financial assurances on BLM land.* 2008. GAO-08–574T. Testimony before congress March 12, 2008.
7. USGS United States Geological Survey. *Mine and mineral processing plant locations-supplemental information for USGS Map I-2654.* May 16, 2012. http://minerals.usgs.gov/minerals/pubs/mapdata/.
8. EPA United States Environmental Protection Agency. *Liquid Assets 2000.* 2000. EPA-840-B-00-001, Office of Water (4101), Washington, D.C. 20460.
9. Gusek JJ, Figueroa LA. *Mitigation of metal mining influenced water*, vol. 2. Littleton (CO): Society for Mining and Exploration; 2009. p. 164.
10. PIRAMID Consortium. *Engineering guidelines for the passive remediation of acidic and/or metalliferous mine drainage and similar wastewaters.* European commission 5th framework RTD Project no. EVK1-CT-1999–000021. Newcastle Upon Tyne (UK): "Passive in-situ remediation of acidic mine/industrial drainage" (PIRAMID). University of Newcastle Upon Tyne; 2003. p. 166.
11. Rose AW. Advances in passive treatment of coal mine drainage 1998–2009. In: Barnhisel RI, editor. *Proceedings of the 2010 national meeting of the American Society of Mining and Reclamation, Pittsburgh, PA bridging reclamation, science and the community June 5–11, 2010.* 3134 Montavesta Rd., Lexington, KY 40502: ASMR; 2010.
12. Skousen J, Rose A, Geidel G, Foreman J, Evans R, Hellier W, et al. *Handbook of technologies for avoidance and remediation of acid mine drainage.* Morgantown, West Virginia: The National Mine Land Reclamation Center, West Virginia University; 1998. p. 131.
13. Younger PL, Banwart SA, Hedin RS. *Mine water: hydrology, pollution, remediation.* Boston: Kluwer Academic Publishers; 2002. p. 464.
14. INAP (The International Network for Acid Prevention. *Global acid rock drainage guide (GARD guide).* 2009. http://www.gardguide.com/index.php/Chapter_7. [accessed 16.05.12].
15. EPA United States Environmental Protection Agency Engineering Technical Support Center Website, 2011. http://www.epa.gov/nrmrl/lrpcd/etsc/. [accessed 16.05.12].

16. EPA United States Environmental Protection Agency Acid Mine Drainage Website, 2012. http://water.epa.gov/polwaste/nps/acid_mne.cfm. [accessed 16.05.12].

17. Guesk JJ. Passive treatment 101: an overview of the technologies. In: *Presented at 2008 U.S. EPA/national groundwater association's remediation of abandoned mine lands conference, Denver, CO, October 2–3, 2008*; 2008. http://www.golder.com/modules.php?name=Publication&sp_id=135 [accessed 21.07.14].

18. ITRC (Interstate Technology & Regulatory Council). *Biochemical reactors for mining-influenced waste*. BCR-1. Washington, D.C.: Interstate Technology & Regulatory Council, Biochemical Reactors for Mining-Influenced Waste Team; 2012. www.itrcweb.org., http://itrcweb.org/bcr-1/Default.htm.

19. Kepler DA, McCleary EC. Successive alkalinity-producing systems (SAPS) for the treatment of acidic mine drainage. In: *Proceedings of the international land reclamation and mine drainage conference*, vol. 1. U.S. Bureau of Mines Special Publication SP 06B-94; 1994. p. 195–204.

20. Licht S. Aqueous solubilities, solubility products and standard oxidation-reduction potentials of the metal sulfides. *J Electrochem Soc* 1988;**135**(12):2971–5.

21. Muyzer G, Stams AJM. The ecology and biotechnology of sulphate-reducing bacteria. *Nat Rev Microbiol* 2008;**6**(6):441–54.

22. Seyler J, Figueroa L, Ahmann D, Wildeman TR, Robustelli M. Effect of solid phase organic substrate characteristics on sulfate reducer activity and metal removal in passive mine drainage treatment systems. In: *Proceedings of the 2003 national meeting of the American Society of Mining and Reclamation and the 9th Billings land reclamation symposium, Billings MT, June 3–6, 2003*. 3134 Montavesta Rd., Lexington, KY 40502: ASMR; 2003.

23. Huntsman BE, Solch JG, Porter MD. *Utilization of* Sphagnum *species dominated bog for coal acid mine drainage abatement*. GSA (91st annual meeting) abstracts, Toronto, Ontario. 1978.

24. Wieder RK, Lang GE. Modification of acid mine drainage in a freshwater wetland. In: McDonald BR, editor. *Proceedings of the symposium on wetlands of the unglaciated Appalachian region*. Morgantown, WV: West Virginia University; 1982. p. 43–53.

25. Kleinman RL, Brooks PR, Huntsman B, Pesavento B. Constructing wetlands for the treatment of mine water. mini-course notes, Lexington KY as cited in Stark LR; Stevens Jr SE, Webster HJ, Wenerick WR. (1990) Iron loading, efficiency and sizing in a constructed wetland receiving mine drainage. In: *Proceedings of the 1990 mining and reclamation conference and exhibition, Charleston, WV, April 23–26, 1990*; 1986. p. 393–401.

26. Girts MA, Kleinman RL, Erickson PM. Performance data on *Typha* and *Sphagnum* wetlands constructed to treat coal mine drainage. In: *Proceedings of the 8th annual surface mine drainage task force symposium, Morgantown, WV*; 1987. Accessible at http://wvmdtaskforce.com/proceedings/1987.cfm.

27. Brodie GA, Hammer DA, Tomljanovich DA. Constructed wetlands for acid drainage control in the Tennessee Valley. In: *Mine drainage and surface mine reclamation. Mine water and mine waste*, vol. 1. Pittsburgh (PA): Bureau of Mines Circular No. 9183; 1988.

28. Dietz JM, Watts RG, Stidinger DM. Evaluation of acidic mine drainage treatment in constructed wetland systems. In: *Proceedings of international land reclamation and mine drainage conference and the third international conference on the abatement of acidic drainage, Pittsburgh, PA, April 24–29, 1994*; 1994. p. 70–9.

29. Hedin RS, Nairn RW. Sizing and performance of constructed wetlands: case studies. In: *Proceedings of the 1990 mining and reclamation conference and exhibition, Charleston, WV, April 23–26, 1990*; 1990. p. 385–91.

30. McIntire PE, Edenborn HM. The use of bacterial sulfate reduction in the treatment of drainage from coal mines. In: *Proceedings of the 1990 mining and reclamation conference and exhibition, Charleston, WV, April 23–26, 1990*; 1990. p. 409–15.

31. Machemer SD, Reynolds JS, Laudon LS, Wildeman TR. Balance of S in a constructed wetland built to treat acid-mine drainage, Idaho-Springs, Colorado, USA. *Appl Geochem* 1993;**8**:587–603.

32. Mueller RF, Sinkbeil DE, Pantano J, Drury W, Diebold F, Chatham W, et al. Treatment of metal contaminated groundwater in passive systems: a demonstration study. In: *Proceedings of the 1996 national meeting of the American Society for Surface Mining and Reclamation, Knoxville, TN, May 19–25*; 1996. p. 590–8.

33. Klusman RW, Dvorak DH, Borek SL. Modeling of wetlands and reactor systems used for mine drainage treatment. In: *Proceedings of the 1993 national meeting of the American Society for Surface Mining and Reclamation, Spokane, WA, May 16–19*; 1993. p. 685–704.

34. Dietz JM, Stidinger DM. Acidic mine drainage abatement in an anaerobic sub-surface flow wetland environment – case history of the treatment system at Corsica, PA. In: *Proceedings of the 1996 annual meeting of the American Society for Surface Mining and Reclamation (ASSMR), Knoxville, Tennessee, May 15–20, 1996*; 1996. p. 531–40.

35. Sencindiver JC, Bhumbla DK. Effects of cattails (*Typha*) on metal removal from mine drainage. Mine drainage and surface reclamation. In: *Proceedings of the 1988 mine drainage and surface mine reclamation conference, Pittsburgh, PA. Mine water and mine waste*, vol. I. U.S. Bureau of Mines Information Circular 9183; 1988. p. 359–69.

36. Beining BA, Otte ML. Retention of metals and longevity of a wetland receiving mine leachate. In: *Proceedings of the 1997 national meeting of the American Society for Surface Mining and Reclamation, Austin, Texas, May 10–16*; 1997.

37. Venot C, Figueroa L, Brennan RA, Wildeman TR, Reisman D, Sieczkowski M. Comparing chitin and organic substrates on the national tunnel waters in Blackhawk, Colorado for manganese removal. In: Barnhisel RI, editor. *Proceedings of the 2008 national meeting of the American Society of Mining and Reclamation, Richmond, VA, new opportunities to apply our science, June 14–19, 2008.* 3134 Montavesta Rd., Lexington, KY 40502: ASMR; 2008.

38. Zobel BJ, van Buijtenen JP. *Wood variation. Its causes and control.* Heidelberg (Germany): Springer-Verlag; 1989.

39. Desch HE, Dinwoodie JM. *Timber: structure, properties, conversion and use.* 7th ed. London: Macmillan Press Ltd; 1996. p. 306.

40. Fitch M, Burken J, Ye C. Measured and modeled removal mechanisms for bench-scale constructed wetlands receiving lead mine water. In: Barnhisel RI, editor. *2008 national meeting of the American Society of Mining and Reclamation, Richmond VA, new opportunities to apply our science, June 14–19, 2008.* 3134 Montavesta Rd., Lexington, KY 40502: ASMR; 2008.

41. Hemsi PS, Shackelford CD, Figueroa LA. Modeling the influence of decomposing organic solids on sulfate reduction rates for iron precipitation. *Environ Sci Technol* 2005;**39**:3215–25.

42. Qu X, Vavilin VA, Mazeas L, Lemunier M, Duquennoi C, He P-J, et al. Anaerobic biodegradation of cellulosic material: batch experiments and inoculum based on isotopic data and focusing on aceticlastic and non-aceticlastic methanogenesis. *Waste Manag* 2009;**29**(6):1828–37.

43. Wang X, Padgett JM, De la Cruz FB, Barlaz MA. Wood biodegradation in laboratory-scale landfills. *Environ Sci Technol* 2011;**45**:6864–71.

44. Rutkowski T, Waples J, Gusek J, Shipley B. Biochemical reactor media characterization study. In: Barnhisel RI, editor. *Proceedings of the 2010 national meeting of the American Society of Mining and Reclamation, Pittsburgh, PA bridging reclamation, science and the community, June 5–11, 2010.* 3134 Montavesta Rd., Lexington, KY 40502: ASMR; 2010.

45. Ruiz A, Figueroa L, Zaluski M, Bless D. Substrate degradation and metal removal performance of a 1,500-gallon sulfate-reducing bioreactor for mining-influenced water treatment. In: Barnhisel RI, editor. *Proceedings of the 2008 national meeting of the American Society of Mining and Reclamation,*

Richmond, VA, new opportunities to apply our science, June 14–19, 2008. 3134 Montavesta Rd., Lexington, KY 40502: ASMR; 2008.

46. Rabenhorst MC, James BR, Shaw JN. Evaluation of potential wetland substrates for optimizing sulfate reduction. In: *Proceedings of the 1992 national meeting of the American Society for Surface Mining and Reclamations, Duluth, Minnesota, June 14–18, 1992*; 1992.
47. Zhao L, Gu W-M, He P-J, Shao L-M. Biodegradation potential of bulking agents used in sludge bio-drying and their contribution to bio-generated heat. *Water Res* 2011;**45**(6):2322–30.
48. Canfield DE, Thamdrup B, Hansen J. The anaerobic degradation of organic matter in Danish coastal sediments: iron reduction, manganese reduction, and sulfate reduction. *Geochim Cosmochim Acta* 1993;**57**:3867–83.
49. Santamaria BE, Nairn RW. Evaluating longevity of vertical flow bioreactors through water quality analysis. In: Barnhisel RI, editor. *Proceedings of the 2010 national meeting of the American Society of Mining and Reclamation, Pittsburgh, PA, bridging reclamation, science and the community, June 5–11, 2010*. 3134 Montavesta Rd., Lexington, KY 40502: ASMR; 2010.
50. Reisman D, Rutkowski T, Smart P, Gusek J, Sieczkowski M. Passive treatment and monitoring at the standard mine superfund site, Crested Butte, CO. In: Barnhisel RI, editor. *Proceedings of the 2009 national meeting of the American Society of Mining and Reclamation, Billings, MT, revitalizing the environment: proven solutions and innovative approaches May 30–June 5, 2009*. 3134 Montavesta Rd., Lexington, KY 40502: ASMR; 2009.
51. LaBar JA, Nairn RW. Evaluation of first 1.5 years of operation of a passive treatment system in SE Oklahoma. In: Barnhisel RI, editor. *Proceedings of the 2009 national meeting of the American Society of Mining and Reclamation, Billings, MT, revitalizing the environment: proven solutions and innovative approaches May 30–June 5, 2009*. 3134 Montavesta Rd., Lexington, KY 40502: ASMR; 2009.
52. Figueroa L, Miller A, Zaluski M, Bless D. Evaluation of a two-stage passive treatment approach for mining influenced waters. In: Barnhisel RI, editor. *Proceedings of the 2007 national meeting of the American Society of Mining and Reclamation, Gillette, WY, 30 Years of SMCRA and beyond, June 2–7, 2007*. 3134 Montavesta Rd., Lexington, KY 40502: ASMR; 2007.
53. Eger P, Moe T, Engesser J. The use of sulfate reducing bacteria to remove sulfate from iron mining tailings water. In: *Proceedings of the 2004 national meeting of the American Society of Mining and Reclamation and the 25th west Virginia surface mine drainage task force, April 18–24, 2004*. 3134 Montavesta Rd., Lexington, KY 40502: ASMR; 2004.
54. O'Sullivan AD, Murray DA, Otte ML. Ecological development of constructed wetlands built for treating mine water at Tara mines, Ireland. In: *Proceedings of the 2003 national meeting of the American Society of Mining and Reclamation and the 9th Billings land reclamation symposium, Billings MT, June 3–6, 2003*. 3134 Montavesta Rd., Lexington, KY 40502: ASMR; 2003.
55. Neculita C-M, Zagury GJ, Bruno B. Effectiveness of sulfate-reducing passive bioreactors for treating highly contaminated acid mine drainage: I. effect of hydraulic retention time. *Appl Geochem* 2008;**23**: 3442–51.
56. Wu S, Chen Z, Braeckevelt M, Seeger EM, Dong R, Kästner M, et al. Dynamics of Fe(II), sulphur and phosphate in pilot-scale constructed wetlands treating a sulphate-rich chlorinated hydrocarbon contaminated groundwater. *Water Res* 2012;**46**:1923–32.
57. Rose AW, Dietz JM. Case studies of passive treatment systems: vertical flow systems. In: *Proceedings of the 2002 national meeting of the American Society of Mining and Reclamation, Lexington, KY, June 9–13, 2002*. 3134 Montavesta Rd., Lexington, KY 40502: ASMR; 2002.
58. Eger P, Wanger J. Sulfate reduction – decreases in substrate reactivity and the implication for long-term treatment. In: *Proceedings of the 2001 national meeting of the American Society of Mining and Reclamation, Albuquerque, NM, June 3–7, 2001*. 3134 Montavesta Rd., Lexington, KY 40502: ASMR; 2001.

59. Greben HA, Tjatji M, Maree JP. COD/SO_4 ratios using propionate and acetate as the energy source for the biological sulphate removal in acid mine drainage. In: Jarvis AP, Dudgeon BA, Younger PL, editors. *Mine water 2004 – proceedings international mine water association symposium 2*. Newcastle upon Tyne: University of Newcastle; 2004. p. 93–9.

60. Di Luca GA, Maine MA, Mufarrege MM, Hadad HR, Sanchez GC, Bonetto CA. Metal retention and distribution in the sediment of a constructed wetland for industrial wastewater treatment. *Ecol Eng* 2011;**37**:1267–75.

61. Bertolino SM, Veloso TC, Leão VA. Performance of a lactate-fed UASB reactor treating sulfate containing waters. In: Rüde TR, Freund A, Wolkersdorfer C, editors. *Mine water – managing the challenges, Proceedings of 11th International Mine Water Association Congress, Aachen, Germany 2011*, ISBN 978-1-897009-47-5. p. 277–80.

62. Ross C. *Hydraulic conductivity for constructed wetlands substrates*. Undergraduate research report. Rolla (Missouri): University of Missouri-Rolla; 1997. p. 10.

63. Bolis JL, Wildeman TR, Dawson HE. Hydraulic conductivity of substrates used for passive acid mine drainage treatment. In: *Proceedings of the 1992 national meeting of the American Society for Surface Mining and Reclamations, Duluth, Minnesota, June 14–18, 1992*; 1992.

64. Butler BA, Smith ME, Reisman DJ, Lazorchak JM. Metal removal efficiency and ecotoxicological assessment of field-scale passive treatment biochemical reactors. *Environ Toxicol Chem* 2011;**30**: 385–92.

65. Spangler K, Figueroa L, Honeyman B. Ironing out your system: removal of ferrihydrite precipitates from mining influenced water pipelines. In: Barnhisel RI, editor. *Proceedings of the 2008 national meeting of the American Society of Mining and Reclamation, Richmond, VA, new opportunities to apply our science, June 14–19, 2008*. 3134 Montavesta Rd., Lexington, KY 40502: ASMR; 2008.

66. Gusek JJ, Schneider R. Passive Management of Mining Influenced Water at the Haile Gold Mine. In: R.I. Barnhisel, ediror. *Proc. 2010 National Meeting of the American Society of Mining and Reclamation, Pittsburgh, PA Bridging Reclamation, Science and the Community June 5–11, 2010*. Published by ASMR: 3134 Montavesta Rd., Lexington, KY 40502; 2010.

67 Lemke PR. *Analyses and optimization of physical and hydraulic properties od constructed wetlands substrates for passive treatment of acid mine drainage. Master's Thesis No. 3823*. Colorado School of Mines, golden; 1989. 78 pp.

68. Genty Thomas, Bussière Bruno, Zagury Gérald J, Benzaazoua, Mostafa. Passive treatment of high-iron acid mine drainage using sulphate reducing bacteria: comparison between eight biofilter mixtures. In: Wolkersdorfer Ch, Freund A, editors. *Mine Water & Innovative Thinking. Proc. 2010 International Mine Water Association Symposium*; 2010. p. 229–233; Sydney, Nova Scotia (CBU Press).

6

Water—Energy—Food: Our Existence Will Require Natural Gas

David Bromley

PRESIDENT DBE ENGINEERING LTD., VANCOUVER, BRITISH COLUMBIA, SHORT COURSE INSTRUCTOR – TEXAS A AND M UNIVERSITY

CHAPTER OUTLINE

The Problem and the Opportunity

Water and energy have become two limited resources that are affecting both our quality of life and the sustainability of our communities. The impact of these limited resources is affecting the ability of our food supply to grow in sync with the needs of the global population. Today's energy generation infrastructure has caused severe water limitations. Simultaneously, the water used for energy generation has caused severe energy consumption. Today's energy and water infrastructure is a dichotomy of counterinterdependence that is doomed for failure. The present day strategy on energy conservation has been misaligned. However, the good news is that there is a readily available solution.

The use of community-based water and energy generation facilities will have significant impact on the sustainability of communities. The use of readily available

Food, Energy, and Water. http://dx.doi.org/10.1016/B978-0-12-800211-7.00006-5

natural gas will be a key to the development of the community-based energy generation systems. In fact, the development of such an infrastructure to replace traditional centralized facilities is essential. Micro-community electrical and heat generation facilities fueled by natural gas will reduce energy consumption by at least 50% and possibly as high as 80%. Correspondingly CO_2 emissions will decrease even further, freshwater inventories will increase significantly and community food production will become the dominant food supply.

Success for community-based generation facilities will be dependent on the generation, management, and analysis of reliable data. The technology revolution in smart data systems is an important success story. The ability to manage micro-community systems is now well underway with the continuous improvement of identifying inefficiencies and reducing operational risk. There are numerous opportunities to use water, energy, and food resources more efficiently through the use of new design methods, materials, monitoring, and control systems.

The good news is that the technology for the community generation infrastructure is readily available. The barrier is not the technology. The barrier is the existing siloed business approach from utility operations, which provide one service and do not use the benefit of water, energy, and food collaboration. The capital entrenchment in existing assets by the present-day private or public utility entity continues to prevent the evolution of a far more efficient community-based water, energy, and food infrastructure.

The Existing and Forecast Continuance of Energy and Water Interdependence

The typical percentages of water use in North America is shown in Figure 1. As indicated, almost 50% of the water is consumed by thermal electric plants.

The US Energy Information Administration has provided a forecast of energy source and consumption up to the year 2040. These forecasts have been based on the use of existing and new centralized facilities to produce electricity, water and waste water treatment. Table 1 provides an overview of the amount of energy that will be consumed until the year 2040. In this forecast, coal remains a key source of energy up to the end of the forecast period in 2040 where it is 26% of the overall energy source for electricity generation.

Table 2 shows the water being utilized primarily for cooling and steam generation purposes in electrical generation facilities using the fuels specified in Table 2. Both coal and nuclear energy sources are high users of water. To put this in perspective, this quantity of water would be sufficient to supply 45 million people in Western-based countries and a billion people in the rest of the world.

Table 3 provides the quantity of water being controlled by electrical generation facilities based on the different fuel sources. These volumes essentially have been removed from the freshwater inventories available to the world's population. There is

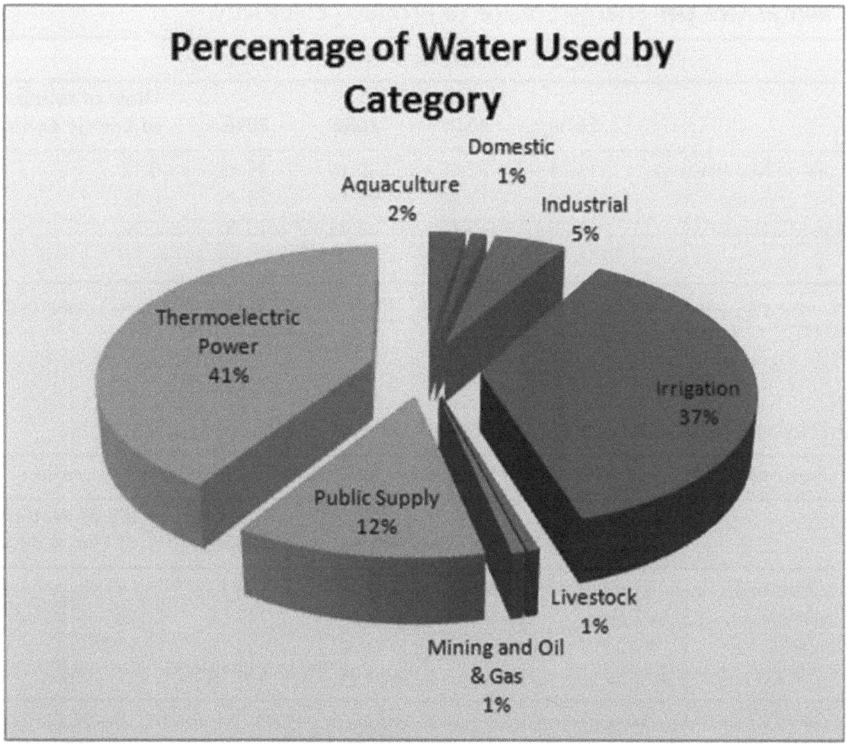

FIGURE 1 Water use in North America.

enough volume in this inventory to supply the entire world 20 times its daily demand for freshwater including all water for agricultural purposes (Bromley, 2014).

Table 4 provides the metric tonne of CO_2 being produced from the fossil fuel energized electrical generation facilities. This amount represents approximately 50% of the overall CO_2 emissions in the world. According to Trends in global CO_2 emissions: 2013 Report,[2] the annual amount of CO_2 emissions in 2012 was 34.5 billion tonnes. The major contributors are China (29%), the United States (16%), and the European Union (11%).

Table 1 Energy Source versus Consumption 2010–2040

Main Sources of Energy Consumed to Create Electrical Power (World Market—Trillion Megajoules Per Year)				
Energy Source/Year	**2010**	**2014**	**2030**	**2040**
Natural gas (Trillion megajoules/year)	123.6	129.2	172.0	202.4
Thermal coal (Trillion megajoules/year)	93.6	101.8	132.0	139.3
Nuclear (Trillion megajoules/year)	28.9	28.4	52.4	60.5
Renewables (Trillion megajoules/year)	59.5	66.0	104.0	126.0
Total	305.5	325.3	460.4	528.3

Reference: http://www.eia.gov/oiaf/aeo/tablebrowser/#release=IEO2013&subject=0-IEO2013&table=2-IEO2013&region=0-0&cases=Reference-d041117

Table 2 Water Use per Energy Source to Produce Electricity

Freshwater Utilized					
Year	**2010**	**2014**	**2030**	**2040**	**Liters of Water/Megajoule of Energy Consumed**
Natural Gas (Billion M3 annually)	15.11	22.65	30.16	35.48	0.12
Coal (Billion M3 annually)	16.40	17.84	23.14	24.43	0.18
Nuclear (Billion M3 annually)	5.20	5.10	9.43	10.90	0.18
Total	36.71	45.59	62.72	70.80	

Reference:http://www.eia.gov/oiaf/aeo/tablebrowser/#release=IEO2013&subject=0-IEO2013&table=2-IEO2013&region=0-0&cases=Reference-d041117

Table 3 Freshwater Withheld from Overall Global Freshwater Inventory

Freshwater Removed from Global Freshwater Inventories for Electricity Generation					
Year	**2010**	**2014**	**2030**	**2040**	**M3 of Water/Megawatt of Electricity Capacity**
Natural Gas (Billion M3)	81.92	85.63	114.04	134.17	51.5
Coal (Billion M3)	147.63	160.55	208.23	219.85	180
Nuclear (Billion M3)	43.04	42.25	78.04	90.18	170
Total	272.59	288.44	400.31	444.19	

Reference:http://www.eia.gov/oiaf/aeo/tablebrowser/#release=IEO2013&subject=0-IEO2013&table=2-IEO2013&region=0-0&cases=Reference-d041117

Table 4 CO_2 Production from Electrical Generation Facilities

CO_2 Generation (Billion Metric Tonnes Per Year)				
Year	**2010**	**2014**	**2030**	**2040**
Natural gas	8.75	9.15	12.18	14.33
Coal	8.53	9.28	12.03	12.70
Nuclear	0.00	0.00	0.00	0.00
Total	17.28	18.42	24.21	27.03

Reference:http://www.eia.gov/oiaf/aeo/tablebrowser/#release=IEO2013&subject=0-IEO2013&table=2-IEO2013&region=0-0&cases=Reference-d041117

Natural Gas Reserves and Water Dependence

Existing Reserves

The amount of natural gas reserves in the world as of 2013 and as per the data provided by US Energy Information Administration is 1.03 Quadrillion m^3. That amount inventory has been increasing annually. Table 5 provides a continental overview as to where the

Table 5 Shale Gas Reserves by Continent

Natural Gas Shale Formation Reserves (Trillion M3)	
North America	263.29
Australia	57.98
South America	181.08
Europe	138.72
Africa	188.85
Asia	184.06
Total	1013.98

Reference:http://www.eia.gov/analysis/studies/worldshalegas/pdf/overview.pdf

reserves are located. The reserves are well distributed throughout the world and are located in areas, which traditionally have not had fossil fuel resources. This even continental distribution will reduce geopolitical tensions significantly.

One cubic meter of natural gas equals 38 MJ of energy. The existing reserves of natural gas equals almost 40 quadrillion megajoules of energy, which is enough energy to replace coal and nuclear power sources for over 200 years based on 2040 forecasted usage as shown in Table 1. This projection is based on the inefficient use of power through centralized mega electrical generation facilities. Needless to say, with the application of energy and water efficient generation facilities, the available Natural Gas in the world would be sufficient for well over 200 years. Key to the development of a highly efficient energy and water infrastructure will be the use of community based micro generation facilities which is counterintuitive to the concept of "economy of scale".

Natural Gas and Impact on Freshwater

There are two main concerns with the extraction of natural gas from shale deposits as a result of the process. To extract gas from the shale formation water is injected into the formation at high pressures to fracture the shale and allow the gas in micropockets or pores to be released. This is called the "fracking" process. Two concerns have evolved from the public's perception of this procedure.

1. The fracking will contaminate groundwater (freshwater) sources.
2. The fracking process uses excessive amounts of freshwater.

With regards to the first concern, there have been numerous studies undertaken to identify locations where groundwater contamination has been caused by the fracturing procedure. The studies have been undertaken by the United States Environmental Protection Agency and individual States in the US including Alabama, Alaska, Colorado, Indiana Kentucky, Louisiana, Michigan, Oklahoma, Tennessee, South Dakota and Wyoming. Some of these States have been using the fracturing processes for 60 years with no incidences of ground water contamination. At the time of publication of this book, there has not been one documented case where the fracturing procedure has caused a contamination impact on freshwater supplies in groundwater aquifers. There have, however, been surface spills, which required clean up. On this basis, this concern is not considered relevant, and no further discussion is warranted.

However, with regards to the second concern, further examination is warranted. Although the oil and gas industry has claimed that the amount of freshwater used in fracking is small compared to other uses of freshwater, there is a severe shortage of freshwater in many of the areas where shale formations are located. Water reuse is essential in these areas, if natural gas extraction from shale formations is to be sustained.

Hydraulic fracturing is performed at depths between 1500 and 3000 m and requires 9500–19,000 m^3 of water per well over a 2–5 day time frame. The water is pressurized down a well into the shale formation to fracture the shale and release the gas and oil. Gas and oil and water return to the surface. Initially much of the water that returns to the surface is the water that was used for the fracturing ("flow-back" water). After 30 days much of the water returning to the surface is "produced" water, which is the naturally high-salt content water in the geological formation. As more and more wells are drilled, the volume of produced water will increase. Eventually, the amount of produced water in a region of shale gas extraction will be higher than the flow-back water volume. For new wells to be drilled and fractured, more and more water will be necessary. The source of this water cannot rely on freshwater supplies. There will be a need to reuse the combination of "flow-back" and "produced" waters. The obvious benefit of this approach is that it will also reduce the disposal of the waters coming from these wells. In North America, the cost of water for fracturing a well and subsequent disposal is in the range US$ 25–US$ 130 per cubic meter of water used for fracturing.[3] This provides a bench mark for treatment of produced and recovered flow-back waters from fracturing, if water is to be reused.

Typical flow-back rates are provided in Table 6.

Water usage for shale fracturing relates to two needs,[3] which are as follows:

1. **Water for the drilling process**: A mixture of clay and water is used to carry rock cuttings to the surface and cool and lubricate the drill bit.
2. **Water for fracturing.** The water is mixed with 0.5% (by volume) of various chemical additives, including cross-linked gelling agents and friction reducers to increase the water's flow and deposition efficiency, a function of rheology. Fracturing jobs utilize highly pressurized fluids between 14 and 56 mpa at an average flow rate of

Table 6 Recovery of Waters Used to Fracture Shale Formations

Flow-Back Waters from Shale Formation Fracturing		
Time	**Flow-Back Rate**	**Flow-Back Recovery %**
1–5 days	16–24 m^3/h	10–25%
5–15 days	3–10 m^3/h	8–12%
15–30 days	1–2 m^3/h	1–5%
30–90 days	0.1 m^3/h	1–2%

Note: After 90 days 15–40% of frac water is recovered.
Ref. [4].

450 m^3/h. These fluids must be slick enough to carry proppants (sand grains and/or ceramic beads) into the microfissures in the rock formation to keep them open and enable the release of the formation resource, which is typically gas and/or oil.[5]

The compositions of fracturing fluids vary from one product to another, and the design of the fluid varies, depending on the characteristics of the target shale gas formation and operational objectives. Figure 2 provides an overview of the additives used in the waters for shale fracturing. The predominant fluids currently being used for fracture treatments in the shale gas formations are as follows:

1. **Water-based fracturing fluids mixed with friction reducing additives (slickwater):**

The fracturing fluid used in slickwater operations typically comprises about 98% water and sand (as a proppant), with additives (e.g., friction reducer, biocide, surfactant, scale inhibitor, breaker, and clay stabilizer) comprising 0.5–2% by volume of the fluid. In addition, polyacrylamide polymers (friction reducers) are used to reduce the tubular friction pressure. Because of its low viscosity (from 2 to 3 cP), a high pump rate is required to transport proppant in the formation.

2. **Linear gel:**

Linear gel is water containing a gelling agent, like guar or xanthan. Other possible additives are biocides, buffers, surfactant, breaker, and clay control. Its viscosity, ranging from 10 to 30 cP, results in improved proppant transport compared to slickwater.

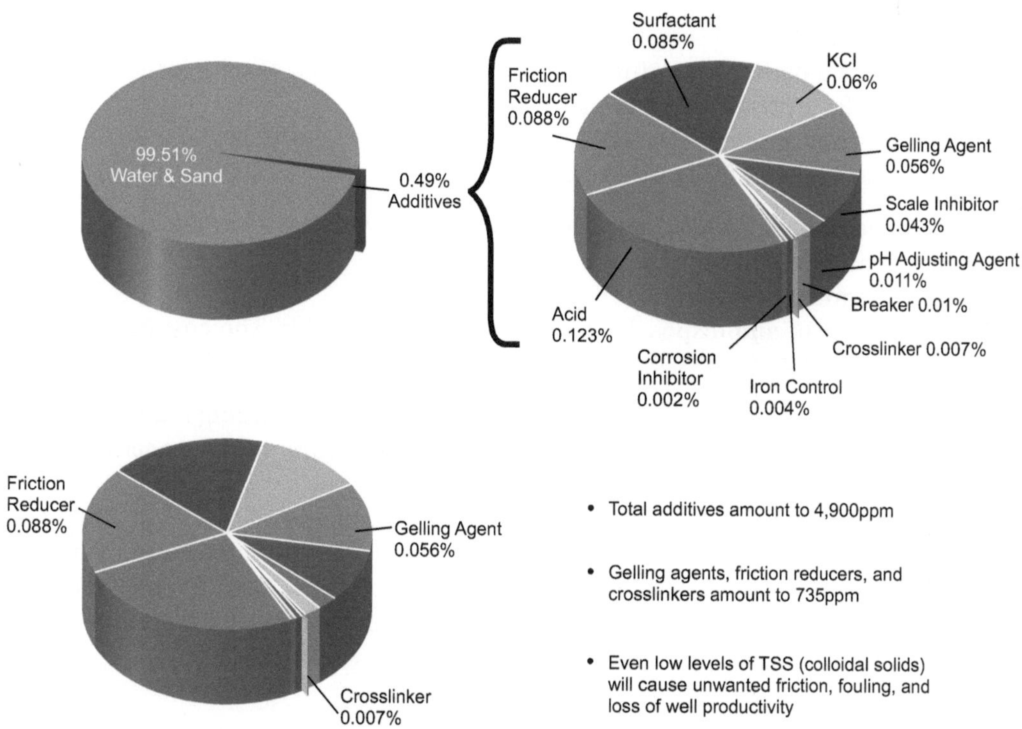

FIGURE 2 Typical frac fluid composition.

3. Crosslinked gel:

Crosslinked gel is water containing any of the gelling agents used in linear gel and a crosslinker like boron, zirconium, titanium, or aluminum. Other possible additives are buffers, biocide, surfactant, breaker, and clay control. This fluid has a high viscosity, ranging from 100 to 1000 cP, which generally results in better proppant transport compared to slickwater and linear gel fracture fluids.

In North America, a shale gas well produces 56.7 million m^3 to 185 million m^3 of natural gas or 2.15 billion megajoules to 7 billion megajoules of energy and uses an average of 12,700 m^3 of water. Flow-back water in shale gas wells should not be confused with produced water in conventional wells. Conventional oil wells produce 7 m^3/m^3 of oil produced. In fractured wells, it is typical to only retrieve back 40% or less of the fractured water used and 15–35% in the first 30 days. The flow rates of fractured water are 10–100 times as great as produced water flow rates. From a chemical perspective, produced water has more minerals than fractured water. The water component that originates from the drilling process for a fractured well will likely have diesels, olefins, and paraffin as well as suspended solids. The flow-back water from the fracturing water will have more suspended solids, oils, and chemicals from the additives. However, a recent research study undertaken by GE[2] indicated that testing of the fractured waters showed no chemical additives. In addition, the quantity of additives is typically less than 1% of the volume. Based on the recovery of the flow-back waters, which can be as low as 10% and never greater than 40%, the impact of the additives on treatment will be small.

In China, the shale gas reserves are in the same zones as conventional oil and gas and as a result this is a perfect opportunity to use produced waters as a major source of the water required for hydraulic fracturing.

In a recent study entitled "Produced and Flowback Water Recycling and Reuse,"[3] four options were considered for North America.

1. Direct reuse of the Produced and Flow-back waters without treatment.
2. On-site treatment and reuse.
3. Off-site treatment and reuse.
4. Off-site treatment and disposal, which was discussed earlier. The cost of this option is from US$ 25 to US$ 130 per cubic meter.

Option 1 was discarded because of the significant concern regarding fouling/plugging the geological formation. Option 3 was considered but the continuing concern with this option is the transportation cost.

A recent publication[6] from the Society of Petroleum Engineers details a study conducted by Halliburton and XTO Energy, which concludes that produced water with total dissolved solids (TDS) levels as high as 285,000 mg/L (28.5% salinity) was shown to generate proper crosslinked rheology for hydraulic fracturing in line with wells that were fracturing with just 20,000 ppm (2% salinity).

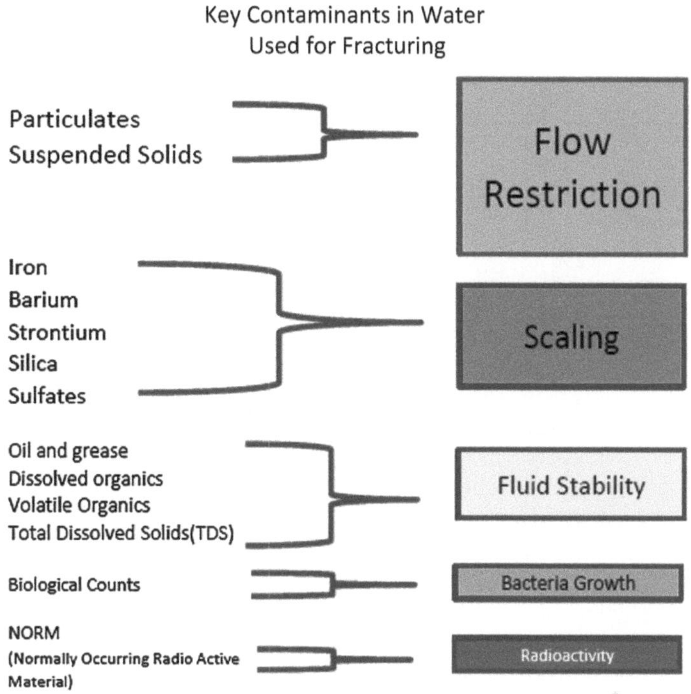

FIGURE 3 Key treatment parameters in order of priority.

In another study by Trican Well Service,[5] which assessed the use of high-TDS waters for fracturing purposes, the study confirmed that with the correct additives, high-salt content (high-TDS) waters could be used without any reduced performance to fracture shale formations in western Canada. In addition, the study confirmed that all additives were commercially available.

The same study assessed the use of freshwater only resulting in disposal of the recovered flow-back waters versus an option where the flow-back water was treated onsite and blended in equal quantities of the treated flow-back water and freshwater. The treatment system removed suspended solids only and the fee charged for the mobile treatment system was US$ 17.50/m^3. Based on this concept it was determined, the on-site treatment of the flow-back water and blending with freshwater provided a cost saving of over US$ 1 million per well.

With the ability to use high-TDS waters (250,000 mg/l), treatment consultants such as CH2M Hill Consultants have confirmed that the focus of treatment for water reuse in fracturing operations is the removal of colloidal solids and scaling parameters. Figure 3 shows the priorities of treatment. Table 7 provides some typical water quality parameters for flow-back waters in fracturing operations.

The recent studies have shown that the key specifications for a water treatment capability to provide water reuse for fracturing operations has to be able to

Table 7 Typical Water Quality

	Typical Flow-Back Water Quality from Fracturing Operations			
(mg/l)	**Sample 1**	**Sample 2**	**Sample 3**	**Sample 4**
Calcium	20,463			29,222
Magnesium	1987			4347
Sodium	37,063	14,100	19,100	70,342
Iron	143	24	73	34.6
Potassium	296	301	340	1660
Barium	203	3	33	5.7
Strontium	2243	70	174	2204
Manganese	43	0.7	1.3	
Bicarbonate	100			
Sulfate	8	134	35	40
Chloride	97,084	23,900	21,000	163,637
Silica reactive	63	67	59	
Carbon dioxide	481			
Hydrogen sulfide	2			
TDS	159,641	39,100	33,800	267,588
TSS	825	841		
Turbidity (NTU)		367	297	182
TOC	150	187	17	
Total hardness	10,000	5170	2310	
Alkalinity		641	495	
Temperature(C)	65			
pH				4.83

TSS, Total suspended solids; TOC, Total organic carbon; TDS, Total dissolved solids.

FIGURE 4 Typical on-site water storage for fracturing water.

1. be mobile and able to move from site to site with ease (see Figure 4),
2. handle a wide variation in flows and quality of water,
3. treat waters at high temperatures (above 40° C),
4. provide a very high removal of colloidal solids to ensure the friction reducer additives are effective in the fractured water and to ensure the formation does not become plugged,
5. reduce scaling parameters, and
6. reduce oil and grease levels.

In summary, the impact of these latest technology developments validates that there is no reason for the oil and gas industry to utilize the volumes of freshwater that they have been using. In fact, economically using freshwater is not viable. Water treatment and reuse of the fractured water flow-back volume and the naturally generated, high-salt (TDS) content water are the logical choices as the source of water for fracturing purposes.

The Future of Energy, Water and Food—A Positive Outlook with Natural Gas

The premise behind the water, energy, and food nexus is the concern that far too much of each resource is being utilized, which will lead to an unsustainable world for humanity. The world's growth has been based on using resources more efficiently and increasing productivity to satisfy demand. The world is entering a new phase where efficiencies are more accessible than ever as a result of data and better understanding of the interrelationships of actions and reactions. Data has evolved from a position where humans obtained data and used the data in a very narrow and siloed analysis to predict outcomes, to the frontier of data analytics where data are machined analyzed at speeds that are unmetaphorical.

Dealing with energy, water, and food for our future will be very dependent on the efficiencies that can be achieved through the management of these resources in microenvironments. Microenvironments depend on data analytics to survive in an overall community of numerous microenvironments. The concept of microenvironments is fundamentally counterintuitive when our economic base has been built on the concepts of economies/efficiencies in scale. The larger the project, the more efficient it will be.

An excellent example of the benefits of efficiency in a microenvironment is the generation of electricity. Most of North America's power generation has been developed on the concept of building a large electrical generation facility in a location where there is a body of water. The body of water is required because these facilities need water for steam generation and cooling waste heat as discussed earlier.

In the development of these facilities, energy is consumed and a typical conversion of energy to electricity is 35%. The remaining 65% is waste heat. From the 35%, a significant amount of this electricity is used to pump and treat water and produce steam or heat water for the purposes of space heating. The electricity also has to be distributed to the end user via transmission systems etc. The end result is that the amount of electrical energy available to the end consumer to satisfy regular commercial, industrial, and residential electricity demands is 11% of the original megajoule of fuel (coal, natural gas).

Tables 8 and 9 show the cost of construction and operation of mega centralized type facilities using various technologies/approaches to generating electricity. As indicated in Tables 8 and 9, the lowest capital cost facility is a natural gas electrical generation facility. Table 9 provides us with an insight into the changes in capital cost from 2010 to 2011. This time period is useful for comparison in that it was immediately following the economic meltdown of 2008. Costs for renewables in 2010 vs 2011 should show the impact of higher market demand in 2010 versus a sluggish demand in 2009. Intuitively one would expect costs to decrease however costs increases in many of the renewable

Table 8 2010 Estimates of Power Plant Capital and Operation Costs (EIA 2010)

	Plant Characteristics		Plant Costs		
	Nominal Capacity (kilowatts)	Heat Rate (Btu/KWH)	Overnight Capital Cost (2010 $/kW)	Fixed O&M Cost (2010 $/kW)	Variable O&M Cost (2010 $/MWh)
Coal					
Single-unit advanced PC	650,000	8800	$3167	$35.97	$4.25
Dual-unit advanced PC	1,300,000	8800	$2844	$29.67	$4.25
Single–unit advanced PC with CCS	650,000	12,000	$5099	$76.62	$9.05
Dual-unit advanced PC with CCS	1,300,000	12,000	$4579	$63.21	$9.05
Single unit IGCC	600,000	8700	$3565	$59.23	$6.87
Dual unit IGCC	1,200,000	8700	$3221	$48.90	$6.87
Single-unit IGCC with CCS	520,000	10,700	$5348	$69.30	$8.04
Natural gas					
Conventional NGCC	540,000	7050	$978	$14.39	$3.43
Advanced NGCC	400,000	6430	$1003	$14.62	$3.11
Advanced NGCC with CCS	340,000	7525	$2060	$30.25	$6.45
Conventional CT	85,000	10,850	$974	$6.98	$14.70
Advanced CT	210,000	9750	$665	$6.70	$9.87
Fuel cells	10,000	9500	$6835	$350	$0.00
Uranium					
Dual unit nuclear	2,236,000	N/A	$5335	$88.75	$2.04
Biomass					
Biomass CC	20,000	12,350	$7894	$338.79	$16.64
Biomass BFB	50,000	13,500	$3860	$100.50	$5.00
Wind					
Onshore wind	100,000	N/A	$2438	$28.07	$0.00
Offshore wind	400,000	N/A	$5975	$53.33	$0.00
Solar					
Solar thermal	100,000	N/A	$4692	$64.00	$0.00
Small photovoltaic	7000	N/A	$6050	$26.04	$0.00
Large photovoltaic	150,000	N/A	$4755	$16.70	$0.00
Geothermal					
Geothermal—dual flash	50,000	N/A	$5578	$84.27	$9.64
Geothermal—binary	50,000	NA	$4141	$84.27	$9.64
MSW					
MSW	50,000	18,000	$8232	$373.76	$8.33
Hydro					
Hydroelectric	500,000	N/A	$3076	$13.44	$0.00
Pumped storage	250,000	N/A	$5595	$13.03	$0.00

http://www.eia.gov/oiaf/beck_plantcosts/

Table 9 Comparison of Updated Plant Costs to *AEO2010* Plant Costs (EIA 2010)

	Overnight Capital Cost ($/kW)			Nominal Capacity kW's[a]	
	AEO 2011	**AEO 2010**	**% Change**	**AEO 2011**	**AEO 2010**
Coal					
Advanced PC w/o CCS	$2844	$2271	25%	1,300,000	600,000
IGCC w/o CCS	$3221	$2624	23%	1,200,000	550,000
IGCC CCS	$5348	$3857	39%	600,000	380,000
Natural gas					
Conventional NGCC	$978	$1005	−3%	540,000	250,000
Advanced NGCC	$1003	$989	1%	400,000	400,000
Advanced NGCC with CCS	$2060	$1973	4%	340,000	400,000
Conventional CT	$974	$700	39%	85,000	160,000
Advanced CT	$665	$662	0%	210,000	230,000
Fuel cells	$6835	$5595	22%	10,000	10,000
Nuclear					
Nuclear	$5339	$3902	37%	2,236,000	1,350,000
Renewables					
Biomass	$3860	$3931	−2%	50,000	80,000
Geothermal	$4141	$1786	132%	50,000	50,000
MSW—landfill gas	$8232	$2655	210%	50,000	30,000
Conventional hydropower	$3078	$2340	32%	500,000	500,000
Wind	$2438	$2007	21%	100,000	50,000
Wind offshore	$5975	$4021	49%	400,000	100,000
Solar thermal	$4692	$5242	−10%	100,000	100,000
Photovoltaic	$4755	$6303	−25%	150,000	5000

[a]Higher plant capacity reflects the assumption that plants would install multiple units per site and that savings could be gained by eliminating redundancies and combining services.
http://www.eia.gov/oiaf/beck_plantcosts/.

options. Solar technology costs, however, did decrease. With natural gas facilities having such a cost advantage, figure 5 shows the impact of natural gas facilities on CO_2 and water demands.

The most important revelation, from figure 5, is that the power generation facilities in North America are expending excessive energy and water to produce energy. The University of Texas has completed two studies that have highlighted the immense wastage of energy and water in traditional steam turbine electricity generation facilities[8]. Figure 5 provides the basis for a major change in the production of electricity and heat. It also provides an overview in the realignment regarding the use of water to generate energy. Figure 5(a) shows the amount of energy that is developed using coal at a centralize electricity generation steam turbine facility. These traditional and entrenched facilities use an opulent amount of energy to make steam and treat water so that energy can be produced. Overall the electrical energy expended on water-related needs for such demands as steam generation, pumping and treating water, space heating, and other

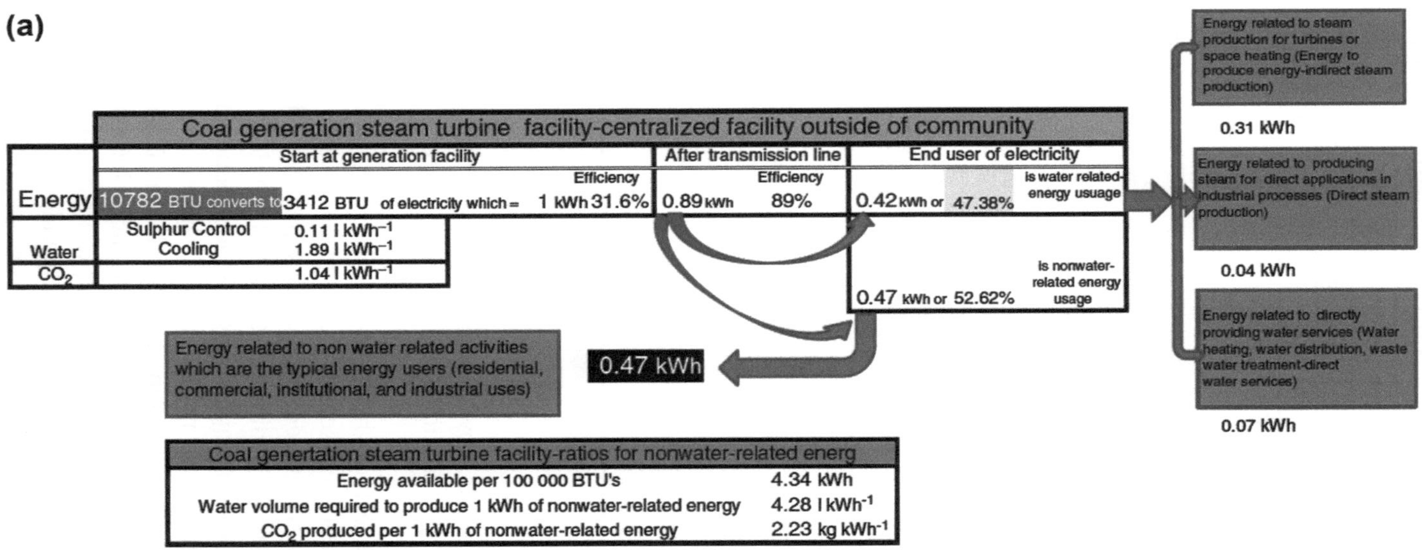

FIGURE 5 (a) Coal fueled—steam generation facility—typical energy usage and efficiency from centralized mega generator. (b) Natural gas-steam generation facility—typical energy usage and efficiency from centralized mega generator. (c) Natural gas/biogas—microgeneration—community-based electricity and heat (CHP) generator—typical energy usage and efficiency.

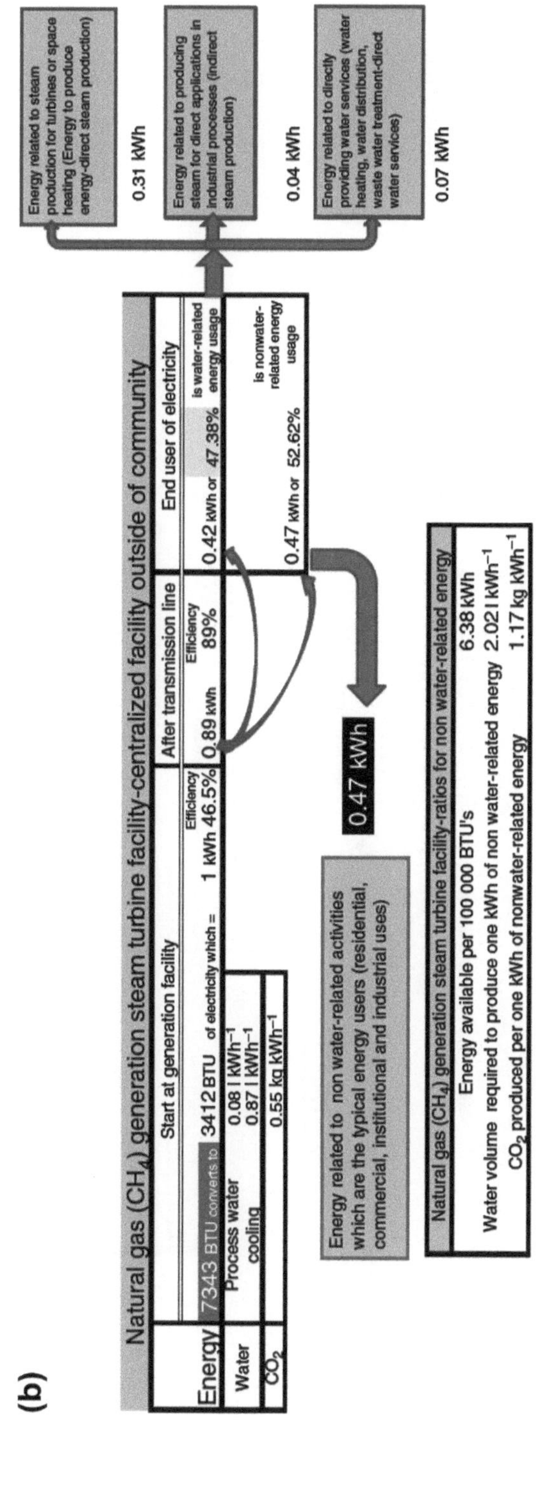

FIGURE 5 Continued

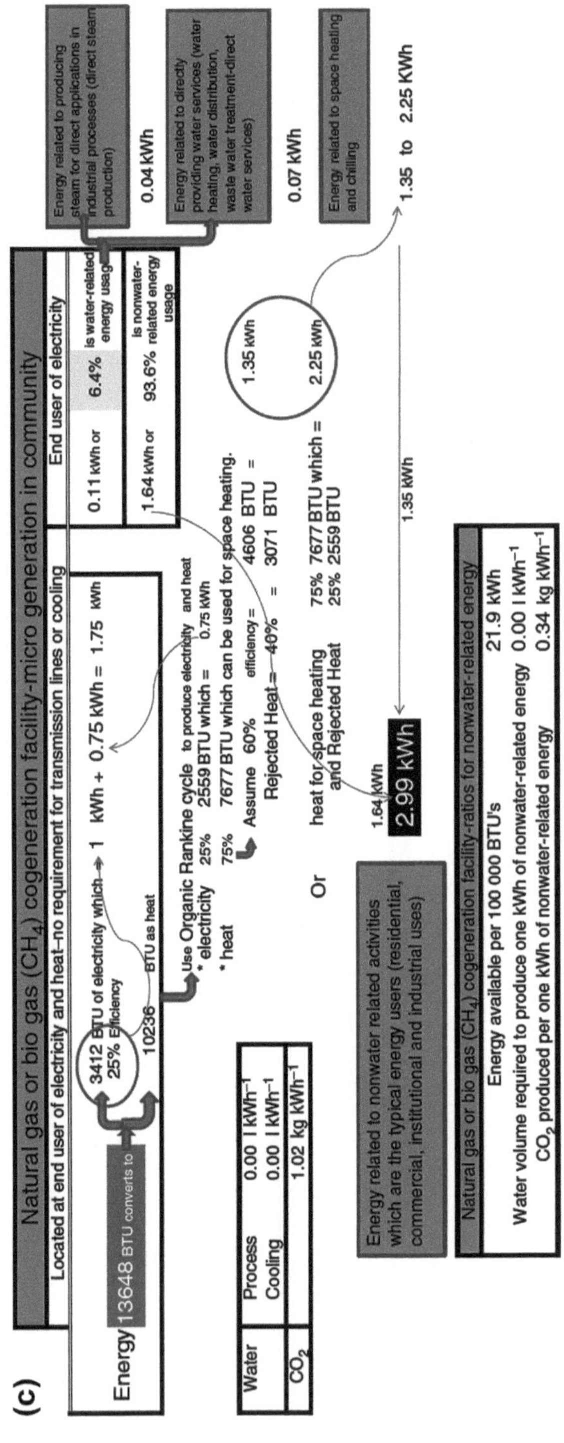

FIGURE 5 Continued

water-related needs represents 47.4% of the electrical energy produced. In addition, 2 litres of water are used for every kilowatt hour produced. CO_2 emissions are high as well. One of the interesting revelations in these figures is the amount of energy available for the nonrelated water uses. In the scenarios where coal and natural gas is the fuel source for a centralized steam turbine operation, only 52.6% of the electrical energy produced is available for all of the energy needs the public requires on a daily basis (residential, commercial, institutional, and industrial uses). These energy uses are termed as the nonwater-related energy uses. Figure 5 has compared the amount of nonwater-related energy that can be developed from 100 MJ. Three scenarios are compared in Figure 5:

1. Coal generation steam turbine facility,
2. Natural gas steam turbine generation facility, and
3. Natural gas or biogas cogeneration—Combined Heat and Power (CHP) microgeneration/community-based facility.

The first two scenarios are the typical centralized facilities that utilities operate today. Fuel is converted to electricity, the excess heat has to be cooled, typically by water, and

Table 10a Comparison of Electrical Energy Produced, Water Requirements, and CO_2

Coal Generation Steam Turbine Facility Ratios for Nonwater-Related Energy		
Energy available per 100 MJ	4.34	KWH
Water volume required to produce one KWH of nonwater-related energy	4.28	liters/KWH
CO_2 produced per one KWH of nonwater-related energy	2.23	kg/KWH

Table 10b Comparison of Electrical Energy Produced, Water Requirements, and CO_2

Natural Gas (CH_4) Generation Steam Turbine Facility Ratios for Nonwater-Related Energy		
Energy available per 100 MJ	6.38	KWH
Water volume required to produce one KWH of nonwater-related energy	2.02	liters/KWH
CO_2 produced per one KWH of nonwater-related energy	1.17	kg/KWH

Table 10c Comparison of Electrical Energy Produced, Water Requirements, and CO_2

Natural Gas or Biogas (CH_4) Cogeneration Facility Ratios for Nonwater-Related Energy		
Energy available per 100 MJ	21.89	KWH
Water volume required to produce one KWH of nonwater-related energy	0.00	liters/KWH
CO_2 produced per one KWH of nonwater-related energy	0.34	kg/KWH

the electricity is distributed to communities many kilometers (miles) from the generation plant via transmission lines. The third scenario, a gas fueled CHP option is built within the community and the resulting heat production is used to produce more electricity, using the Organic Rankine Cycle, and/or for space heating or chilling (air conditioning). As a result, waste heat is significantly reduced.

Table 10 provides a summary of the amount of usable electrical and heat energy available from a base consumption unit (100MJ) of energy as well as CO_2 emissions and fresh water consumption.Table 10a and Table 10b assess the impact of centralized mega coal fired and natural gas fired steam generation facilities respectively

Table 10c shows the very compelling benefit of the microgeneration-based power systems, and in this case, the significant benefit in the use of gas in a CHP facility. The kilowatt hour per megajoule of fuel is 3 and 5 times better than a natural gas steam turbine facility and the coal steam turbine facility, respectively. This significant improvement in efficiency provides a much greater longevity and value in the use of gas as a fuel source. In Section 1.3.1, a review showed that by replacing all existing coal fired and nuclear electrical generation facilities with natural gas fueled generation facilities, the existing reserves of natural gas would last 200 years. This life expectancy of the reserves of natural gas can be extended to 400 years through the use of community-based microgeneration facilities due to the significant efficiency improvements in the use of gas to produce usable electricity and heat.

In addition, there are significant savings in the use of water. The water usage shown in Table 10 relates to nonwater-related energy usage whereas Figure 5 shows the water usage per kilowatt hour for all power demands. Table 3, however, shows the real impact of the centralized mega power generation facility on the freshwater available for human and economic-related activities. Approximately 300 billion m^3 of freshwater is not available for human and economic needs because this volume of freshwater has been set aside for use as cooling water and steam production for the mega sized centralized steam/electricity generation facilities. This volume of water is more than sufficient for a population twice the size of the existing global population to live at standards similar to any Western country and have sufficient water for all aspects of human and economic needs. Typically a centralized electrical generation facility is located near a water body and in reasonable distance to transmit the energy to a large population. As a result, the water withheld from the freshwater inventory for this centralized power facility is also in close proximity to a population and economic base that could use the freshwater inventory. An electrical utility based on a series of microgeneration facilities located within a community similar to the process outlined in Figure 5(c), would free the significant freshwater inventories now set aside in the conventional electrical utility infrastructure.

Finally, the use of the microgeneration facility and its resulting significant reduction in energy consumption would correspondingly reduce CO_2 production. Table 4 shows the amount of CO_2 emitted as a result of conventional coal and natural gas mega steam generation facilities. Approximately, 18 billion metric tonnes emanate from these facilities. Using the microgeneration type concept located within communities should facilitate a CO_2 reduction of 50% or 9 billion m^3. With a total annual CO_2 production anticipated at 36 billion m^3 in 2014, the impact of this change in generating electricity would reduce CO_2 generation by 25%.

Creating the Next Generation of Efficiency and Connectivity with Energy, Water, and Food Through the Development of Data Utilities

The ability to implement microgeneration facilities for electricity and heat production is well within the realm of technological capability. Cogeneration technology has been available for many years. The hardware is well established and readily available. In fact, the hardware could be as simple as a combustion engine. In addition, the combustion engine would have to be equipped with a jacket to capture the waste heat. The heat transfer medium in the jacket could be as simple as water but more likely a glycol-type product would be used.

The hardware for microgeneration systems is not a barrier. There are, however, two important barriers that could impede any progress in the development of community-based utilities and specifically the microgeneration systems.

Firstly, and probably more difficult to overcome than any other barrier is the resistance to change. For the development of microgeneration facilities in a community environment, the resistance will be significant. Major utilities throughout the developed world have operated their utilities in a regulated environment where the utility is compensated based on the amount of capital investment. As a result of the years of capital asset development, it would be presumptuous to expect these utilities to give up these assets and their existing regulated business operations.

The second major obstacle is the coordination of numerous microgeneration facilities. The world's ability to have communication connection between the smallest of instruments to unlimited networking will be the conduit to a reconfigured utility and community service structure. Figure 6 shows the exceptional growth in the connectivity of instruments, monitors, applications, and data management systems that manage our infrastructure and our environs everyday. All of this connectivity is driving continuous efficiency improvement in the application of data. The data levels are such that no human being could assess such data volumes. As a result, machine data assessment is the key to driving the efficiency gains in the utilization of resources. If the concept of microgeneration and community-based utilities is going to be successful in providing improved resource efficiencies, the acquisition, management, and analysis of data will be paramount. Electrical distribution systems have initiated the change through the use of smart grid-type concepts. Electrical distribution systems are only a small part of an overall data driven infrastructure. The overall integration and coordination of all services is essential to develop the significant improvements in resource utilization and

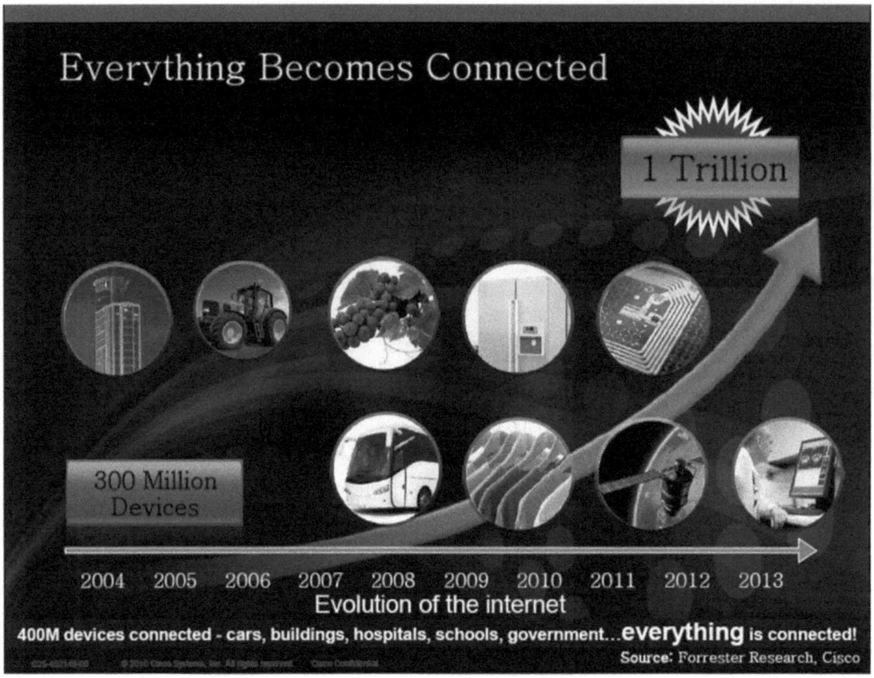

FIGURE 6 The growth of connectivity.

efficiency. This overall data infrastructure has been defined as the data utility. The data utility will become as essential as our water, energy, and food services. The data utility will create the nexus of these three essential requirements of life.

Conclusions

A very disruptive opportunity exists today as a result of the evolution in technology to extract natural gas from shale geological formations. Inventory of natural gas reserves is not only significant but it is well distributed throughout the world. The extraction of shale gas has generated unwarranted concerns that local water resources will be impacted. Technology in the extraction process has improved significantly since inception. Advancements such as 100% water reuse are a reality. In addition, there have been no incidents of groundwater contamination as a result of fracturing procedures.

Some very significant opportunities will result through the use of local community-based microgeneration systems, which can be fueled by the existing natural gas infrastructure and application of existing well proven CHP generation technology.

Firstly, energy utilization can be decreased significantly. The replacement of conventional steam turbine electrical generation facilities fueled by coal, natural gas, or nuclear reactors with natural gas microgeneration facilities could reduce energy requirements by as much as 80%.

Secondly, the existing natural gas reserves combined with the efficiency of microgeneration facilities would be sufficient to supply all of the world's electricity needs for the next 400 years. The transition over this time would be to substitute natural gas with biogas and other renewable energy sources as well as further levels of efficiency in operating microgeneration facilities.

Thirdly, the conversion from the existing electrical generation facilities to the microgeneration concept would increase freshwater inventories dramatically resulting in water availability in all of the areas of high population growth.

Fourthly, CO_2 generation can be reduced by approximately 25% worldwide if the microgeneration concept was adopted globally.

Finally, the development of the microgeneration concept within communities would catapult the development of data utilities and the interconnection of data generating devices with data analytic applications. This data infrastructure has been defined as the data utility. The importance of such a utility is the opportunity to interconnect electricity generating systems within a community to the overall infrastructure including water, food, and other key services that impact the quality of life in a community. It is through this concept of developing microcommunity systems managed by machine-based data analytics that will lead to the next revolution in macroefficiency improvements.

References

1. Bromley D. Water and its controlling dominance in energy consumption. In: Ahuja S, editor. *Comprehensive water quality and purification*, vol. 2. United States of America: Elsevier; 2014. p. 34–59.
2. Olivier JGJ, Janssens-Maenhout G, Muntean M, Peters JAHW. *Trends in global CO_2 emissions: 2013 Report, 2013.* The Hague: PBL Netherlands Environmental Assessment Agency; 2013. ISBN: 978-94-91506-51-2 PBL publication number: 1148:JRC Technical Note number: JRC83593:EUR number: EUR 26098 EN.
3. Boschee Pam. *Produced and flowback water recycling and reuse–economics limitations and technology.* February 2014.
4. Acharya H, et al. Cost effective recovery of low TDS frac flow back water for reuse. *GE Global Res.* June 2011.
5. Paktinat J, et al. *Case studies: impact of high salt tolerant friction reducers on fresh water conservation in Canadian shale fracturing treatments.* SPE 149272. November 2011.
6. Lebas R, et al. *Development and use of high-TDS recycled produced water for crosslinked-gel-based hydraulic fracturing.* SPE 163824. February 2013.
7. Gunderson J. Oil and gas, dynamic treatment challenges. *Ind Water World.* April 2014.

8 Saunders K, Webber M, University of Texas. Evaluating the Energy Consumed for Water Use in the United States , *Environmental Resource Letter* 7, August 2012.

Relevant Websites

1. http://www.eia.gov/forecasts/ieo/coal.cfm.
2. Table on BTU production from various energy sources http://www.eia.gov/oiaf/aeo/tablebrowser/#release=IEO2013&subject=0-IEO2013&table=2-IEO2013&region=0-0&cases=Reference-d041117.
3. Table on the BTU's required to produce 1 kwh (3412 BTU) of electricity http://www.eia.gov/electricity/annual/html/epa_08_01.html.
4. Shale gas reserves http://www.eia.gov/analysis/studies/worldshalegas/pdf/overview.pdf.

7

The Role of Water in Unconventional In Situ Energy Resource Extraction Technologies

Tanya J. Gallegos[1], Carleton R. Bern[2], Justin E. Birdwell[3], Seth S. Haines[3], Mark Engle[4]

[1]U.S. GEOLOGICAL SURVEY, EASTERN ENERGY RESOURCES SCIENCE CENTER, RESTON, VA, USA; [2]U.S. GEOLOGICAL SURVEY, CRUSTAL GEOPHYSICS AND GEOCHEMISTRY SCIENCE CENTER, DENVER, CO, USA; [3]U.S. GEOLOGICAL SURVEY, CENTRAL ENERGY RESOURCES SCIENCE CENTER, DENVER FEDERAL CENTER, DENVER, CO, USA; [4]U.S. GEOLOGICAL SURVEY, DEPARTMENT OF GEOLOGICAL SCIENCES, UNIVERSITY OF TEXAS AT EL PASO, EL PASO, TX, USA

CHAPTER OUTLINE

Food, Energy, and Water. http://dx.doi.org/10.1016/B978-0-12-800211-7.00007-7

Introduction

In situ energy resource exploitation techniques have emerged as the world's most readily accessible and technically recoverable uranium deposits and conventional accumulations of oil and gas have grown scarce. The so-called conventional accumulations of oil and gas, those having a delineated, downdip, water–hydrocarbon contact within high-permeability rock (porosity greater than ~8% and permeability greater than 0.1 milliDarcy (mD)[1]), can be economically produced using conventional technology. On the other hand, hydraulic fracturing is needed to economically liberate high volumes of in-place continuous petroleum resources (oil, gas, and natural gas liquids) trapped in low-permeability reservoirs with generally less than ~10% porosity and permeability less than 0.1 mD (e.g., tight gas, shale gas, tight oil (or shale-hosted oil), and coalbed methane reservoirs).[1] Similarly, the capital costs of labor-intensive conventional underground uranium mining has prompted advances in in situ recovery (ISR) of uranium to improve the technical accessibility and the economics of extracting uranium from deep, low-grade sandstone-hosted uranium deposits. Finally, oil shale retorting is emerging as an in situ method for producing synthetic crude oil from thermally immature kerogen, the precursor of crude oil and natural gas.

Fluids are injected into the subsurface during ISR of uranium, in situ retorting of oil shale, and to enhance hydrocarbon production through hydraulic fracturing. Although these in situ unconventional extraction technologies are likely to continue to increase the national endowment of oil, gas, and uranium resources, they rely on the use of water (often fresh) as an input to each extraction process. In turn, these methods produce a wastewater effluent that is usually of lesser quality, typically containing elevated concentration of salts, organic compounds, metals, metalloids, and/or radionuclides. Many wastewater effluents from energy resource production are disposed of in deep injection wells. In the United States, the >650,000 injection wells are regulated through the US Environmental Protection Agency (EPA) Underground Injection Control (UIC) program, and wells are categorized by type. For discussion in this chapter, the most notable types of UIC wells include class I (hazardous waste, nonhazardous liquids, and municipal wastewater, >600 wells), class II (oil and gas, >170,000 wells), class III (solution mining (e.g., ISR), >22,000 wells), class IV (shallow, hazardous, and radioactive waste, ~35 wells), class V (shallow, nonhazardous waste, ~400,000 to 650,000 wells), and class VI (geological carbon sequestration, currently no permitted class VI wells). Ultimately, the EPA oversees the permitting of these wells, but many states have accepted primacy and administer their own programs for one or more well classes, making for some variance in permitting guidelines across state borders. Recent concerns of fluid injection triggering earthquakes (so-called induced seismicity) have shed light on potential environmental issues related to deep well injection.[2,3]

Because these in situ processes intersect or take place within or in proximity to fresh groundwater aquifers, there remain uncertainties regarding other potential environmental subsurface impacts such as (1) releases of chemical additives, mining fluids, trace metals, organic compounds, and radionuclides from minerals and formation waters outside of the subsurface extraction zones; (2) alterations or perturbations to geochemical equilibrium such that restabilizing organics, radionuclides, and trace metals released during production is difficult; and (3) uncertainties in groundwater flow paths and connections to drinking water sources. These impacts are related to the large volumes of water used during the energy resource life cycle (production and subsequent restoration) and the resulting wastewaters sent to evaporation ponds or injected into the deep subsurface for disposal.

It has been estimated that acid-based uranium ISR (not used in the United States) uses on the order of 8520 m^3 of water per ton of uranium oxide or "yellowcake" (U_3O_8) and 0.1–0.2 L of water is used to produce a cubic meter of shale gas using hydraulic fracturing.[4–6] Water recycling and reuse can be instrumental in reducing the overall consumptive water use,[7] and returning water to the environment that is equal or better condition than was withdrawn,[8] in some cases. "Water recycling" and "reuse" are often used as two interchangeable terms; however, "water recycling" is defined herein as the on-site use of water for the same purpose (either directly or after treatment) and "water reuse" as the use of water elsewhere, possibly for another purpose (e.g., agriculture). As such, the water balance needs to include an understanding of water use in each stage of the energy resource extraction life cycle from the production to groundwater restoration steps, including injected volumes of water, wastewater streams, water treatment processes, and ultimate disposal options. This understanding allows for a better delineation of consumptive use and the potential for recycling and beneficial reuse, especially for agriculture.

The goal of this work is to contribute to an understanding of the role of water in the life cycle of unconventional in situ energy resource extraction technologies and stimulation methods (namely, hydraulic fracturing for oil and gas production, in situ retorting of oil shale, and ISR of uranium). The objectives are to provide an overview of (1) the use of water during these situ oil, gas, and uranium extraction life cycles, (2) groundwater concerns and restoration methods, (3) disposal of wastewaters, and (4) the challenges of and opportunities for recycling and beneficial reuse of wastewater for agricultural purposes. This evaluation will help inform energy- and water-resource extraction decisions and provide a context for beneficial reuse of wastewaters.

Water Issues and In Situ Oil Shale Retorting

Petroleum resources come in many forms. One of the largest, mostly untapped resources is oil shale, with an estimated 2.8 trillion barrels of in-place synthetic crude oil in various basins worldwide.[9] Some of the thickest and richest oil shale deposits are located in the United States, specifically in Colorado, Utah, and Wyoming in the Eocene Green River Formation.[10,11] Significant resources are also found in Russia, Brazil, China, Australia, Israel, Morocco, and Jordan.[9] Oil shales are sedimentary rocks rich in kerogen, the precursor of crude oil and natural gas.[12] Unlike petroleum source rocks, which have been buried to a sufficient depth for long enough to generate petroleum and charge conventional and unconventional reservoirs, oil shales are thermally immature and do not contain readily extractable petroleum in significant quantities. In order to obtain useful products from oil shale, it must be heated to convert kerogen to oil and/or gas or burned to generate electricity.

The history of oil shale development has been turbulent. Although very organic-rich oil shale can be used like coal as a feedstock for power plants, as it is in Estonia, the primary goal of oil shale utilization has been the extraction of synthetic crude oil. Therefore, the lower cost associated with development of conventionally—and more recently unconventionally—produced crude oil has been the critical factor in preventing the oil shale industry from developing into a large-scale commercial enterprise throughout most of the world. When oil supplies have been strained and crude oil prices have increased significantly, however, the prospect of bringing additional supply to market has made oil shale prospective, leading to several boom-and-bust cycles in the United States over the last century.[13]

Synthetic crude oil is generated from oil shale using a process called retorting, an industry term for pyrolysis or heating in the absence of oxygen. During oil shale retorting, kerogen is thermochemically converted into smaller compounds, including saturated and aromatic hydrocarbons, polar compounds (e.g., resins and asphaltenes), and various gases (e.g., methane, ethane, and hydrogen). The temperature range for retorting is typically between 660° and 1100 °F (350°–600 °C). The retorting process can be conducted ex situ by mining and heating the shale in large industrial reactors or in situ by heating the shale in place and collecting the oil and gas generated using traditional petroleum production techniques.[10] The key factors in determining the applicability of ex situ over in situ development methods are related to mining, specifically the amount of overburden rock that must be removed and the thickness and richness of the underlying oil shale interval. Surface disturbance and related environmental impacts are also important considerations.

In many places, oil shale resources are too deeply buried to be economically mined. This is the case for much of the resource in the Piceance Basin of northwestern Colorado,[14] central Jordan,[15] and the HaShefela region of Israel.[16] In these places, in situ retorting techniques are being developed or have been tested by companies at the pilot scale using approaches designed to address the specific geological and environmental issues related to development of the particular resource.[17,18] To date, the only successful in situ retorting operations have been pilot-scale or small commercial ventures (<1000 bbl/day) and no technology has yet been demonstrated to be viable at large scale (>10,000 bbl/day) anywhere in the world. A key issue for oil shale development is water use. In places where oil shale is abundant, in Colorado, Jordan, and Israel, this is especially true due to their arid climates and corresponding limits on freshwater availability. The water needs of an oil shale industry developing in any of these regions are a potential limiting factor, and development is expected to be contingent on minimizing water consumption and other impacts on water supplies.

Water Needs for In Situ Retorting

An important distinction for water use in energy production is between what is withdrawn and what is consumed during the process.[19] Consumed water is that which is removed and not directly returned to the system from which it was taken. Examples of consumptive water use are evaporation and product incorporation. Withdrawn water is used and returned to the source; a common example would be nonevaporated cooling water for a power plant. Water consumption is always less than or equal to water withdrawn. The focus of most studies on the water use issue related to oil shale development is on consumed water, and all values reported herein will be for consumptive use.

The focus in examining water use for in situ retorting has generally been on processes that use downhole electric heaters, as electric heating is the most firmly established technology currently being considered for large-scale in situ development of oil shale resources.[10,17] Such processes will need significant expansion of the electricity-generating infrastructure in regions that are being targeted for development, potentially leading to concomitant increases in water demand for power plants. Other heating approaches include the use of downhole burners and more exotic technologies that may use less water, such as radio frequency, microwave, or solid-oxide fuel-cell heating.

Water-intensive activities associated with in situ oil shale retorting operations are generally well understood, even though there is some uncertainty in the absolute amount of water that any particular aspect of production will need. Water use categories for an in situ process include construction and drilling, process water (e.g., dust mitigation and other on-site needs), power generation, product upgrading, and site reclamation. Additional external water uses include those needed to accommodate any population growth associated with a large-scale oil shale industry.[20] The single largest component of water use relates to how electricity is generated. Power plant consumption contributes between one (combined-cycle, air-cooled natural gas power plant) and five (nuclear power plant) barrels of water per barrel of oil produced (bbl-w/bbl-o) to overall water needs.[20,21] Water for site reclamation is potentially another major factor. Estimates for reclamation water use have focused on the proposal by Shell that oil shale retorted in situ be flushed with water to remove labile organic and inorganic contaminants and in doing so, reduce the potential for and long-term impact of groundwater pollution.[21–23]

A range of 0–5.5 bbl-w/bbl-o for reclamation water has been reported by the US Government Accountability Office (GAO).[20] The high end of this range was based on low-probability scenarios incorporated into an analysis of water needs by the Colorado School of Mines and Idaho National Laboratory[23] that was updated with a narrower range of potential values from 0–1.2 bbl-w/bbl-o based on reclamation experiments and modeling studies conducted by Shell.[21]

The estimated range of values for total water used to produce synthetic crude oil from oil shale by in situ methods is 1–3 bbl-w/bbl-o.[21,23–25] This range is based, in part, on assumptions regarding what the most probable sources of electricity will be for powering in situ operations (e.g., natural gas is considered to be a more likely fuel than coal or nuclear)[21,23] and the most recent estimates of water needed for reclamation of in situ retorts. The US GAO reported a wider range of 1–12 bbl-w/bbl-o due to the inclusion of low-probability estimates of reclamation water needs.[20] Contributions to the total from other water use categories are very similar throughout the available literature.

Some water is generated during the retorting process. It has been suggested that at least a fraction of the water coproduced with shale oil, which is similar in some ways to wastewater generated during crude oil refining, could be recycled to meet various on-site water needs for an in situ retorting operation.[26] The degree to which oil shale produced water can meet process needs depends heavily on the water content of the particular oil shale being targeted and the extent to which the produced water needs treatment. Coproduced water typically has a pH greater than 8.5 and contains high total dissolved solids (TDS) of 10–20 g/L, dissolved organic carbon of 1–10 g/L, and an array of hazardous inorganic and organic compounds, including arsenic, barium, mercury, various ketones, and aromatic hydrocarbons.[27–29] The quality of produced water can be upgraded for on-site use (e.g., boiler water) or discharged following a combination of distillation, froth removal, biological treatment, and reverse osmosis (RO) to remove volatiles, dissolved organic compounds, and salts.[26] A recent study of Piceance Basin oil shales estimated that the typical ratio of water to oil produced is significantly less than one, particularly for high-grade oil shale (oil shale yielding >25 gallons of synthetic crude oil per ton of rock). Therefore, produced water would be unlikely to satisfy more than a fraction of water requirements.[30] Oil shales in other parts of the world are expected to be better sources of retort-produced water.[9] Leachate water from spent in situ retorts typically contains the same types of compounds as oil shale produced waters but with lower concentrations.[31–33] It is expected that water used for spent retort reclamation could be recycled multiple times with less treatment than coproduced water would need or could be used in a variety of off-site applications.

Comparison of Retorting Water Use to Conventional Production of Other Fuels

Comparing the water use for production of oil shale-derived transportation fuels to those of other resources provides additional perspective on the relative impact of oil shale development. Although these estimates do not take into account the regional differences in water availability, they provide a consistent basis for comparing the full life cycle water use in fuel production from various resources. Analysis by King and Webber[34] compared a range of different petroleum- and agriculturally derived transportation fuels by combining water use values reported for production of the initial feedstock (crude oil, tar sand or oil shale synthetic crude, corn, etc.) with that needed to refine the feedstock into transportation fuel and the utilization of the fuel itself. Values were reported for both withdrawn and consumed water on a gallons-of-water-per-mile-driven (gal/mile) basis. Estimated consumptive water intensity of oil shale-derived gasoline ranged from 0.15 to 0.37 gal/mile, which was higher than that of traditional crude oil (0.07–0.14 gal/mile), but comparable to that of gasoline produced from Canadian tar sands (0.2–0.46 gal/mile). Some of the most water-intensive fuels were ethanol based and derived from irrigated crops like corn and soybeans (average values between 12 and 28 gal/mile). Electricity from the US grid utilized by electric and plug-in hybrid vehicles had an estimated water intensity of 0.24 gal/mile. With values between 0.03 and 0.07 gal/mile, compressed natural gas vehicles had some of the lowest water intensities. Although this study did not specifically address fuels derived from in situ oil shale retorting operations, the range of water use values for production of synthetic crude oil from oil shale that was incorporated into their analysis is consistent with estimated values for in situ retorting; therefore, the comparison is still illustrative of how fuels derived from in situ oil shale retorting operations will compare with other resources.

The Role of Water in Hydraulic Fracturing and Oil and Gas Resource Development

Water has played an integral role throughout the history of petroleum (oil, natural gas, and natural gas liquids) production. Well drilling uses drilling fluid, often water-based mud that is pumped down the borehole to cool the drill bit, clear cuttings from the borehole, and maintain hydrostatic pressure to prevent formation fluids from entering the borehole. Water is also a main component of the cement used to seal the annular space between the metal well casing and the surrounding geologic formation; together, the casing and cement are intended to prevent flow of fluids between the borehole and geologic strata, including aquifer units. Once oil or gas production begins, oftentimes water from the geologic formation is coproduced along with the oil and/or gas. In some geologic settings, the volume of produced water can be substantially larger than the volume of oil or gas, particularly for wells that have been in production for years or decades. This "produced water" is generally defined as all water that is extracted from a petroleum well.[6,35] Historically, the volumes of water used to drill and cement petroleum wells have been relatively small, whereas the disposal of produced waters has been a main water-related consideration in oil and gas production.

During the last two decades, technological advances in horizontal drilling and hydraulic fracturing (Figure 1) have greatly expanded the production of oil and gas from shale reservoirs and other formations previously considered unsuitable for oil and gas

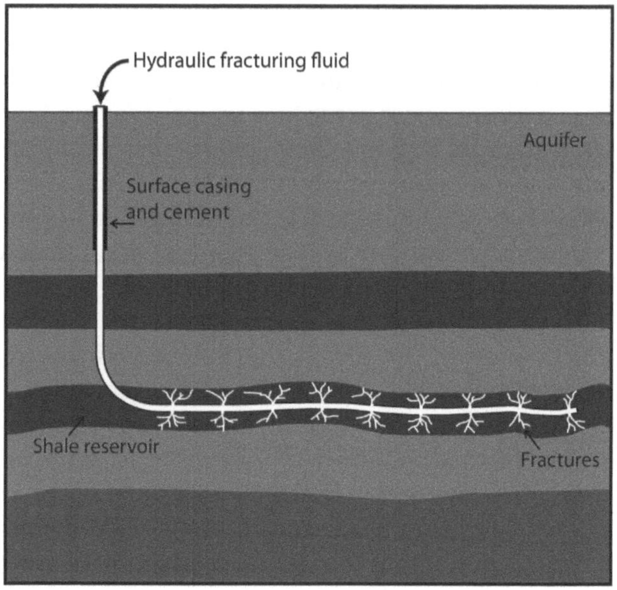

FIGURE 1 High-precision horizontal drilling and multistage hydraulic fracturing have enabled oil and gas production from shale formations and other low-permeability rock units. Wells are commonly drilled to depths as great as ~6400 m (~21,000 ft) and extend 3000 m (~9843 ft) or more laterally.

development due to low yields resulting from low porosity and permeability; these advances have also led to greater water use during development. Horizontal drilling has been conducted since the 1930s, but its use has expanded considerably since the 1990s, when techniques for high-precision directional and horizontal drilling[36] were developed in the Barnett Formation in Texas.[37] Horizontal boreholes result in greater contact area between the well bore and the oil or gas reservoir rock than a vertical well can provide, and they also use greater quantities of water for drilling and cementing well bores. Hydraulic fracturing has been used since the late 1940s[38] as a form of well stimulation treatment, and its use and the quantity of water per treatment have both grown in recent years. Hydraulic fracturing along a horizontal borehole enables the stimulation of a large volume of reservoir rock, as compared to fracture stimulation in vertical wells. A fluid, generally water based, is pumped into the borehole under high pressure to fracture the reservoir rock around the borehole to facilitate the flow of oil or gas. A propping agent or "proppant," most often sand, is carried by the hydraulic fracturing fluid into the fractures that are formed, to prop open the fracture, maintaining these fluid-flow paths. A variety of chemicals may be included in the hydraulic fracturing fluid to optimize the transport of the proppant into the fractures, to prevent swelling of clays that could close the recently opened fractures, and to inhibit the growth of microbes that might otherwise clog the resulting fluid-flow pathways. Modern approaches typically involve separation of the borehole into multiple sections that are sequentially hydraulically fractured, beginning with the segment farthest from the vertical portion of the borehole and moving inward; this is referred to as multistage hydraulic fracturing. The quantity of water per hydraulic fracturing treatment varies with the type of treatment being performed and the characteristics of the reservoir. For example, slick water treatments achieve the desired flow of proppant into fractures by pumping the fluids at very high rates (often utilizing a friction reducer), whereas gel treatments and similar approaches (e.g., cross-linked gel) attempt to move proppant using higher viscosity fluids to overcome reservoir pressures inhibiting proppant migration. Slick water is a water-based hydraulic fracturing fluid type including friction reducing agents, commonly a surfactant, electrolyte (e.g., potassium chloride), and/or polymers; it is particularly well suited to shale gas production.[39] Historically, hydraulic fracturing has also sometimes used other fluids such as diesel fuel and oil, but non-water-based approaches are not discussed here.

Hydraulic Fracturing Water Needs and Sources

At the present time, horizontal drilling and hydraulic fracturing are regularly conducted throughout the United States and around the world. Specific practices vary regionally and between operating companies, and the water needs vary accordingly (Figure 2). Modern horizontal wells commonly extend 1500–3000 m (4921–9843 ft) or more laterally, and they use between 65,000 gal (~246 m^3) and 600,000 gal (~2271 m^3) of water for drilling and cement.[5,40] Hydraulic fracturing treatments can include as many as 20 or more stages and can use between 500,000 gal (~1892 m^3) and greater than 10 million gallons (~37,854 m^3) of water, with a median of roughly 4 million gal (~15,141 m^3) for recent wells in several areas in the United States.[40–42] These advances have been integral to the expansion of US production of oil and gas. Crude oil production in the United States increased by 30% between 2008 and 2012, following a downward trend from 1986 to 2008, and net import of oil and other petroleum liquids has fallen since 2005.[43] Total US proved reserves (i.e., known resources that are producible with current technology) of

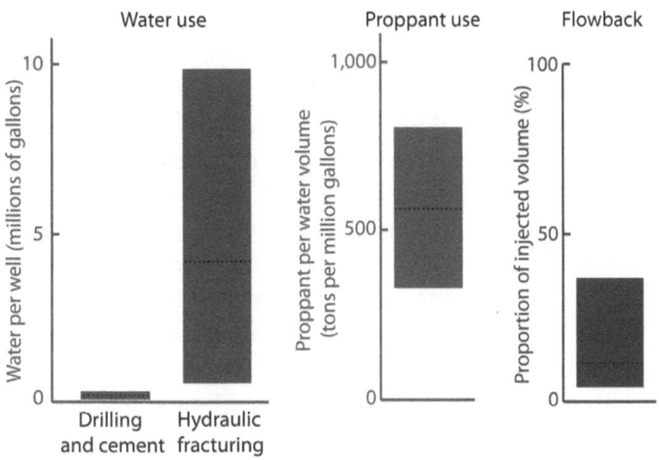

FIGURE 2 Relative water use for drilling/cement and hydraulic fracturing, proppant use for hydraulic fracturing, and the percentage of injected water that returns as flowback. *After Ref. 35.*

all natural gas have increased every year since 1999, and the proportion made up by shale gas has risen from less than 10% in 2007 to 38% in 2011.[44]

Water-per-well usage quantities can be combined with estimates of future well drilling to yield estimates of water demand that can be linked to regional water budgets to evaluate supply/demand considerations. Estimates of possible future water use can be based on drilling projections derived from recent well drilling rates and counts of available drill rigs, often including scenarios for high, medium, and low rates of drilling.[41,45,46] Alternatively, a geology-based approach, linked to probabilistic assessments of technically recoverable petroleum resources, can provide a range of possible future water needs and water production.[35] In general, studies of past and on-going water use for oil and gas development and estimates of possible future water use show oil- and gas-related water consumption to be a small percentage of regional water budgets and suggest that shortages and conflicts will mainly arise in localized areas and in cases of severe regional water shortage.[41,42,45,47]

Traditionally, water for hydraulic fracturing has been drawn from freshwater sources to avoid possible detrimental effects on equipment and on the hydraulic fracturing fluid design due to salts and other dissolved substances. The chemical makeup of hydraulic fracturing fluid is carefully controlled to maximize effectiveness and to minimize microbial growth. Information regarding preferred starting water chemistry is generally not reported, but estimates for maximum allowed TDS are in the range of 25,000 to 30,000 mg/L.[47] In recent years, water scarcity and technological advances in hydraulic fracturing methods have led to increased use of lower quality water from various sources. In areas such as British Columbia and drought-impacted Texas, brackish and saline waters are increasingly used for hydraulic fracturing.[42,48] And in areas such as the Appalachian Basin, where inexpensive disposal options are extremely limited, operators increasingly recycle produced water for hydraulic fracturing of subsequent oil and gas wells.[49]

Storage, Recycling, and Disposal of Flowback and Produced Water

At the start of oil and gas production from a hydraulically fractured well, a fraction of the hydraulic fracturing fluid returns to the surface at a rate that is high initially and then drops off exponentially over time. This initial slug of water, typically lasting a few days to weeks, is commonly referred to as "flowback water," and it represents one constituent of produced water. The percentage of hydraulic fracturing water that returns to the surface as flowback varies among geologic provinces, and it also varies with stimulation and production techniques. Estimates for several areas across the United States range from 5% to 20% (Figure 2).[40,49] Flowback water contains the chemical additives that were included with the hydraulic fracturing fluid, and it also includes any additional substances that were originally present in the geologic formation. The chemical makeup of produced water varies considerably across the United States, with TDS ranging between 25,000 mg/L and 180,000 mg/L, and sometimes including various toxic and radioactive elements (reviewed by Vengosh et al.[47]).

Historically, produced water has been used for injection into oil and gas reservoirs to maintain pressure and improve production (e.g., waterflooding) or disposed of by injection into deep formations that contain nonpotable water. As modern oil and gas production has increased both water demand and water production, the recycling of produced water has become more common, though disposal by deep injection remains a standard practice in most areas. Trends regarding disposal versus recycling vary regionally. For example, in the area of the Bakken Formation, where surface water is relatively plentiful and disposal wells are readily available, virtually all produced water is disposed of and little is recycled. In contrast, in the areas of production from the Marcellus Formation, where disposal wells are scarce, as much as 90% of produced water may be recycled for subsequent petroleum production.[50,51] Typically most of the total suspended solids, oil, and grease and some fraction of solutes are removed from the produced water prior to recycling, and it may also be diluted with freshwater to reach the desired characteristics for hydraulic fracturing. Recycling of produced water helps to reduce demands on local water sources, but low flowback percentages (5–20%) dictate that even with 100% recycling, produced water can fill only a fraction of demand at current rates of development.

It has long been known that the subsurface injection of fluids can cause induced seismicity (e.g., Refs 52,53). In fact, hydraulic fracturing itself represents the application of this principle in a controlled and small-scale manner. Larger seismic events (earthquakes that can be felt at the surface) may occur when large volumes of fluid are injected over long periods of time and/or under high pressure, particularly in locations where the earth's crust is already stressed nearly to the point of fault slip (e.g., Ref. 54). The disposal of produced water by deep injection has been linked with felt earthquakes in a handful of cases, a small number relative to the tens of thousands of active wastewater disposal wells in the United States. Ongoing research aims to provide a more thorough understanding of this issue.[2,54,55]

Impacts of Hydraulic Fracturing on Ground and Surface Water

As hydraulic fracturing has become more common in populated areas, public concern regarding petroleum development-related contamination of freshwater sources has increased. Historically, contamination of freshwater sources by oil and gas production activities has primarily been limited to spills of produced water into surface water bodies. Fluids can by spilled due to leakage from unlined storage pits or during transportation by pipe or by truck. Modern shale gas and shale oil development often involves considerable numbers of wells yielding produced water composed of brines and possibly elevated concentrations of trace metals, organic compounds, and radionuclides. As such, concerns regarding spills and wastewater disposal into surface waters remain. A review of recent evidence regarding surface spills of produced waters suggest a possible linkage between produced water and observed surface water contamination, based on limited data.[47]

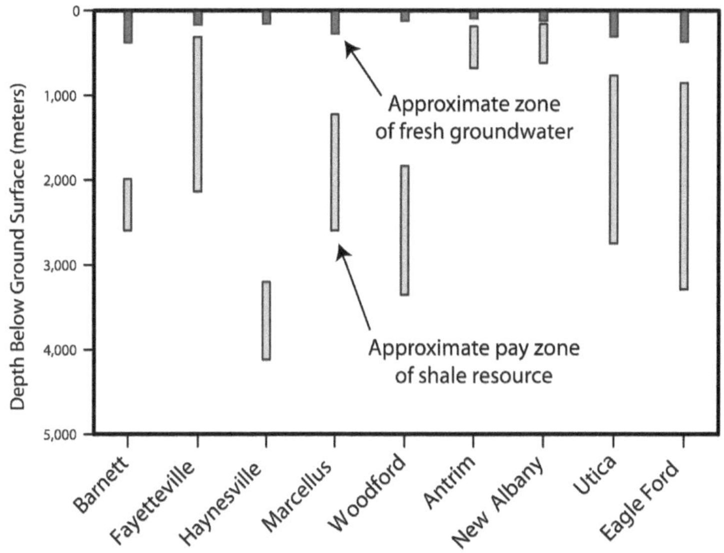

FIGURE 3 Drinking water aquifers are typically within the upper 300 m (~1000 ft) below the ground surface, whereas oil and gas reservoirs can be as deep as 6400 m (~21,000 ft), or as shallow as 150 m (~500 ft) below the ground surface. This plot shows aquifer and gas reservoir depths for selected shale gas areas in the United States. *After Healy et al.*[6]*; based on data from the National Energy Technology Laboratory.*[56]

The process of hydraulic fracturing, involving large volumes of water and chemicals pumped into the subsurface specifically to create fractures, has led to concern that injected hydraulic fracturing fluid could reach drinking water sources. The vertical distance between oil and gas reservoirs and freshwater aquifers (Figure 3) is clearly a factor in whether individual fractures can enable fluid flow between the two; however, geologic heterogeneity complicates prediction of this risk and, regardless of depth, no evidence for direct reservoir–aquifer fluid flow has been reported in the literature. The vertical borehole represents a possible fluid flow path, and thus well development practices may factor into the risk of subsurface hydraulic fracturing fluid migration. For example, the cement in the annular space of the well bore is in part intended to prevent fluid flow along this pathway, and improper or damaged well cement has been identified as a possible conduit for fluid migration from reservoir depths to surface aquifer layers.[47,49] As of early 2014, there have been no confirmed cases of contamination of drinking water aquifers by hydraulic fracturing fluids, but evidence of "stray" natural gas in aquifers suggests that gas flow may be possible along boreholes and, thus, that fluid contamination could potentially also occur via improperly cemented boreholes.[47,49,57,58]

Water for Uranium ISR Mining in the United States

ISR mining of uranium, also known as in situ leach, is a solution-mining technique whereby an ISR-amenable deposit is mined and milled in place underground through a series of injection and extraction wells. In 2011, ISR was used to produce about 45% of uranium in the world.[59] Although there are some 14 major uranium deposit type categories based on geologic setting in the world,[60] ISR was developed in the 1960s specifically to exploit sandstone-hosted uranium deposits, in aquifers confined by impermeable layers (e.g., clay, mudstone, shale). ISR allows access to lower grade deposits (typically 0.04–0.2% U_3O_8) hosted in weakly lithified sandstone located below the water table at depths nearly 2000 ft below the ground surface,[61] conditions that make access using conventional underground mining too dangerous or uneconomic.

Sandstone deposits originally formed as dissolved uranium was transported in oxidized waters through permeable sedimentary formations and then precipitated as coatings on sand grains when it contacted reductants, such as pyrite, hydrogen sulfide gas, or organic matter. Uranium is typically found as U^{+4} in minerals such as pitchblende, uraninite, coffinite, and minor brannerite.[61] Zones of uranium mineralization tend to follow the general trend of paleodrainage channels (usually narrow, often less than 300 ft long, and only tens of feet wide and thick).[62] The four main groups of sandstone deposits include roll type (or rollfront), tabular, basal channel, and structurally controlled (or the tectonic lithologic type). These can be further subdivided by the age of mineralization, intensity of post-ore metamorphism, distribution of the uranium minerals, associated metals, and nature of reductants.[61] In the United States, an estimated 135,000 metric tons (tU) of uranium reserves in Tertiary-aged sandstones in Wyoming, Nebraska, and Texas and the shallow parts of the 120,500 tU of reserves in Mesozoic sandstone formations in the Colorado Plateau may be mined by ISR.[63]

At the end of 2012, five US uranium ISR plants were operating with a combined capacity of 10.8 million pounds U_3O_8 per year (Crow Butte Operation in Nebraska; Alta Mesa Project, Hobson ISR Plant/La Palangana in Texas; Smith Ranch-Highland Operation and Willow Creek Project in Wyoming). Kingsville Dome and Rosita ISR plants in Texas were on standby with a total capacity of 2.0 million pounds U_3O_8 per year. In 2012, Lost Creek Project and Nichols Ranch ISR Project were under construction in Wyoming and several ISR plants/mines were licensed and either partially or fully permitted, including the Ross and Moore Ranch Projects in Wyoming, the Crownpoint and Church Rock projects in New Mexico, and the Goliad Project in Texas.[64] Other projects in the planning and permitting stages include the Dewey-Burdock site in South Dakota.[65]

Water Use in the ISR Life Cycle

Water and hydraulics are vital to the ISR life cycle, especially during the uranium production and the groundwater restoration stages. During ISR production, a leaching solution, called a "lixiviant," is injected into an ISR-amenable uranium deposit to dissolve the uranium into the groundwater, allowing for the uranium-enriched groundwater to be pumped to the surface. Although both acid- and alkaline-based lixiviants are used worldwide, only alkaline solutions are currently used in the United States and likely to be used in the near future.[66] Historical ISR operations used ammonium carbonate lixiviants, but ammonium ions trapped within clay matrices proved problematic during post-ISR groundwater restoration.[67] Current lixiviant chemistry is typically fresh groundwater that has been fortified with oxygen gas (oxidant) to oxidize the reduced uranium (as U^{+4}) to uranyl (UO_2^{2+}), and carbon dioxide, sodium bicarbonate, or sodium carbonate that complexes the uranium in solution, as aqueous uranyl carbonate complexes, mainly $UO_2(CO_3)_2^{2-}$ and/or $UO_2(CO_3)_3^{4-}$, depending on pH. The uranium-rich groundwater is then pumped to the surface and sent to a central processing plant where it is further concentrated via ion exchange and precipitated to produce yellowcake (UO_3 or U_3O_8), the final solid product of uranium mining and milling. Leaching of the uranium ore is greatly enhanced if a sandstone exhibits good porosity and permeability and if the majority of the uranium mineral is in accessible locations in the sandstone, such as along the sand grain boundaries of a porous and permeable sandstone.[68] ISR, however, can be inhibited if preferential flow paths form or uranium grains are coated by other minerals such as clays, which can limit access of the ore to the leaching solutions.[68]

A typical ISR mine consists of a series of well fields along the uranium mineralized zone. Typically, each well field contains an extraction (i.e., production) well surrounded by four equally spaced injection wells (generally measuring from 30 to 150 ft (9–15 m) per side to form a square) (i.e., a "five-spot pattern")[65,69] completed within the target uranium mineralization zone located within a saturated, confined aquifer. The injection wells are used to inject the lixiviant into the deposit to dissolve the uranium, and production wells are used to extract the uranium-rich water. All injection wells in the United States are permitted through the EPA's UIC Program. All ISR sites require a UIC class III well permit, which is a classification specifically for wells used in mineral extraction. Additionally, before ISR begins, the portion of the aquifer containing the uranium ore zone is permanently exempted by the EPA as an underground source of drinking water (USDW), as part of the permitting process, once it has been established that the aquifer containing the uranium deposit has water quality that is not acceptable as a drinking water source. Because the mining takes place within a confined aquifer, often in proximity to other aquifers used as underground sources of drinking water, there are concerns regarding "excursions," or the escape of fluids enriched in uranium and other elements from the mining zone into adjacent, underlying, or overlying aquifers or regional aquifers, whose water may be consumed by livestock or humans.[70] Excursions can be caused by improper water balance between injection and recovery rates, preferential flow paths, improperly plugged and abandoned boreholes, breach of the confining layers, or compromised well integrity.[71]

In general, hydraulics are used to minimize the potential for excursions by maintaining a higher extraction rate than injection rate, resulting in excess water withdrawal (called an aquifer "bleed"), to create a cone of depression that will maintain an inward hydraulic gradient toward the well field/mining zone. Baseline groundwater and surface water sampling of premining water quality within the mining zone aquifer and in adjacent aquifers is conducted as part of the permitting process to (1) qualify that water within the mining zone is not usable as a drinking water source and may therefore be exempted; (2) determine postmining groundwater restoration water quality standards within the mining zone; (3) establish a record of premining baseline water quality that can be compared to water quality during and after mining to determine if ISR activities have impacted nearby drinking water aquifers or surface waters through excursions, spills, or leakage of mining fluids; (4) identify parameters that would indicate excursions; and (5) establish the upper control limits (UCLs) for monitoring wells. A monitoring well ring is constructed about 400 ft (~122 m) outside of the injection/extraction ISR wells with monitoring wells completed in aquifers above and below the mining zone to monitor both horizontal and vertical excursions. If two or more chemical excursion indicators in a monitoring well, such as chloride, bicarbonate, and specific conductivity, exceed their UCLs, the flow rates of the nearby injection and production wells are adjusted to increase process bleed in the area of the excursion to retrieve the horizontal excursions back into the production zone. Vertical excursions are thought to be more difficult to retrieve, however.[67] If the excursions cannot be retrieved, the ISR operator may be forced to shut down.[72]

Following mining, groundwater concentrations of uranium, radium-226 (Ra-226), chloride, alkalinity, pH, and sometimes ancillary gangue mineral elements and compounds (such as Mo, Se, V, As, and sulfate among others) within the mining zone are often elevated and need remediation to restore groundwater quality to premining baseline quality or an EPA-established maximum contaminant level (MCL), whichever is higher. In some cases, if this cannot be achieved using the best available technologies, relaxed standards are requested, called "alternate concentrations levels" (ACLs). Historically, remediating impacted groundwater resources after mining has proven difficult for acid-based lixiviant chemistry[66] and ammonium carbonate-based lixiviant chemistry used until around 1980.[73,74] Likewise, groundwater restoration following current-day oxygen/carbonate lixiviant-based ISR operations is challenging.

A common method of groundwater remediation is the "groundwater sweep," whereby several pore volumes of water from the mining zone are extracted and disposed of in order for fresh native groundwater to flow into the impacted area. Due to heterogeneities in the aquifers among other reasons, groundwater sweep alone is insufficient and uneconomical.[65,74] Furthermore, groundwater sweep may cause oxic conditions from upgradient waters to enter the ore zone, making it more difficult to reestablish equilibrium and possibly the chemically reducing conditions that might be helpful in immobilizing uranium.[67] Alternatively, a groundwater sweep of a single pore volume from the mining zone followed by surface treatment by electrodialysis or RO of multiple pore volumes is thought to be a more efficient method of remediation,[65,70] especially if best practices are followed. These best practices include daily balancing of injection and extraction flow rates during production, starting restoration operations as soon as is reasonably possible following the cessation of recovery operations, and maintaining aquifer bleeds.[65] During RO, the contaminated groundwater is pumped to the surface and passed through pressurized semipermeable membranes, resulting in clean water or permeate that can be recirculated in the mining zone while brine water with concentrated ions/contaminants is sent to disposal.[67] Additionally, the groundwater may be passed through a decarbonation unit to remove residual carbon dioxide remaining in groundwater after uranium recovery prior to reinjection.[72] Calcium, barium, strontium, carbonate, sulfate, silica, and residual uranium present in postmining zone water can cause membrane fouling problems that increase the amount of brine for waste disposal. The problem, however, is greatly reduced by the initial groundwater sweep of one pore volume prior to RO.[74,75]

Wastewater Disposal Options in ISR

Wastewater streams are generated from both the uranium production and the groundwater restoration phases of the ISR life cycle. During production, wastewater consists of process bleed water pumped from the ore zone and effluents from the central processing plant (e.g., process drains, elution circuit bleed, and washdown water), with the process bleed being the largest contributor. During restoration, wastewaters generated also include the restoration bleed, the well field purge water extracted during the groundwater sweep, and ion exchange or RO reject concentrate and brines produced during aquifer restoration activities, with the majority of liquid waste coming from the groundwater sweep and RO processes. Historically, ISR facilities have employed several methods of wastewater disposal from ISR production and groundwater restoration processes, including deep wells, solar evaporation ponds, land application, or surface water discharge.[76] The most common disposal method for recent and future proposed ISR sites involves using deep wells and/or land application.[65,71,72,76–79]

The underground injection of wastewater (i.e., either with or without treatment) into either class I or class V deep disposal wells (DDWs) requires permits through the EPA's UIC Program. Generally, the DDWs at ISR sites are classified as class I for the injection of untreated waste (i.e., radioactive, hazardous, or nonhazardous fluids) into deep isolated formations beneath the lowermost USDW within $^1/_4$ mile of the ISR site. The wastewater does not require treatment prior to injection into class I wells. In states where class I wells are banned (such as in South Dakota), DDWs are class V. Class V DDWs are generally shallower wells used to place a variety of nonradioactive, nonhazardous fluids into or above USDWs. Treatment of the wastewater is needed prior to discharge to a class V well to ensure that all toxic substances remain at concentrations less than the class-of-use standards or any federal primary drinking water standards, whichever is more stringent. Before ISR begins, a class III UIC permit is also required to provide an estimate of the amount of water to be disposed of in a class I or class V well over the life of a project in order to ensure that the wells have sufficient disposal capacity for the volume of wastewater generated from ISR. Depending on the availability and capacity of DDWs, ISR mines may exclusively employ land application in lieu of deep well disposal or augment deep-well disposal with land application.

Land application also involves treatment of wastewaters via ion exchange for removal of uranium and other dissolved species followed by radium removal through addition of barium chloride and coprecipitation with existing sulfate as radiobarite in lined radium settling ponds prior to seasonal application of treated wastewater on designated land surface areas through center-pivot sprinklers. Land application would be permitted under a groundwater discharge plan. Radium removal tests conducted on sample mine water indicate that more than 99% of radium in solution would be removed resulting in radium concentrations below 1% of Federal limits for releases to waterways.[78] Radium settling pond sludge along with other solid wastes (i.e., by-product material), such as production equipment and piping, soils contaminated by spills or leaks, spills of loaded or spent ion exchange resin, filter sand or other process media, debris, and personal protective equipment that cannot be decontaminated for unrestricted release, are considered Nuclear Regulatory Commission (NRC)-regulated wastes. As such, these solid wastes require disposal at an NRC or NRC agreement state-licensed mill tailings facility or other licensed facility in accordance with NRC license requirements.[65]

Water Recycling in the Production and Restoration Phases of ISR

Figure 4 illustrates possible water balance scenarios based on several ISR sites currently under development.[65,71,72,76–79] Generally, at most ISR sites, a large portion of water withdrawn from the production wells during the ISR uranium production phase is recycled ($Q_{recycle}$) by reinjecting treated water (i.e., "permeate") as part of the lixiviant following either ion exchange and/or RO such that the net withdrawal rate is only a fraction of the gross pumping rate (Figure 4(a)). Typically the wastewater to be disposed during ISR uranium production is equal to the aquifer bleed (Q_{bleed}), which varies by site (typically 0.875–4% of the well field production flow (Q_{ISR})), plus any wastewater from the central processing plant (10–25 gallons per minute, gpm) (Q_p). Well field production flow rates estimated for several proposed ISR sites range from 1000 to 8000 gpm, depending on the ore grade, hydrologic properties, target uranium annual production, and solubility of uranium minerals, among other factors.[65,71,72,76–79]

Compared to the ISR uranium production phase, the largest contribution to the total wastewater stream is derived from the groundwater restoration steps following the uranium production phase. Although the volume of water necessary to achieve restoration standards depends on the geochemical environment and the reactions that may occur during groundwater restoration,[74] the amount of water consumed or wasted during restoration varies according to the groundwater restoration method. The water balances for three restoration options in Figure 4(b)(i) show that the groundwater sweep and disposal of ion exchange effluents via class I deep wells and/or land application (following radium removal) results in 100% consumptive use with the wastewater flow (Q_{waste}) equaling the restoration flow rate (Q). RO coupled with recycling or reuse of permeate and disposal of reject brines into class I wells, is the preferred treatment and results in greater water conservation[65] (Figure 4(b)(ii)). The amount of wastewater (i.e., mainly reject brines) going into the DDWs depends on the

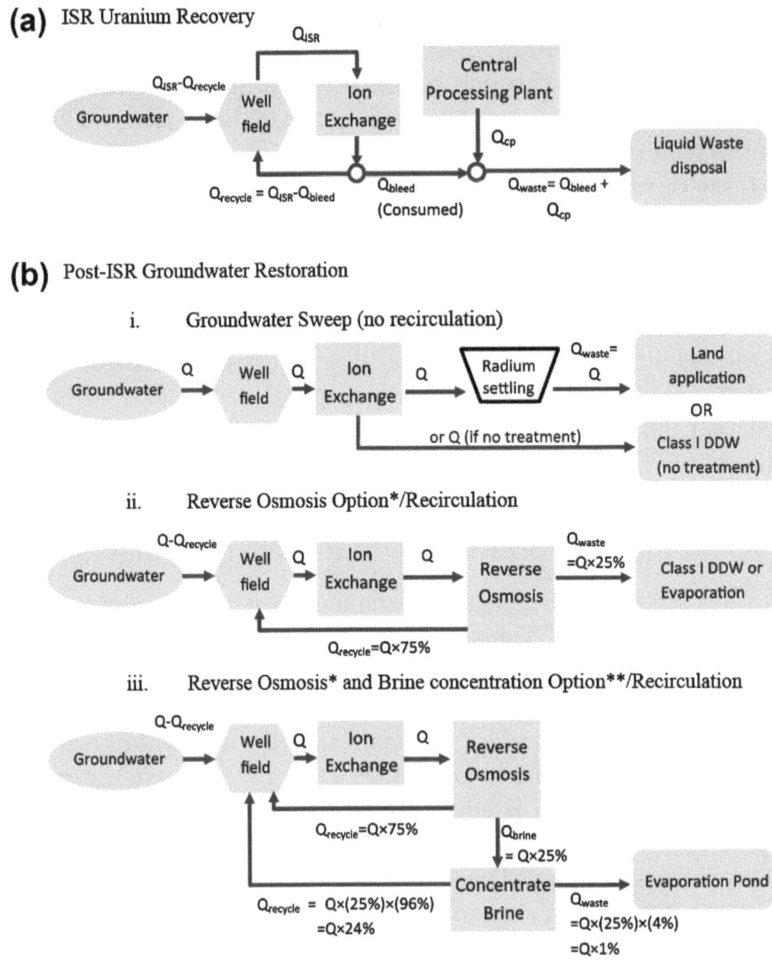

FIGURE 4 Water balance for in situ recovery (ISR) mine (a) recovery phase and (b) restoration/disposal options including (i) groundwater sweep, (ii) reverse osmosis (RO), and (iii) RO and brine concentration adapted from Ref. 65 and Ref. 78. *RO efficiency assumed to be 75%, and **brine concentrator estimated to be 96%. DDW, deep disposal well.

efficiency of the RO treatment system; generally 70–90% of the clean water permeate from the RO unit is reinjected (i.e., recycled) into the mining zone after the RO treatment.

Another modification alternative to the water balances (Figure 4) is the "groundwater transfer" between a well field starting groundwater restoration and another beginning ISR production, which is also advantageous in reducing the amount of disposed wastewater. During groundwater transfer, it is projected that the TDS of the well field being restored is theoretically lowered by dilution with water of baseline quality, thus reducing the volume of water needed to be treated and the amount of reject brine produced.[72] The proposed Crownpoint Uranium Project also evaluated the use of a brine concentrator following RO (Figure 4(b)(iii)), which heats and evaporates the water in the brine, then condenses the water vapor as pure water, resulting in a waste stream of 1–2 gpm of water during recovery and restoration.[78] Although brine concentration is an ideal option for reduction of wastewaters and water conservation, costs make this option the least feasible. The proposed Moore Ranch ISR project evaluated alternative disposal options including mechanical evaporation of liquid wastes using either gas or electric power and liquid waste volume reduction by chemical precipitation/RO, and also found these options to be uneconomic.[72]

Figure 4 demonstrates that the degree of recycling during groundwater restoration is closely dependent upon the ultimate wastewater disposal options available. For example, if there are plentiful and readily accessible class I DDWs available to accept untreated ISR wastes, then groundwater sweep may be more cost effective than RO. Although historical applications of the groundwater sweep option suggest that it is an unnecessary, ineffective step in the groundwater restoration process and results in large consumptive use of groundwater,[65,80] if there is plenty of deep disposal well capacity, then groundwater sweep may likely be the most attractive option.

Reuse of Wastewaters from Uranium ISR Operations

The untreated wastewaters from ISR processes are classified as by-product. As such, reuse of untreated wastewaters in agriculture is problematic due to radionuclides and other potentially toxic constituents. Table 1 lists typical water qualities of various waste streams and treatment effluents expected at future ISR sites.[65,77] Table 1 shows an expected typical range of Ra-226 concentrations of waste streams from <5 to 300 pCi per liter (pCi/L Ra-226)[65] with some future ISR sites projecting as high as 3000 pCi/L Ra-226 in wastewaters.[72] Wastewater extracted during post-ISR groundwater restoration slated for disposal represents the largest potential water volumes for reuse in agriculture or other applications, but some challenges remain. For example, reuse would need advanced treatment to meet reuse standards and suitable and (or) economic methods for treatment of reject brines such as (1) land available for evaporation of waste brines, (2) wells available for disposal of waste brines, or (3) economic methods to concentrate waste brines into smaller volumes. Furthermore, even with advanced treatment, a major restriction to reuse could stem from high Ra-226 concentrations similar to the projected concentrations of 30 pCi/L Ra-226 that persist in the "clean" RO permeate water in Table 1. This concentration is on the order of six times higher than the acceptable MCL established by EPA of 5 pCi/L Ra-226.[81] So, not only is Ra-226 a problem for reuse of untreated waters in agriculture, but it could also be a problem in RO-treated waters. Since Ra-226 decays to radon-222 (Rn-222) gas, this further suggests that Rn-222 may also pose a significant barrier to both recycling and reuse. Some other issues related to reuse of ISR wastewater are similar for beneficial reuse of produced waters discussed in detail in the next section.

Table 1 Projected Wastewater Quality for Disposal via Class I Deep Disposal Wells and Land Application and Reinjection

	Class I Deep Wells (Untreated)[a]						Reinjection
Analyte	**Water Softener Brine**	**Resin Rinse**	**Elution Bleed**	**Yellowcake Wash Water**	**Restoration Wastes**	**Land Application (Following IX/ Radium Removal)[a]**	**RO Permeate[b]**
pH						6.5–7.5	6.5
Total dissolved solids, TDS						1000–5000	200
Conductivity, μS/cm						1500–6000	300
Alkalinity as $CaCO_3$							100
Bicarbonate		600–900			400–700	50–300	150
Calcium	3000–5000					200–1000	0
Carbonate		500–800			300–600	<1	
Chloride	15,000–20,000	10,000–15,000	12,000–15,000	4000–6000		300–1300	15
Magnesium	1000–2000					50–300	
Potassium						10	
Sodium	10,000–15,000	6000–11,000	6000–8000	3000–4000	380–720	100–500	50
Sulfate					100–200	500–2000	15
Sodium adsorption ratio						2–6 (Unitless)	
Arsenic					0.1–0.3	0.01	0
Barium						0.4	
Cadmium						0.3	
Chromium						0.4	
Copper						0.3	
Iron						0.2	
Molybdenum						<0.1	
Nickel						0.3	
Selenium					0.05–0.15	<0.2	0
Uranium	<1	1–3	5–10	3–5	<1		0
Vanadium						<10	
Lead-210, pCi/L						<10	
Radium-226, pCi/L	<5	100–200	100–300	20–50	50–100	<60	30
Thorium-230, pCi/L	<5	50–100	10–30	10–20	50–150	<100	
U-natural, pCi/L						<300	
Gross alpha, pCi/L					2000–3000		
Gross beta, pCi/L					2500–3500		

Values in milligrams per liter, unless otherwise noted. RO, reverse osmosis.

[a]Ref. 65.

[b]Ref. 77.

Beneficial Agricultural Reuse of Produced Waters

Produced waters from petroleum energy development are derived from waters coexisting naturally with the oil or gas resource in the subsurface and are transported to the surface as a by-product of resource extraction. Depending upon volume and composition, produced water management and disposal can be a significant burden on the extraction process. Beneficial agricultural use of produced waters is a relatively new endeavor that faces numerous and significant challenges and thus currently occupies select but expanding niches worldwide. The challenges can be broadly divided between physical, water chemistry, and regulatory issues. With the right circumstances and innovation, however, the challenges can be met economically, and a substantial boon to agriculture can be achieved, particularly in arid to semiarid regions. Beneficial agricultural use is typically irrigation, although uses such as stock watering also exist. As climate fluctuations negatively impact freshwater resources and water demands increase, the motivation to derive beneficial use from produced waters will only grow. While the list of hurdles faced by this linkage between energy development and agriculture is substantial, pessimism is unwarranted. The section ends with examples of beneficial agricultural use of produced waters that meet the challenges described.

Physical Challenges to Agricultural Reuse of Produced Waters

Geography presents one of the first challenges to using produced waters for agriculture, because of distance between points of generation and points of potential use. Oil and gas wells are located based upon access to productive regions of basins, while agriculture is generally concentrated in areas of productive soil, and the two do not necessarily coincide. Conveying water from one area to the other entails infrastructure and operational costs. Unlike other produced water disposal options, both collection and final handling of the water occur at dispersed locations.

The seasonality of agricultural production and crop water demand are often out of sync with the cycles of water production from oil and gas wells. Fluctuations of climate can also generate uncertainty in the capacity of irrigation as a disposal mechanism, as cooler or wetter than average conditions reduce or temporarily eliminate demand for irrigation. Such conditions indicate a need for some flexibility in timing of using produced waters for irrigation. Surge ponds and small storage structures allow for short-term delay between water production and use. Larger reservoirs or aquifer storage, however, might be needed if produced waters are to be retained through the nongrowing season for later use. Ideally, irrigation can be one of a menu of produced water disposal options that can be engaged and disengaged as conditions change.

The transient nature of water production from energy development needs consideration from the agriculturalist's point of view. Energy resources each have their own specific water-production curves that vary depending on the resource, basin characteristics, well configuration, and development strategy. Coalbed methane wells tend to produce less water over time, while conventional oil wells produce more (Figure 5).

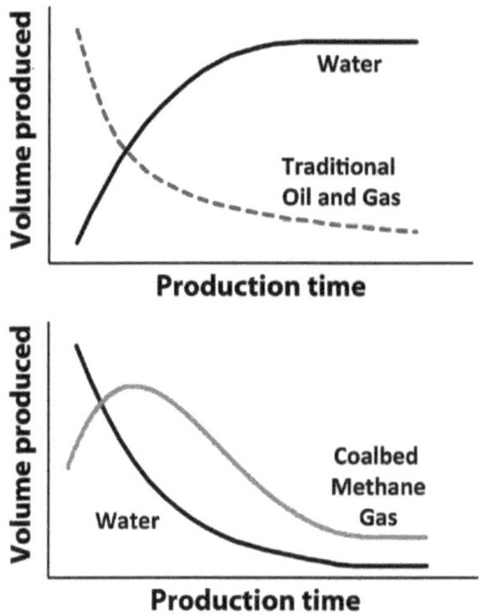

FIGURE 5 Production curves of water and petroleum products from traditional oil and gas wells (top panel) and coalbed methane wells (bottom panel).

Energy development also tends to go through boom and bust cycles, often driven by economics outside the region of production. Predictions of water availability are therefore difficult to make. Perhaps the only certainty is that the produced water will not be available in perpetuity.

Water Chemistry Challenges to Agricultural Reuse

Salinity, the dissolved inorganic constituents in water, can be the greatest obstacle to beneficial agricultural use of produced waters, due to negative effects on both crop growth and soil properties. Salinity is the sum of dissolved ions and is most directly measured by evaporating to dryness; it is reported as TDS in units of milligrams per liter. The TDS in produced waters varies widely from around >500,000 mg/L measured in some oil field brines, down to <1000 mg/L found in certain coalbed methane produced waters.[82,83] By comparison, drinking water typically contains <1000 mg/L TDS and seawater around 35,000 mg/L, making some produced waters >10 times saltier than seawater. Because ions in solution contribute fairly straightforwardly to the electrical conductivity (EC) of water at lower concentrations, this more rapid determination is often used in agricultural applications and the results reported in milliSiemens per centimeter (mS/cm). The EC of relatively pure water is approximately 0.002 mS/cm, that of drinking water is typically <0.5 mS/cm, and seawater has an EC of around 58 mS/cm.

Crops can suffer direct negative impacts from salinity in two ways: through osmotic effects and direct ion toxicity. Osmotic effects occur when solute concentrations in soil water exceed those in root cells of a plant, decreasing the favorability of soil-to-root water transfer. To compensate, plants take up salts or synthesize organic compounds to restore the osmotic gradient. Such processes need energy that is diverted from plant growth. For each crop there is a threshold soil salinity below which normal yields can be achieved. Above that threshold, yields generally decrease in linear fashion with slopes that vary from crop to crop. Crops vary widely in their tolerance of soil salinity, with threshold tolerance values ranging more than an order of magnitude. Certain crops, termed halophytes, actually grow better at relatively high salinity.

Direct ion toxicity effects on plants have been reported for sodium and chloride. Such effects can be difficult to distinguish from osmotic effects of salinity, however, even in test settings, if other ions are not also tested.[84] Nevertheless, sodium is not essential for plant growth, and leaf injuries are observed when it accumulates to higher levels. High sodium levels in soil may impair plant uptake of calcium, resulting in nutrient imbalances and impaired growth. Chloride is an essential micronutrient, but like sodium, higher concentrations can damage leaves and impair growth. Crop tolerances to chloride vary by over an order of magnitude. Boron is another essential micronutrient, but for many plants there is a relatively narrow difference between levels optimum for growth and those that cause toxicity. Natural boron levels can be high in soils of arid climates, compounding issues with its presence in produced waters.

Depending upon the proportions and concentrations of major ions in a produced water, its use for irrigation has the potential to negatively impact soil physical properties. The ability of water and air to permeate soil, and for mechanized equipment to manipulate the soil, is influenced positively when mineral particles form stable aggregates. Aggregation in turn is influenced by the degree to which clay particles are drawn together to agglomerate or pushed apart to disperse. Clay particles generally have a net negative charge, attracting positively charged ions (cations) to their surfaces. Such ions are loosely held by electrostatic forces, allowing them to exchange as the composition of the surrounding solution changes. Sodium ions have a larger hydrated radius than calcium or magnesium ions and are less attracted to clay surfaces. As the proportion of sodium ions in the soil solution increases, the thickness of the layers of ions on clay surfaces increases, and as the positively charged layers repel each other, the clays are forced apart and disperse.

The percentage of exchangeable sodium on clays, relative to calcium and magnesium, correlates most directly with clay dispersion, but has a strong relationship with the sodium adsorption ratio (SAR) of water in contact with the soil. The SAR is defined as

$$\mathrm{SAR} = \frac{[\mathrm{Na}]}{\sqrt{\frac{[\mathrm{Ca}]+[\mathrm{Mg}]}{2}}}$$

where [Na], [Ca], and [Mg] are concentrations of sodium, calcium, and magnesium, respectively, expressed in units of milliequivalents per liter (meq/L). Units of SAR are $\mathrm{meq}^{1/2}/\mathrm{L}^{1/2}$, but common convention in reporting data is to omit the units. In addition to SAR, the overall concentration of solutes in water also influences clay dispersion. At lower concentrations, more water moves in between clay particles, drawn by the higher concentrations of ions on the clay surfaces. Thus, both the salinity (measured via EC) and SAR of irrigation water can influence clay dispersion and the ability of water to infiltrate the soil. Guidelines have been published to suggest how combinations of those parameters will or will not influence infiltration.[85] While guidelines are useful, the particle size distribution, clay mineralogy, and other properties particular to a given soil will also influence how it responds to irrigation waters of a given EC and SAR. In particular, soils with a higher proportion of swelling clays like smectites are more likely to suffer from problems with dispersion and permeability.

Soil salinity and the proportion of sodium on exchange sites of clays must be managed for irrigation with produced waters to be successful. A time-tested means of addressing problems with sodium is by boosting the calcium supply via addition of gypsum ($CaSO_4 \cdot 2H_2O$), a comparatively soluble mineral and relatively inexpensive agricultural amendment. Removal of salts from soil can only be accomplished by flushing them down deeper through application of water in excess of evapotranspiration demand.[86]

Hydrocarbons, hydraulic fracturing additives, toxic trace metals, and radionuclides are other components of produced waters that can limit their use for irrigation. Some produced waters, like those associated with coalbed methane, tend to have lower hydrocarbon content compared to those from conventional oil and gas production. The sheer diversity of hydrocarbon compounds and the variety of their occurrence in produced waters limits the generalizations that can be made. Higher concentrations of some compounds are likely to generate toxic effects in organisms from plants to soil microbes. At less extreme concentrations, it may be the diversity of the microbial community that is primarily affected, possibly by favoring hydrocarbon-degrading microbes and suppressing others. Such shifts may alter enzyme activity, soil respiration and microbial biomass, all changes that might influence the cycling of nutrients to crops. Organic components such as alkanes may be decomposed rapidly in soil, while others, like asphaltenes, may persist much longer.

Produced waters often need multiple treatment steps to make them suitable for irrigation. Oil can be removed by various devices that coalesce the small drops of suspended oil ($<100\ \mu m$) into larger volumes and physically separate them. Dissolved organic components can be removed by pH adjustment or adsorption onto activated carbon or organoclays, whereas the more volatile components can be removed by aeration. A variety of technologies are also available for removal of inorganic solutes, including RO, ion exchange, capacitive deionization, and thermal distillation. The energy needed to operate each of these technologies is a primary consideration in their employment, and desalinization costs are perhaps the biggest barrier to beneficial reuse of produced waters. Additional costs come from disposal of the concentrated organics or solutes (brine) that are generated as by-products of purification. Many technologies for removing inorganic solutes produce water that is too poor in ions and nutrients for irrigation use, and blending with untreated water or water from another source is needed. Direct amendment of produced water with gypsum or magnesium sulfate can also be used to adjust both the EC and SAR to values more suitable for irrigation.

Additionally, special organic compounds such as polyacrylamide can be added to improve soil aggregation.

Examples of Irrigation Using Produced Waters

One of the older examples of produced water use for agriculture exists in Kern County, California, USA.[87] Since 1996, produced water from the Kern River oil field has been sold to the Cawelo Water District, which then delivers it to growers of crops including almonds, pistachios, grapes, and citrus. The oil field produces heavy oil and currently employs enhanced oil recovery techniques, including steam flooding. The produced water is of unusually good quality by oil field standards, having an EC around 0.9 mS/cm and SAR around 7,[87] but such chemistry is still likely to cause infiltration problems. Boron is a constituent of concern, with the waters containing a concentration around 1 ppm. Oil and other organics are removed from the produced water using skimmers and walnut shell filters. Blending of produced water with surface water and groundwater in a reservoir reduces EC to about 0.5 mS/cm, SAR to about 5, and boron to about 0.5 ppm. From 2001 to 2005, produced water cost the district and growers one-tenth as much as freshwater and accounted for about 25% of the Water District's supply.[87] Gypsum is often added through irrigation systems to boost EC to 1.7 mS/cm and lower SAR to <1 and thus reduce water penetration problems in the sandy loam soils. Water is generally applied via microsprinklers to prevent salinity damage to leaves. Problems encountered with buildup of soil salinity have been combated using winter irrigation to flush solutes out of the rooting zone.

A variety of strategies have been employed to use coalbed methane produced waters for irrigation in the Powder River Basin, Wyoming, USA. Of these, the most sophisticated is a system of deep subsurface drip irrigation.[88] Drip tubing is buried 0.92 m deep to prevent freezing during winter and limit salt migration to the surface. Deep-rooted perennial crops including alfalfa and grasses are planted that can access the applied water. Produced water is applied at relatively steady monthly rates year-round, with annual rates slightly in excess of potential evapotranspiration. Produced waters used in such systems have an EC around 2.1 mS/cm and SAR around 54, indicating severe risks of infiltration problems. Treatment by sulfuric acid addition reduces water pH from 8.0–8.5 to 5.5–6.0 and converts the sodium bicarbonate water to a sodium sulfate water, thus preventing loss of soil calcium to carbonate precipitation.[89] The systems derive substantial benefit from native gypsum present in the subsurface at around 2.4 wt.%. Gypsum dissolution decreases applied water SAR to <7 and increases EC to >6 mS/cm, preventing serious permeability problems in the clay loam soils. Compared to irrigation strategies elsewhere in the basin that applied produced water at the surface, soil salinity and sodium levels were lower in surface soil.

In Queensland, Australia, coalbed methane waters are being used for irrigation in a project that illustrates the advantages of flexibility in such endeavors[90] (personal communication from David Macfarlane, Santos Ltd., on 3/21/14). Produced waters have an EC in the range of 0.7–11 mS/cm, and SAR ranging up to 260. Treatment usually consists of RO, blending with untreated water, and amendment with micronized gypsum, although magnesium sulfate has also been used. The different treatment components allow water chemistry to be adjusted depending upon fluctuations in costs and disposal options. Irrigation is one disposal option, alongside direct discharge and quite limited reinjection capacity. Flexibility in disposal options was key in 2010, which saw above-average rainfall and cooler temperatures in Queensland[90] (personal communication from David Macfarlane, Santos Ltd., on 3/21/14), severely limiting irrigation to prevent water logging of fields. Currently, the project irrigates leucaena via center-pivot to provide forage for cattle and drip irrigates Chinchilla white gum trees to produce timber.

These examples illustrate niches where produced waters have been beneficially reused for irrigation. Each overcame the technical challenges outlined above and refined its methodology over time. Additional projects exist or are being explored elsewhere in the world, indicating growing acceptance of the practice and expertise to carry it out. Ultimately, economics will always be the overriding factor in a decision to reuse produced waters for irrigation. The infrastructure and operation costs for any agricultural use will be compared to those of other disposal methods, which may contain fewer uncertainties. The less tangible benefits of meeting innovation challenges, reusing a valuable resource, and positive public perceptions, however, should not be discounted. Particularly as costs and energy needs for desalinization decrease, it is easy to predict that this linkage between energy production and agriculture will expand in the future.

Opportunities and Regulatory Challenges

As shown in the previous sections, a clear linkage exists between in situ resource extraction methods, water use, and wastewater disposal. As applications of these advanced and innovative in situ techniques grow in number to exploit previously inaccessible oil, gas, and uranium resources, water use will concomitantly increase. In fact, water consumption is a function of each stage of the in situ extraction technology life cycle. Specifically, wastewater effluents resulting from the groundwater restoration phase of the oil shale retorting and uranium ISR life cycles, and the flowback and produced waters evolved following hydraulic fracturing for oil and gas production, represent the largest untapped volumes of wasted water and thus the greatest opportunity for advancement of technologies for water conservation through both recycling and beneficial reuse. However, regulatory challenges need to be addressed before recycling and reuse can become commonplace.

Regulatory Challenges in Recycling and Reuse of Wastewaters from Uranium Production Using ISR

Because wastewaters from uranium ISR production and restoration processes commonly contain high concentrations of trace metals, dissolved organic compounds, radionuclides, and/or salts, both recycling and reuse hinge upon additional treatment.

During uranium production, recycling entails treatment of wastewaters to meet both process chemistry needs as well as environmental regulatory requirements for reinjection of nonfreshwater into or adjacent to near-surface aquifers. Additional treatment is also necessary for some wastewater disposal options including surface water discharge, land application, and class V disposal options. In many cases, however, it is more economical to directly dispose of untreated wastewater into class I DDWs, especially in geographic areas that have greater water availability, limited land availability, and large class I disposal well capacity. Alternatively, ISR wastewaters could potentially be reused for agriculture, but only if their treatment produces a high-quality effluent, most notably free of radium and other toxic elements. In order for this option to become feasible, however, the following need to be available for dealing with reject brines: class I DDW capacity, large tracts of land for solar evaporation, or a low-cost method for enhanced concentration/evaporation. Class I wells thus both prevent and enhance the impetus for wastewater treatment and reuse, to some degree. Furthermore, until freshwater becomes prohibitively expensive and/or advanced technologies for wastewater treatment evolve and become more cost-effective, water recycling and/or reuse of ISR wastewaters in agriculture may remain at a minimum.

Regulatory Challenges in Recycling and Reuse of Wastewaters from Oil and Gas Production

Recycling of produced waters for use in hydraulic fracturing and other oil field operations is an area with public support, but it can be difficult to initiate depending on locality. In the United States, produced waters and drilling fluids are not considered hazardous wastes (CFR § 261.4(b)(5)), meaning that treatment of produced waters is not subject to federal regulation. Legal petitions requesting reconsideration of this exemption, however, wherein such treatment would be federally regulated, have caused concern by some operators and wastewater managers.[91] State and local regulations have greatly impacted the ability of operators to use brackish and reclaimed produced waters. In Texas, the Rail Road Commission permits recycling operations but has no authority over fresh or brackish water use.[92] By comparison, Pennsylvania has remarkably different rules depending on whether the operation takes place in the Ohio, Susquehanna, or Delaware River basins. Hydraulic fracturing is considered a consumptive water use by the Susquehanna River Basin Commission, requiring an operator to obtain a permit to transfer the flowback or produced water from another well site (interpreted as a water diversion) and a second permit to allow for its consumption during hydraulic fracturing. Additional oversight regarding produced water recycled in Pennsylvania is provided by the Department of Environmental Protection's Residual Waste Division, which requires a permit for recycling operations in addition to permits required by some of the Water Basin Commissions. Such potentially involved regulatory regimes tied together with an interest in updating and modifying permitting processes has made managing recycling operations challenging.

Regulation and the legal issues of liability and risk can pose a substantial challenge to the beneficial reuse of produced waters for irrigation. The novelty of using produced waters for irrigation often means that projects may not fit existing regulatory structures or permitting processes. Water law can also be complex in regions where water is scarce. Fears of liability can forestall projects even where technical challenges are relatively limited. In some cases, there may be benefit to having a specialized operator with both technical and regulatory expertise function as an intermediary between an energy company and agriculturalists.

The Future of Reuse and Recycling

Perceptions of any given project by all of the factions involved with differing goals (including the public) are important and likely to change through time. The details of the energy and agricultural industries are often not understood by outsiders, and it takes time to build understanding. One factor in favor of any proposed project is that using what might otherwise be wastewater to benefit agriculture fits within the ideas of efficiency and recycling that help define "green" technology. Although projects may initially be viewed with skepticism and may not accommodate all the demands of stakeholders on one side or the other of the water equation, early establishment and continued use of smaller scale test plots can be a crucial means of validating technology in the eyes of stakeholders, regulators, and the public.

Conclusions

Demands on water and energy supplies will only increase in the coming decades as the world population grows and economies in low to middle income countries continue to develop. Efficient use of water, through conservation, reuse and recycling, where possible, will be a key issue in the continuing development and implementation of established and prospective in situ extraction technologies. Only by continuing to improve the efficiency of water use in the extraction of petroleum and uranium resources and finding ways to re-purpose water supplies degraded by use in energy development will we be able to meet both the energy and water needs of future generations.

Disclaimer

Any use of trade, firm, or product names is for descriptive purposes only and does not imply endorsement by the US Government.

Acknowledgments

Gary Kitzrow, David Macfarlane, Blake Sanden, and John Zupancic contributed to this subsection through commentary and interviews. We appreciate the valuable insights of Valois Shea of the US EPA's Underground Injection Program and James Clay of Cameco Resources, Smith-Ranch Highland ISR site.

References

1. Coleman J. Tight-gas sandstone reservoirs: the 200-year path from unconventional to conventional gas resource and beyond. In: Carr T, D'Agostino T, Ambrose W, Pashin J, Rosen NC, editors. *Unconventional energy resources: making the unconventional conventional: 29th annual GCSSEPM foundation Bob F. Perkins research conference, Houston, Tex., 2009*; 2009. p. 31.
2. Ellsworth WL. Injection-induced earthquakes. *Science* 2013;**341**(6142).
3. Block LV, Wood CK, Yeck WL, King VM. The 24 January 2013 ML 4.4 earthquake near Paradox, Colorado, and its relation to deep well injection. *Seismol Res Lett* 2014;**85**(3):609–24.
4. Mudd GM. The future of yellowcake: a global assessment of uranium resources and mining. *Sci Total Environ* 2014;**472**:590–607.
5. Mantell ME. Deep shale natural gas and water use, part two: abundant, affordable, and still water efficient, water/energy sustainability symposium at the 2010 ground water protection council annual forum, Pittsburgh, PA; 2010.
6. Healy RW, Alley WM, Engle MA, McMahon PB, Bales JD. The energy-water nexus: an earth-science perspective, U.S. Geological Survey Circular 1407, in press.
7. Hoekstra AY, Chapagain AK, Aldaya MM, Mekonnen MM. *The water footprint assessment manual: setting the global standard.* p. 228. London; Washington, D.C.: Earthscan; 2011.
8. Schneider E, Carlsen B, Tavrides E, van der Hoeven C, Phathanapirom U. A top-down assessment of energy, water and land use in uranium mining, milling, and refining. *Energy Econ* 2013;**40**:911–26.
9. Dyni JR. *Geology and resources of some world oil-shale deposits.* U.S. Geological Survey Scientific Investigations Report 2005–5294; 2006. p. 42.
10. Bartis JT, LaTourrette T, Dixon L, Peterson DJ, Cecchine G. *Oil shale development in the United States: prospects and policy issues.* Santa Monica (CA): RAND Corporation; 2005. p. 68.
11. Birdwell JE, Mercier TJ, Johnson RC, Brownfield ME. *In-place oil shale resources examined by grade in the major basins of the Green River Formation, Colorado, Utah, and Wyoming.* U.S. Geological Survey Fact Sheet 2012–3145; 2013. p. 4.
12. Tissot BP, Welte DH. *Petroleum formation and occurrence.* pp. 148–184. Germany: Springer, Verlag; 1984.
13. Hanson JL, Limerick P. *What every westerner should know about oil shale: a guide to shale country.* University of Colorado at Boulder Center of the American West; 2009. Report #10. p. 58.
14. Mercier TJ. Calculation of overburden above the Mahogany zone in the Piceance Basin: U.S. Geological Survey Digital Data Series DDS-69-Y. In: Johnson RC, et al., editors. *Oil shale and nahcolite resources of the Piceance Basin, Colorado*; 2010.
15. Terres R, Omari W, Al-Najjar I. Progress in characterizing Jordan's in situ oil shale resource. In: *32nd Oil Shale Symposium at Colorado School of Mines, Golden, CO*; 2013.
16. Minster T, Sladek T. Oil shales potential and prospects in Israel – an update. In: *32nd Oil Shale Symposium at Colorado School of Mines, Golden, CO*; 2013.
17. Ryan RC, Fowler TD, Beer GL, Nair V. Shell's in situ conversion process – from laboratory to field pilots. In: Ogunsola OI, et al., editors. *Oil shale: a solution to the liquid fuel dilemma.* (Washington, D.C.): American Chemical Society; 2010. p. 161–84.
18. Birdwell JE, Mercier TJ, Johnson RC, Brownfield ME, Dietrich JD. *In-place oil shale resources in the saline-mineral and saline-leached intervals, Parachute Creek Member of the Green River Formation, Piceance Basin, Colorado, U.S.* Geological Survey Fact Sheet 2013–3115; 2014. p. 6.
19. Kenny JF, Barber NL, Hutson SS, Linsey KS, Lovelace JK, Maupin MA. *Estimated use of water in the United States in 2005.* U.S. Geological Survey Circular 1344; 2009. p. 52.

20. Gaffigan M, Mittal A. *Energy-water nexus: a better and coordinated understanding of water resources could help mitigate the impacts of potential oil shale development: report to Congressional Requesters, GAO-11-35*. 2010. p. 75.
21. Boak J. A second look at water use for in situ shale oil production. In: *32nd Oil Shale Symposium at Colorado School of Mines, Golden, CO*; 2013.
22. Ryan RC. Oil shale pyrolysis water treatability study. In: *29th Oil Shale Symposium at Colorado School of Mines, Golden, CO*; 2010.
23. Boak J, Mattson E. Water use for in situ production of oil from oil shale. In: *29th Oil Shale Symposium at Colorado School of Mines, Golden, CO*; 2010.
24. AMEC. Energy development water needs assessment phase II – final report prepared for the Colorado River Basin Roundtable and Yampa/White River Basin Roundtable. 2012. p. 196.
25. Gleick PH. Water and energy. *Annu Rev Energy Environ* 1994;**19**:267–99.
26. Shell Frontier Oil and Gas Inc. *Plan of operation – e-ICP test project*. Prepared for the Bureau of Land Management; 2006. p. 70.
27. Leenheer JA, Noyes TI, Stuber HA. Determination of polar organic solutes in oil-shale retort water. *Environ Sci Technol* 1982;**16**:714–23.
28. Stephenson M, Dobson KR, Greenfield PF, Bell PRF. Characterisation of process waters from Fischer assay retorting of Rundle oil shale. *Environ Technol Lett* 1982;**3**:241–6.
29. Hill R, Birdwell JE, Lewan MD. *Evaluation byproducts of environmental concern in water generated during hydrous pyrolysis of oil shale*. 2010. Geological Society of America Abstracts with Programs, 2010, p. 340.
30. Johnson RC, Mercier TJ, Brownfield ME. *Spatial and stratigraphic distribution of water in oil shale of the Green River Formation using Fischer assay, Piceance Basin, Northwestern Colorado*. 2010. Geological Society of America Abstracts with Programs, 2010, p. 339.
31. Amy GL, Hines AL, Thomas JF, Selleck RE. Groundwater leaching of organic pollutants from in situ retorted oil shale – a mass transfer analysis. *Environ Sci Technol* 1980;**14**:831–5.
32. Bethea RM, Barber DR, Parker HW, Guven N. Effect of groundwater and pH modification on quality of leachate from simulated in-situ retorted Utah oil shale. *Soc Pet Eng J* 1983;**23**:809–18.
33. Birdwell JE, Lewan MD. *Leaching of hazardous substances from retorted oil shale during high temperature water flushing*. 2010. Geological Society of America Abstracts with Programs, Denver, CO, 2010, p. 340.
34. King CW, Webber ME. Water intensity of transportation. *Environ Sci Technol* 2008;**42**:7866–72.
35. Haines SS, Cook TA, Thamke JN, Davis KW, Long AJ, Healy RW, et al. *A framework for assessing water and proppant use and flowback water extraction associated with development of continuous petroleum resources*. U.S. Geological Survey 2014–3010; 2014. Available at http://dx.doi.org/10.3133/fs20143010 [accessed 14.04.14], p. 6.
36. Soeder DJ, Kappel WM. *Water resources and natural gas production from the Marcellus Shale*. U.S. Geological Survey Fact Sheet 2009–3032; 2009. Available at: http://pubs.usgs.gov/fs/2009/3032/ [accessed 14.10.30], p. 6.
37. Airhart M. *The Barnett shale gas boom igniting a hunt for unconventional natural gas resources*. Geology.com; 2007. http://geology.com/research/barnett-shale-gas.shtml [accessed 14.04.14].
38. Andrews A, Folger P, Humphries M, Copeland C, Tiemann M, Meltz R, et al. *Unconventional gas shales: development, technology, and policy issues*. Congressional Research Service; 2009. Available at https://www.fas.org/sgp/crs/misc/R40894.pdf [accessed 14.04.14], p. 53.
39. Arthur JD, Bohm B, Layne M. *Hydraulic fracturing considerations for natural gas wells of the Marcellus shale*. Cincinnati (OH): Ground Water Protection Council Annual Forum; 2008.

40. Clark CE, Horner RM, Harto CB. Life cycle water consumption for shale gas and conventional natural gas. *Environ Sci Technol* 2013;**47**(20):11829–36.

41. Nicot JP, Scanlon BR. Water use for shale-gas production in Texas. *US Environ Sci Technol* 2012; **46**(6):3580–6.

42. Nicot JP, Scanlon BR, Reedy RC, Costley RA. Source and fate of hydraulic fracturing water in the Barnett shale: a historical perspective. *Environ Sci Technol* 2014;**48**(4):2464–71.

43. U.S. Energy Information Administration. *Annual energy outlook 2013 with projections to 2040. DOE/EIA-0383(2013).* (Washington, D.C.): Department of Energy; 2013. Available at http://www.eia.gov/forecasts/archive/aeo13/index.cfm [accessed 14.04.14].

44. U.S. Energy Information Administration. *U.S. crude oil and natural gas proved reserves 2011.* (Washington, D.C.): Department of Energy; 2013. Available at http://www.eia.gov/naturalgas/crudeoilreserves/pdf/uscrudeoil.pdf [accessed 14.04.14].

45. Belanger L. *Fracking our future: measuring water & community impacts from hydraulic fracturing.* Western Resource Advocates; 2012. Available at http://www.westernresourceadvocates.org/frackwater/fracking_our_future_july_2012.pdf [accessed 14.04.14], p. 27.

46. Colorado Division of Water Resources and the Colorado Water Conservation Board and the Colorado Oil and Gas Conservation Commission. *Water sources and demand for the hydraulic fracturing of oil and gas wells in Colorado from 2010 through 2015.* 2012. Available at http://cogcc.state.co.us/Library/Oil_and_Gas_Water_Sources_Fact_Sheet.pdf [accessed 14.04.14], p. 9.

47. Vengosh A, Jackson RB, Warner N, Darrah TH, Kondash A. A critical review of the risks to water resources from unconventional shale gas development and hydraulic fracturing in the United States. *Environ Sci Technol* 2014;**48**(15):8334–48.

48. Rivard C, Lavoie D, Lefebvre R, Séjourné S, Lamontagne C, Duchesne M. An overview of Canadian shale gas production and environmental concerns. *Int J Coal Geol* 2014;**126**:64–76.

49. Vidic RD, Brantley SL, Vandenbossche JM, Yoxtheimer D, Abad JD. Impact of shale gas development on regional water quality. *Science* 2013;**340**(6134).

50. Veil JA. *Water management technologies used by Marcellus shale gas producers. ANL/EVS/R-10/3.* (Washington, D.C.): prepared for the U.S. Department of Energy and the National Energy Technology Laboratory; 2010. Available at http://www.veilenvironmental.com/publications/pw/Water_Mgmt_in_Marcellus-final-jul10.pdf [accessed 14.04.14], p. 56.

51. Maloney KO, Yoxtheimer DA. Production and disposal of waste materials from gas and oil extraction from the Marcellus shale play. *Pa Environ Pract* 2012;**14**(04):278–87.

52. Raleigh CB, Healy JH, Bredehoeft JD. An experiment in earthquake control at Rangely, Colorado. *Science* 1976;**191**:1230–7.

53. Nicholson C, Wesson RL. Earthquake hazard associated with deep well injection – a report to the U.S. Environmental Protection Agency. *US Geol Surv Bull* 1990;**1951**:74.

54. National Research Council. *Induced seismicity potential in energy technologies.* p. 263. Washington, D.C.: The National Academies Press; 2013.

55. Horton S. Disposal of hydrofracking waste fluid by injection into subsurface aquifers triggers earthquake swarm in central Arkansas with potential for damaging earthquake. *Seismol Res Lett* 2012;**83**:250–60.

56. NETL. *Modern shale gas development in the United States: an update (DE-FG26–04NT15455).* (Washington, D.C.): U.S. Department of Energy, National Energy Technology Laboratory Strategic Center for Natural Gas and Oil; 2013. Available at http://www.netl.doe.gov/File%20Library/Research/Oil-Gas/shale-gas-primer-update-2013.pdf [accessed 02.05.14].

57. Osborn SG, Vengosh A, Warner NR, Jackson RB. Methane contamination of drinking water accompanying gas-well drilling and hydraulic fracturing. *Proc Natl Acad Sci USA* 2011;**108**(20):8172–6.

58. Jackson RB, Vengosh A, Darrah TH, Warner NR, Down A, Poreda RJ, et al. Increased stray gas abundance in a subset of drinking water wells near Marcellus shale gas extraction. *Proc Natl Acad Sci* 2013;**110**(28):11250–5.

59. World Nuclear Association. *In situ leach (ISL) mining of uranium.* 2014. http://www.world-nuclear.org/info/Nuclear-Fuel-Cycle/Mining-of-Uranium/In-Situ-Leach-Mining-of-Uranium/ [accessed 10.06.14].

60. McCammon RB. Uranium resources/reserves lecture: U.S. Geological Survey Open-File Report 97-701. 1997. p. 38.

61. Abzalov MZ. Review: sandstone-hosted uranium deposits amenable for exploitation by in situ leaching technologies. *Appl Earth Sci (Trans Inst Min Metall B)* 2012;**121**(2):55–64.

62. IAEA. In situ leaching of uranium: technical, environmental and economic aspects, IAEA-TECDOC-492. In: *Technical committee meeting organized by the International Atomic Energy Agency, International Atomic Energy Association: Vienna, Austria*; 1989. p. 168.

63. Finch WI. Uranium—fuel for nuclear energy 2002. *US Geol Surv Bull* 2003. 2179-A.

64. U.S. Energy Information Administration. *2012 domestic uranium production report.* 2013. Washington, D.C., p. 21.

65. Powertech Inc. *Dewey-Burdock project Fall River and Custer Counties, South Dakota large scale mine permit application.* Prepared for: South Dakota Department of Environment and Natural Resources; 2012. p. 505.

66 Mudd GM. Acid in situ leach uranium mining – 1. USA and Australia. *Environ Geol* 2001;**41**(3-4): 390–403.

67. U.S. Environmental Protection Agency. *Draft technical report: considerations related to post-closure monitoring of uranium in-situ leach/in-situ recovery (ISL/ISR) sites.* Washington, D.C.: U.S. Environmental Protection Agency; 2011. p. 127.

68. Stewart CL, Reimann LJ, Swapp SM. Mineralogic considerations for uranium in situ leach mining: a preliminary study of uranium and associated mineralogy of roll-front uranium deposits in Wyoming and Nebraska. *Cim Bull* 2000;**93**(1045):89–96.

69. Way S-C. Well-field mechanics for in situ recovery of uranium. *Southwest Hydrol* 2008;**7**(6):30–1.

70. Borch T, Roche N, Johnson TE. Determination of contaminant levels and remediation efficacy in groundwater at a former in situ recovery uranium mine. *J Environ Monit* 2012;**14**(7):1814–23.

71. U.S. Nuclear Regulatory Commission. *Environmental impact statement for the Nichols Ranch ISR project in Campbell and Johnson Counties, Wyoming supplement to the generic environmental impact statement for in-situ leach uranium milling facilities: final report.* Washington, D.C.: U.S. Nuclear Regulatory Commission; 2011. p. 749.

72. U.S. Nuclear Regulatory Commission. *Environmental impact statement for the Moore Ranch ISR project in Campbell county, Wyoming supplement to the generic environmental impact statement for in-situ leach uranium milling facilities: final report.* Washington, D.C.: U.S. Nuclear Regulatory, Commission; 2010. p. 570.

73. Grant DC. *In-situ leaching studies of uranium ores – phase V. 80–9E3-INSIT-R3.* 1980. Salt Lake City, UT, p. 591.

74. Davis JA, Curtis GP. *Consideration of geochemical issues in groundwater restoration at uranium in-situ leach mining facilities. NUREG/CR-6870.* Washington, D.C.: U.S. Nuclear Regulatory Commission. Division of Fuel, Engineering, and Radiological Research & U.S. Geological Survey; 2007.

75. Montgomery AH. Adapting uranium in situ mining technology for new commercial applications. In: IAEA, editor. *In situ leaching of uranium: technical, environmental and economic aspects, IAEA-TECDOC-492 technical committee meeting organized by the International Atomic Energy Agency.* Vienna (Austria): International Atomic Energy Association; 1989. p. 75–96.

76. U.S. Nuclear Regulatory Commission. *Environmental impact statement for the Lost Creek ISR project in Sweetwater County, Wyoming supplement to the generic environmental impact statement for in-situ leach uranium milling facilities: final report.* Washington, D.C.: U.S. Nuclear Regulatory Commission; 2011. p. 735.
77. U.S. Nuclear Regulatory Commission. *Environmental impact statement for the Ross ISR project in Crook County, WY: supplement to the generic environmental impact statement for in-situ leach uranium milling facilities. NUREG-1910, Supplement 5.* Washington, D.C.: U.S. Nuclear Regulatory Commission; 2014. p. 805.
78. U.S. Nuclear Regulatory Commission. *Final environmental impact statement to construct and operate the Crownpoint uranium solution mining project, Crownpoint, NM. NUREG-1508/BLM NM-010-93-02/BIA EIS-92-001.* Washington, D.C.: U.S. Nuclear Regulatory Commission; 1997. p. 452.
79. Uranium Energy Corp. *Goliad project: draft environmental assessment for a proposed in situ uranium mining facility – license application No. R06064.* 2011.
80. COGEMA Mining Inc., Petrotek Engineering Corporation. *Wellfield restoration report: Christensen Ranch Project, Wyoming. Mills, WY.* 2008. p. 206.
81. U.S. Environmental Protection Agency. National primary drinking water regulations. In: *40 CFR Parts 9, 141, and 142*, vol. 65. Fed. Regist.; 2000.
82. Breit G, Otton J. *Produced waters database.* U.S. Geological Survey; 2002. http://energy.cr.usgs.gov/prov/prodwat/data.htm [accessed 04.04.14].
83. Blondes M, Gans K, Thordsen J, Reidy M, Thomas B, Engle M, et al. *U.S. Geological Survey national produced waters geochemical database v2.0.* 2014. http://energy.usgs.gov/EnvironmentalAspects/EnvironmentalAspectsofEnergyProductionandUse/ProducedWaters.aspx#3822349-data [accessed 18.04.14].
84. Hu Y, Schmidhalter U. Drought and salinity: a comparison of their effects on mineral nutrition of plants. *J Plant Nutr Soil Sci* 2005;**168**(4):541–9.
85. Ayers R, Westcot D. *Water quality for agriculture: FAO irrigation and drainage paper no. 29 Rev. 1.* Rome (Italy): Food and Agriculture Organization of the United Nations; 1985.
86. Ayars J, Hoffman G, Corwin D. Leaching and rootzone salinity control. In: Wallendar W, Tanji K, editors. *Agricultural salinity assessment and management.* 2nd ed. Reston (VA): ASCE; 2012. p. 371–403.
87. Sanden B, Ansolabehere D, Le H. In: Wickramasinghe R, Waskom R, editors. *Conjunctive use of oilfield produced water for irrigation in the Southern San Joaquin Valley of California – California case study. Produced waters workshop: Colorado Water Institute Information Series No. 102.* Fort Collins (CO): Colorado State University; 2006. p. 81–6. Available at http://www.cwi.colostate.edu/publications/is/102.pdf [accessed 02.04.14].
88. Zupancic J, Roenbaugh J, Mull G, Nicholson C. System and method for dispersing of coal bed sodic water. BeneTerra, LLC (Pratt, Kansas). 2008. Patent: 7,320,559.
89. Bern CR, Breit GN, Healy RW, Zupancic JW. Deep subsurface drip irrigation using coal-bed sodic water: part II. geochemistry. *Agric Water Manag* 2013;**118**:135–49.
90. Santos Ltd. *Coal seam gas water.* 2014. http://www.santos.com/coal-seam-gas/coal-seam-gas-water.aspx [accessed 06.06.14].
91. Romo CR, Janoe JS. Regulatory regimes for recycling produced and frac flowback water, paper 2012-A-453-AWMA. Air & Waste Management Association, 2012. Available at SSRN: http://ssrn.com/abstract=2124696 or http://dx.doi.org/10.2139/ssrn.2124696 [accessed 05.05.14], p. 12.
92. Bath CR. A commentary on Texas water law and policy. *Nat Resour J* 1999;**39**:121–8.

8

Water Management in Unconventional Oil and Gas Development—The Issues and Their Optimization

Neha Mehta[1], Francis O' Sullivan[2]

[1]CIVIL AND ENVIRONMENTAL ENGINEERING, MASSACHUSETTS INSTITUTE OF TECHNOLOGY, CAMBRIDGE, MA, USA; [2]MIT ENERGY INITIATIVE, MASSACHUSETTS INSTITUTE OF TECHNOLOGY, CAMBRIDGE, MA, USA

CHAPTER OUTLINE

Food, Energy, and Water. http://dx.doi.org/10.1016/B978-0-12-800211-7.00008-9

Introduction

The US natural gas industry, and by extension the industry globally has witnessed tremendous change over the past decade. During this period, the US natural gas production levels have risen from a 20-year low of 18 Tcf[a] (510 Bcm[b]) in 2005, to an all-time high of 24 Tcf (680 Bcm) in 2012[c]. At the same time, natural gas prices have fallen to levels not seen since the period immediately following the US gas market deregulation in the mid-1990s. The underlying driver of these dynamics has been the very rapid growth in the production of unconventional natural gas resources, and in particular shale gas resources—historically considered unrecoverable.

Technical advances in the areas of drilling and reservoir stimulation have been the key to unlocking shale gas. Today's shale gas is largely sourced from wells drilled with horizontal bores that have been subjected to large-scale hydraulic fracture stimulation. The combined efficacy of these technologies in enabling gas production from shale formations (and indeed other low-permeability reservoir settings) is such that their development has led to enormous upward revisions in the assessed scale of the total recoverable natural gas resource in the United States. Today, analysis by organizations including the U.S. Energy Information Administration and the Potential Gas Committee suggest that the likely recoverable US shale gas resource is in the 800–1000 Tcf (23–28 Tcm) range. This contrasts with the National Petroleum Council 2003, assessment that estimated the shale resource at 35 Tcf (1 Tcm).

The prospect of a much larger and indeed lower-cost domestic natural gas resource in the US is having major impacts on both the US and international energy sectors. In the United States, unsurprisingly, many are now projecting a more gas-centric future than was envisioned even a few years ago. For example, in the power generation sector, gas-fired generation is now expected to make up a much higher proportion of total US output over the coming 20–30 years than was anticipated before the full extent of the so-called shale gas revolution became apparent. On the international front, the impacts of shale gas on the global energy systems are likely to be significant. Already, shale gas has essentially removed the United States from the global market for Liquefied Natural Gas (LNG) imports. LNG is liquid form of natural gas that is formed by cooling natural gas to −260 F. Under this liquefied state, natural gas can easily stored and shipped. There are now a range of proposed projects at various levels of maturity planning to export LNG from the United States. How these developments ultimately alter the shape of global gas markets remains to be seen; however, it is undeniable that US shale gas is leading to a shift in both global gas market structures and altering energy-related geopolitical balances.

[a]Tcf stands for US trillion cubic feet.
[b]Bcm stands for US billion cubic meters.
[c]Marketed production as reported by U.S. Energy Information Administration, June 2013.

Hydraulic Fracturing—A Key Enabler of Unconventional Gas Production

The shale resource is really a collection of many hydrocarbon-prone mud rock formations with a diverse set of geological, geomechanical, geochemical, and petrophysical characteristics. One unifying feature of these formations is that they generally do not produce gas at economically feasible rates unless artificially stimulated, owing to very low rock matrix permeability.[1] Hydraulic fracturing of these formations results in a significant enhancement in well productivity, and today it is a ubiquitous reservoir stimulation technique across shale and other low and not so low-permeability formations.

Although certainly synonymous with shale gas development, the process of hydraulic fracturing has a much wider utility. Its use is widespread across the oil and gas sectors and its adoption predates large-scale shale gas production by decades. Some specific nonshale applications of hydraulic fracturing include production acceleration from conventional formations, and the management of skin and near-wellbore damage issues. Regardless of what type of formation is being stimulated, the fundamental processes involved in hydraulic fracturing remains the same. A fluid is injected (water is the commonly used) into the wellbore at a sufficient rate to generate a pressure differential between the wellbore and the reservoir. This causes stresses around the wellbore to increase beyond the tensile strength of the rock, at which point it splits or "fractures." Assuming a sufficient pumping rate is maintained, it is possible to "grow" these fractures both in terms of width and length, such that they propagate away from the wellbore into the surrounding formation. Generally speaking, once pumping ceases the induced fractures would close. So in order to keep fractures open, a proppant material must be placed in the fractures. This is achieved by transporting the proppant down hole as part of the slurry comprising of the proppant and the fracturing fluid itself.

Environmental Impacts of Hydraulic Fracturing

The widespread use of hydraulic fracturing technology to extract hydrocarbons from shale formations, has resulted in a number of potentially harmful environmental and public health consequences. One of the key concerns is the procurement of water for fracturing a well. Wells are often not located adjacent to water resources, thus the process water is either piped or trucked to the well site. The increasing cost of transportation of water to well site has made the local water withdrawals as the first choice for natural gas producers to meet their water demand. Although, water withdrawals in major shale plays represent a relatively minor volume of the total water used in the specific area; the problem is that large volumes of water are procured over a relatively short period of time, which could compete with the availability of water for other local users and create a transient stress on the water source. Furthermore, rapid and concentrated procurement of water could result in regional shortages and altered flow regime, impacting the habitat of aquatic biota.[2]

A related concern is how to manage the postfracturing produced fluids (aka flowback water). These fluids are highly contaminated and can be expected to contain high levels of salts, scaling ions, oil and grease, and other organics, naturally occurring radioactive materials (NORMs), etc. The management of these fluids on surface present handling risks, including surface spillage from vehicle accidents and potential contamination of freshwater source.[3,4]

Finally, the potential impacts of fracking activities on the air quality can pose hazardous health consequences of the residents. Assessment of potential air quality degradation from fracking will require a separate chapter in itself. Therefore, the sources of air pollution are given below:

- Fuel emissions form trucks is a major source of air pollution. Large number of trucks are commonly seen on a well pad for carrying fluid, piping, etc., back and forth. It its estimated that for completing an entire fracturing operation on a well pad, requires 600–1000 truck trips.[5]
- The storage of postfracturing fluids on-site in open pits can result in release of volatile organic compounds in the air. Inhalation of some of these organic compounds can lead to respiratory problems.
- Pipeline leaks can result in the release of fugitive methane, which is at least 20 times more potent greenhouse gas than CO_2 over a 100-year period.

To address these environmental challenges, various steps are being taken jointly by industry and regulatory agencies. On a regulatory level, Environmental Protection Agency (EPA) is currently studying the impacts of fracturing operations on water life cycle. The study is targeted to be completed shortly and will be critical for developing a regulatory framework for hydraulic fracturing activities. On the industry front, there is an increasing focus on developing fracturing additives that would enable fracturing operations to be performed using high salinity water. Overcoming the limited salt tolerance of these chemicals would allow use of high salinity water sources. This could dramatically reduce demand on freshwater resources, particularly in regions where such resources are limited.[6] Simultaneously, there has been an increasing emphasis on recycling or reusing flowback water to fracture subsequent fracturing processes. This does help reduce freshwater demand, however, economics and the regulatory environment will play an important role in determining the extent to which flowback water reuse and recycling occurs. In summary, the combined and coordinated efforts of the industry and regulatory agencies are essential to execute fracking operations in a sustainable fashion.

Fracture Fluid Systems

For a typical fracturing treatment to be successful, the fracture fluid should have the following characteristics:

- It should be able to effectively transport the proppant particles between the induced fractures—This is necessary in order to keep the fracture open after the treatment is completed.

- It should be compatible with the formation and reservoir fluid chemistries.
- It should have a low fluid loss in order to avoid low "leak off" of fluid into the formation, thereby displacing hydrocarbons.
- It should have low flow resistance.
- It should be able to clean up the fluid from the fractures so that there is no obstruction to the flow of hydrocarbons.
- It should be inexpensive.

To perform abovementioned functions, the composition of fracture fluid has evolved significantly, ranging from simple water and sand mixtures to complex polymeric compounds and hydrocarbon mixtures. In the early stages of hydraulic fracturing development, oil was a preferred base fluid due to its perceived minimal damaging effects on the hydrocarbon formations. However, oil-based fluids are not only expensive, but also pose significant environmental and safety issues. In fact, the use of diesel for fracturing is regulated under the Safe Drinking Water Act. Acid-based fluids fracture the rock by etching mechanism, thereby they do not require proppant. Acid fracture systems are associated with high fluid loss and used mostly in conjunction with fluid loss additives.[7] A major application of oil- and acid-based fluids is in high-temperature and water-sensitive formations. The majority of the formations uses water-based preparation, mainly due to its' low cost, ease of handling, expert execution, and widespread accessibility. There is also occasional use of waterless fluid system approaches such as gas-based fluids, which uses propane-based liquefied petroleum gas (LPG) as base fluid instead of water.[8]

Finally, the choice of a fracture fluid is dominated by the nature of the reservoir being treated. A brief description of the chemistry of commonly used fracture fluid systems is discussed below. These systems mainly include slickwater fluids, gelled fluids, viscoelastic fluids, energized fluids, and nonaqueous fluids.

Gelled Fluids

These fluids consist of a water-soluble polymer compound to increase the viscosity of the base fluid at borehole temperatures. The earliest water-soluble polymer to be used in water-based fluids was guar gum or galactomannan. Guar Gum is a naturally occurring polysaccharide composed of galactose and mannose units. Mannose (M) units are joined together via β-1,4-glycosidic linkages whereas galactose (G) units are bonded to mannose via α-1,4-glycosidic linkages.[9] The ratio of M/G is critical to the solubility of the polymer in water. High ratio implies low solubility in water. For instance, guar derived from locust bean has low galactose content and is sparingly soluble in water.[10] The performance of natural guar is sensitive to pH, temperature, shear, and salinity.[9] Chemically modified guar derivatives such as hydroxypropyl guar and carboxymethyl hydroxypropyl guar (CMHPG) exhibit improved performance and chemical and thermal stability in fracturing operations. Fluids using low molecular weight guar and its derivatives are also referred to as "linear gels".

The viscosity of the gelled fluids is significantly influenced by temperature and shear. Guar gels degrade at high temperatures resulting in premature reduction of viscosity.

Early applications used higher polymer concentrations in an effort to offset this viscosity reduction. However, increased polymer loading can result in a reduced fracture conductivity. Fracture conductivity is a measure of how conductive the fracture is compared to the formation. To improve fluid performance at lower polymer loading, crosslinking agents can be used. These fluid systems are also referred as "crosslinked gels". Such gels exhibits a stable viscosity at elevated temperature as well as increased viscosity at low polymer loading.[7] Salts of metal ions such as borate, zirconium, titanium, and aluminum are commonly used as cross-linkers.[11] These salts are sensitive to pH and selective for a specific polymer. The metal ions combine with the reactive group (in case of CMHPG it is the carboxyl group) present in the polymer and produce a dense network of cross-linked chains.

Fracture fluid experiences a wide shear variation during injection. The high shear rates experienced by the fluid occur when the fluid is being pumped through the wellbore tubing. Once the fluid enters the fracture, the shear rate of the fluid is substantially reduced.[12] Gelled fluids are known to exhibit shear-thinning properties, which mean that the viscosity of the fluid decreases with increasing shear stress.[9,13,14] As a result, gelled fluids show a low viscosity in the wellbore tubing. However, in the low-stress region, the fluid may recover its viscosity depending on the relaxation time. Relaxation time is the characteristics of a polymer matrix and is a measure of the time required for the polymer molecule to relax and regain its viscosity.[15] It is a function of the gel network structure, shear rates, etc. In case, where the viscosity is partially recovered under low-stress regimes, proppant transport can be compromised. The use of borate cross-linkers enhances the shear tolerance of the fluid, however, zirconium-based cross-linkers have good temperature stability, but low-shear tolerance.[12,16]

Slickwater Fluids

In recent years, slickwater fluids have become a standard technique in fracture stimulations of the US shale plays. It is estimated that slickwater fluids comprise more than 30% of the US fracturing stimulations in 2004.[17,18] Such fluids are low-viscosity fluids and mainly composed of water and a friction reducer. Due to low viscosity, the proppant suspension properties of these fluids are inferior to gelled fluids. Insufficient suspension of proppant particles in the fluid can result in premature dropping out of proppant from the fluid (also referred to as screenout) and settling of the proppant particles near the wellbore. As a consequence of the inefficient proppant transport, well productivity can be severely impaired. To compensate for these concerns, slickwater fluids are pumped at high injection rate so that the turbulence generated by the use of higher injection rates can carry the proppant effectively in the fracture with minimal proppant screening and settling.

Injection rates, such as 60–100 barrels per minute are very common in areas stimulated using slickwater fluids.[17] The turbulent motion under these flow rates results in significant frictional losses. In order to minimize these losses, friction reducers are added to the slickwater fluids. Friction-reducing agents minimize the pressure head loss in tubing by converting the flow regime from turbulent to laminar flow.[19] They are described in more detail later in the chapter.

A modification of slickwater fracture fluids is hybrid fluid systems. Hybrid fluids are a combination of slickwater fluids and gelled fluids. The fluids are mainly composed of a low loading of the linear guar gel, friction reducer, and water. The reservoir is first fractured using slickwater fluids to maximize fracture complexity, followed by a fracturing with a gel to increase fracture aperture and improve proppant placement. The advantages of using hybrid fluids include, low cost of fracture treatment, less water consumption, enhanced fracture conductivity, and improved proppant transport.

Viscoelastic Surfactants

These fluids are polymer-free systems and use inorganic compounds such of quaternary ammonium salts to achieve a sufficient viscosity for transporting proppant into the fracture.[20] These surfactants are composed of lyophilic and lyophobic parts. Upon addition to the water, lyophilic parts align itself in contact with the base fluid and lyophobic parts are appended away from the base fluid. Such a structure is also called as micelles and their aggregated growth is responsible for imparting viscosity to the base fluid.[7] The shear tolerance of these fluids is better than the gelled fluids. These fluids are not sensitive to shear because of the elasticity of the micelles structure. The micelles quickly combine and recover back after the shear stress ceases. The application of surfactant systems is currently limited to deep and high-temperature reservoirs.[21]

Nonaqueous Fluid Systems

Still in the early stage of development, waterless fracture systems are being explored as a method to address the increasing water demand of the fracturing operations and improve fracture performance. Gasfrac Energy Services, a Canadian company has developed a technology for using gelled propane-based LPG fracture fluid system as a substitute for water-based fluid systems.[8] The fluid is composed of around 90% propane and 10% gelling agents.

There are several benefits, which are offered by a propane-based fluid. Firstly, it has a favorable rheology to achieve efficient proppant transport and generate longer and complex fractures, thereby enhancing the reservoir performance. Secondly, there is no water used in the fluid. This implies that all the environmental concerns related to water and fracking wastewater are mitigated. Lastly, unlike water-based systems where only a fraction of injected water is recovered to the surface, propane-based fluid result in a high percentage of the fracture fluid. The reason for this high recovery is that once the gas gel is broken down, propane flashes and mixes with reservoir hydrocarbon and is recovered at the surface with the reservoir gas flow.

Despite numerous benefits, there are certain safety issues, which could limit the widespread use of the gas fluids in fracking operations. Propane is an explosive chemical and requires stringent safety protocols to be followed at the well site. Furthermore, the water savings so generated by the use of waterless systems can in turn be less significant if one takes into account the amount of water used to produce and liquefy propane. A thorough life cycle analysis of waterless fluids can provide a better picture about potential water savings it can generate. Nevertheless, a number of operators in Canada have successfully utilized propane-based fluids for fracking the rocks and is being considered a promising alternative to water-based fracture fluids in the United States.

Energized Fluids

Energized fluids are multiphase fluids and composed of either nitrogen (N_2) or carbon dioxide (CO_2). Nitrogen is a chemically inert gas and immiscible with formation fluids, whereas carbon dioxide is reactive and adversely impact the additive performance. This type of fluid not only minimizes the volume of water injected into the well but also enhances well performance by efficient transport of proppants. These systems are characterized by their quality, which is the ratio of volume percent of energized medium to the total fluid volume.[7] Quality of systems above 53% is referred to as foams. Foam is a stable mixture of liquid in gas. Foamed fluids are very well suited for formations with marginal water compatibility.

Fracture Fluid Additives

Fracture fluid additives are chemicals mixed with the base fracture fluid to maintain fluid integrity in the subsurface environment. For most water-based fluids, the additives comprise about 1% or less of the fluid by volume. However, given the large volume of fluid used in hydraulic fracturing treatments, a 1% additives' volumetric concentration on an absolute scale can represent a very significant volume. The nature of the fracture fluid additives varies based on the type of gelling agents used as well as geochemistry of the formation being treated.

An increasing number of states require operators to disclose fracture additives being used, with Wyoming being the first state to implement it. Depending on the states, the exemptions in the state regulations defer, but the most important exemptions granted across the states is the exclusion of proprietary additives from the disclosure. These disclosures are publicly reported either on a national hydraulic fracturing chemical registry, managed by the Ground Water Protection Council and the Interstate Oil and Gas Compact Commission www.fracfocus.org or state Web site. It is not possible here to document an exhaustive list of additives, but commonly used additives and their functions are noted in Table 1.[5]

Breakers

After proppant particles are successfully placed into the fracture, the gelled fracturing fluid must be broken down (or degraded). This is necessary as otherwise the gel would

Table 1 Commonly Used Additives in Fracture Fluid

Additive Type	Typical Chemical Compound	Function
Breaker	Ammonium persulfate	Allows a delayed breakdown of the polymer chains
Friction reducer	Polyacrylamide/mineral oil	Minimizes friction between the fluid and pipe
Corrosion inhibitor	N,N, dimethyl formamide	Prevents the corrosion of pipe
Scale inhibitor	Ethylene glycol	Prevents scale deposits in the pipe
Biocides	Glutaraldehyde	Eliminates bacteria in the water that produce corrosive by-products
Cross-linker	Borate salts	Maintains fluid viscosity as temperature increases
Iron control	Citric acid	Prevents precipitation of metal oxides
Clay stabilizer	Potassium chloride	Prevents clay from swelling in presence of water

reduce the conductivity of the induced fracture and ultimately reduce gas production. Breaker compounds are used to achieve this. Ammonium persulfate is a commonly used breaker due to its strong oxidizing potential. The gel degradation is a free radical mechanism. Figure 1 shows a schematic of degradation of guar gel using ammonium persulfate.[22,23]

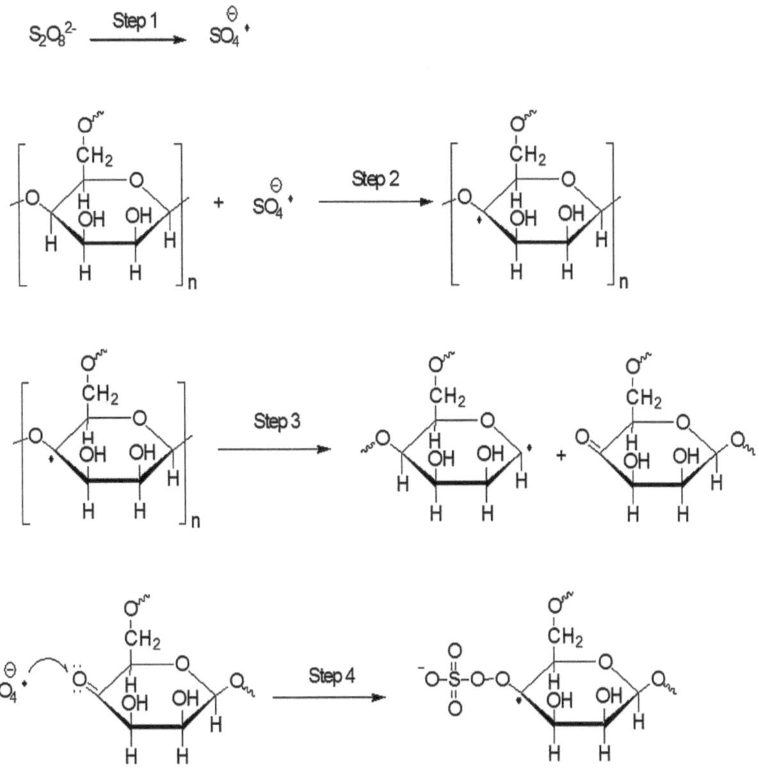

FIGURE 1 A schematic of gel degradation using ammonium persulfate.

There are two locations where the gel degradation can take place in a guar chain: (1) the main mannose chain and (2) the side galactose chain. Ammonium persulfate can attack at both locations, but an attack on the main chain results in greater viscosity reduction than a side chain attack. As shown in Figure 1, the free radical activity can be inhibited by some of the ketonic fragments formed during the reaction. The limited free radical activity negatively impacts the extent of the degradation reaction and results in formation damage as previously mentioned.

At low temperatures, reaction rates are low and gel degradation is very slow. To enhance the rate of reaction at low temperatures initiators such as tertiary amines, organic esters, and transition metal salts are often added in the fluid mixture.[23] In contrast, in high-temperature reservoirs, the reaction rate can be very fast and uncontrolled, resulting in premature thinning of the gelled fluid. In such cases, breakers are encapsulated in a semipermeable coating for slow release into the solution. Examples of such breaker formulations include rosin or paraffin wax coats and elastomeric sulfonates coating.[24]

Enzyme-based breakers are also used to degrade the gels. Enzymes are selective for a specific polymer gel and degrade the gel by the same mechanism as chemical breakers. However, enzymes are not consumed in the reaction like the chemical breakers and can achieve equivalent performance to their chemical counterparts at lower concentrations.[25] Widespread application of enzyme breakers is often constrained due to high cost and high sensitivity to process parameters such as pH and temperature.

Biocide

High reservoir temperatures and an abundance of biodegradable polymeric gel means microbial growth can take place within a fracture fluid. This microbial growth can rapidly degrade the polymeric gel, and this can result in loss of fluid viscosity in only a few hours. In addition, acid-producing bacteria and sulfate-reducing bacteria can cause localized corrosion in the equipment. The latter can also be responsible for well souring and iron sulfide precipitation. Biocides are mixed into the injected fluid to prevent growth of bacteria in the wellbore environment. Biocides are regulated by the EPA under the Federal Insecticide, Fungicide, and Rodenticide Act. The most common biocides are typically either oxidizing or nonoxidizing in nature. Oxidizing biocides are faster than nonoxidizing biocides, but faster activity could result in interaction of biocide with other components of the fracture fluid, which could compromise the fluid stability.[26] Nonoxidizing biocides in contrast, are generally passive to other fluid components, but are intrinsically hazardous and toxic, thus requiring special handling.

Commonly adjust spacing used biocides include glutaraldehyde, DBNPA (2, 2, dibromo-3-nitrilopropionamide), BNPD (bromonitropropane-1, 3 diol), THPS (tetrakishydroxymethylphosphonium sulfate), Isothiazolone.[7] Most of these biocides function effectively within a wide pH range; however a few (THPS, DBNPA) may show limited stability at high pH.

Friction Reducer

As mentioned earlier, friction reducer is one of the key components of the slickwater fluids. Partially hydrolyzed polyacrylamide (PAM) is commonly used as a friction-reducing agent. The mechanism of friction reduction by polymers is postulated by several theories. Two prominent theories are proposed by Lumley and De Gennes.[27] According to Lumley, the stretching of polymeric chains outside the viscous sublayer can induce a large increase in the effective viscosity near the wall (because the extensional strain rates are highest at the pipe fluid interface), which suppresses the turbulent fluctuations and thereby reduces the frictional losses.[28] On the other hand, De Gennes postulated that when the polymer chains are moderately stretched they do not produce a measurable change in viscosity and in fact, it is the elastic behavior of the stretched polymeric molecules, which absorbs the turbulent kinetic energy leading to reduction in turbulent friction.[29] Both these theories constitute two fundamental concepts in the area of polymeric friction reduction.

Regardless of the mechanism of friction reduction, the degree of hydration of PAM is a decisive factor in determining its effectiveness as a friction reducer. The rate of hydration is sensitive to fluid temperature and concentration of divalent ions. Polyacrylamide gel undergoes a phenomenon known as "syneresis" in the presence of high concentration of scaling ions. Syneresis is a process where the polyacrylamide chains excessively hydrolyze (causing precipitation) in solution to carboxylate polyions, resulting in the gel collapse (see Figure 2). The degree of hydrolysis is dependent on pH, temperature, and divalent ion concentration.[30–32] In the absence of scaling ions, syneresis occurs at a temperature of approximately 200 F and greater.[32] High concentration of scaling ions can result in increased hydrolysis sufficient to precipitate the polymer at lower temperatures.[33] Due to these factors the source water quality must be carefully controlled in order to limit the precipitation of the polymer.

Over time, various alternative friction-reducing formulations have been commercialized, which have higher salt and ionic tolerance, improved stability at high temperature, and easier handling and storage characteristics. For example, aqueous dispersions containing either organic solvents or surfactants are fast dissolving to reach the desired activity[34]; copolymers of acrylamide and quaternary salts are effective in freshwater as well as high salinity brines.[35]

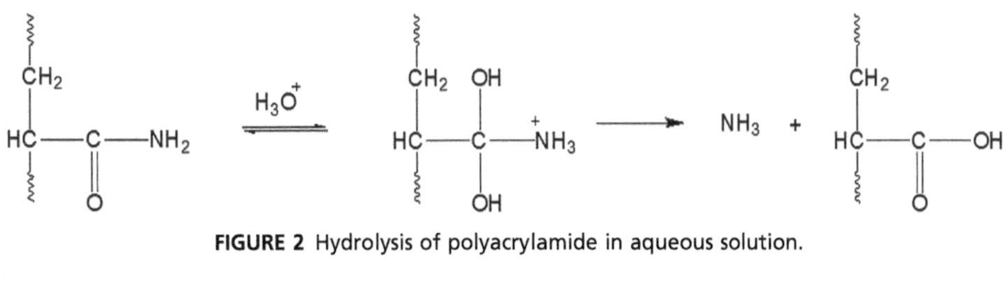

FIGURE 2 Hydrolysis of polyacrylamide in aqueous solution.

Scale Inhibitors

Mineral scaling is a major issue for hydraulic fracturing operations. The formation of scales downhole can cause reduction in formation permeability, block pipes, and have other negative consequences. To avoid scale development, scale inhibitors are used in fracture fluid. Some of the scale inhibitors are not directly mixed with the fracture fluid for several reasons.[36] Firstly, scale inhibitors are pH sensitive. The different additives used in fracture fluid can result in large swings in pH levels of the fluid, thereby compromising the effectiveness of the scale inhibitor. Secondly, scale inhibitors can quickly destroy cross-linkings, resulting in diminished proppant carrying capacity. To overcome these issues, those scale inhibitors are preferred whose optimal functioning pH range is similar to that of the cross-linking agent's functioning pH. Commonly used scale inhibitors include polyphosphates, phosphate esters, polyacrylic acid derivate, and chelating agents such as EDTA. More recently, scaling inhibitors have been encapsulated in a material coating for slow release in the solution in order to provide improved control over the reaction rate as well as to minimize the effect of varying pH levels.[6,37]

Understanding Flowback Water

In the days immediately following a hydraulic fracturing treatment, large volumes of the fluid flow from a well. This fluid is generically referred to as "flow back" or "flowback water" and the volumes involved can often be 10–20% of the originally injected volume.[38,39] The chemical constituents of flowback water are highly dependent on various water–rock interactions, the chemicals used in the fracture fluid and the point during the flowback period when the fluid is sampled. No typical chemical profile for flowback water exists, however, it can generally be expected to contain elevated levels of salts, scaling ions, oil and grease, and other organics, NORM, and derivative compounds of those used as additives in the originally injected fluid.[5,39–41] Because of the large volume of flowback water that can be produced from individual wells, and the high concentration of solids and other pollutants that can be contained in the flowback water, there are very significant environmental concerns surrounding flowback water management, treatment, and ultimate disposal.

A thorough understanding of the chemical and physical composition of flowback water is fundamental to predicting the environmental impacts that may arise from the mismanagement of the flowback water. An exemplar chemical profile of flowback water of the Marcellus shale region is shown in Table 2.[42] The analytical characterization of flowback water quality parameters is a challenging job and requires insight into how different chemical interferences can limit the accuracy of conventional testing and analysis methods. For instance, from Table 2, it can be seen that the chemical oxygen demand (COD) values of the flowback water increase drastically with time whereas the biochemical oxygen demand (BOD) decreases. This dynamic implies increased toxicity with time. Nevertheless, this conclusion can be incorrect since large COD values are representative of not only organic pollutants, but also high concentrations of inorganic oxidizable

Table 2 An Example of Flowback Water Quality

Parameters	Fracture Fluid (mg/L)	Flow Back (mg/L)	
		Day 5	Day 14
pH	7.2	6.6	6.2
Total alkalinity	–	138	85.2
Total suspended solids	–	99	209
Total hardness as $CaCO_3$	130	17,700	34,000
Total dissolved solids	735	67,300	120,000
Total organic carbon	226	62.8	38.7
Chemical oxygen demand (COD)	1730	4870	8530
Biochemical oxygen demand (BOD)	<2–2220	144	39.8
Oil and gas	–	6.3	ND
Calcium	–	4950	ND
Barium	–	686	ND
Strontium	–	1080	ND
Sulfate	–	–	–

pollutants, especially chloride.[43] Since the conventional COD testing procedure is restricted in its scope to prevent interference from high concentration of chloride and other inorganic oxidants, it is essential to properly mask any inorganic species before measuring the COD of a flowback water sample.

Despite the analytical challenges, some general analysis is presented here to develop an understanding of flowback constituents. From Table 2 it is seen that the flowback water is mildly acidic with moderate alkalinity.[42] Low alkalinity coupled with a high degree of hardness observed in flowback water implies that the majority of hardness is caused by the noncarbonate salts of divalent ions, such as calcium, barium, strontium, and magnesium. Such noncarbonate scales are difficult to remove and can be very abrasive to equipment surfaces. Among the noncarbonate scales, $BaSO_4$ is of special importance in flowback water as it has the lowest solubility product among sulfate salts of divalent ions and is among the first scale to precipitate in the formation. The flowback water is supersaturated with respect to barium with an ionic product of $BaSO_4$ being on the order of 10^{-3}. This is 100,000-fold higher than the solubility product of barium sulfate ($1.05*10^{-10}$ at 25 C) assuming activities in solution for both ions to be unity.[44] The apparent activity coefficients are likely much lower than 1 due to possible complex-forming reactions of barium with low molecular weight acids, such as aliphatic acids, dicarboxylic acids, aromatic acids, and cyclic acids.[45,46] Such reactions are more common in low salinity flow backs that contain high concentrations of organic matter.[47] In addition, barite solubility increases with temperature, pressure, and salinity. These factors substantially increase the dissolution of barium in flowback water.[48,49] Flowback water with elevated concentrations of barium also contains elevated concentrations of strontium and radium in the form of Ba–Sr–Ra complexes.[50] These interactions further decrease the activity of barium in solution and lead to increased leaching of barium from formation rocks into flowback water.

Suspended solids in flowback water are colloidal particles with sizes varying between 1 and 10 μm.[51] There is no distinct composition of suspended solids. Some of XRD studies have found barite to be the primary constituent of suspended solids in the Marcellus region.[41] On the other hand, analytical characterization of suspended solids in the Daqing oil fields in China has found that the major suspended solid constituents were organics, iron, and barium.[52] The presence of suspended solids in flowback water increases its turbidity. High turbid water acts as a barrier to sunlight and provides favorable conditions for growth of bacteria, which ultimately damages the biological profile of the flowback water.

Contemporary Flowback Management

Three generic pathways exist for flowback water management. They are injection into dedicated wastewater wells, treatment for surface discharge, and reuse or recycling of flowback water for use during a subsequent fracture treatment. This final pathway may or may not involve some form of treatment prior to reuse.[5,38,53] Injection may not be a feasible management pathway in all plays due to the nonavailability of suitable injection wells near to a well site. For example, in Pennsylvania, there were only seven operating class II wastewater disposal wells in 2008, whereas Texas had over 11,000 class II wastewater disposal wells in 2008.[39] This means that flowback water from the Marcellus play in some cases had to be transported to out-of-state disposal wells, often located in Ohio and West Virginia. As such the cost of injection can be very significant. For instance, in Bakken shale play, cost of deep-well injection range from \$1–\$11bbl^{-1}, out of which transportation costs represent 50–80% of total injection costs.[54]

Treatment of flowback water for the surface discharge is to a great extent determined by the regulatory discharge limits prevalent in a neighborhood. In the early stages of Marcellus Shale development (2008–09), the majority of flowback water was sent to domestic wastewater treatment plants (WWTP) for treatment and dilution followed by the subsequent surface discharge. However, WWTPs are designed to handle municipal wastewaters and so cannot remove contaminants such as dissolved salts, barium, and other potentially harmful substances present in the flowback water.[55] As a consequence, in Pennsylvania, WWTPs have limited the intake of flowback water to remain in compliance to the discharge limits.[2] Centralized wastewater treatment plants (CWT) are better fitted to handle flowback pollutants by using advanced treatment technologies. Cost estimates for a surface discharge range between \$2.50–\$3.50bbl^{-1} in Wyoming, whereas in Pennsylvania, the range lies between \$2.25–\$2.75 \$bbl^{-1}.[56] The trade-off between sending the flowback water to an injection well or to a CWT facility depends on the cost of transportation. In regions, where CWT facilities are not located in close vicinity of the well, injection becomes an economically viable option and vice versa.

With increasing shale gas development, reuse and recycling of flowback water have emerged as a promising option for flowback management. It not only reduces the water intensity of fracturing operations by partially making up the process water demand, but also reduces the impacts associated with trucking large volumes of water to and from a well site. Flow back can be treated either on-site or off-site. The feasibility of the flowback water reuse pathway is dependent on the volume of flowback produced over time. Wells producing large volumes early in the flowback period prefer to reuse their flow back due to the logistics involved in storing and transporting flowback water and the relatively lower levels of pollutants seen during the initial flowback period. For example, Barnett, Fayetteville, and Marcellus shales all produce approximately 10–15% of flowback water over a period of 2 weeks, enabling the effective reuse process, whereas Haynesville wells produce approximately 5% of flow back over the same period, limiting reuse opportunities.[38] The technologies incorporated for recycling can range from dilution to the use of desalination technologies such as reverse osmosis, mechanical vapor compression, and multieffect distillation.[57,58] Seldom, desalination technologies are utilized for treating flow back at a commercial scale due to unfavorable economics of these technologies. Major treatment schemes are either on-site dilution of flow back without treatment or dilution of flowback water after passing it through primary treatment processes (treat the water for everything other than dissolved salts) to achieve desired water quality.

Target Contaminants to Be Removed Before Recycling and Their Handling

For successfully recycling of flowback water, it is necessary that the water quality is compatible with fracturing additives. High concentration of mineral scales, dissolved salts, and colloidal particles can render the fracture fluid ineffective. As noted earlier, mineral scales tend to precipitate the polymeric gel and result in its collapse. Lime softening, ion exchange, and/or membrane filtration are used to remove scales prior to reusing flow back. However, the sludge generated in these processes can be potentially hazardous and its disposal can pose another management issue. Some companies have handled scales in flowback water by mixing it with sulfate-rich water, such as acid mine drainage (AMD).[59] Upon mixing, the divalent ions rapidly combine with the sulfate present in AMD water and precipitate in the solution. Similar effect has been achieved by mixing barium-rich flowback water with freshwater and using barium seeding to precipitate barite.[41]

The salt tolerance of the gelling agents and other fracture additives have substantially improved over the years. A few successful fracturing treatments have been executed using saline water with total dissolved salts (TDS) as high as 270,000 ppm (parts per million).[60] Selection of a high saline-based fracture fluid formulation is in accordance to the geology of the formation and overall economics of the fracture treatment.

Alternatively, high saline water could be treated using desalination technologies, such as reverse osmosis, multieffect distillation, crystallization, and freeze thaw to reduce the salt concentration to a desired level. Often a combination of these technologies is required to completely manage the salinity. For example, 212 resources have developed treatment train, which combines vapor compression and steam distillation for treating high TDS flow back.[56] Flow back needs to be preconditioned before using these technologies to prevent fouling. However, in practice, it is often seen that dilution of flowback water either at a WWTP or with freshwater is more commonly practiced for managing flowback salinity.

Flowback water rich in suspended solids can result in premature biological degradation of the polymeric gel, resulting in gel instability and poor performance. Suspended solids with high concentration of organic content can further damage the formation permeability. Depending on the size and type of the colloidal particles, different removal technologies are used. Coarse solids can largely be removed by sedimentation, media filtration whereas coarse solids require coagulation/flocculation followed by sedimentation and filtration. Suspended solids rich in organic content can be managed by aerobic/anaerobic oxidation methods, ozonation, or ultraviolet radiations.

Irrespective of method of recycling, ideal reusable flowback water should have low levels of hardness, salinity, and suspended solids. The current standard for water quality of recycled flowback water is uncertain due to different fracture fluid formulations used by the operators. For instance, flowback water in Marcellus has very high content of barium and is a prime pollutant to be removed before recycling. On the other hand, Barnett shale flowback water is low on barium, but has a high level of dissolved organic content and suspended solids, which must be removed before recycling. Thus, the water quality guideline for recycled water will be different for different regions. A general guideline has been established to facilitate the baseline design of wastewater treatment systems and shown in Table 3.[61] As shown in the Table 3, dissolved solids in treated flowback water range between 9000 and 16,000 ppm and hardness levels are maintained at 125–625 mg/l as $CaCO_3$.

Table 3 An Example Target Reuse Water Quality

Parameter	Range
Total dissolved solids, mg/l	9000–16,000
Turbidity, NTU	0–5
pH	6.5–8
Iron, mg/l	1–10
Chloride, mg/l	5000–10,000
Potassium, mg/l	100–500
Calcium, mg/l	50–250
Magnesium, mg/l	10–100
Sodium, mg/l	2000–5000
Boron, mg/l	0–20

Decision-Making Process for Managing Flowback Water

The management of flowback water is a complex technoeconomic issue. Different management approaches are selected depending upon a plethora of factors, including the relative economics of management pathways, the local availability of water, and the regulatory environment within which they operate. Widely formulated frameworks are targeted to capture as many of these dimensions affecting flowback management. Some noteworthy formulations include the integrated multicriteria decision-making model, Produced Water Management Information System (PWMIS), and water decision tree.

Integrated multicriteria decision-making models are based on analytical hierarchy processes that integrate subjective and relative preferences in performing analysis of feasible options.[62] As the first step in the process, the flowback management options and the criteria under which these options are evaluated are organized into a hierarchical structure. A judgment matrix is developed by means of a pairwise comparison between any two criteria and quantified by using a scale. The options are assigned scores based on the comparison matrix and further ranked according to scores. The major advantages of the approach include a user-friendly interface, the inclusion of verbal and expert judgment, and the structured analysis of the problem. Despite being a broad-spectrum approach, the aggregation of scores that are from scales of different units is often not easily interpretable.[63]

PWMIS is an online resource for technical and regulatory information for managing the produced water. It was developed by Argonne National Laboratory for Department of Energy. The practices and technologies are organized in a three tier system, namely minimization, recycle/reuse, and disposal. The user is asked for a series of questions regarding the well characteristics, regulatory framework, company policy, etc. Based on these user inputs, the system advances management options, which have less environmental impact. Currently, the system is unavailable for application.

Water decision tree was jointly developed by Petroleum Technology Alliance Canada, Schlumberger, and Science and Community Environmental Knowledge. The decision-making process screens the viability of various treatment options based on series of operational and water quality questions. Operational questions include information about the fracturing fluid treatment type, additive types, bottom hole pressure and temperature, and geology of the formation, whereas the water quality question includes information about blending ratios, target water quality, etc. Based on these data, the fate of flow back is determined. There is a limited flexibility to directly incorporate regulatory and soft factors in the decision process.

These approaches are based on a common principle of encompassing the different dimensions impacting flowback management, but seldom do any of them provide a stand-alone solution to the problem. Furthermore, they merely reiterate the complexities and challenges involved in the integration of diverse sets of interests in managing flowback water.

System Dynamics Approach to Flowback Management

A system dynamics based framework to capture various complexities in flowback management is presented here in the form of a linear optimization model. A conceptual layout of the system is shown in the Figure 3. The framework identifies the following technical elements for investigation: (1) flowback water volume and quality; (2) on-site versus off-site management; (3) treatment options; and (4) impact on water supply within the local watersheds. Each region has specific treatment endpoints denoted by X1, X2, X3... Xn in the Figure 3. Some example of these endpoints could be underground injection, on-site dilution, primary treatment, and desalination. The recovered flow back is distributed across these endpoints depending on the model inputs and constraints. After any dilution as required, the flow back is ready to be used for the subsequent fracturing process.

Quantitative relationships are derived within each endpoint to relate process variables to the overall cost of flowback management. Each treatment endpoint is characterized by three parameters: (1) transportation costs, (2) treatment cost, and (3) avoided cost. The transportation cost includes driver cost and fuel cost, treatment cost includes operating and capital cost of the plant and, avoided costs signify the cost savings resulting from a reduction in water demand for subsequent fracturing operations owing to any reuse and recycling of streams. Avoided cost is equivalent to water cost and the transportation cost of hauling the water to the well site from the freshwater source. The output stream of the model is calibrated at a certain water quality level (example, water quality is shown in Table 2) after necessary dilution, such that it could be further reused in the process. The framework is bounded by constraints set on the intake capacities of endpoints, water withdrawal for any dilution if necessary, and, lastly, the mass balances across the model boundary. Finally, based on these inputs and parameters, it is desired to determine the optimal management strategy at a minimum cost to

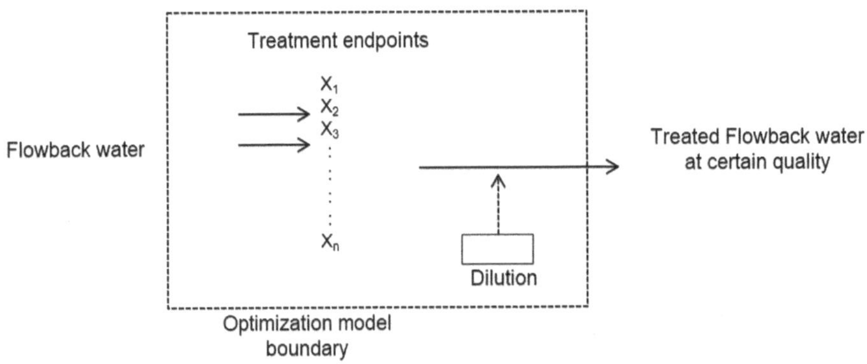

FIGURE 3 Conceptual schematic of the optimization framework.

various stakeholders. The general mathematical formulation of the schematic is as follows

$$\begin{aligned}
&Minimize \sum_{t=1}^{15}\sum_{i=1}^{n}[\alpha_i(1+\varepsilon_i)+b_i-c_i\varepsilon_i-d\varepsilon_i]v_{it}\\
&s.t. \sum_{t=1}^{15} v_{it}\varepsilon_i(\alpha_{it}-1)\le w_o \quad i=1,2,3\ldots n\\
&\sum_{i=1}^{15} v_{it}=V_t \quad t=1,2,3\ldots15\\
&v_{it}\le y_i \quad i=1,2,3\ldots n,\ t=1,2,3\ldots15\\
&v_{it}\ge 0 \quad i=1,2,3\ldots n,\quad t=1,2,3\ldots15
\end{aligned}$$

where,

a_i = cost of transportation to ith endpoint, b_i = cost of treatment for ith endpoint, d = price of water in the region, c_i = cost of transportation of water from a surface stream to well site, ε_i = treatment efficiency of ith endpoint, v_{it} = flow back entering in the ith endpoint at time t, α_{it} = dilution factor for ith endpoint at time t, w_o = daily freshwater withdrawal limit, V_t = flowback water recovered in time t, y_i = intake capacity of ith endpoint.

Analysis and Discussion

A scenario is studied here to illustrate the functionality of the framework. Let us assume a well that recovers 15% of the injected water volume as flow back over a period of 2 weeks. To manage the flowback water, the options available in this scenario for evaluation are on-site dilution, off-site primary treatment, on-site primary treatment, off-site desalination, and underground injection. The primary treatment plant is a two-stage softening system designed to treat the flowback water for contaminants such as hardness, suspended solids, and other organic matter whereas off-site desalination plant is a multieffect distillation plant. The sludge/residue produced in the processes is managed by a conventional sludge thickening train and sent to landfill. The salinity of the treated water in the primary treatment plant is managed by mixing freshwater with treated water. In this base case, it is assumed that the off-site plants are located at a distance of 15 miles from the well, whereas injection wells are located at a distance of 50 miles. These values will exhibit a large variation depending on the location of the wells. Freshwater in the process is procured locally. The water quality and quantity data used for running the model is shown in Table 4.[42]

For running this scenario, the cost associated with treatment endpoints is determined based on the input specifications. The transportation cost of trucking water to and from the well site is dependent on the distance between treatment endpoints and well site. For the base case, the transportation costs range between \$1–\$3bbl^{-1}, and increases proportionally with distance. The water treatment costs are based on EPA cost curves for wastewater treatment processes,[64] with an amortization rate of 10% for a period of

Table 4 Flowback Water Quality and Quantity Data Set used for the Analysis of Base Case

Time, Days	Hardness*, ppm	TSS†, ppm	TDS‡, ppm	Volume, gpd
1	2200	237	30,995	68,339
2	2585	251	33,627	66,386
3	3037	266	36,481	64,488
4	3568	282	39,578	62,645
5	4191	299	42,938	60,854
6	4924	317	46,583	59,115
7	5785	336	50,538	57,425
8	6796	356	54,828	55,784
9	7984	377	59,482	54,189
10	9380	400	64,532	52,640
11	11,019	424	70,010	51,136
12	12,945	449	75,953	49,674
13	15,208	476	82,401	48,254
14	17,867	505	89,396	46,875
15	20,990	535	96,985	45,535

*Hardness here denotes the amount of divalent ions present in the flowback water. Higher hardness results in scale formation on the equipments.

†TSS is total suspended solids that include colloidal particles and any particulate matter.

‡TDS is total dissolved salts that reflect the salinity of flowback water by measuring the salt (NaCl) amount in water.

30 years and 10 years for off-site and on-site plants. Depending on the level of contaminants removed, the cost of treatment ranges between $\$0.05\text{bbl}^{-1}$–$\$1.5\text{bbl}^{-1}$. Some regions have price of water owing to the water rights system prevalent in their regions. As a result, in this scenario we assumed a flat water price of \$0.025/gal. It is important to point out that these cost estimates are an approximation of the actual cost as there is large uncertainty around the available water treatment infrastructure, volume of water, and its water quality. The model provides the user to customize the model based on their cost specifications.

Based on these inputs, the framework determines that reusing 40% of flow back by dilution, 26% by treating it in off-site primary treatment plant, 20% by treating in on-site primary treatment plant, 4% by treating it at the off-site desalination plant, and sending 10% to deep-well injection is the optimal strategy. The majority of the early day flow back is captured for reuse as it is relatively cleaner in comparison to later day flow back. The implementation of the strategy will cost $\$\ 0.75\text{bbl}^{-1}$. The overall environmental scorecard of this strategy results in recycling of nearly 90% of flowback water, which is equivalent to a reduction in 100 trucking round-trip. Using this as the base case, sensitivity analysis could be performed across different dimensions as per the user requirement. For instance, in the current example scenario, Figure 4 and Figure 5 show the impact of flowback water quality and distance to injection wells on the model results. For a given distance from an injection well and increasing pollution level of flowback

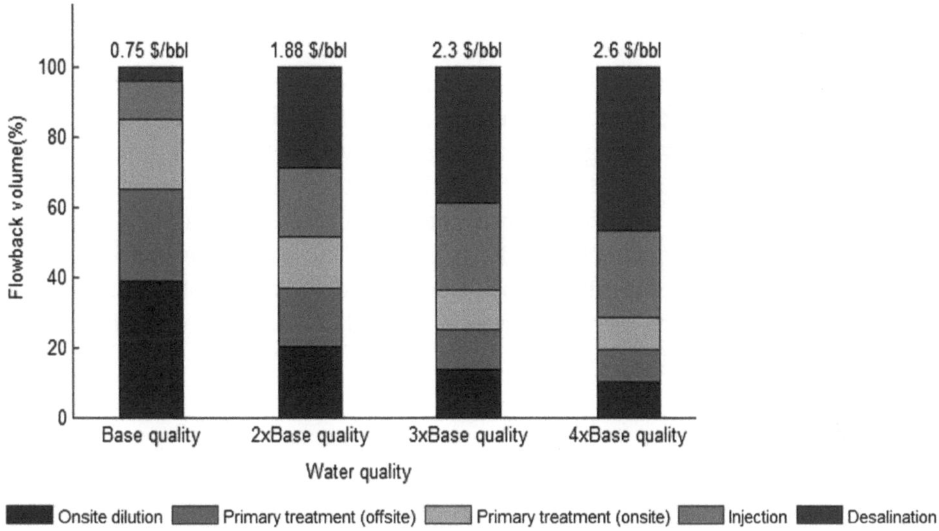

FIGURE 4 Effect of varied water quality on flowback management and its cost. The x-axis denotes different water quality as n times base quality, where n = 1, 2, 3, 4. The y-axis shows the volumetric percentage of flow back received in a particular endpoint over a period of 2 weeks.

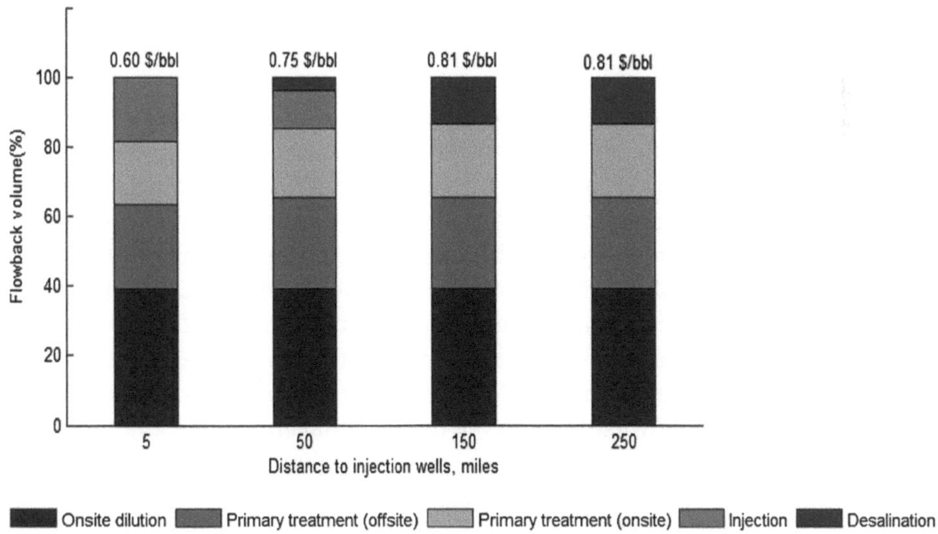

FIGURE 5 Impact of varying injection well distance on flowback management. The y-axis shows the volumetric percentage of flow back received in a particular endpoint over a period of 2 weeks and x-axis shows the distance to injection well.

water, the majority of water recycling is achieved via desalination plants as its poor water quality makes it difficult to achieve desired reuse water quality level using non-desalination technologies. In this specific case, at least 70 trucking round-trip can be reduced by using model strategy and 70–80% flowback water can be reclaimed for process reuse at costs less than \$2.3 bbl^{-1}. On the other hand, for a given flowback water profile and varying injection well distance, desalination technologies could substitute underground injection if the high costs of desalination treatment can offset the high transportation costs of underground injection. In this specific case, at least 100 trucking round-trips are reduced by using the model strategy and 80–100% flowback water is reclaimed for process reuse at costs less than \$1 bbl^{-1}. It is important to point out that the economics of management and treatment strategy will vary on a case by case basis and there is no standard solution, which could apply to all. The use of the example scenario in the above analysis is purely for the purpose to demonstrate the application of the model and its capabilities in providing a comprehensive method to manage the flowback water systematically and sustainably.

Conclusions

Recent developments in hydraulic fracturing have enabled a dramatic increase in the production of unconventional oil and gas resources. Contemporary hydraulic fracturing treatments are water intensive, requiring several million gallons of water per well. Environmental externalities associated with shale gas development, particularly related to flowback water quality and quantities have drawn significant public and regulatory attention. Because of the large flowback water volumes and high concentrations of solids and other pollutants, there are significant concerns about the management of this water and its environmental impacts, potentially with the release of inadequately treated or untreated wastewater to the environment. Furthermore, the variability of the chemical profile of flow back and its dependence on the geology, fracture fluid composition, and sampling time, aggravates the problem of managing it. With technological advances in multipad drillings and refracturing of wells, these environmental concerns will increase unless an effective flowback management framework is developed, which encompasses the temporal and spatial dynamics in fluid composition and quantity, economics of technology, the availability of physical infrastructure, and impact of water withdrawal rates on the sustainability of local surface water bodies. Previous attempts to integrate the various dimensions in flowback management provide a consistent approach to decision-making, but complete integration of the diverse interests of multiple stakeholders is challenging. In this chapter, an alternative approach based on system dynamics is presented. Using this approach, flowback management strategy can be tailored to the specific spatial and temporal data. This type of holistic approach is essential for the long-term responsible shale gas development and for secure availability of water and energy at large.

References

1. Sondergeld C, Newsham K. *SPE 131768 Petrophysical considerations in evaluating and producing shale gas resources*. Society of Petroleum Engineers; 2010.
2. Entrekin S, Evans-White M, Johnson B, Hagenbuch E. Rapid expansion of natural gas development poses a threat to surface waters. *Front Ecol Environ* 2011;**9**(9):503–11.

3. Entrekin S, Evans-White M, Slutz J. *SPE 147264 shale Gas water treatment value chain – a review of technologies, including case studies.* Society of Petroleum Engineers; 2011.
4. Vidic RD, Brantley SL, Vandenbossche JM, Yoxtheimer D, Abad JD. Impact of shale gas development on regional water quality. *SCIENCE* 2013;**340**(6134).
5. NYS DEC. *Draft supplemental generic environmental impact statement on the oil, Gas and solution mining regulatory program, well permit issuance for horizontal drilling and high-volume hydraulic fracturing to develop the Marcellus shale and other low-permeability G.* New York State Department of Environmental Conservation (NYS DEC); 2011.
6. Kaufman P, Penny G, Paktinat J. *SPE 119900 critical evaluations of additives used in shale slickwater Fracs.* Society of Petroleum Engineers; 2008.
7. Economides MJ, Martin T. *Modern fracturing.* Houston, TX: ET Publishing; 2007.
8. Lestz R. *WaterLess fracturing technology.* 2011. Available at, http://www.siltnet.net/documents/lestz_condensed_version_2_15_11.pdf.
9. Aqualon. *Guar and guar derivatives oil and Gas Field applications.* 2007. Available at, http://www.ashland.com/Ashland/Static/Documents/AAFI/PRO_250-61_Guar.pdf.
10. Weaver J, Gdanski R, Karcher A. *SPE 80226 guar gum Degradation : a kinetic study.* Society of Petroleum Engineers; 2003.
11. Conway MW, Almond SW, Briscoe JE, Lawrence HE. *SPE 9334 chemical model for the rheological behavior of crosslinked fluid systems.* Society of Petroleum Engineers; 1980.
12. Kalg R, Patil P. *SPE 151890 A Novel, shear-stable fracturing fluid for high-temperature applications.* Society of Petroleum Engineers; 2012.
13. Prabhanjan H, Gharia MM, Srivastava HC. Guar gum derivatives. Part I: preparation and properties. *Carbohydr Polym* 1989;**11**(4):279–92.
14. Wu M. Shear-thinning and viscosity synergism in mixed solution of guar gum and its etherified derivatives. *Polym Bull* 2009;**63**(6):853–63.
15. Sheng J. *Polymer viscoelastic behavior and its effect on field facilities and operations," modern chemical enhanced oil recovery.* Gulf Professional Publishing; 2011. pp. 207–238.
16. Bjornen HK, Hodge MR, Cawiezel EK, England WK. *SPE 143962 shear sensitivity of borate fracturing fluids.* Society of Petroleum Engineers; 2011. pp. 1–17.
17. Palisch TT, Vincent MC, Handren PJ. Slickwater fracturing: food for thought. *SPE Prod Operations* 2010;**25**(3):327–44.
18. Schein G. *SPE108807 the application and technology of slickwater fracturing.* Society of Petroleum Engineers; 2005.
19. White CM, Mungal MG. Mechanics and prediction of turbulent drag reduction with polymer additives. *Annu Rev Fluid Mech* 2008;**40**:235–56.
20. Stewart BR, Mullen ME, Howard WJ, Norman W. *SPE 30114 Use of a solids-free viscous carrying fluid in fracturing Applications: an economic and productivity comparison in shallow completions.* Society of Petroleum Engineers; 1995.
21. Rahim Z, Al-Kanaan AH, Al-Kanaan AA. Selecting optimal fracture fluids, breaker system, and proppant type for successful hydraulic fracturing and enhanced gas production—case studies. In: *Proceedings of 2013 SPE middle East unconventional Gas conference & exhibition.* Society of Petroleum Engineers; 2013.
22. Hsu SC, Don TM, Chiu WY, Hsu Shih C, Don Trong M. "Free radical degradation of chitosan with potassium persulfate. *Polym Degrad Stab* 2001;**75**(1):73–83.
23. Rae P, Lullo G. *SPE 37359 fracturing fluids and breaker systems - a review of the state-of-the-art.* Society of Petroleum Engineers; 1995.

24. McDougall. *Encapsulated breaker chemical.* 1992. US Patent Number 5,102,558.
25. Sarwar MU, Cawiezel KE, Nasr-El-Din AH. *SPE 140520 gel degradation studies of oxidative and enzyme breakers to optimize breaker type and concentration for effective break profiles at low and medium temperature ranges.* Society of Petroleum Engineers; 2011.
26. Rimassa SM, Howard P, MacKay B, Blow K, Coffman N. *SPE 141211 Evaluation of an oxidative biocide during and after a hydraulic fracturing job in the Marcellus shale.* Society of Petroleum Engineers; 2011.
27. Fu Z, Kawaguchi Y. A short review on drag-reduced turbulent flow of inhomogeneous polymer solutions. *Adv Mech Eng* 2013:1–12.
28. Lumley LJ. Drag reduction by additives. *Annu Rev Fluid Mech* 1969;**1**(1):367–84.
29. De Gennes PG. Towards a scaling theory of drag reduction. *Phys A* 1986;**140**(1–2):9–25.
30. Kabir AH. *SPE 72119 chemical water & Gas shutoff technology - an overview.* Society of Petroleum Engineers; 2001.
31. DiGiacomo PM, Schramm CM. SPE 11787 mechanism of polyacrylamide gel syneresis determined by C-13 NMR. In: *Proceedings of Spe oilfield and geothermal chemistry symposium, Denver, CO*; 1983.
32. Moradi-Araghi A, Doe P. *"SPE 13033 hydrolysis and precipitation of polyacrylamides in hard brines at elevated temperatures.* Society of Petroleum Engineers; 1987.
33. McCormik CL, Hester RD, Wildman GC. *Improved polymers for enhanced oil recovery-synthesis and rheology.* 1984. Available at, http://www.netl.doe.gov/kmd/cds/disk44/C-Chemical%20Flooding/BC10321-16.pdf.
34. Blair CC, Chang KT, Trybig DS. *Use of dispersion polymers as friction reducers in aqueous fracturing fluids.* 2004. US Patent Number 6,787,506, 2(12).
35. Phillips KG, Hunter WE. *Method for reducing friction loss in a well fracturing process.* 1979. US Patent Number 4152274.
36. Vetter OJ, Lankford S, Nilssen T, Shelton M. SPE-17284 well stimulation and scale inhibitors. In: *Proceedings of Permian Basin oil and gas recovery conference, Midland, Texas*; 1988.
37. Ronald PJ, Fischer AR, Gdanski RD, McCabe MA. *SPE 30700 encapsulated scale inhibitors for use in fracturing treatments.* Society of Petroleum Engineers; 1995.
38. Mantell M. Produced water reuse and recycling challenges and opportunities across major shale plays. In: *Proceedings of EPA hydraulic fracturing study technical workshop on water resource management, Washington D.C*; 2011.
39. Gregory KB, Vidic RD, Dzombak DA. Water management challenges associated with the production of shale Gas by hydraulic fracturing. *Elements* 2011;**7**(3):181–6.
40. Kimball RJ. Key considerations for Frac flowback/produced water reuse and treatment. In: *Proceedings of New Jersey water environment association conference, Atlantic city*; 2012.
41. Minnich K. A water chemistry perspective on flowback reuse with several case studies. In: *Proceedings of EPA hydraulic fracturing study technical workshop on water resource management*; 2011.
42. Hayes T, Severin BF. *Barnett and Appalachian shale water management and reuse technologies* [Project report by Gas Technology Institute for Research Partnership to Secure Energy for America (RPSEA)]. 2012.
43. Wolford R, Dempsey B. *Characterization of organics in the Marcellus shale flowback and produced waters.* Department of Civil and Environmental Engineering, The Pennsylvania State University; 2011.
44. Haarberg T, Selm I, Granbakken BD, Ostvoid T, Read P, Schmidt T. *SPE 19449 scale formation in reservoir and production equipment during oil Recovery: an equilibrium model.* Society of Petroleum Engineers; 1990.

45. Neff J, Theodor S. *Barium in produced water-fate and effects in the Marine environment.* American Petroleum Institute; 1995.
46. Palmer AD, Drummond SE. Thermal decarboxylation of acetate. Part II. Boundary conditions for the role of acetate in the primary migration of natural gas and the transportation of metals in hydrothermal systems. *Geochim Cosmochim Acta* 1986;**50**(5):825–33.
47. Shock EL, Koretsky CM. Metal-organic complexes in geochemical processes: calculation of standard partial molal thermodynamic properties of aqueous acetate complexes at high pressures and temperatures. *Geochim Cosmochim Acta* 1993;**59**(8):1497–532.
48. Blount CW. Barite solubilities and thermodynamic quantities up to 300 C and 1400 bars. *Am Mineral* 1977;**62**(9–10):942–57.
49. Templeton CC. Solubility of barium sulfate in sodium chloride solutions from 25 to 95 C. *J Chem Eng Data* 1960;**5**(4):514–6.
50. Hanor HS, Chan Lui-Heung. Non-conservative behavior of barium during mixing of Mississippi river and Gulf of Mexico Water. *Earth Planet Sci Lett* 1977;**37**(2):242–50.
51. Ziemkiewicz P, Hause J, Lovett R, Locke D, Johnson H, Patchen D. *Zero discharge water management for horizontal shale Gas well development.* Project report by West Virginia Water Research Institute to U.S. Department of Energy. Available at. 2012., http://www.netl.doe.gov/File%20Library/Research/Oil-Gas/Natural%20Gas/shale%20gas/FE0001466_TSA.pdf.
52. Deng S, Yu G, Shubo D, Gang Y, Zhongxi C, et al. Characterization of suspended solids in produced water in Daqing oilfield. *Colloids Surfaces A: Physicochem Eng Aspects* 2009;**332**(1):63–9.
53. Veil J. *Water management technologies used by Marcellus shale Gas producers.* 2010. Available at, http://fracfocus.org/sites/default/files/publications/water_management_in_the_marcellus.pdf. July.
54. Stepan JD, Shockey ER, Kurz AB, et al. *Bakken water opportunities assessment-phase 1.* 2010. Available at, http://www.undeerc.org/bakken/pdfs/FracWaterPhaseIreport.pdf.
55. Ferrar KJ, Michanowicz DR, Christen CL, Mulcahy N, Malone SL, Sharma RK, et al. Assessment of effluent contaminants from three facilities discharging Marcellus shale wastewater to surface waters in Pennsylvania. *Environ Sci Technol* 2013;**47**(7):3472–81.
56. RPSEA, Colorado School of Mines. *An integrated framework for treatment and management of produced water.* 2009. Available at, http://aqwatec.mines.edu/produced_water/treat/docs/Tech_Assessment_PW_Treatment_Tech.pdf. November.
57. Nathan K, Rusty Weline, Ken N. Water reuse and recycling in the oil and Gas Industry: Devon's water management Success. In: *Proceedings of 2nd annual Texas reuse conference, Texas*; 2012.
58. Vasiliu CC, Pierce D, Bertrand K. *SPE 157615 challenging wastewater treatment.* Society of Petroleum Engineers; 2012.
59. Keister TE. *Treatment of Gas well production wastewaters.* 2013. US Patent Number 20130048562.
60. Lebas R, Lord P, Luna D, Shahan T. *SPE 163824 development and use of high-TDS recycled produced water for crosslinked-gel-based hydraulic fracturing.* Society of Petroleum Engineers; 2013.
61. Shafer L. *SPE 141448 water recycling and purification in the Pinedale anticline Field: results from the anticline disposal project.* Society of Petroleum Engineers; 2011.
62. Lee K, Neff J. *Produced water.* New York: Springer; 2011. New York.
63. Ishizaka A, Labib A. Analytic hierarchy process and expert choice: benefits and limitations. *Operation Res Insight* 2009;**22**(4):201–20.
64. Clark RM. *Estimating water treatment costs: cost curves applicable to 2500 gpd to 1 mgd treatment Plants.* 1979. Available at, http://nepis.epa.gov/Exe/ZyNET.exe/300009IH.TXT?ZyActionD=ZyDocument&Client=EPA&Index=1976+Thru+1980&Docs=&Query=&Time=&EndTime=&SearchMethod=1&TocRestrict=n&Toc=&TocEntry=&QField=&QFieldYear=&QFieldMonth=&QFieldDay=&IntQFieldOp=0&ExtQFieldOp.

9

Water Management in the Oil Sands

Neera Chawla
CONSULTANT, ENERGY & PETROLEUM, CHICAGO, IL

CHAPTER OUTLINE

Introduction

The global demand for oil is high and with the projected growth in world population, this demand is expected to multiply. It is important to understand that the production of oil comes at the cost of energy and water. Energy is required to produce oil including the treating and transporting of water; while water is required to extract the oil from the reservoir. Increased production rates continue to drive the consumption of water and

Food, Energy, and Water. http://dx.doi.org/10.1016/B978-0-12-800211-7.00009-5

energy. Moreover, higher energy consumption yields an increase in greenhouse gases (GHGs) released into the atmosphere. It is believed that GHGs contribute to a climate change that negatively impacts the supply of fresh water. Actions continue to be taken in the form of regulations and policies to promote water conservation and mitigate the release of GHGs.

There has been a surge in unconventional production over the last decade, which signals a shift in oil production from accessible reserves to those that are less accessible. One such example is the Alberta oil sands, which is home to the third largest proven oil reserve in the world. The importance and relationship of energy and water is perhaps most apparent in this region. Oil sands in situ methods use 0.5 barrels of water per barrel of bitumen while conventional oil uses 0.1–0.3 barrels of water per barrel of oil produced.[1,2] With the current technologies, large amounts of water are required to extract oil from Alberta's hydrocarbon reserves, which creates a significant demand for water in the region and sets up a natural competition between advancing the region's economy and protecting and conserving its limited source of fresh water. Statistics show that over the last three decades, stream flows in the Athabasca River, which is the main supply of fresh water in the region, have declined by 30%.[3] It could be possible that the future of the oil sands may ultimately be limited by water supply in the region. The projected growth of the industry coupled with climate changes could potentially reduce the river's flow significantly. Eventually, the rate of growth of the oil sands will be dependent on the availability of water to meet future production demands while preserving the environment.

Alberta Oil Sands

As conventional sources of crude are being depleted, supply is being replaced by unconventional crudes such as the bitumen found in Alberta's oil sands. This creates a global importance for the oil found in this region. Alberta is a province situated in the Western region of Canada with deposits located in the northeast area of the province covering approximately 140,000 square kilometers. They are currently the largest provider of oil to the United States. Figure 1 depicts the geographical position of the oil sands deposits.

Alberta's oil reserves play a very strategic role in the global context with a proven 170 billion barrels, which stand as the third largest in the world. This consists of approximately 168 billion barrels of bitumen and the remaining of conventional crude oil. They provide more crude to the United States than any other nation including Saudi Arabia, Venezuela, and Mexico. Total production stands at roughly 1.9 million barrels of oil per day of which 1.4 million barrels per day are exported via pipeline to the United States.[4] As such, part of America's energy security is dependent on the ability of the oil sands to remain sustainable. See Figure 2 below for Alberta's stance in the global reserve context.

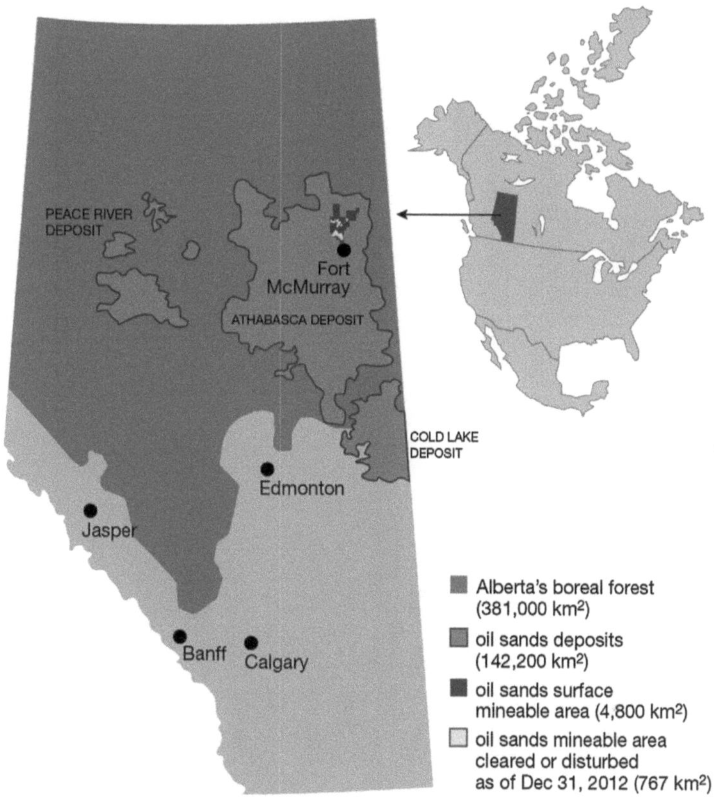

Note: 1 km^2 = 1 square kilometre = 0.39 square miles

FIGURE 1 Alberta's oil sands deposits.[4]

The oil sands contain bitumen, which is a type of heavy oil and is characterized by high viscosity, relatively low amounts of organic hydrogen, and high levels of impurities such as sulfur, nitrogen, and heavy metals. In addition, the bitumen resides in formations containing large amounts of sand and clay. The bitumen must be separated from the sand, clay, and water in order to obtain feedstock that can then be upgraded or diluted for transportation to refineries. The dilution requires the addition of some condensate material referred to as diluent to the bitumen in order to be shipped via pipeline. This material is often pentanes and lighter liquid hydrocarbons that come from natural gas production. The refineries then upgrade or refine the oil into finished fuels like gasoline and diesel, and petrochemical feedstocks.

There are generally two methods used for extraction in the oil sands: surface mining and in situ techniques. The method used depends on the location of the reservoir relative to the surface. Surface mining may be used for shallow reserves that have less than 75 m in overburden, while in situ methods are used to extract from deeper reserves.[6] Over 80% of oil sands deposits can only be accessed through in situ methods as they lie more than

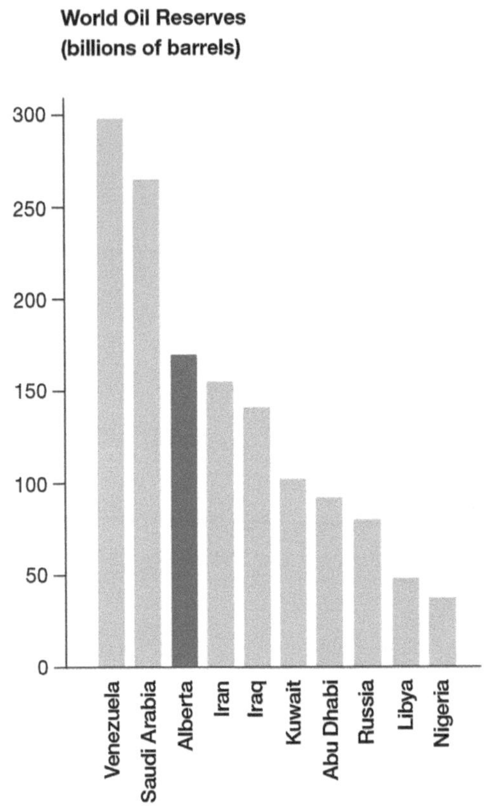

FIGURE 2 Comparative oil reserves.[5] *Oil & Gas Journal, AER.*

400 m below the surface.[7] As such, these methods have the most significant growth potential and are expected to be a growing proportion of future production. Currently, two main in situ techniques are commercially available to produce bitumen from the deeper reservoirs. The technologies are steam-assisted gravity drainage (SAGD) and cyclic steam stimulation (CSS). As indicated by their names, both require water in the form of steam to act as the motive force to extract the oil. The next section discusses SAGD in more detail and will be the focus of this chapter. A brief outline of CSS is also provided for comparative purposes.

Thermal In Situ Oil Production Methods

Steam-Assisted Gravity Drainage

SAGD is an enhanced oil recovery method that uses thermal energy to produce oil. This technique was developed after more than a decade of research and continues to play an important part in the development of Alberta's oil sands. It represents a critical step forward in in situ production technology as improvements are being made on an ongoing basis in order to reduce GHG emissions and water use. The process requires two horizontally parallel wells drilled into an oil reservoir. Steam is generated at the surface facilities and is injected into the upper wellbore. The steam forms a steam chamber that provides an area for steam to condense and provide the heat to lower the viscosity of the bitumen lying within the permeable rock and sand formations. The effect is a "melting" of the bitumen, which then bleeds into the lower wellbore along with the condensed steam and a small amount of produced gas forming an emulsion that is transported to the surface either by natural or artificial lift. Figure 3 shows the positioning of the horizontal wells and provides an overall basic picture of how a SAGD system operates below ground.

At the surface, the oil, the produced gas, and the condensed steam are separated. The oil is either upgraded or diluted to be sold, the produced gas is used to supplement natural gas to produce steam, and the condensed steam, more commonly termed produced water, is treated and reused to produce steam.

The efficiency of the SAGD process is typically measured by the steam-to-oil ratio (SOR). This equates to the volume of steam required on a cold water equivalent basis to generate one unit volume of oil. The oil sands require large amounts of water for the generation of steam. The SOR is indicative of the operation's technical and environmental performance. Higher SOR operations require more water and larger amounts of natural gas to generate the steam. The combustion of the natural gas produces GHG emissions. Moreover, higher steam rates lead to larger water treatment facilities and increased capital and operating costs. The SOR is a function of a variety of factors; these can be the reservoir's characteristics, extraction technology, and operating conditions.

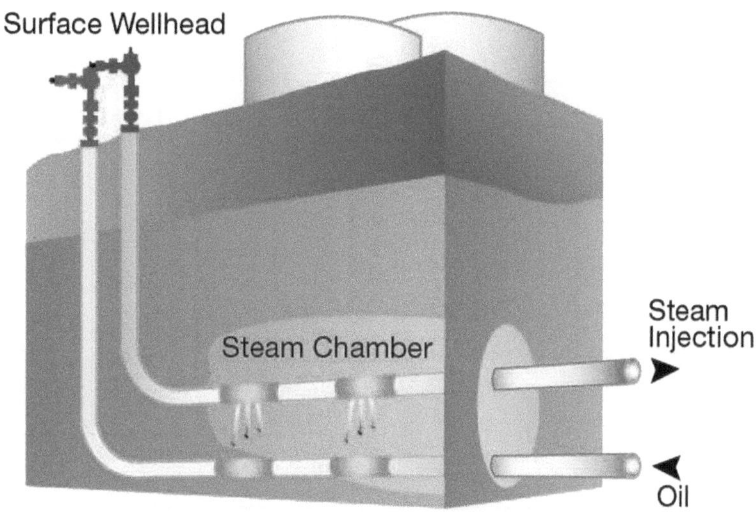

FIGURE 3 Steam-assisted gravity drainage (SAGD) process.[8]

Generally, the SOR, or the amount of steam needed can be anywhere between two and five times the oil produced for a SAGD operation. Because a large amount of steam is required to produce the oil, it is important for operators to efficiently use and recycle water in the system. The proper use and treating of the water is critical to the environmentally sustainable oil sand production operation. Due to the treatment of water, the Alberta in situ producers cumulatively recycle between 80% and 95% of the water required for production.[1]

Water Treatment Technologies

The produced water contains silica, organics, carbonates, and dissolved salts, which are also known as total dissolved solids (TDS). Lime softening and evaporation are the only commercially available water treatment technologies currently used in oil sands facilities.

Lime Softening

Figure 4 shows a typical SAGD configuration with lime softening used as the water treatment method. Emulsion from the reservoir is sent for oil and water separation. Following this, the de-oiled water is combined with makeup water and both are processed though a warm or hot lime softener depending on the temperature of the operations. A temperature between 80 and 85 °C is the target for warm lime softening

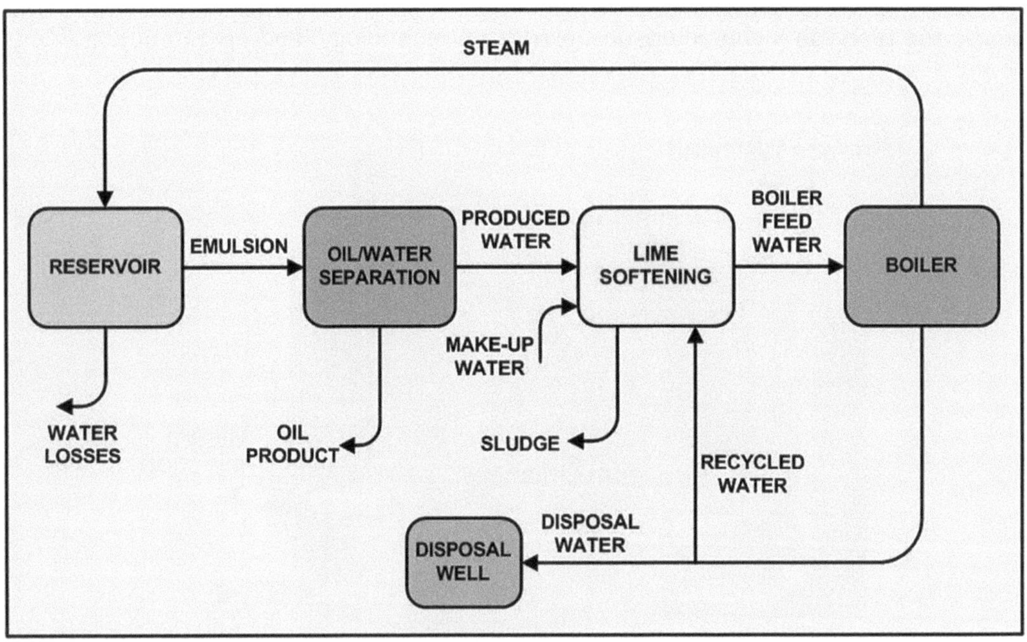

FIGURE 4 Steam-assisted gravity drainage (SAGD) configuration with lime softening water treatment.

while a hot lime softener operates between 105 and 120 °C. In this process, chemicals are used to reduce the hardness and silica in the waters. Silica is removed by adsorption on a magnesium oxide hydrate matrix. The TDS is not impacted since the ions present in the water are substituted rather than removed. The water then enters a weak acid cation unit, which is an ion exchange unit to reduce hardness. In this unit, a regenerable resin is used to replace the calcium and magnesium ions with sodium. The water exiting the water treatment, generally referred to as boiler feed water, is then sent to boilers to be heated for steam generation. The blowdown from the boilers, which is the water separated from the steam, is recycled back to water treatment typically to a boiler limit. The remaining blowdown is disposed either via deep well injection or through further processing prior to disposal.[9]

Evaporation

Figure 5 shows a SAGD facility with the only difference being the change in water treatment technology. Here the produced and makeup waters are combined and sent to an evaporator for treatment. The evaporator produces a clean distillate stream and disposal stream. The distillate stream is virtually contaminant free except for lighter residual organics that can vaporize from the feed stream. The disposal stream is a concentrated stream of contaminants containing almost all the TDS, silica, and hardness in the feed stream. Due to the cleaner nature of the distillate, either a different type of boiler can be used or higher water recycle rates are achievable. The contaminant-laden

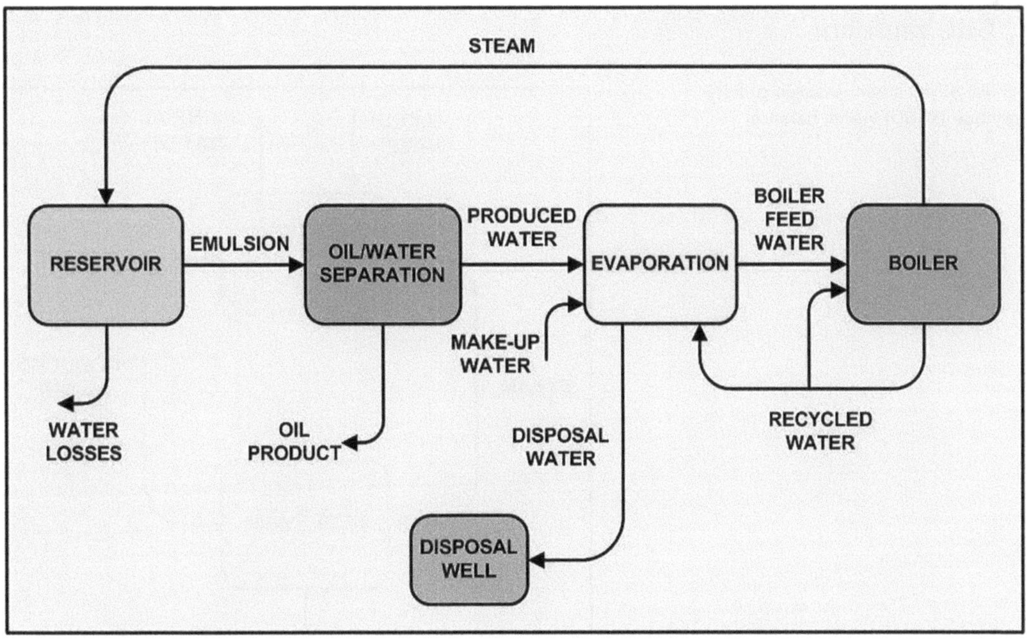

FIGURE 5 Steam-assisted gravity drainage (SAGD) configuration with evaporation water treatment.

stream from the evaporator typically contains large amounts of silica and must be further treated for silica removal prior to disposal.[9]

It is important to note that the most common evaporation process consumes power. As a result, evaporator-based water treatment facilities have higher GHG emissions than chemical-based lime softening water treatment facilities.

The Water Balance

The steam that is injected and condensed downhole typically ends up in one of three places; it is recycled back into the system, lost to the reservoir, or disposed of as a waste product following water treatment. Since a particular SOR is maintained for an operation, the water lost must be supplemented through a makeup water source in order to maintain a constant steam volume. There is limited control over how much of this water is lost to the reservoir; however, there is much that can be done to limit the disposal quantities. This is mainly a factor of the makeup water quality and the water treatment technology used. Figure 6 depicts the overall water balance. Produced water and makeup water qualities play a key role in the overall picture, with the main difference being that while the produced water quality is a function of the reservoir, the makeup water quality is dependent on the source. The availability of the type of water used directly impacts the following[9]:

- Disposal water volumes
- Makeup water volumes
- Technology selection
- GHG emissions

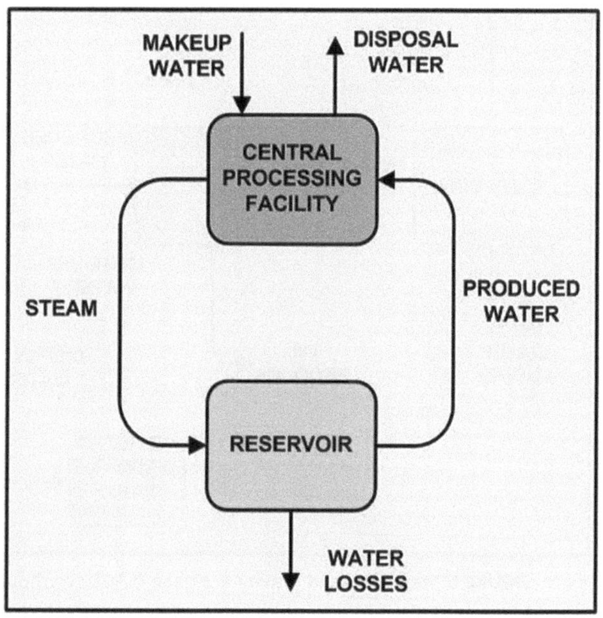

FIGURE 6 The steam-assisted gravity drainage (SAGD) water balance.

Cyclic Steam Stimulation

An alternate form of production used by some operators in the Alberta region is CSS. This type of in situ production uses repeating cycles of steaming, soaking, and production to produce the bitumen. Higher pressures are often used and therefore, this method requires a caprock and overburden of more than 400 m.[7] CSS differs from SAGD in that it only involves the use of one wellbore. Steam is injected into the reservoir through a combination of horizontal and vertical wells and is left there to soak for a few weeks. The flow is then reversed allowing the water/bitumen emulsion to drain into the same wellbore and brought up to the surface. This cycle is repeated until it becomes uneconomical to recover the oil. Another major difference is that while SAGD makes use of dry steam injection, CSS is operated on a wet steam injection basis. From a water treatment standpoint, SAGD and CSS are similar.

Water Use in In Situ Production

As mentioned above, somewhere between 80% and 95% of all the water used for in situ production is recycled and some amount of makeup water is required. Most of the makeup water currently used is groundwater, specifically the Athabasca River as the primary water source.[1] Consequently, protecting the Athabasca River and preventing the depletion of groundwater sources are important. It is starting to become more common to use higher TDS brackish water found in underground aquifers as makeup water.

The quality of water is inversely proportional to the amount of TDS in the water, which is mainly made up of salts and silica. Fresh water is classified by having less than 4000 mg L^{-1} TDS, while brackish water is classified as having TDS levels above 4000 mg L^{-1}.[10]

Understandably, with more water being recycled, the result becomes less makeup water intake. Reports indicate that the cumulative withdrawal rate from the average annual flow of the Athabasca River is less than 1%. For in situ operations, the total water use per barrel of oil produced is 0.5.[1] In general, operators are encouraged to maximize water recycle rates, which would minimize the intake of new water into their facilities and when water supplementation is needed, to make use of high salinity waters.

The oil sands industry in conjunction with the local government strives to maintain water conservation through limited and efficient use of fresh water sources in the area. This is enforced through regulations that are outlined in the next section.

Government Regulations for Water Conservation

It is believed that the world will be highly dependent on carbon-based fuel in the years to come. The Alberta oil sands must be able to sustainably meet global demand while conserving its water sources in the region. Alberta works with the industry to ensure that the development of the oil sands occurs responsibly through improved regulations, environmental monitoring, and technology innovation.[1]

The Alberta Energy Regulator (AER) is an independent agency of the government responsible for regulating the development of energy resources in Alberta. A regulation known as Directive 081: Water Disposal Limits and Reporting Requirements for Thermal In Situ Oil Sands Scheme has been compiled in order to regulate water management within the industry. Its main purpose is to set disposal limits and a methodology for reporting water injection streams in order to meet water balance requirements. This provides a medium to monitor performances and provides the ability to compare water use among thermal operations in the Alberta region. The disposal limits enforce efficient recycle of produced water and effective use of makeup water. Optimal use of water and efficient use of energy can be achieved through efficient water treatment, recycle, and disposal.[10]

Allowed Water Disposal Quantities

Water disposal of any particular production facility is impacted by the following:

- Technique to recover the bitumen
- Water treatment technology
- Water qualities and quantities

Directive 081 has established a formula for disposal limits in order to meet the objective of efficient produced water recycle and effective makeup water use. This formula sets forth a unique limit for each operator. As shown below, the equation differs depending on the type of thermal technique used and is based on annual volumes in cubic meters.[10]

For SAGD,

$$\textit{Disposal Limit}\ \% = \frac{(0.10 * \textit{Produced Water}) + (0.03 * \textit{Fresh Water}) + (0.35 * \textit{Brackish Water})}{(\textit{Produced Water} + \textit{Makeup Water})} * 100\%$$

For CSS,

$$\textit{Disposal Limit}\ \% = \frac{(0.07 * \textit{Produced Water}) + (0.03 * \textit{Fresh Water}) + (0.35 * \textit{Brackish Water})}{(\textit{Produced Water} + \textit{Makeup Water})} * 100\%$$

It is expected that the AER will continually revisit these limits and may make amendments to them by assessing changes in the following key factors[10]:

- Impacts on water treatment, recycle, and disposal through changes in bitumen recovery technology
- Change of field operations and water handling by operators
- Water treatment, recycle, and steam generation technology improvements
- Conformity with other AER directives
- Changes in regulations and policies by government agencies

Water management is achieved through a rigorous regulatory system by the province. Its primary responsibility is to oversee the water withdrawals and disposal water released into Alberta's rivers. In addition, frameworks are set in place through regional environmental groups to monitor the collective effects of oil sands developments in the Lower Athabasca region. The Athabasca River is maintained at natural flow conditions. Weekly caps on the amount of water withdrawals are placed in order to maintain the river's flow. Water qualities of surface and groundwater of the regional aquifers and Athabasca River are routinely monitored to maintain overall qualities. In addition to consultants and several multistakeholder groups, staff also monitor, approve, and ensure projects are in compliance. This includes the auditing of data provided by operators. Furthermore, players in the industry are required to submit progress reports, which are made public on their Web site. These reports serve the purpose of fully disclosing the details of their operations.[1]

Greenhouse Gas Emissions and Energy Efficiency

Many believe that the global fresh water supply is threatened by climate change caused by increased GHG emissions. The global distribution of water must be considered in order to understand this issue. The supply of water is either salty or fresh. Salty water makes up 98% of the world's supply, which consist of the oceans and seas. That leaves only 2% of fresh water sources. Of this 2%, 70% comes from snow and ice, 30% from ground water, less than 0.5% comes from surface water, which consists of lakes and rivers, and finally less than 0.05% comes from the atmosphere. As emissions rise yielding in higher average temperatures, fresh water could potentially become a diminishing resource.[11]

In keeping pace with North America's changing climate policies, the Alberta government has regulations in place to control GHG emissions by investing in emerging technologies that aid in reducing the emissions at the source. Alberta currently contributes to one-third of the total amount of emissions from Canada. These are mainly from coal-fired electricity and oil and gas activity. In Alberta, the oil sands account for approximately 21% of the emissions. This equates to 8% of Canadian emissions and less than 0.15% on a global level. Figure 7 below shows Alberta's GHG emissions and how the oil sands stack up against the other sources; Figure 8 illustrates where the oil sands' emissions stand relative to the rest of the nation, while Figure 9 shows how the oil sands' emissions compare to the rest of the world.[12]

The total GHG per barrel of oil produced otherwise termed the GHG intensity continues to decrease with increasingly efficient operations or lower SORs; however, the total GHG emissions have increased due to higher production rates. The oil sands are subject to mandatory GHG targets by the Alberta government and as such are working to reduce their emissions to meet these regulations. Figure 10 depicts Alberta's upcoming targets. This shows that GHG mitigation is planned through improving the energy efficiency of oil sands operations, capturing and storing carbon, and using greener methods for production.[12] A more in-depth look at energy-efficient SAGD operations is to follow.

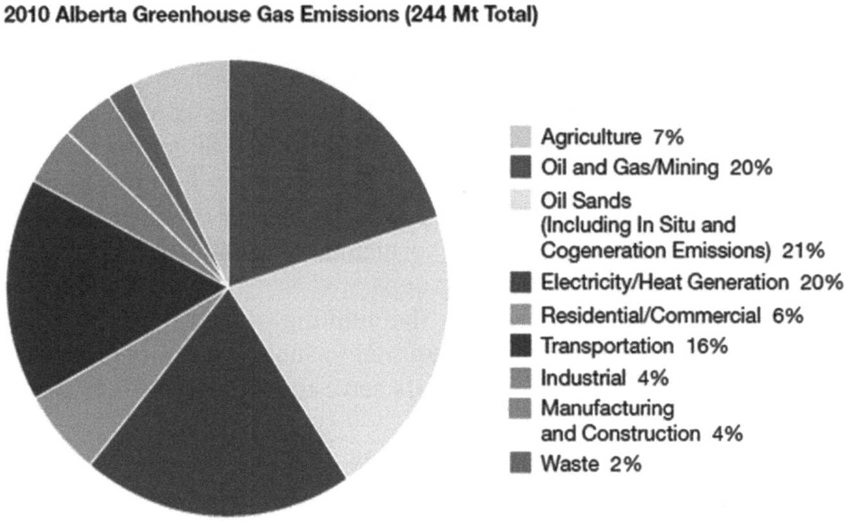

FIGURE 7 Alberta greenhouse gas emissions (2010).[12]

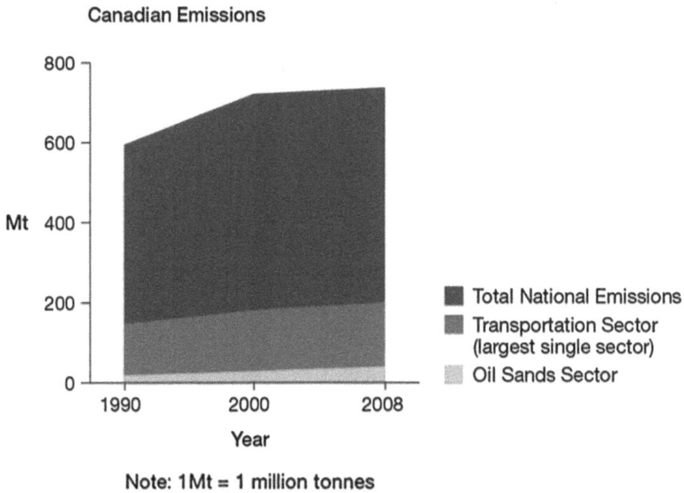

FIGURE 8 Canadian emissions (2008).[13]

Life Cycle of Well-to-Wheels

GHG emissions are generated from the stage at which oil is produced through to the refining process and ends with the gasoline consumption in vehicles. This process is generally termed the life cycle of the well-to-wheels process. It can be broken down into the following:

- Oil extraction, for example, through the SAGD process.
- Upgrading of bitumen or separation of light ends from heavy material.

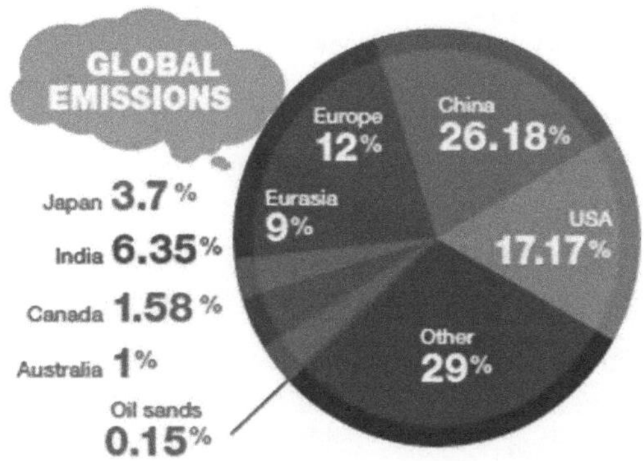

FIGURE 9 2010 Global sources of emissions.[16]

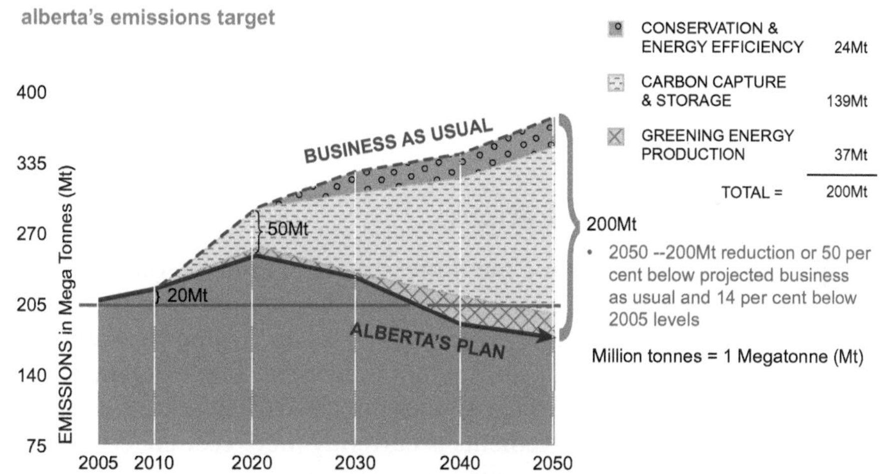

FIGURE 10 Alberta's emissions target.[12]

- Transportation of diluted bitumen to refineries.
- Processing of diluted bitumen into transportation fuels such as kerosene, gasoline, and diesel.
- Transportation of products to market for vehicle consumption.
- Vehicle combustion of oil.

It has been determined that the final use of the oil that is combusted and emitted through the tailpipes of vehicles accounts for 70–80% of the total emissions in the life cycle. These emissions are independent of the type of crude used to produce the gasoline. Figure 11 portrays how the full life cycle of GHG emissions from the oil sands compares to other sources in the world. It emphasizes the fact that the tank to wheels

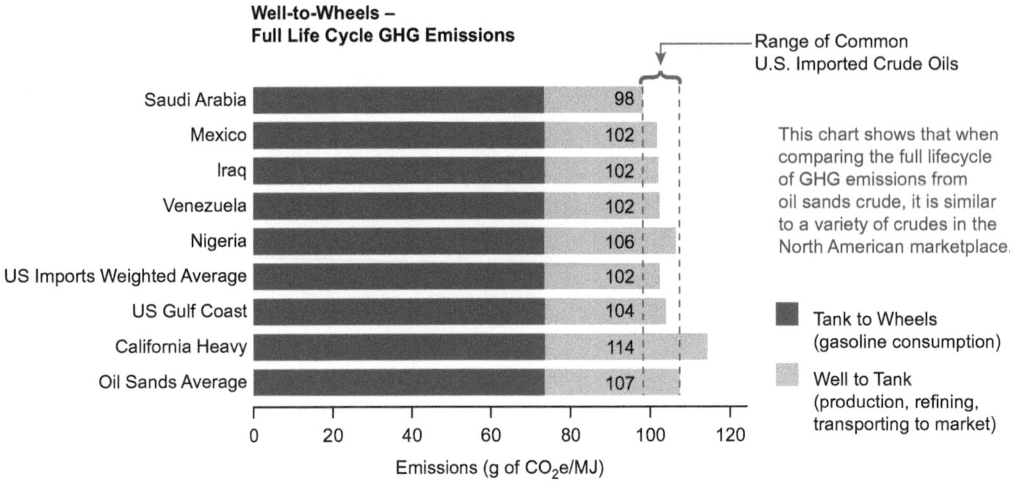

FIGURE 11 Well-to-wheels full life cycle GHG emissions.[13]

holds the greater carbon footprint while the well to tank has a relatively smaller footprint and varies with crude type.[13]

Reducing GHGs in SAGD Operations

GHG mitigation continues to become more pressing with the maturing of the oil sands industry. As energy prices continue to rise and more importantly as the affects of these emissions are felt in other industries, companies will be required to achieve greater energy-efficient operations through technology innovation and operational improvements. An efficiently run plant translates to minimized GHG emissions. Reducing GHG emissions through efficiency improvements is currently more cost-effective and feasible from a technology standpoint than carbon capture and storage technology. First and foremost, reducing the SOR, or the amount of steam required to produce each barrel of oil, helps in decreasing energy consumption and GHG emissions as this results in less natural gas usage and smaller water treatment facilities.

For the SAGD process, GHG emissions are a result of the energy required to produce the oil; this includes lifting the oil from the reservoir, separating and treating the oil and water, and generating steam to inject downhole. An investigation at a more granular level reveals that there are several areas within the SAGD operation that require energy or from where energy in the system may be lost therefore causing energy inefficiencies.

If the operation runs on downhole pumps, then the energy consumed by these must be considered. The power required for these pumps is strongly dependent on the reservoir pressure, depth of the reservoir, and the amount of material that must be pumped to the surface. The energy needed is supplied by a source of electricity that may be generated on-site through a cogeneration facility or electricity imported from the grid.

The separation of the oil, water, and gas mixture does not require a significant amount of energy. It consists of a separator from where the produced gas is sent to be treated for hydrogen sulfide removal. The gas can then be used as fuel to the steam boilers. Post treatment, the oil, which is now a mixture of diluent and bitumen, is stabilized, which involves the removal of light ends, mainly in the methane to butane range, before being sent to tanks to be transported to market. The de-oiled water requires further treatment before being heated and injected back into the wells. If the water treatment technique being used is evaporation, then the electricity requirements of this process must be accounted for. Additionally, the pumping required in these processes is the major source of energy consumption as they run on electricity.

Finally, the major source of emissions comes from the generation of steam. Fuel gas, which is generally a mixture of produced gas with natural gas supplementation, is burned in the boilers to create the steam required for the operation.

In addition to energy consumers, there are energy losses that occur at various points in the process. The majority of the energy may be lost downhole as heat lost to the Earth. This is a result of a heat imbalance between the heat injected as steam and the heat exiting the reservoir with the emulsion. Additionally, water losses in the reservoir represent the heat that is not returned into the system. As a result, more water lost to the reservoir denotes higher rates of makeup water to be supplemented to maintain steam rates.[14] Since this makeup water comes in cold, it requires heat to sustain temperatures for optimal water treatment conditions. Furthermore, there are losses at the well pad that occurs with pressure change across the wellbore, which result in flashing at the well pads. Heat is lost at the stack of the steam boilers that increases with higher stack temperatures. Lastly, there exist inefficiencies of using glycol as a heat exchange media in place of direct heat exchange within the process streams that ultimately results in heat lost or rejected to the atmosphere via an air cooler located in the glycol circuit.

Energy losses can be minimized by applying some operational and technological changes. To begin with, maintaining a lower reservoir pressure minimizes the heat lost to the Earth. Furthermore, the reservoir pressure has a direct correlation with the SOR of the reservoir. As the reservoir pressure decreases, there is lower water leakage and a lower SOR.[14] Moreover, achieving a lower SOR results in the need for smaller water treatment facilities and less natural gas fired to the boilers. Additionally, using downhole pumps in the reservoir can help to deliver an emulsion at a higher temperature at the well pads, thus resulting in less vaporization.

Recovery of the heat lost in the stack by exchanging it to preheat air for the combustion process or other possible process streams is another technique.[14] Not only does recovering some of this heat reduce the losses through the stack, but also by using this heat to increase the combustion air temperature, the boiler can be operated more efficiently. A reduction in the stack temperature also leads to natural gas savings, which equates to lower GHG emissions.

Additionally, optimizing the heat integration of the overall system reduces dependencies on the glycol system and therefore reduces the glycol air cooler duty, which signifies a reduction in heat rejected to the atmosphere.[14]

Direct and Indirect Carbon Dioxide Emissions

Carbon dioxide is a primary GHG. The GHG emissions from a SAGD facility can be broken up into direct and indirect sources of carbon dioxide. The direct emissions are mainly a result of the natural gas burned in the boilers to generate the steam needed for injection. In some cases, where the plant has a heat source deficit, a glycol heater may be needed, which contributes to these direct emissions. The indirect sources constitute a smaller quantity relative to the direct sources; these are the emissions generated from the production of power, trucking of solid waste from site, and the production of imported chemicals used for oil and water treating.

Cogeneration Plants for Self-Sustaining Power Demands

Some operators have incorporated a cogeneration facility on-site to provide the power needs of the plant. A cogeneration unit has the ability to produce power and steam simultaneously. This system generates steam by a heat recovery steam generator that uses the exhaust gas from the gas turbine as fuel. The power generated is used for the needs of the plant and the net power is then sent to the power grid, which replaces the use of coal-fired generation. There are several benefits for integrating a cogeneration plant into a SAGD facility. These can be summed up as follows[14]:

- Improved power reliability.
- Improved overall energy efficiency. The waste heat from the gas turbines is used to generate steam.
- Improved electrical generation efficiency when compared to stand-alone combined cycle facilities or coal generation.
- Improved economics of selling power into the grid.

Power generation credits may be awarded based on the efficiency of incorporating a cogeneration facility when compared to a stand-alone combined cycle facility. This efficiency would be greater when compared to coal power generation since coal is a higher carbon content fuel and would thus generate greater GHGs. This credit is given since the power produced on-site would offset power that would have been produced at another location, which eliminates the GHGs that would have otherwise been generated.[14]

Energy savings can also be achieved through this process as it allows a plant to generate its own steam and electricity needs. This is through efficient fuel consumption versus independent generation of steam and electricity. Less fuel burned equates to less GHG emissions released to the atmosphere.

Conclusions

The world's demand for energy is expected to rise over the next several decades. The depletion of conventional crude reserves is giving way to a surge in unconventional crude projects to satisfy global demand. One of the largest reserves of unconventional crude and one that holds strong financial value is the Alberta oil sands. The oil sands stand strong in the global market as some of the largest energy companies in the world are competing to develop and invest in this resource· There are several methods of extraction used in this region, and as discussed above, the SAGD method requires significant amounts of water and energy to produce oil. As such, this resource must be sustainably developed while minimizing environmental impacts.

Many Canadian oil sands producers have joined hands in an effort to sustainably grow their operations in an environmentally responsible manner. This is achieved through an organization known as the Canadian Oil Sands Innovation Alliance (COSIA). The focus lies on bringing together companies with similar goals to share best practices that will drive innovation and operational improvements with respect to the environment.

In addition to COSIA, the Alberta government has continually worked to encourage sustainable development of the oil sands resource in the interest of all the people of Alberta. With these two organizations working together, the oil sands are making strides in meeting the growing global demands of energy while keeping a close eye on water use and GHG emissions. In the end, only environmentally sound oil sand resource development is sustainable.

References

1. Government of Alberta. *Oil sands water.* November 2013. http://oilsands.alberta.ca/FactSheets/Water_FSht_Nov_2013_Online.pdf [Retrieved December 2013].
2. Lightbrown V. *Oil, gas mining.* Issue 2. *New SAGD technologies show promise in reducing environmental impact of oil sand production,* vol. 1; March 2014. http://albertainnovates.ca/media/20420/sagd_technologies_ogm_lightbown.pdf [Retrieved June 2014].
3. Pacific Institute. *A ceres report. Water scarcity and climate change.* February 2009. http://www.pacinst.org/wp-content/uploads/sites/21/2013/02/full_report30.pdf [Retrieved April 2014].
4. Government of Alberta. *Alberta's clean energy story.* January 2014. http://oilsands.alberta.ca/images/CEnergy_Brochure_Jan_2014_Online.pdf [Retrieved January 2014].
5. Government of Alberta. *Oil sands the resource.* September 2013. http://oilsands.alberta.ca/FactSheets/Resource_FSht_Sep_2013_Online.pdf [Retrieved December 2013].
6. SAGD drilling parameters evolve for oil sands. *Oil Gas J.* June 7, 2004.
7. Centre for Energy. *Canada's oil sands.* 3rd ed. November 2011. http://www.centreforenergy.com/shopping/uploads/12.pdf?1662014174741 [Retrieved June 2014].
8. Canadian Association of Petroleum Producers. *Oil Sands Today. What are Oil Sands?* http://www.oilsandstoday.ca/whatareoilsands/Pages/RecoveringtheOil.aspx [Retrieved August 2014].
9. Hill R. *Thermal in situ water conservation study, a summary report.* May 2012. http://www.ai-ees.ca/media/6868/thermal-in-situ-water-summary-report.pdf [Retrieved Janurary 2014].

10. *Directive 081: water disposal limits and reporting requirements for thermal in situ oil sands schemes.* November 2012. http://www.aer.ca/documents/directives/Directive081.pdf [Retrieved June 2014].

11. Mcintyre N. Department of Civil and Environmental Engineering, Grantham Institute for Climate Change, Imperial College of London. *How will climate change impact on fresh water security?* December 2012. http://www.theguardian.com/environment/2012/nov/30/climate-change-water [Retrieved March 2014].

12. Government of Alberta. Climate change. September 2013. http://oilsands.alberta.ca/FactSheets/CChange_FSht_Sep_2013_Online.pdf [Retrieved December 2013].

13. Government of Alberta. *Oil sands greenhouse gases.* March 2014. http://oilsands.alberta.ca/FactSheets/GHGs_FSht_March_2014.pdf [Retrieved March 2014].

14. Buchanan I, Westphalen D, Davies S, Chawla N, Gieseman J. *SAGD energy efficiency study.* September 2009. http://www.ai-ees.ca/media/12516/aeri-jacobs_consultancy_sagd_study-1009c-jg_final_rev1.pdf [Retrieved January 2014].

15. Technologies lower in situ bitumen recovery costs. *Oil Gas J.* September 6, 2010.

16. Government of Alberta. *Alberta's Oil Sands: The Facts.* January 2014. http://www.energy.alberta.ca/OilSands/pdfs/AlbertasOilSandsFactsJan14.pdf [Retrieved October 2014].

10

Coal Use as a Cause of Water Quality Impairment

Lawrence B. Cahoon
DEPARTMENT OF BIOLOGY AND MARINE BIOLOGY, UNC WILMINGTON, WILMINGTON, NC, USA

CHAPTER OUTLINE

Introduction

Coal arguably drove the early stages of the Industrial Revolution from the late eighteenth century onward. This energy-dense, solid fuel could be mined, transported, stored, and burned (usually to fire boilers, but also as a source of direct heating) relatively easily with the technological capabilities of the times. The advent of far more efficient internal combustion engines fueled by liquids partially displaced coal, particularly as technical capabilities and infrastructure grew to match the needs of liquid fuel technologies. One key moment in the transition from coal-fired power to petroleum fuels was First Admiralty Lord Winston Churchill's decision in World War I to convert the British navy from coal-fired engines, with their strategically critical needs for coaling stations and colliers and crew-exhausting recoaling chores, to oil-fired engines, with more power and more flexible fuel storage and handling capabilities.[1] Coal has remained a significant industrial fuel, however, particularly for firing power plants and large industrial heating systems. Its advantages for such uses include large supply, easy storage, and relatively low price to consumers.

Some of coal's environmental impacts, particularly air pollution, have been well known for many years. Coal is an impure substance; its biological and geological origins

Food, Energy, and Water. http://dx.doi.org/10.1016/B978-0-12-800211-7.00010-7

cause inclusions of many elements, notably sulfur, but including a long list of others. Many of those elements are vaporized and/or oxidized by the combustion process, and carbon combustion is never 100% efficient, so coal smoke contains many pollutants, including black soot, other particulates, and gases. London's historically famous smogs were largely caused by widespread use of coal for heating homes. Coal combustion's role in causing acid rain via production of SO_x and NO_x compounds has been recognized and addressed by a variety of measures, including scrubbers and emissions cap and trade programs. More recently, the accumulation of large quantities of coal ash and occasional but massive spills of that material, which contains a variety of toxicants, has attracted more public and regulatory attention.

Carbon dioxide emissions from coal and other fossil fuels combustion are now thought to be a significant driver of climate change and ocean acidification.[2] Owing to coal's carbon–carbon bond structure, as opposed to the hydrocarbon chemistry of petroleum and natural gas, coal generates approximately twice as much carbon dioxide per unit energy derived as natural gas. Consequently, climate change mitigation policies often focus on reduced coal use in particular. The Obama Administration's recent climate change initiative, for example, encourages energy utilities to switch from coal-fired power plants to natural gas-powered plants, among other options to reduce carbon dioxide emissions (http://www.epa.gov/climatechange/impacts-adaptation/fed-programs.html). Market forces have separately favored this transition as well, at least in the United States, as the "fracking" boom has enhanced national supplies of natural gas.

Coal continues to be heavily used in other industrialized and developing countries, notably China, with its extensive domestic supplies of coal. Global supplies of coal are estimated to be sufficient to support several centuries' projected demand, as opposed to much more limited and uncertain projections for petroleum and natural gas reserves. Consequently, barring significant shifts away from coal as an energy source, its use for the foreseeable future is likely to remain a large, if not dominant, factor in the global economy.

Coal Production Impacts on Water

Coal mining has interacted with groundwater from the beginning of its large-scale development. Prior to the advent of the steam engine mining was limited to shallow depths by flooding by groundwater and inflows from rain. Steam-powered pumps were much more powerful than muscle-driven pumping systems, allowing much more effective removal of water from what then became deeper and deeper mines. As more coal became accessible to mining, better steam-powered pumps became cheaper and more widely available, creating a positive feedback that supported even more use of coal. Not surprisingly, the rise of the industrial powers in the nineteenth century was powered by coal, and so groundwater removal from mines and disposal off-site was the first interaction of water with coal use.

Closely linked to the initial interaction of groundwater and rain water with coal mines was the exposure of mine tailings to rain water. Not all materials extracted from coal mines were marketable coal, and this unwanted material was usually discarded on the surface, exposed to air and rainfall. Coal was typically not the only mineral found in the ground by mining for coal, so other minerals, some containing various metals, were then exposed to oxidation and dissolution. The result often was an acidic mixture, derived primarily from oxidation of metallic sulfides and sulfur residues in coal dust to sulfuric acid, which enhanced leaching of other materials, including various metals and their compounds. So-called "acid mine drainage" is, of course, derived from many types of mining, not just coal mining, but is one kind of water quality issue resulting from coal mining. Such drainage poses a multidimensional threat to surface waters: low pH and high concentrations of sometimes toxic solutes.

Mining and disposal of mine tailings poses another potential threat to water quality: sedimentation. Land disturbance activities such as mining move large quantities of soil aside to reach coal deposits, particularly large-scale surface forms of mining like "contour" (strip) mining and the more recently adopted mountaintop removal mining. Older style shaft mining generated smaller volumes of mine debris and tailings, but the surface mining practices generate truly significant volumes of unwanted material that is moved away from the active mining site and generally disposed nearby. Frequently, in areas with high relief, such disposal practices generate large-scale downslope movement of sediment and rock, which can bury small hillside watersheds altogether and allow erosion of sediments downslope and downstream. Sedimentation chokes or buries stream channels, fills natural and artificial impoundments, severely impacts the biota, and degrades the ecosystem services performed by these aquatic ecosystems.

Coal Combustion Impacts on Water

The major use of coal currently is for fueling large industrial boilers and heating systems, especially for generating electricity, but also for calcining limestone in cement manufacture and for other industrial purposes. Residential use of coal for heating and cooking, once widespread, is now more commonly found in developing countries, notably China. These uses drive combustion of a very large amount of coal and thereby generate very large quantities of combustion by-products. Those by-products may be broadly categorized as airborne particulates, gaseous emissions, and solid waste.

Airborne particulates include soot or black carbon from incomplete combustion, as well as volatilized materials that condense upon cooling when they leave the combustion chamber, forming "fly ash," which is mostly silica (SiO_2), but includes a variety of other compounds and elements that coprecipitate with the silica. Some of the gaseous emissions can also generate atmospheric aerosols, notably sulfur oxides (SO_x), which can further oxidize in air to sulfate ion (SO_4^{2-}), which is an important atmospheric aerosol.

Inorganic gas emissions from coal combustion include materials vaporized by the high temperatures at which coal burns, such as mercury, and materials oxidized by the combustion process, such as sulfur, yielding sulfur oxides, (SO_x), nitrogen oxides (NO_x), and, of course, carbon (primarily as carbon dioxide, CO_2, but also carbon monoxide, CO). NO_x compounds are formed by the high-temperature oxidation of nitrogen (N_2) and oxygen (O_2) in the combustion chamber, but SO_x compounds are formed by oxidation of sulfur contained in coal itself. Some coals, notably "hard coal" or anthracite, contain relatively low levels of sulfur, whereas "soft" or bituminous, subbituminous, and lignite coals often contain relatively more sulfur and generate correspondingly more SO_x emissions.

SO_x and NO_x emissions from coal-fired power plants and other sources are well known to have been largely responsible for so-called "acid rain" and "acid fogs." Atmospheric oxidation of SO_x and NO_x gases and reactions with water vapor yield a mixture of sulfurous (H_2SO_3) and sulfuric (H_2SO_4) acids and nitrous (HNO_2) and nitric (HNO_3) acids, respectively. Aerosols like sulfate ion (SO_4^{2-}) act as cloud condensation nuclei that form liquid droplets of water vapor, which yield precipitation with much higher acidity (lower pH) than normal rainfall. Normal atmospheric conditions yield rainfall with a pH no lower than about 5.6 (versus neutrality at pH 7.0) as a result of the reaction of water vapor with carbon dioxide (CO_2) to form the relatively weak carbonic acid (H_2CO_3). The much stronger nitrogen and sulfur acids found in polluted air can drive rainfall pH levels as low as $\sim$2, with important and severe consequences for ecosystems experiencing such acidic precipitation.

Acid precipitation is well known to have direct impacts on surface waters as well as effects mediated through processes in surrounding watersheds. Direct deposition of acidic precipitation lowers the pH of surface water, particularly oligotrophic waters in regions with poorly buffered soils, such as the granite-derived soils of the Adirondack Mountains in the northeast region of United States. Eutrophic waters are more resistant to acidification, owing to several biologically mediated processes. Nitric acid is a source of the primary plant nutrient, nitrate (NO_3^-), uptake of which by plants drives stoichiometric excretion of hydroxyl ion (OH^-), which then neutralizes acid. Plant production also involves uptake of CO_2, which alters the carbonic acid–bicarbonate equilibrium to favor more alkaline conditions. Sulfuric acid, although not a source of plant nutrients, yields sulfate ion in solution. Sulfate ion can be used by anaerobic bacteria in the absence of molecular oxygen (O_2) as a source of atomic oxygen for cellular respiration, yielding hydrogen sulfide (H_2S). Hydrogen sulfide can then react to form insoluble metallic sulfides or be oxidized upon contact with molecular oxygen in water to reform sulfuric acid/sulfate ion. This sulfate reduction cycle also plays an important role in the cycling of mercury, which will be discussed later.

Acid deposition on landscapes is well known to have a wider range of sometimes more subtle effects on receiving waters. Again, the buffering capacities of the habitat, in this case soils and parent rock, matter significantly. Soils with significant concentrations of labile cations, particularly calcium (Ca^{2+}) and magnesium (Mg^{2+}), can yield significant quantities of the salts of these elements, with relatively minor impacts on the acidity of receiving surface waters. Soils with poorer ion exchange capacities, however, can be less effective at protecting surface waters from acidification. Moreover, acidic soil solutions are much more effective at solubilizing metals in soils and driving their export to surface waters. Various studies have shown that enhanced metals loading and resulting toxicity has been a major factor in acid rain-induced impacts on surface water ecosystems (see review by Schindler[3]).

Mercury emissions from coal combustion have also received considerable attention. Coal contains small but significant amounts of mercury, which is volatilized by combustion and emitted as gaseous mercury, typically in elemental form (Hg^0). Mercury is oxidized in the atmosphere to Hg^{2+}, which dissolves more readily in condensed water vapor and falls in precipitation. Under certain circumstances (anaerobic conditions with high levels of organic matter and sulfate, as found in some wetlands and sediment habitats in surface waters) mercury can be methylated by sulfate-reducing anaerobes to methylmercury (MeHg). Methylmercury is fat soluble, so it is biomagnified, becoming concentrated in organisms toward the tops of aquatic food chains. This is the single most important cause of fish consumption advisories in many aquatic ecosystems, including freshwater and coastal fisheries, as well as some oceanic pelagic fisheries for top carnivores like tunas and sharks. Methylmercury is a potent neurotoxin with particular ability to damage the developing nervous systems of fetuses and young children. Coal combustion accounts for the majority of anthropogenic mercury emissions to the atmosphere (http://www.epa.gov/hg/about.htm); total global emissions include significant emissions from volcanic activity as well.

Gaseous emissions from coal combustion also include various organic compounds, as well, as no combustion process is 100% efficient. Some of these compounds are simply volatilized from coal, e.g., n-alkanes, while others are reaction products.[4] Notable among these products as potential hazards to surface waters are polynucleated aromatic hydrocarbons (PAHs), some of which are formed during combustion by pyrolysis (breakdown of larger molecules) and some by pyrosynthesis (reactions during combustion that form larger molecules from small ones).[5] Many PAHs are water soluble although as a class they are strongly lipophilic, some are toxic and/or carcinogenic, and some can be biomagnified and thereby accumulate in higher trophic level organisms.[6,7] PAHs tend to adsorb strongly to particles, which renders them an inhalation risk for public health purposes, and are also thereby more readily deposited downwind of emission sources and transferred into aquatic and terrestrial food chains.[5,8] Coal combustion is by no means the only source of atmospheric PAH pollution, but can be a significant source of some PAHs.[9] Moreover, the relatively low temperature combustion of coal in residential heating applications may generate higher total quantities and a more diverse set of PAHs than the high temperature combustion typical of industrial processes,[9,10] making such use and the consequent intimate exposure particularly hazardous.

Related categories of organic coal combustion by-products include the nitroaromatics, formed by reactions of nitrogen, oxygen, and hydrocarbons during combustion, along with polychlorinated biphenyls (PCBs), polychlorinated naphthalenes (PCNs), dibenzo-*p*-dioxins (PCDDs) and dibenzofurans (PCDFs). Many of these compounds are also toxic and carcinogenic, acting as mutagens in various bioassays.[11] Low temperature combustion of coal, as in residential heating uses, seems particularly capable of generating these compounds.[10] As with PAHs, these persistent organic pollutants (POPs) can be strongly lipophilic, enabling them to accumulate in fatty tissues and biomagnify through food chains.

Some studies have found that coal combustion is a major source of POP loading to the environment in industrialized settings (e.g., PAHs[8]). Sourcing of these POPs specifically to coal combustion from ambient air sampling, however, can be deceptively difficult,[12] as multiple sources of these compounds, variability in emission factors within source type, and atmospheric reactions can alias the compound ratio signatures that some studies have used as source identification methods. Consequently, the most reliable information about coal combustion as a source of POPs comes from studies of stack emissions.

The largest gaseous emissions from coal combustion, of course, are not derived from its impurities but from its fundamental chemistry: carbon emissions as carbon dioxide. Coal use as fuel is one of the most important sources of anthropogenic CO_2 emissions globally, owing to its large-scale use and to coal's carbon–carbon composition (vs. petroleum and natural gas with predominantly carbon–hydrogen composition), which generates relatively more CO_2 per energy unit (http://www.eia.gov/tools/faqs/faq.cfm?id=73&t=11). Although climate change *per se* is beyond the scope of this discussion, several aspects of carbon dioxide loading to the atmosphere are relevant to consideration of coal impacts on aquatic ecosystems. First, carbon dioxide is relatively soluble in water and dissolves through a series of equilibrium reactions that generate carbonic acid (H_2CO_3), which disassociates to acid (H^+) and bicarbonate ion (HCO_3^-), thus lowering pH. The primary impact of this process manifests as ocean acidification; seawater is typically buffered at slightly alkaline pH values (~8.0–8.2) by reactions with carbonate, but anthropogenic carbon dioxide invasion of ocean waters is shifting oceanic pH values measurably.[13] Ocean acidification threatens the physiological and ecological health of oceanic species using carbonate for shell and skeleton formation, e.g., hard corals, bivalves,[14] Acidification-induced shifts in carbonate solubility relations are potentially quite significant, with implications for fisheries, habitat integrity, and ecosystem function throughout the world ocean.

Consideration must also be given to effects on the hydrological cycle driven by CO_2-induced warming. Warming can alter precipitation patterns, driving geographical shifts in rainfall patterns, increased drying in already dry areas, heavier precipitation in other areas, and alterations of the seasonality of precipitation, among other effects.[15] Given the importance of hydrology (the balance of water inputs and outputs) on the function of aquatic ecosystems, these are important considerations. For example, the effects of nutrient loading and other pollutant loadings are magnified when hydrological residence times are long in dry periods, whereas runoff and other watershed inputs are greater in wetter circumstances. Of course, global warming is driven by more than simply coal-sourced greenhouse gases and effects, so hydrological impacts must be considered in a broader context of multiple causes and effects.

Coal Waste Impacts on Water

Coal combustion also creates solid wastes, owing to the impurities in coal itself. Formation of "fly ash" was mentioned above in the context of aerial emissions, but fly ash is effectively a solid waste resulting from condensation of vaporized mineral constituents, and is managed as a solid waste. Another significant form of solid waste created by coal combustion is so-called bottom ash—material that does not combust or vaporize and that remains as residue in coal furnaces, requiring periodic removal. Standard practice is to remove fly ash from smokestacks and bottom ash from furnaces concurrently, as both processes require the temporary shutdown of a coal furnace, so in most situations the two kinds of solid residue would be handled simultaneously and become mixed. Henceforth, they will be referred to collectively as "coal ash," which may in practice also include solids generated by precipitation in scrubbers and other air quality management systems.

Coal ash has become a major waste product of coal use owing to the very large quantities of coal burned over centuries, with coal use rising particularly rapidly in recent decades. Estimates of global annual coal waste production were on the order of 5×10^8 tons two decades ago,[16] and have surely increased. The sheer quantities of this waste material have resulted in accumulations of large amounts of it with no comparably voluminous beneficial uses; use as a filler for cement has been estimated to utilize only about one-third of global production.[16] Several other uses have been exploited and will be described below, but the supply of coal ash now far exceeds demand for it, leaving large accumulations of it, on the order of 10^9 tons in just the United States.

Storage practices for coal ash typically utilize simple methods—creating piles of it on the ground or placing it into pits. Above-ground storage exposes coal ash to leaching by rain water and dispersal by wind, as much of the ash consists of fine-grained particles. The smallest coal ash particles are particularly subject to aeolian dispersal and are also the most dangerous to human health as pulmonary disease agents.[17,18] Consequently, industry practice has been to store coal ash in large pits; rain water and groundwater moisten the ash, limiting its wind dispersal problems, but posing potential for leaching, leaking, and groundwater contamination. In addition, there have been a number of breaches of wet coal ash storage pits, with catastrophic local effects. The worst such accident in the United States occurred in 2008 near Kingston, Tennessee, when a coal ash storage facility owned by the Tennessee Valley Authority breached, spilling an estimated 4,200,000 m^3 of coal ash into the Emory and Clinch River valleys, necessitating a cleanup estimated to approach $1 billion when completed (Figure 1; http://en.wikipedia.org/wiki/Kingston_Fossil_Plant_coal_fly_ash_slurry_spill). More recently (February. 2, 2014), a coal ash spill in North Carolina, United States, released an

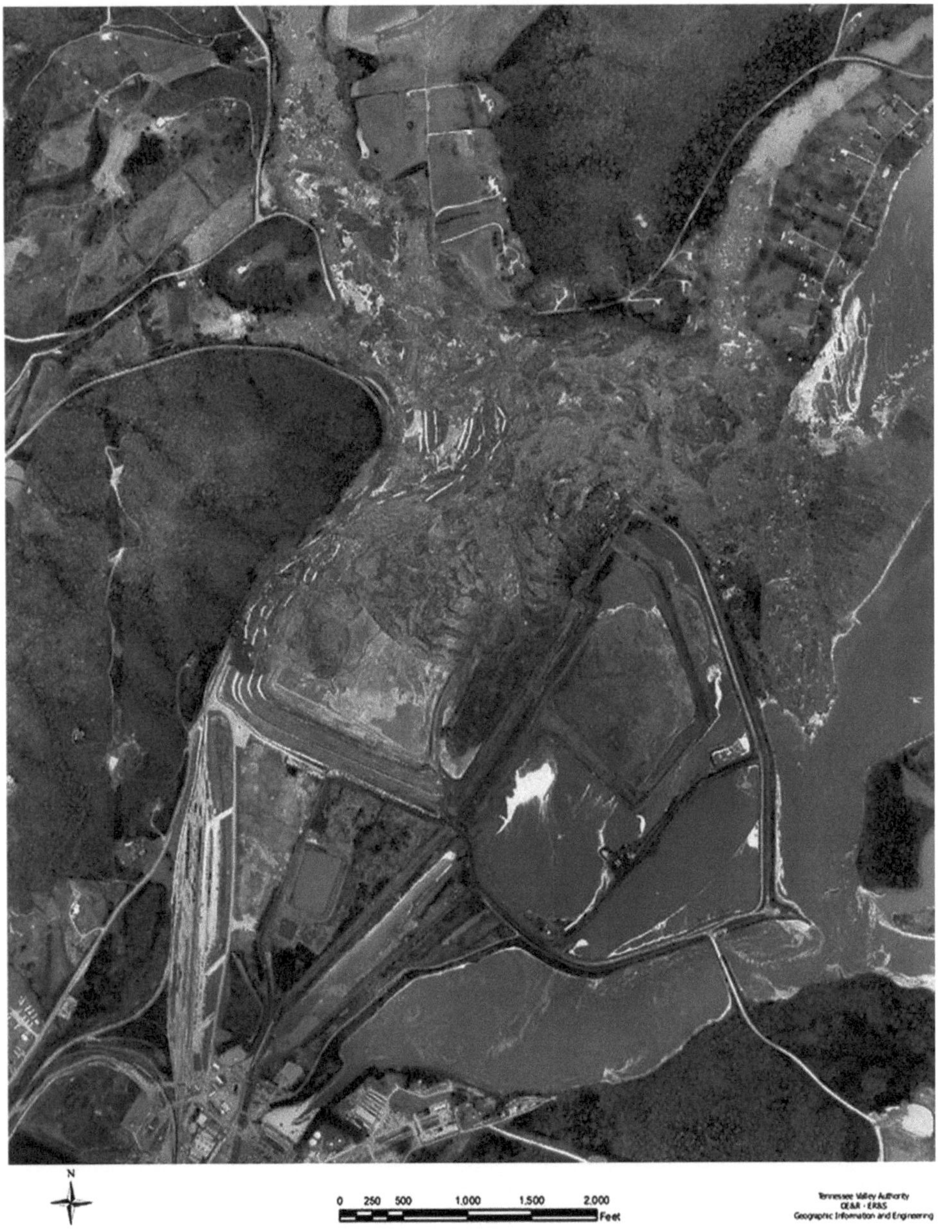

FIGURE 1 Aerial image of Tennessee Valley Authority coal ash spill in December, 2008 near Kingston, Tennessee, United States; the largest coal ash spill in the United States to date. *Tennessee Valley Authority.*

estimated 39,000 tons of coal ash into the Dan River, prompting federal investigations and nationwide focus on issues surrounding storage of coal ash wastes.

Coal ash contains a long list of elements,[19,20] e.g., a variety of potentially hazardous constituents, many of which are leachable and therefore able to cause water pollution problems.[21] The term, "hazardous," has two distinct connotations in this discussion. The common use of the term implies the potential to cause harm to human or ecological health. Legally, however, the term implies a formal designation of a given material as hazardous, for example, under the provisions of the U.S. Resource Conservation and Resource Recovery Act (RCRA) or the Toxic Substances Control Act (TSCA). Coal ash *per se* is not currently designated as a hazardous waste, so its handling and disposal is not subject directly to RCRA or TSCA. Many constituents of coal ash, including most of the "RCRA Nine" metals, e.g., arsenic, cadmium, chromium, lead, selenium, and mercury, however, are legally designated as hazardous substances. The U.S. Environmental Protection Agency (US EPA) declined to designate coal ash as a hazardous material in December, 2014, partly as such a designation would have made beneficial uses of coal ash much more problematic.

Potentially hazardous elements in coal ash, reported from analyses of coal ash from several countries and thereby spanning different sources and types of coal, are listed in Table 1. Some of these elements are simply hazardous and have no known biological function, e.g., mercury and lead. Others function as micronutrients at low concentrations, e.g., selenium, but become toxic at high concentrations. Some, e.g., chromium, are micronutrients in one valence state (Cr III) but toxic in another (Cr VI). All can react in the environment to form a variety of compounds, some more soluble than others. Controlling conditions include presence of reactants, redox conditions, and in some cases, e.g., mercury cycling, biological processes. Thus, the overall hazards posed by these contaminants in coal ash are multidimensional and require consideration of particular circumstances.

Variation in coal sources, characteristics, combustion technologies, and waste handling practices drive substantial variation in the concentrations of these constituents in coal ash. Moreover, two factors control the degree to which specific toxicants pose hazards to aquatic ecosystems: "leachability" and bioavailability. "Leachability" may be defined as the tendency for toxicants like metals to enter solution, either in elemental form or as compounds. Solution chemistry, redox reactions, complexation reactions, and hydrology act to regulate leaching. For example, elemental lead is rather less soluble than some lead compounds, but acidity can enhance leaching of elemental lead. Once in solution, additional reactions can transform compounds, in some cases rendering them more or less soluble and available for uptake by the local biota. Consequently, quantitative analyses of toxicant concentrations in coal ash must be put into more specific context before meaningful risk assessments can be performed. For the purposes of this review, therefore, presentations of quantitative concentration data have limited value.

Leaching of toxic elements, especially metals, is one of the premier environmental concerns posed by coal ash, given the large amounts of this waste material present, its exposure to rainfall and groundwater, and the general lack of strict regulation to mitigate leaching problems. Many studies have examined the chemistry of coal ash itself and ability of typical environmental conditions to support metals leaching. Huggins[26] reported that arsenic in coal ash typically occurs as readily leachable arsenate ion

Table 1 Elements Reported from Analyses of Coal Ash with Summary of Potential Hazards (Hazards are Dependent on Dosage, Route of Exposure, Chemical form, and Other Variables; "Minor Toxicants" are Considered Toxic only at High Dose or by Specific Route of Exposure)*

Element[references]	Hazard (http://www.lenntech.com/periodic/elements/
Aluminum[a,c]	Pulmonary hazard, aquatic ecosystem toxicant
Antimony[c,d]	Pulmonary hazard, skin irritant
Arsenic[b,c,d,f,g]	Major toxicant and carcinogenic (RCRA 9)
Barium[c,d]	Minor toxicant
Beryllium[a,c,d]	Pulmonary hazard, carcinogen
Bismuth[d]	Minor hazard (kidneys)
Boron[a,c,d]	Phytotoxic
Cadmium[a,b,c,d,e]	Major toxicant and carcinogen (RCRA 9)
Calcium[a,c,e]	None
Cesium[c,d]	Toxic via potassium displacement
Chromium[a,c,d]	Major toxicant and carcinogen (RCRA 9)
Copper[a,b,c,d,e]	Minor toxicant
Gallium[c,d]	Minor toxicant
Germanium[c,d]	Toxic to diatoms (displaces SiO_2 as GeO_2)
Hafnium[d]	None
Iron[b,c]	None
Lanthanum[d]	Minor toxicant
Lead[b,d]	Major toxicant and neurotoxin (RCRA 9)
Magnesium[b]	None
Manganese[b,c,d]	Pulmonary hazard
Molybdenum[a,c,d]	Minor toxicant
Nickel[c,d]	Toxicant and carcinogen
Rubidium[c,d]	Minor, pulmonary irritant
Scandium[d]	Minor toxicant
Selenium[a,c,d]	Major toxicant (RCRA 9)
Silicon[a,g]	Pulmonary irritant as SiO_2–silicosis
Silver[d]	Toxicant
Strontium[c,d]	Nontoxic
Tellurium[d]	Toxic and teratogenic
Thallium[c]	Toxic
Titanium[a,c]	Nontoxic
Uranium[c,d]	Minor toxicant
Vanadium[a,c,d]	Minor toxicant
Zirconium[d]	Nontoxic
Zinc[a,b,c,d,g]	Minor toxicant

*None of these studies reported mercury.
[a]Dreesen et al.[21].
[b]Alberts et al.[22].
[c]Queros et al.[19].
[d]Vassilev and Vassileva.[20].
[e]Fytianos et al.[23].
[f]Matjie et al.[24].
[g]Pandey and Singh.[25].

FIGURE 2 Coal ash spill residue from a pipe breach at a coal ash storage location near Eden, North Carolina, USA, February, 2014. *Photo credit: WaterKeepers Alliance/Southeast_CoalAsh.org.*

(AsO_4^{3-}: As V), but arsenic also leaches from coal ash as the more toxic reduced form (As III,[27]). Wadge and Hutton[28] demonstrated in laboratory experiments that selenium (as Se IV) was the most leachable of four toxic metals (As, Cd, Pb, and Se) found in coal ash, whereas Fytianos[23] found that lead and cadmium were most readily leachable in their analyses, and that lowered pH enhanced leaching generally. Similarly, Wang et al.[29] found lead and arsenic relatively leachable from coal ash under acidic conditions. Dutta et al.[30] argued that one potential use for large quantities of coal ash waste, backfilling old coal mines, would present leaching problems for such toxic elements as arsenic, chromium and lead unless pH modification was employed. Similarly, use of coal ash for a soil amendment in agriculture poses a risk of leaching under acidic soil conditions,[31] but can be managed with soil pH controls. Cox et al.[32] also showed significant leaching of boron, a potentially phytotoxic element, from fly ash, although bottom ash posed much less risk in this regard. Not surprisingly, many of these leachable elements and compounds, including such toxicants as arsenic, cadmium and chromium, are detectable in solution in coal ash ponds and streams receiving pond effluents.[22] US EPA has identified almost 650 wet storage sites for coal ash in the United States, most of them in the more heavily populated eastern half of the United States, (Figure 2; http://www.epa.gov/solidwaste/nonhaz/industrial/special/fossil/ccrs-fs/index.htm), so the potential for contamination of drinking water supplies and impacts on other human uses of surface waters is not trivial.

Bioavailability refers to the ability of organisms to take up materials from solution. Various studies have examined biological uptake and biomagnification of coal ash-derived pollutant metals in aquatic ecosystems. Invertebrates exposed to coal ash solutions bioaccumulate various constituents in species- and constituent-dependent fashion.[33] Guthrie and Cherry[34] demonstrated uptake of cadmium and mercury by crayfish and of selenium by mosquitofish. Reash et al.[35] examined selenium uptake by fish, which appears to be a common problem in waters receiving coal ash effluents. Selenium can bioaccumulate in fishes, causing them physiological harm and rendering them unsafe for human consumption, as can arsenic,[36,37] although chronically exposed fish populations appear to develop resistance to harm. Coal ash toxicants may also manifest effects ecologically, through impacts on exposed populations that then affect their interactions in food webs.[38] Nagle et al.[39] documented uptake of arsenic, cadmium, chromium, and selenium by turtles, but transfer of selenium as well to eggs and young, implying potential for transgenerational effects.

Alternative Uses of Coal Ash

Waste products generated in large quantities offer possibilities for other uses, both as a means of avoiding waste storage and disposal costs and deriving income. Coal ash has found such uses in the past, and new research has identified more potential alternative uses.[40,41] Some of the largest volume uses for coal ash have included use as fill for roadbeds and as an additive to concrete mixes[42,43] and have been mentioned above. These uses are driven by demand for and local availability of coal ash as an inexpensive filler material, however, not from any particularly desirable aspect of coal ash for this purpose. Consequently, these more traditional uses of coal ash have certainly not kept pace with the supply of coal ash, as large quantities of coal ash remain in storage at many locations. Moreover, concerns have been raised about leaching of toxicants from coal ash used as road fill, as groundwater can interact with this material, albeit relatively slowly owing to compaction. Additional uses for coal ash that might divert some of it from simple waste storage have been investigated, and include synthesis of ceramics,[44] in flue gas desulfurization,[45] as adsorbents for a variety of waste compounds,[46,47] synthesis of zeolites,[45,46] a source of potassium silicate fertilizer,[45] extraction of useful metals such as aluminum, iron, and titanium,[24] and a soil amendment for agriculture,[46] although not without risk of toxicant leaching at high application rates[25,48] and possibly sequestration of soil phosphorus,[49] as well as for other bulk uses in construction and engineering applications.[50] It remains to be seen, however, if these uses can be expected to reduce safely the very large volumes of coal ash now in storage.

Conclusions

Coal use has supported industrialization and economic growth for centuries, but has also generated a wide variety of environmental impacts. This review focuses on coal impacts on aquatic ecosystems, whereas most considerations of coal use impacts have addressed air quality concerns. Recent attention to coal ash spills, however, has prompted concerns about water pollution effects from coal use, particularly the long-term accumulation and storage of truly immense quantities of solid wastes generated thereby. Coal ash contains a long list of elements and compounds, some of which pose serious threats of environmental pollution, and many of which are potent human toxicants. Interaction of coal ash with rainfall and groundwater clearly poses threats to the integrity of aquatic ecosystems and, ultimately human health, if these effects are not adequately mitigated. Cleanup costs may ultimately represent one of the largest costs of coal use, an aspect of coal pricing not currently reflected in energy markets.

References

1. Dahl EJ. Naval innovation: from coal to oil. *Jt Forces Q Winter* 2001;(2000-2001):50–6.
2. IPCC. Summary for Policymakers. In: Stocker TF, Qin D, Plattner G-K, Tignor M, Allen SK, Boschung J, Nauels A, Xia Y, Bex V, Midgley PM, editors. *In: climate change 2013: the physical science basis. Contribution of working group I to the Fifth assessment report of the intergovernmental panel on climate change.* Cambridge, United Kingdom and New York, NY, USA: Cambridge University Press; 2013.
3. Schindler DW. Effects of acid rain on freshwater ecosystems. *Science* 1988;**239**:139–240.
4. Zhang Y, Schauer JJ, Zhang Y, Zeng L, Wei Y, Liu Y, et al. Characteristics of particulate carbon emissions from real-world Chinese coal combustion. *Environ Sci Technol* 2008;**42**:5068–73.
5. Liu G, Niu Z, Van Niekerk D, Xue J, Zheng L. Polycyclic aromatic hydrocarbons (PAHs) from coal combustion: emissions, analysis, and toxicology. *Rev Environ Contam Toxicol* 2008;**192**:1–28.
6. Mastral AM, Callén MS, Garcia T. Toxic organic emissions from coal combustion. *Fuel Process Technol* 2000;**67**:1–10.
7. Chen Y, Sheng G, Bi X, Feng Y, Mai B, Fu J. Emission factors for carbonaceous particles and polycyclic aromatic hydrocarbons from residential coal combustion in China. *Environ Sci Technol* 2005;**39**:1861–7.
8. Simcik MF, Eisenreich SJ, Lioy PJ. Source apportionment and source/sink relationships of PAHs in the coastal atmosphere of Chicago and Lake Michigan. *Atmospheric Environ* 1999;**33**:5071–9.
9. Ravindra K, Sokhia R, Van Grieken R. Atmospheric polycyclic aromatic hydrocarbons: source attribution, emission factors and regulation. *Atmospheric Environ* 2008;**42**:2895–921.
10. Lee RGM, Coleman P, Jones JL, Jones KC, Lohmann A. Emission factors and importance of PCDD/Fs, PCBs, PCNs, PAHs and PM10 from the domestic burning of coal and wood in the U.K. *Environ Sci Technol* 2005;**39**:1436–47.
11. Purohit V, Basu AK. Mutagenicity of nitroaromatic compounds. *Chem Res Toxicol* 2000;**13**:673–92.
12. Galarneau E. Source specificity and atmospheric processing of airborne PAHs:Implications for source apportionment. *Atmospheric Environ* 2008;**42**:8139–49.
13. Sarmiento JL, Gruber N. *Ocean biogeochemical dynamics.* Princeton, NJ: Princeton University Press; 2006. 503 pp.
14. Talmadge SC, Gobler CJ. The effects of elevated carbon dioxide concentrations on the metamorphosis, size, and survival of larval hard clams (*Mercenaria mercenaria*), bay scallops (*Argopecten irradians*), and Eastern oysters (*Crassostrea virginica*). *Limnol Oceanogr* 2009;**54**:2072–80.
15. Oki T, Kanae S. Global hydrological cycles and world water resources. *Science* 2005;**313**:1068–72.
16. Manz OE. Worldwide production of coal ash and utilization in concrete and other products. *Fuel* 1997;**76**(8):691–6.
17. Linak WP, Miller CA, Seames WS, Wendt JOL, Ishinomori T, Endo Y, et al. On trimodal particle size distributions in fly ash from pulverized-coal combustion. *Proc Combust Inst* 2002;**29**:441–7.
18. Linak WP, Yoo J-I, Wasson SJ, Zhu W, Wendt JOL, Huggins FE, et al. Ultrafine ash aerosols from coal combustion: characterization and health effects. *Proc Combust Inst* 2007;**31**:1929–37.

19. Querol X, Juan R, Lopez-Soler A, Fernandez-Turiel JL, Ruiz CR. Mobility of trace elements from coal and coal combustion wastes. *Fuel* 1996;**75**:821–38.

20. Vassilev SV, Vassileva CG. Geochemistry of coals, coal ashes and combustion wastes from coal-fired power stations. *Fuel Process Technol* 1997;**5I**:19–45.

21. Dreeson DR, Gladney ES, Owens JW, Perkins BL, Wienke CL, Wangen LE. Comparison of levels of trace elements extracted from fly ash and levels found in effluent waters from a coal-fired power plant. *Environ Sci Technol* 1977;**11**:1017–9.

22. Alberts JA, Newman MC, Evans DW. Seasonal variations of trace elements in dissolved and suspended loads for coal ash ponds and pond effluents. *Water, Air Soil Pollut* 1985;**26**:111–28.

23. Fytianos K, Tsaniklidi B, Voudrias E. Leachability of heavy metals in Greek fly ash from coal combustion. *Environ Int* 1998;**24**:477–86.

24. Matjie RH, Bunt JR, van Heerden JHP. Extraction of alumina from coal fly ash generated from a selected low rank bituminous South African coal. *Miner Eng* 2005;**18**:299–310.

25. Pandey VC, Singh N. Impact of fly ash incorporation in soil systems. *Agriculture, Ecosyst Environ* 2010;**136**:16–27.

26. Huggins FE, Shah N, Zhao J, Lu F, Huffman GP. Nondestructive determination of trace element speciation in coal and coal ash by XAFS spectroscopy. *Energy & Fuels* 1993;**7**:482–9.

27. Turner RR. Oxidation state of arsenic in coal ash leachate. *Environ Sci Technol* 1981;**15**:1062–6.

28. Wadge A, Hutton M. The leachability and chemical speciation of selected trace elements in fly ash from coal combustion and refuse incineration. *Environ Pollut* 1987;**48**:85–99.

29. Wang Y, Ren D, Zhao F. Comparative leaching experiments for trace elements in raw coal, laboratory ash, fly ash and bottom ash. *Int J Coal Geology* 1999;**40**:103–8.

30. Dutta BK, Khanra S, Mallick D. Leaching of elements from coal fly ash: assessment of its potential for use in filling abandoned coal mines. *Fuel* 2009;**88**:1314–23.

31. Chaudhuri D, Tripathy S, Veeresh H, Powell MA, Hart BR. Mobility and bioavailability of selected heavy metals in coal ash- and sewage sludge-amended acid soil. *Environ Geol* 2003;**44**:419–32.

32. Cox JA, Lundquist GL, Przyjazny A, Schmulbach CD. Leaching of boron from coal ash. *Environ Sci Technol* 1978;**12**:722–3.

33. Cherry DS, Guthrie RK, Sherberger FF, Larrick SR. The influence of coal ash and thermal discharges upon the distribution and bioaccumulation of aquatic invertebrates. *Hydrobiologia* 1979;**62**:257–67.

34. Guthrie RK, Cherry DS. Trophic level accumulation of heavy metals in a coal ash basin drainage system. *J Am Water Resour Assoc* 1979;**15**:244–8.

35. Reash RJ, Lohner TW, Wood KV. Selenium and other trace metals in fish inhabiting a fly ash stream: Implications for regulatory tissue thresholds. *Environ Pollut* 2006;**142**:397–408.

36. Lohner TW, Reash RJ, Willet VE, Fletcher J. Assessment of tolerant sunfish populations (*Lepomis* sp.) inhabiting selenium-laden coal ash effluents. 3. Serum chemistry and fish health indicators. *Ecotoxicol Environ Saf* 2001a;**50**:225–32.

37. Lohner TW, Reash RJ, Willet VE, Rose LA. Assessment of tolerant sunfish populations (*Lepomis* sp.) inhabiting selenium-laden coal ash effluents. 1. Hematological and population level assessment. *Ecotoxicol Environ Saf* 2001b;**50**:203–16.

38. Rowe CL, Kinney OM, Fiori AP, Congdon JD. Oral deformities in tadpoles (*Rana catesbeiana*) associated with coal ash deposition: effects on grazing ability and growth. *Freshwater Biol* 1996;**36**: 723–30.

39. Nagle RD, Rowe CL, Congdon JD. Accumulation and selective maternal transfer of contaminants in the turtle *Trachemys scripta* associated with coal ash deposition. *Arch Environ Contam Toxicol* 2001;**40**:531–6.

40. Ahmaruzzaman M. A review on the utilization of fly ash. *Prog Energy Combust Sci* 2010;**36**:327–63.
41. Siddique R. Utilization of coal combustion by-products in sustainable construction materials. *Resour Conservation Recycl* 2010;**54**:1060–6.
42. Toth PS, Chan HT, Cragg CB. Coal ash as structural fill, with special reference to Ontario experience. *Can Geotechnical J* 1988;**25**:694–704.
43. Churchill E, Amirkhanian S. Coal ash utilization in asphalt concrete mixtures. *J Mater Civ Eng* 1999; **11**:295–301.
44. Cumpston B, Shadman H, Risbud S. Utilization of coal-ash minerals for technological ceramics. *J Mater Sci* 1992;**27**:1781–4.
45. Kikuchi R. Application of coal ash to environmental improvement: transformation into zeolite, potassium fertilizer, and FGD absorbent. *Resources, Conservation Recycl* 1999;**27**:333–46.
46. Iyer RS, Scott JA. Power station fly ash—a review of value-added utilization outside of the construction industry. Resources. *Conservation Recycl* 2001;**31**:217–28.
47. Wang S, Wu H. Environmental-benign utilisation of fly ash as low-cost adsorbents. *J Hazard Mater* 2006;**B136**:482–501.
48. Jaya S, Goyal D. Fly ash as a soil ameliorant for improving crop production—a review. *Bioresour Technol* 2006;**97**:1136–47.
49. Adriano DC, Woodford TA, Ciravolo TG. Growth and elemental composition of corn and bean seedlings as influenced by soil application of coal ash. *J Environ Qual* 1977;**7**:416–21.
50. Kumar R, Kumar S, Mehrotra SP. Towards sustainable solutions for fly ash through mechanical activation. *Resources, Conservation Recycl* 2007;**52**:157–79.

11

Biodiesel from Plant Oils

Nikul K. Patel[1], Shailesh N. Shah[2]

[1]MECHANICAL ENGINEERING DEPARTMENT, FACULTY OF TECHNOLOGY & ENGINEERING, THE M S UNIVERSITY OF BARODA, VADODARA, INDIA; [2]CHEMISTRY DEPARTMENT, FACULTY OF SCIENCE, THE M S UNIVERSITY OF BARODA, VADODARA, INDIA

CHAPTER OUTLINE

Food, Energy, and Water. http://dx.doi.org/10.1016/B978-0-12-800211-7.00011-9

Introduction

Fossil fuels are fuels formed by natural processes such as anaerobic decomposition of buried dead organisms. The age of the organisms and their resulting fossil fuels are typically millions of years in age, and can sometimes exceed 650 million years.[1] The US Energy Information Administration in 2007 estimated that the primary sources of energy consisted of petroleum 36.0%, coal 27.4%, and natural gas 23.0%, amounting to an 86.4% share for fossil fuels in primary energy consumption in the world. Nonfossil sources in 2006 included hydroelectric 6.3%, nuclear 8.5%, and others (geothermal, solar, tidal, wind, wood, and waste) amounting to 0.9%.[2] World energy consumption was growing about 2.3% per year. Strictly speaking, fossil fuels are renewable resources. They are continuously being formed via natural processes; for example, when plants and animals die and decompose, they become trapped beneath sediments. However, fossil fuels are generally considered to be nonrenewable resources because they take millions of years to form, and known viable reserves are being depleted much faster than new ones are being made. Biomass is organic matter derived from decomposition of plants and animals, available on a renewable basis. Biomass includes wood and agricultural crops, herbaceous and woody energy crops, municipal organic wastes, and manure. Biofuels are fuels derived from biomass or waste feedstocks, including ethanol and biodiesel.

For economic development of any country, there should be a large enough supply of electricity and transportation capability. Both electricity and transportation are dependent on fuels available from Earth's crust, known as fossil fuels. The fossil fuels used are not renewed rapidly and because of their rapid use, they will be depleted in the near future, and there will be an energy crisis in the world. Moreover, important fossil fuels like oil and gas are concentrated in a small number of countries, resulting in a large number of countries dependent on oil-producing countries to provide fossil fuel. Because of globalization, in developing countries like India and China the consumption of petroleum products and natural gas increases year by year at unbelievable rates, growing their import bills and adversely affecting their economy. Fossil fuels' dominance will remain unchallenged for at least the next four decades, even if countries pursue environment policies, according to a report by the World Energy Council (WEC). Tectonic shifts are taking place, with China replacing the United States as the world's leading crude importer, even as the United States reinvents itself as the world's largest producer of oil liquids. The WEC report predicts two scenarios for the future: "Jazz," in which energy accessibility and affordability trump sustainable development, and "Symphony," in which countries move toward a more harmonious and environment-friendly path.[3] In both scenarios, fossil fuels remain extremely dominant.

"The future primary energy mix in 2050 shows that growth rates will be highest for renewable energy sources. In absolute terms, fossil fuels (coal, oil, gas) will remain dominant up to 2050. The share of fossil fuels will be 77% in the Jazz scenario and 59% in the Symphony scenario – compared to 79% in 2010." This shows that renewable sources will grow slowly and steadily to help replace conventional sources of energy. Of the total contribution of renewable sources, biofuel will make a significant contribution.

While renewable energies are making a dent, they will remain a periphery energy source. Their share of energy sources could increase from around 15% in 2010 to almost 20% in the Jazz scenario in 2050 and almost 30% in the Symphony scenario in 2050, according to various reports.

Renewable energy will play an important role in our future energy mix. However, a number of challenges for renewable sources remain. Christopher Frei, Secretary General of the WEC, made the following statement:

"There is huge unexploited hydropower potential, especially in Africa, Asia, and Latin America, but a number of large projects are facing local resistance. There is significant potential of biomass energy, particularly in Latin America, but concerns about the energy–water–food nexus have to be carefully managed. Other technologies, such as marine energy, still require a lot of efforts in R&D."

A predominant source of energy, bioenergy, is explored in this chapter. Bioenergy is renewable energy made available from materials derived from biological sources. Biomass is any organic material that has stored up sunlight in the form of chemical energy. As a fuel, it may include wood, wood waste, straw, manure, sugarcane, plants, seeds from fruits, and many other by-products from a variety of agricultural processes. The main focus is on biofuels produced from the seeds of plants that can be grown specifically for their seeds. Oil is extracted from the seeds; the oil undergoes a process known as esterification or transesterification. In this process, oil is converted to bio-diesel, which can be blended or used directly as fuel in diesel engines.

Plants Catalog

In light of energy and environmental problems associated with the utilization of fossil fuels in transportation and for power generation, increasing attention is being paid worldwide by engineers and other scientists alike for the utilization of renewable energy sources. Considering the demand of fossil fuels and their effect on climate, it is important to consider alternate sources of energy as a long-term solution to Earth's pollution, while substituting conventional energy sources.

There are many types of renewable energy sources—solar, wind, hydropower, ocean thermal, geothermal, wave, biomass, and bioenergy. It is necessary to develop the source of renewable energy that increases its utilization to reduce dependence on fossil fuels and environmental pollution-related issues. In the future, when fossil fuels are depleted, these renewable energy sources will be essential. Biomass and bioenergy are among the important renewable energy sources. One of the key sources of bioenergy is biodiesel.

Biodiesel can be produced from various sources such as edible or nonedible plant oils.[4] Biodiesel is a promising alternative fuel resource for diesel engines. It is defined as the monoalkyl esters of long-chain fatty acids derived from vegetable oils or animal fats and alcohol, with or without a catalyst. It is renewable, biodegradable, environment friendly, nontoxic, readily available, transport friendly, and an eco-friendly fuel.[5–8] Biodiesel can be produced from edible or nonedible seeds of plants. The edible oils that can be used for biodiesel production are mahua, palm, tobacco seed, rice bran, sesame, sunflower, barley, coconut, corn, used cooking oil, linseed, mustard, soybean, rapeseed, groundnut, cottonseed, pumpkin, peanut, olive, etc. Utilizing edible oil biodiesel has raised numerous concerns and ignited food-versus-fuel debates all over the world, possibly causing shortages of food, especially in developing countries. Moreover, environmental problems causing the utilization of cultivable land for production of biofuels may lead to serious issues like deforestation for planting of edible oil seeds, resulting in tremendous damage to the environment. Biodiesel can also be produced from nonedible seeds that do not compete with edible oils. An additional necessity of such nonedible seeds is the ability to cultivate the crop on a large scale on fallow marginal lands and wastelands. Nonedible oils such as neem (*Azadirachta indica*), *Jatropha curcas*, jojoba (*Simmondsia chinensis*), kusum (*Schleichera oleosa*), karanj (*Millettia pinnata*), punnakka (*Calophyllum inophyllum*), kapok (*Ceiba pentandra*), soapnut (*Sapindus*), desert date (*Balanites roxburghii*), wild mustard and other such nonedible seeds can be used for extracting oil and producing biodiesel fuel.[9]

Moringa oleifera[10–14]

Moringa oleifera is the most widely cultivated species of the genus *Moringa*, which is the only genus in the family Moringaceae. *M. oleifera* is a fast-growing, deciduous tree that grows best in *semiarid*, tropical, and subtropical regions at an altitude of 2000 m. Flowers and pods of *M. oleifera* are shown in Figures 1 and 2, respectively. It can reach a height of 10–12 m,[10] and the trunk can reach a diameter of 45 cm.[11] The bark has a whitish-gray color and is surrounded by thick cork. The tree grows best in dry *sandy soil* and tolerates poor soil, including coastal areas. It is found in parts of Africa, Asia, and Latin America.

FIGURE 1 Flowers of *Moringa Oleifera*.

FIGURE 2 Pods of *Moringa Oleifera*.

The plant is grown from cuttings, and the first harvest can take place 6–8 months after planting. Often, the fruit yield is generally low in the first year. By the beginning of the second year, the plant produces around 300 pods; by the third year, around 400–500 pods. A healthy mature tree can yield 1000 or more pods.[12] In India, a hectare can produce 31 tons of pods per year.[13] In north India, the fruits ripen during the summer, with a single harvesting. In *south India*, harvesting takes place twice a year: July to September and March to April.[14] To reduce the acid value of the *M. oleifera* oil, it is treated with acid and from which biodiesel is obtained by a standard transesterification procedure, with methanol and an alkali catalyst at 60 °C and an alcohol/oil ratio of 6:1. Biodiesel obtained from this oil exhibits a high cetane number (CN) of approximately 60, one of the highest for a biodiesel fuel.

Glycine max (L.) Merr.—Soybean[15–19]

The soybean (United States), or soya bean (United Kingdom) (*Glycine max*) is a species of *legume* native to East Asia, widely grown for its edible *bean* and also which has numerous uses. The soybean plant and seeds are shown in Figures 3 and 4, respectively.

FIGURE 3 Soybean plant.

FIGURE 4 Soybean seeds.

The plant is classed as an *oilseed* by the UN *Food and Agricultural Organization*. The plant varies in growth and habit, varying from less than 0.2 to 2.0 m (0.66–6.56 ft). The pods, stems, and leaves are covered with fine brown or gray hairs. The *leaves* have three to four leaflets per leaf, and the leaflets are 6–15 cm (2.4–5.9 in) long and 2–7 cm (0.79–2.76 in) broad. The leaves fall before the seeds are mature. Inconspicuous, self-fertile flowers are borne in the axil of the leaf and may be white, pink, or purple. Hairy *pods* grow in clusters of three to five fruits, each containing two to four seeds.

Cultivation is successful in climates with hot summers, with optimum growing conditions in mean temperatures of 20–30 °C (68–86 °F); temperatures of below 20 °C and over 40 °C (68 °F, 104 °F) stunt growth significantly. They can grow in a wide range of soils, with optimum growth in moist alluvial soils with a good organic content. Soybeans are widely used as edible foods in the form of oil, meal, flour, or infant formulas. In addition, soybeans are utilized in industrial products such as soap, cosmetics, resins, plastics, inks, crayons, and also as biodiesel.[16–18] Engines, when tested with different proportions of soybean oil, varying from 5% to 50% of the oil in diesel fuel, along with variations in intake temperature of biodiesel, showed no changes in the running of the engine, consumption, or the supply of water. On the other hand, it constituted changes to CO, HC, NO, and smoke too. CO emissions are reduced; the exhaust temperature is not affected either by the percentage of soy oil in diesel or by the fuel temperature, with the engine operating without any modification.[19]

Rapeseed (*Brassica napus*)[20,21]

According to Wikipedia, rapeseed oil was produced in the nineteenth century as a source of a lubricant for steam engines. It is less preferred as a food for animals and humans because it has a bitter taste. The production of this oil takes place in the European Union, Canada, United States, Australia, China, and India. It comes from the black seeds of the rapeseed plant, *Brassica napus*, from the same *Brassica* family as the health-enhancing vegetables broccoli, cabbage, and cauliflower. The plant produces sunny, yellow flowers in the springtime, producing golden fields brightening our beautiful landscape, as shown in Figure 5. On conducting a diesel engine performance test using rapeseed oil, the brake-specific fuel consumption (a measure of the *fuel efficiency* of any prime mover that burns fuel and produces rotational, or shaft, power) at the maximum torque and rated power is correspondingly higher by 12.2% and 12.8% than that of diesel fuel, while its brake thermal efficiency remains almost the same of 0.37–0.38 for diesel and 0.38–0.39 for rapeseed.[20] Fuel properties of rapeseed biodiesel are comparable with petrodiesel and when its blend with petrodiesel is used as fuel in a diesel engine, its engine performance parameters show good results in accordance with petrodiesel. Thus, rapeseed biodiesel can be partially substituted for petrodiesel under most operating conditions, regarding both performance parameters and exhaust, without modifications made to the engine.[21]

FIGURE 5 Flowering of rapeseeds.

FIGURE 6 Tree of dessert date.

Desert Date[22–24]

Balanites aegyptiaca is a dicotyledonous flowering plant that is popularly known as "desert date" in English. It is a widely grown desert tree, as shown in Figure 6, with a multitude of uses. The fruits containing seeds are shown in Figure 7. The plant is found throughout the Sudano-Sahelian region and in other arid areas of Africa, the Middle East, India, and Burma. It is one of the most drought-resistant tree species in arid regions.[22] It is highly resistant to hazardous conditions such as sandstorms and heat waves and grows extensively even when neglected. In Nigeria, *Balanites* are found mostly in the northern and western part of the country[23]; although the fruit pulp is bitter-tasting, it is still edible. Pounded fruits make a refreshing drink, which becomes alcoholic if left to ferment. The seed has a low moisture content of 8.73%. Low moisture content is an indication of a reasonable shelf life for the seed, because there is little or no water for the hydrolysis of the oil to take place. The average oil content obtained from *B. aegyptiaca* seeds is 37.2%.[24]

Pongamia pinnata L. (karanja)[25–30]

Pongamia pinnata (L.), also known as karanja, is an evergreen tree that grows in India, Southeast Asia, Australia, New Zealand, China, and the United States.[25–27] Assuming that

FIGURE 7 Fruits of dessert date.

one tree has a yield of 9–90 kg seeds, a yield of 900–9000 kg of seeds/hectare with a planting of 100 trees per hectare is obtained.[26] The plant can be grown by various methods, such as direct sowing, transplanting, and root or shoot cuttings. Its maturity comes after 5–7 years and Figures 8 and 9 show the tree and the fruits of *P. pinnata*, respectively. The plant grows rapidly, requires little water, and is highly tolerant of salinity. It has been recognized as a viable source of oil for the biofuel industry. The tree should be planted with a spacing of $3 \times 3\ m^2$, and the oil content ranges between 30 and 40 wt%.[28,29] The oil is reddish brown in color, and it is rich in unsaponifiable matter and oleic acid.[30] This plant has been used in India as a source of traditional medicines, animal fodder, green manure, timber, water–paint binder, pesticide, fish poison, and fuel.

FIGURE 8 Tree of karanj.

FIGURE 9 Fruits of karanj.

Jatropha curcas[31]

Jatropha curcas Linnaeus is a species of flowering plant in the spurge family, Euphorbiaceae; the tree is shown in Figure 10 and its seeds are shown in Figure 11. It is a tree or shrub that is native to the American tropics, mostly Mexico and Central America, but it grows under a variety of agroclimatic conditions and is commonly found in most of the tropical and subtropical regions of the world. Thus, it ensures a reasonable production of seeds with very minor care. The oil content of *Jatropha* seed ranges from 30% to 35% by weight. The common by-products produced while processing the biodiesel are glycerol and oilseed cake.[31] The plant can grow on *wastelands* and almost any terrain, including *gravelly*, *sandy*, and *saline soils*. It can thrive in poor and stony soils, although new research suggests that the plant's ability to adapt to these poor soils is not as extensive as had been previously discussed. Complete germination is achieved within 9 days and harvesting can occur from 9 to 12 months' time, but the best yields are obtained after 2–3 years.

Neem Tree[32–34]

Azadirachta indica (neem) tree belongs to the Meliaceae family; the tree is shown in Figure 12, and its seeds are shown in Figure 13. It is a multipurpose, evergreen tree, 12–18 m tall, and can grow in almost any kind of soil, including clay, saline, alkaline, dry, stony, shallow soils, and even on highly calcareous soil. It is native to India, Pakistan, Sri Lanka, Burma, Malaysia, Indonesia, Japan, and the tropical regions of Australia. It thrives well in arid and semiarid climates with a maximum shade temperature of 49 °C and rainfall as low as 250 mm/year. It can be raised by directly sowing its seed or by transplanting nursery-raised seedlings in monsoon rains. It reaches maximum productivity after 15 years and has a life span of 150–200 years. Planting is usually done at a density of 400 plants per hectare. The productivity of neem oil varies from 2–4 ton/ha/year

FIGURE 10 Tree of Jatropha curcas.

FIGURE 11 Fruits and seeds of Jatropha curcas.

FIGURE 12 Neem tree.

FIGURE 13 Fruits bearing seeds of neem.

and a mature neem tree produces 30–50 kg fruit. The seed of the fruit contains 20–30 wt% oil, and kernels contain 40–50% of an olive green to brown color oil.[32–34]

Madhuca indica[35–37]

Madhuca indica is found mainly in India.[28,31,35,36] It belongs to the Sapotaceae family and grows quickly to approximately 20 m in height, possesses evergreen or semi-evergreen foliage, and is adapted to arid environments. The tree and their fruit are shown in Figures 14 and 15, respectively. *M. indica* oil is a forest-based tree-borne nonedible oil

FIGURE 14 Tree of Madhuca indica.

FIGURE 15 Fruits bearing seeds of Madhuca indica.

with large production potential of about 60 million tons per annum in India. The *M. indica* tree starts producing seeds after 10 years and continues up to 60 years. The kernel constitutes about 70% of the seed and contains 50% of the oil.[28,33,37] Each tree yields about 20–40 kg of seeds per year, depending upon the maturity and size of the tree, and the total oil yield per hectare is 2.7 tons per year. The tree's seed contains about 35–40 wt% of *M. indica* oil.[35]

Plants are shown from which biodiesel can be produced. Biodiesel is made from the seeds of plants that are renewable in nature and thus can be good alternatives for the production of biodiesel. There are a number of such seeds that are identified as potential feedstocks for the production of biodiesel as shown in Table 1.[29]

Production of Biofuels

Chemical processes involved in production of biofuels and economics involved in the processes are discussed below.

Extraction of Oil

Seeds that are available in nature are in solid form and cannot be directly used in engines as a fuel. Hence, the primary requirement is to convert this solid form of seeds into a liquid form. Oil is extracted by crushing seeds using different methods. Three methods have been identified for extraction of the oil from seeds, namely, mechanical extraction, solvent extraction, and enzymatic extraction. Oil from the seeds cannot be extracted when there is water in the seeds, so before oil extraction takes place, seeds have to be dried in an oven at a required temperature, or sun-dried. This process normally ranges from 3–4 weeks.

Mechanical Expeller

An expeller press is a screw-type machine that presses oilseeds through a caged barrellike cavity as shown in Figure 16. Raw materials enter one side of the press and waste products exit the other side. The machine uses friction and continuous pressure from the screw drives to move and compress the seed material. The oil seeps through small openings that do not allow seed fiber solids to pass through. Afterward, the pressed seeds are formed into a hardened cake, which is removed from the machine. Pressure involved in expeller pressing creates heat in the range of 140–210 °F (60–99 °C). Some companies claim that they use a cooling apparatus to reduce this temperature to protect certain properties of the oils being extracted.

The technique of oil extraction using mechanical expellers is the most conventional practice. In this type, either a manual ram press or an engine-driven screw press can be used. It has been found that an engine-driven screw press can extract 68–80% of the available oil while the ram press achieves only 60–65%.[38–42] Further treatment such as filtering and degumming of oil is required when the oil is obtained using a mechanical expeller. Pretreatment such as cooking the seeds can increase the oil yield of screw pressing up to 89% after a single pass and 91% after dual passes.[38,43,44]

Solvent Extraction

Liquid–liquid extraction, also known as solvent extraction or partitioning, is a method to separate compounds based on their relative *solubilities* in two different *immiscible*

Table 1 List of Non-Edible Plants Used for Biodiesel Production

1. Anacardiaceae *Rhus succedanea* L.	**28.** *Hydnocarpus wightiana* Blume Guttiferae	**55.** *Myristica malabarica* Lam. Papaveraceae
2. Annonaceae *Annona reticulate* L. Apocynaceae	**29.** *Calophyllum apetalum* Wild	**56.** *Argemone Mexicana* Fabaceae
3. *Ervatamia coronaria* Stapf	**30.** *Calophyllum inophyllum* L.	**57.** *Pongamia pinnata* Pierre Rhamnaceae
4. *Thevetia peruviana* Merrill	**31.** *Garcinia combogia* Desr	**58.** *Ziziphus mauritiana* Lam. Rosaceae
5. *Vallaris solanacea* Kuntze Balanitaceae	**32.** *Garcinia indica* Choisy	**59.** *Princepia utilis* Royle Rubiaceae
6. *Balanites roxburghii* Planch Basellaceae	**33.** *Garcinia echinocarpa* Thw	**60.** *Meyna laxiflora* Robyns Rutaceae
7. *Basella rubra* L. Cannabinaceae	**34.** *Garcinia Morella* Desr	**61.** *Aegle marmelos correa* Roxb. Salvadoraceae
8. *Canarium commune* L. Cannabinaceae	**35.** *Mesua ferrea* L. Icacinaceae	**62.** *Salvadora oleoides* Decne
9. *Cannabis sativa* L. Celastraceae	**36.** *Mappia foetida* Milers Illiciceae	**63.** *Salvadora persica* L. Santalaceae
10. *Celastrus paniculatus* L.	**37.** *Illicium verum* Hook Labiatae	**64.** *Santalum album* L. Sapindaceae
11. *Euonymus hamiltonianuis* Wall Combretaceae	**38.** *Saturega hortensis* L.	**65.** *Nephelium lappaceum* L.
12. *Terminalia bellirica* Roxb	**39.** *Perilla frutescens* Britton Lauraceae	**66.** *Sapindus trifoliatus* L.
13. *Terminalia chebula* Retz Asteraceae	**40.** *Actinodaphne angustifolia*	**67.** *Schleichera oleosa* Oken Sapotaceae
14. *Vernonia cinerea* Less–Herb Corylaceae	**41.** *Litsea glutinosa* Robins	**68.** *Madhuca butyracea* Mac
15. *Corylus avellana* Cucuribitaceae	**42.** *Neolitsea cassia* L.	**69.** *Maduca indica* JF Gmel
16. *Momordica dioica* Rox Euphorbiaceae	**43.** *Neolitsea umbrosa* Gamble Magnoliaceae	**70.** *Mimusops hexendra* Roxb Simaroubaceae
17. *Aleurites fordii* Hemsl	**44.** *Michelia champaca* L. Malpighiaceae	**71.** *Quassia indica* Nooleboom
18. *Aleurites moluccana* Wild	**45.** *Hiptage benghalensis* Kurz Meliaceae	**72.** *Ximenia Americana* L. Sterculaceae
19. *Aleurites Montana* Wils	**46.** *Aphanamixis polystachya* Park	**73.** *Pterygota alata* Rbr. Ulmaceae
20. *Croton tiglium* L.	**47.** *Azadirachta indica*	**74.** *Holoptelia integrifolia* Urticaceae
21. *Euphorbia helioscopia* L.	**48.** *Melia azadirach* L.	**75.** *Urtica dioica* L. Verbenaceae
22. *Jatropa curcas* L.	**49.** *Swietenia mahagoni* Jacq Menispermaceae	**76.** Tectona grandis L.
23. *Joannesia princeps* Vell	**50.** *Anamirta cocculus* Wight & Hrn Moraceae	
24. *Mallotus phillippinensis* Arg	**51.** *Broussonetia papyrifera* Vent	
25. *Putranjiva rosburghii*	**52.** Moringaceae	
26. *Sapium sebiferum* Roxb Flacourtiaceae	**53.** *Moringa concanensis* Nimmo	
27. *Hydnocarpus kurzii* Warb	**54.** *Moringa oleifera* Lam Myristicaceae	

FIGURE 16 Mechanical expeller.

liquids, usually water and an *organic solvent*. It is an extraction of a substance from one liquid into another liquid phase. Liquid–liquid extraction is a basic technique in chemical laboratories, where it is performed using a *separatory funnel*. This type of process is commonly performed after a chemical reaction as part of the *work-up*. The term partitioning is commonly used to refer to the underlying chemical and physical processes involved in liquid–liquid extraction but may be fully synonymous. The term solvent extraction can also refer to the separation of a substance from a mixture by preferentially dissolving that substance in a suitable solvent. In that case, a soluble compound is separated from an insoluble compound or a complex matrix. This process is also called leaching. There are many factors influencing the rate of extraction, such as particle size, the type of liquid chosen, temperature, and agitation of the solvent. The liquid chosen should be a good selective solvent, and its viscosity should be sufficiently low to circulate freely. Temperature also affects the extraction rate, and solubility of the material increases with an increase in temperature. There are three methods that are used for extraction[38,43,44]: (1) hot water extraction, (2) Soxhlet extraction, and (3) ultrasonication technique.

HOT WATER EXTRACTION

Water, which is cheap, safe, nontoxic, nonflammable, and recyclable, is one of the alternatives for organic solvents. Subcritical water, or pressurized hot water, provides liquid water under pressure at temperatures between the boiling point (100 °C) and the critical temperature (374 °C). The polarity of superheated water significantly decreases on increasing temperature and pressure, and the properties of superheated water appear to be those of an organic solvent, such as methanol or ethanol. Over the superheated temperature range, the extensive hydrogen bonds break down, changing the properties more than usually expected by increasing temperature alone. Solubility of organic solutes increases by several orders of magnitude and the water itself can act as a solvent, as in extractions and separations.[45]

SOXHLET EXTRACTION

Soxhlet extraction is a chemical term that means the separation process of compounds using solvents by dissolving the mixture in another soluble solvent. The product of the extraction is called a distribution ratio. The separations that can be achieved by this method are simple, convenient, and rapid. The procedure is applicable to both trace and macro levels. A further advantage of solvent extraction method lies in the convenience of subsequent analysis of the extracted species.[46]

ULTRASONICATION TECHNIQUE

An ultrasonication method is a procedure for extracting nonvolatile and semivolatile organic compounds from solids such as soils, sludges, and wastes. It ensures intimate contact of the sample matrix with the extraction solvent. This method is divided into two procedures based on the expected concentration of organic compounds. The low concentration procedure is for individual organic components expected at less than or equal to 20 mg/kg and uses large sample sizes and three serial extractions. The medium- or high-concentration procedure is for individual organic components expected at greater than 20 mg/kg and uses a smaller sample and a single extraction. It is highly recommended that the extracts be subject to some form of cleanup prior to analysis. Because of the limited contact time between the solvent and the sample, ultrasonic extraction may not be as rigorous as other extraction methods for soils and solids. Therefore, it is critical that the method be followed explicitly, in order to achieve the maximum extraction efficiency.[47]

Enzymatic Oil Extraction

Enzymatic oil extraction has emerged as a promising technique for extraction of oils. Conventional processes for the extraction of oil involve mechanical treatment that submits the oil cake for further extraction with *n*-hexane.[48] Although these technologies are economically justifiable, they have certain well-known drawbacks: damage to the environment and quality loss of finished products (e.g., high free fatty acids and lower resistance to rancidity). On the other hand, enzymatic oil extraction is friendly to the environment and does not produce volatile organic compounds. Several studies have been carried out on aqueous enzymatic oil extractions.[49,50] The enzymatic extraction of vegetable oils, developed at Embrapa Food-Agro Industry[51] reaches the high-yield extraction of tropical fruit oils (avocado, pupunha, pequi, tucumã) when the incubation process is based on a combination of pectinolytic enzymes. Furthermore, the high-yield extraction of seed oils is obtained with a combination of cellulose and protease enzymes.

Production of Biodiesel

Oil extracted from the seeds of plants is highly viscous, preventing their direct use in engines as a fuel. There are many techniques used to produce biodiesel from various feedstocks. Different processes included are pyrolysis, microemulsification, dilution, and transesterification. The following sections discuss these technologies.

Pyrolysis (Thermal Cracking)

Pyrolysis is a *thermochemical decomposition* of *organic material* at elevated temperatures in the absence of *oxygen*. It involves the simultaneous change of *chemical composition* and physical phase, and it is irreversible. Pyrolysis is a type of *thermolysis*, and it is most commonly observed in *organic* materials exposed to high temperatures. It also occurs in fires where solid fuels are burning or when vegetation comes into contact with lava in *volcanic eruptions*. In general, pyrolysis of organic substances produces gas and liquid products and leaves a solid residue richer in carbon content, *char*. In extreme cases of pyrolysis, the residue is carbon and the process is known as carbonization. In the case of biofuels, pyrolyzed material can be vegetable oils, animal fats, natural fatty acids, or methyl esters of fatty acids. Thermal decomposition of triglycerides produces alkanes, alkenes, alkadienes, aromatics, and carboxylic acids. Liquid fractions of the thermally decomposed vegetable oils are likely to display characteristics of diesel fuel.

Microemulsification

Microemulsions are clear, thermodynamically stable, *isotropic* liquid mixtures of oil, water, and *surfactant*, frequently in combination with a *cosurfactant*. The aqueous *phase* may contain *salt*(s) and/or other ingredients, and the "oil" may actually be a complex mixture of different *hydrocarbons* and *olefins*. In contrast to ordinary *emulsions*, microemulsions form upon simple mixing of the components and do not require the high *shear* conditions generally used in the formation of ordinary emulsions. The three basic types of microemulsions are direct (oil dispersed in water), reversed (water dispersed in oil), and bicontinuous. Microemulsions can be made from vegetable oils with an ester and dispersant (cosolvent), or hexanol and a surfactant and a cetane improver, with or without diesel fuels. Microemulsification has been considered a reliable approach to solve the problem of high viscosity of vegetable oils.[5,6,52]

Dilution

To improve the performance of an engine it is a "must" to reduce the viscosity of the oil, which can be achieved by diluting with diesel fuel. This method does not require any chemical process: vegetable oil is mixed with diesel oil. It has been reported that substitution of 100% vegetable oil for diesel fuel is not practical.[6] Therefore, a blend of 20–25% vegetable oil to diesel oil gives good results for diesel engines.[5,6,52–54]

Transesterification

Transesterification is the process of exchanging the organic alkyl groups (R_1, R_2, R_3) of vegetable/plant oil – an ester) with the methyl group methyl alcohol as shown in

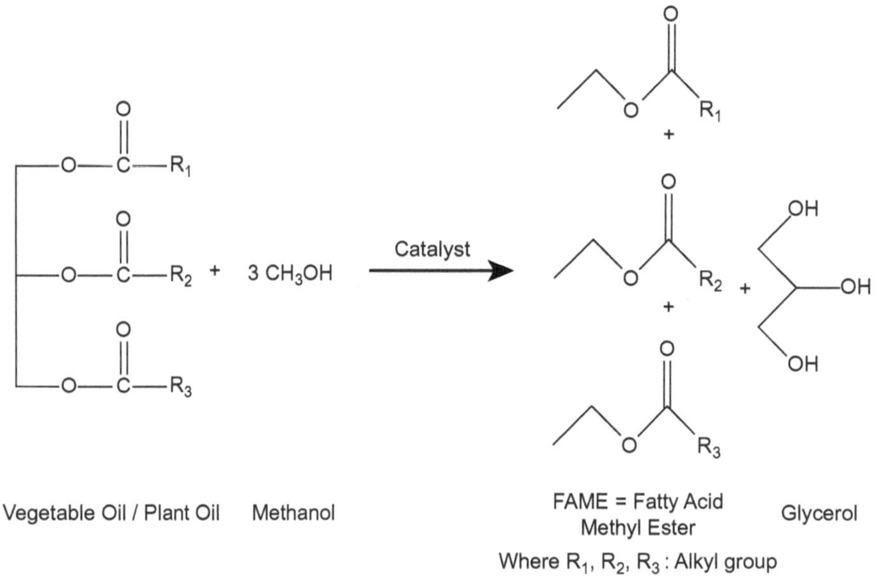

FIGURE 17 Chemical structure of transesterification process.

Figure 17. These reactions are often *catalyzed* by the addition of an *acid* or *base* catalyst. In the transesterification mechanism, the carbonyl carbon of the starting ester of Vegetable/Plant oil undergoes nucleophilic attack by the incoming alkoxide from catalyst to give a tetrahedral intermediate, which either reverts to the starting material or proceeds to the transesterified product and glycerol as shown in Figure 17. In this reaction, methanol and ethanol are the most commonly used alcohols because of their availability and low cost. This reaction has been widely used to reduce the viscosity of vegetable oil and conversion of the triglycerides into esters.

Transesterification can be carried out by two ways: (1) catalytic transesterification and (2) noncatalytic transesterification. It is widely known that catalytic transesterification has two problems. The main problem is that the process is relatively time consuming and needs separation of the vegetable oil/alcohol/catalyst/saponified impurities mixture from the biodiesel. Furthermore, the wastewater generated during biodiesel purification is not environment-friendly. Under such conditions, supercritical alcohol transesterification is an option to solve the problems by employing two-phased methanol/oil mixtures by forming a single phase as a result of the lower value of the dielectric constant of methanol in the supercritical state. As a result, the reaction is less time consuming. Moreover, purification of biodiesel is much easier, as no catalyst is required during the supercritical transesterification process, thus preventing soap formation or saponification. However, the drawbacks of supercritical alcohol transesterification process are due to the high temperature and pressure that result in high cost of the apparatus.[52]

Economics for Production of Biodiesel

The economics of using biodiesel are still unclear for a number of reasons. Although biodiesel fuel has been proven to be a more environmentally beneficial fuel source in diesel engines than standard petroleum-based products, biodiesel fuel may still not be as cost-effective as traditional fuels. It is expected that over time, the cost of mass-producing biofuels will eventually fall so that they more closely match the price of other, more readily available fueling products. To understand the production and economics estimation of biodiesel from nonedible seeds, let us take the case study of jatropha, a novel feedstock for biodiesel production.[55] By-products of the transesterification process are glycerin, deoiled cake/press cake, husk, and hulls. Glycerin is about 95% pure, a product that can be sold to refiners, deoiled cake is a good source of manure such that 1 ton is equivalent to 200 kg of mineral fertilizer, and husk can be used as fuel in cook stove (chulhas) for rural areas or in the boiler in an oil mill. The annual yield of jatropha after the first, second, third, fourth, and fifth years will be 2.5, 3.0, 5.0, 8.0, and 12.5 tons per hectare, respectively. The factors involved in carrying out analyses are capacity utilization of production plant; raw material purchase price; project cost such as land, machinery, and equipment; power rate to run the plant; water cost; salaries; taxes; etc. Calculations are based on the above parameters. Suppose 10,000 L of jatropha-based biodiesel is required, where the cost of 1 L is calculated based on recent costs of materials available in India. Its corresponding values are converted from liter to gallon as 3.78 L equal 1 gallon. Next, rupees to dollars is converted using Rs 60 = US$1.

	Cost Components	
(A)	Raw material cost	
(i)	Amount of jatropha seeds (kilograms) for production of 2646 gallons of biodiesel	35,000
	Cost of seeds $0.2 per kg	$7000
(ii)	Amount of methanol (gallons) for production of 2646 gallons of biodiesel	556
	Cost of methanol at $1.5 per gallon	$33
(iii)	Amount of NaOH (kilograms) for production of 2646 gallons of biodiesel	250
	Cost of NaOH at $0.4 per kg	$100
(iv)	Miscellaneous material costs	$50
	Total cost of raw materials	$7983
(B)	Processing cost	
(i)	Cost of oil extraction at $0.1 per gallon	$320
(ii)	Transesterification cost at $0.125 per gallon	$400
(iii)	Miscellaneous costs per gallon at $0.01	$32
	Total processing cost	$752
(C)	Total production cost (A) + (B)	$8736
(D)	Revenue from sale of by-products	
(i)	Cake (24,500 kg × 0.9 {loss factor} × $0.03)	$661.5
(ii)	Glycerin (2495 kg × 0.9 {loss factor} × $0.6)	$1347
	Total revenue from by-products	$2009
	Cost of 1 gallon of biodiesel: {(C) − (D)}/2646	$2.54

Table 2 Variation in Cost of Biodiesel with Change in Cost of Plant Seeds

Cost of Seeds in US$	Raw Material Cost in US$	Process Cost in US$	Total Production Cost in US$	Revenue from By-Product in US$	Biodiesel Cost per Gallon in US$
0.2	7983	752	8736	2009	2.54
0.25	9733	752	10,486	2009	3.20
0.3	11,483	752	12,236	2009	3.87
0.35	13,233	752	13,986	2009	4.53

If the cost of seeds is US$0.2 per kg, the cost of 1 gallon of biofuel is US$2.54. Table 2 shows the variation in the cost of 1 gallon of biodiesel with the variation in the cost of seeds.

Properties of Biofuels

It is very important to understand the behavior of fuel properties. Considerable effort has been made to develop nonedible oil-based biodiesel fuels that approximate properties and performances of petrodiesel. Finding out different properties of biodiesel fuels is important because alternative fuels are being created worldwide. Subsequently, the international standards as given in Table 3 define the use of biodiesel in internal combustion engines (ICEs) and give a definite value of various properties.

Research has shown that the properties of oil from which biodiesel has been prepared may vary significantly, depending on their chemical composition and fatty acid composition, which give obvious effects on engine performance and emissions. Discussed below are 11 properties that are essential with respect to use in engines.

Table 3 Important Biodiesel Fuel Property and Different International Standard

Property	Unit	ASTM D6751 (United States)	EN 14214 (Europe)	Australian Standard	Brazilian Standard
Flash point	°C	130 min	120 min	120 min	100 min
Water and sediment	% Volume	0.050 max	0.050 max	0.050 max	0.020 max
Kinematic viscosity, 40 °C	mm^2/s	1.9–6.0	3.5–5.0	3.5–5.0	3.5–5.0
Sulfated ash	%Mass	0.020 max	0.020 max	0.020 max	0.020 max
Sulfur	%Mass	0.0015 max	0.0010 max	0.0050 max	0.0010 max
Cetane number	–	47 min	51 min	51 min	45 min
Carbon residue	%Mass	0.050 max	0.030 max	0.030 max	0.050 max
Acid number	mg KOH/g	0.80 max	0.50 max	0.80 max	0.80 max
Free glycerine	%Mass	0.020 max	0.020 max	0.020 max	0.020 max
Total glycerine	%Mass	0.240 max	0.250 max	0.250 max	0.380 max
Phosphorous content	%Mass	0.001 max	0.001 max	0.001 max	0.001 max
Oxidative stability, 110 °C	Minute	NR	4.0	6.0	6.0

NR, not reported.

Flash Point

The flash point (FP) of a fuel is the temperature at which it will ignite when exposed to a flame or spark, i.e., it is the lowest temperature at which fuel emits enough vapors to ignite. The FP varies inversely with the fuel's volatility. This property is measured in accordance with ASTM (American Standard for Testing Materials) D93, which is tested by the Pensky–Martens closed-cup method. In this type, the cup is sealed with a lid through which the ignition source can be introduced. Closed-cup testers normally give lower values for the FP than open-cup testers (typically 5–10 °C lower, or 9–18 °F lower) and are better temperatures at which the vapor pressure reaches the *lower flammable limit*. The FP is an empirical measurement rather than a fundamental physical parameter.

Water and Sediment Content

In a mixture of water and sediments, water has either of two forms: dissolved water or suspended water droplets. While biodiesel is generally considered to be insoluble in water, it actually takes up a considerably greater amount of water than does diesel fuel. On the other hand, the water content of biodiesel reduces the heat of combustion and causes corrosion of vital fuel system components. Moreover, the sediment may consist of suspended rust and dirt particles, or it may originate from the fuel as insoluble compounds formed during fuel oxidation.[45] The standards of water content are ASTM D2709, which limit the amount of water to be maximum of 0.05 (v%).[46]

Kinematic Viscosity

Viscosity is defined as thickness, or a measure of how resistant a liquid is to flowing. It refers to the thickness of the oil and is determined by measuring the amount of time taken for a given measure of oil to pass through an orifice of a specified size. Kinematic viscosity is the most important property of biodiesel since it affects the operation of fuel-injection equipment, particularly at low temperatures when an increase in viscosity affects the fluidity of the fuel. Moreover, high viscosity may lead to the formation of soot and engine deposits because of insufficient atomization. It is observed that the viscosity of oil methyl esters decreases sharply after transesterification processes of biodiesel. Viscosity is measured with various types of *viscometers* and *rheometers*. The kinematic viscosity in biodiesel is determined using ASTM D445 (1.9–6.0 mm^2/s).

Sulfated Ash Content

Ash content describes the amount of inorganic contaminants such as abrasive solids and catalyst residues, and the concentration of soluble metal soaps contained in a fuel sample. The sulfated ash test uses a procedure to measure the amount of residual substance not volatilized from a sample when the sample is ignited in the presence of sulfuric acid. The test is usually used for determining the content of inorganic impurities in an organic substance, and the procedure is as follows. Accurately weigh about 1 g of the substance into a suitable crucible (usually platinum) and moisten with sulfuric acid. Heat gently to remove the excess acid and ignite at about 800 °C until all the black particles disappear; again, moisten with sulfuric acid and reignite; then add a small amount of ammonium carbonate and ignite to constant weight. Unless otherwise specified, if the amount of residue so obtained exceeds the limit specified, repeat the moistening with sulfuric acid, heat and ignite as before, using a 30-min ignition period, until two consecutive weighings of the residue do not differ by more than 0.5 mg or the percentage of residue complies with the limit.

Sulfur Content

Combustion of fuel containing sulfur causes emissions of sulfur oxides.[56] Most of the vegetable oil- and animal fat-based biodiesel have very low levels of sulfur content. However, specifying this parameter is important for engine operability. Ultra-low-sulfur diesel (ULSD) is a *diesel fuel* with substantially lowered *sulfur* content. As of 2006, almost all of the petroleum-based diesel fuel available in the *European Union* and North America is of the ULSD type. The move to lower sulfur content is expected to allow the application of newer emissions-control technologies that should substantially lower emissions of particulate matter from diesel engines. This change occurred first in the *European Union* and is now happening in *North America*. New emissions standards, which are dependent on the cleaner fuel, have been in effect for automobiles in the *United States* since the 2007.

Cetane Number

The CN is a measure of a fuel's ignition delay, the time period between the start of injection and the first identifiable pressure increase during combustion of the fuel. In a particular diesel engine, higher cetane fuels will have shorter ignition delay periods than lower cetane fuels. Fuels with a low CN tend to cause diesel knocking and show increased gaseous and particulate exhaust emissions because of the occurrence of incomplete combustion. Generally, diesel engines operate well with a CN from 40 to 55. This property is measured in accordance with ASTM D613, in which the CN is obtained by burning the fuel in a rare diesel engine called a Cooperative Fuel Research (CFR) engine, under standard test conditions. The operator of the CFR engine uses a hand wheel to increase the compression ratio (and therefore the peak pressure within the cylinder) of the engine until the time between fuel injection and ignition is 2.407 ms. The resulting CN is then calculated by determining which mixture of *cetane* (*hexadecane*) and *isocetane* (2,2,4,4,6,8,8-heptamethylnonane) results in the same ignition delay.

Carbon Residue

A carbon residue test is used to indicate the extent of deposits resulting from the combustion of a fuel. Carbon residue, which is formed by decomposition and subsequent pyrolysis of the fuel components, can clog the fuel injectors. This property is measured in accordance with ASTM D 4530, in which 4 g of the sample are put into a weighed glass bulb. The sample in the bulb is heated in a bath at 553 °C for 20 min. After cooling, the bulb is weighed again and the difference is recorded. This method is popularly known as the Ramsbottom carbon residue, and is well known in the *petroleum industry*.

Acid Number

In chemistry, acid value (or "neutralization number" or "acid number" or "acidity") is the mass of potassium hydroxide (KOH) in milligrams that is required to neutralize 1 g of chemical substance. The acid number is a measure of the amount of carboxylic acid groups in a chemical compound, such as a fatty acid, or in a mixture of compounds. This property is measured in accordance with ASTM D664, in which a known amount of sample dissolved in an organic solvent (isopropanol is often used) is titrated with a solution of potassium hydroxide with a known concentration and with phenolphthalein as a color indicator. The acid number is used to quantify the amount of acid present, for example, in a sample of biodiesel. It is the quantity of base expressed in milligrams of potassium hydroxide that is required to neutralize the acidic constituents in 1 g of sample. Higher acid content can cause severe corrosion in the fuel supply system and in ICEs.

Free and Total Glycerine

Free and total glycerin is a measurement of how much triglyceride remains unconverted into methyl esters. Total glycerin is calculated from the amount of free glycerin, monoglycerides, diglycerides, and triglycerides. Structurally, triglyceride is a reaction product of a molecule of glycerol with fatty acid molecules, yielding three molecules of water and one molecule of triglyceride.[57,58] This property is measured in accordance with ASTM D6584.

Phosphorus, Calcium, and Magnesium Content

Phosphorus, calcium, and magnesium are minor components, which are typically associated with phospholipids and gums that may act as emulsifiers or cause sediment, lowering yields during the transesterification processes.[59] The specification from ASTM D6751 states that the phosphorus content in biodiesel must be less than 10 ppm, and calcium and magnesium combined must be less than 5 ppm. Phosphorus is determined using ASTM D4951 and EN (European Nations) 14107; calcium and magnesium are determined using EN 14538.[56]

Oxidative Stability

Oxidative stability is a chemical reaction that occurs with a combination of the lubricating oil and oxygen. The rate of oxidation is accelerated by high temperatures, water, acids, and catalysts such as copper; the rate of oxidation increases with time. The service life of a lubricant is also reduced with increases in temperature. Oxidation will lead to an increase in the oil's viscosity and in deposits of varnish and sludge. The rate of oxidation is dependent on the quality and type of base oil as well as the additive package used. Some synthetics, such as polyalphaolefins, have inherently better oxidation stability than do mineral oils. This improved oxidation stability accounts for the slightly higher operating temperatures that these synthetic oils can accommodate. Several methods may be used to determine or evaluate the oxidation stability of oil, which is usually regarded as the number of hours until a given increase in viscosity is noted or until there is a given increase in the acid number.

Applications of Biofuels

Diesel engines are used to power automobiles, locomotives, trucks, ships, and irrigation pumps. They are also widely used to generate electric power. Diesel engines offer high thermal efficiency and durability; using biodiesel as fuel in petrodiesel engine helps in understanding of engine performance with respect to conventional petrodiesel. Despite these advantages, the environmental pollution caused by diesel engines becomes a major concern throughout the world. Diesel engines produce smoke, particulate matter, oxides of nitrogen (NO_x), oxides of carbon (CO and CO_2), and unburned hydrocarbon (HC).[60] Thus, the application of biofuels can be divided into two classes: (1) engine performance and (2) engine emissions.

Engine Performance

The ICE is an *engine* in which the *combustion* of a *fuel* (normally a *fossil fuel* and now also a biofuel) occurs with an oxidizer (usually air) in a *combustion chamber* that is an integral part of the working fluid flow circuit. In an ICE, the expansion of the high-*temperature* and high-*pressure* gases produced by combustion applies a direct *force* to some components of the engine. The force is applied typically to *pistons*. This force moves the component over a distance, transforming chemical energy into useful mechanical *energy*. The conversion of this chemical energy into mechanical energy helps determine different parameters such as mechanical efficiency, brake thermal efficiency, brake-specific fuel consumption, and brake power; measuring all of these gives the performance of an engine. Thus, performance of an engine using different biofuels will help determine the extensive use of biofuels as alternative fuels.

Oil from vegetable seeds can be used as fuel without any modification in compression ignition (CI) engines. Comparative studies show that nonedible oils are good alternative fuels. Without changing the engine technology, blending of biodiesel is an interim approach to overcoming the problems of short supply and emission. Experimental findings proved that nonedible biodiesel, when blended with diesel from 10% to 20%, shows similar engine performance as engine fueled with conventional diesel.[61] Experimentation was conducted on a four-stroke, four-cylinder indirect injection water-cooled CI engine to evaluate the performance of engines. The fuel used was B20 (a blend of 20% neem biodiesel and 80% petroleum diesel by volume). The load was varied from no-load conditions to a maximum of 12 kW at a constant speed of 1500 RPM. It was found that the brake thermal efficiency was higher for the biodiesel blend than diesel. The FP of B20 was higher, enhancing safety during storage and transportation.[62] A performance study was carried out with an engine using karanj oil (K100) and blended karanj oil with diesel fuel, named K10, K15, or K20. Specific fuel consumption increases with the increase in blend, but K15 has minimum brake-specific fuel consumption because of the properties similar to diesel.[63] The effect of neem oil and its methyl ester on a direct-injected, four-stroke, single-cylinder diesel engine shows that at full load, the peak cylinder pressure is higher, the peak heat release rate during premixed combustion phase is lower. In this situation, ignition delay is lower for neat neem oil and neem oil methyl ester when compared with diesel at full load. The combustion duration is higher; the brake thermal efficiency is slightly lower.[64]

Engine Emissions

Basically, CI engines use diesel as fuel and are used mainly in industrial, transport, and agricultural applications because of their reliability, durability, and high fuel efficiency. Despite diesel's extensive use in industry, high smoke and NO_X emissions are major issues related to it. Unburned hydrocarbons, because of lower availability of oxygen in the combustion chamber, lead to emissions. An extensive study has been done to determine the increase in the oxygen content in the fuel by the means of additives to diesel. Biodiesel itself has a large oxygen content compared to petrodiesel, making it less emissive compared to petrodiesel.

Emissions study is important with respect to stringent standards followed by different countries as to decrease greenhouse gas emissions. Performance and emission study on a single-cylinder diesel engine using preheated mahua oil was carried out. Because of preheating to 130 °C of the mahua oil, the viscosity of the oil decreases, which not only enhances the heat release rate but also improves the engine performance and emissions. NO_X emission marginally increased, but preheated mahua oil can be used as diesel substitute in an emergency as well as running situations.[60] Experimentation on a four-stroke, four-cylinder indirect injection water-cooled CI engine was carried out to understand engine emissions. The fuel used was B20 (blend of 20% neem biodiesel and 80% petroleum diesel by volume), and it was observed that emission of CO, NO_X, and SO_2 is less with B20 fuel compared to diesel.[62] The effect of emissions using neem oil and its methyl ester on a direct-injected, four-stroke, single-cylinder diesel engine shows a reduction in emission in NO_X for neem oil and its methyl ester, along with an increase in CO, HC, and smoke emissions.[64]

Work was recently reported that reveals the use of additives in petrodiesel. An oxygenated additive is a chemical compound containing oxygen. It enhances the combustion process and reduces emissions by volume, and it reduces the amount of petrodiesel consumption. Dimethyl carbonate ($C_3H_6O_3$), often abbreviated DMC, has an oxygen content up to 53.3 wt%, with a lower heating value of 15.78 MJ/kg and a boiling point of 90 °C, which are much lower than that of diesel fuel. Ethylene glycol monoacetate ($C_4H_8O_3$), often abbreviated EGM, is a clear, colorless liquid with a boiling point of 187 °C, which is an ideal solvent for oil, cellulose ester, etc. DMC and EGM addition to diesel changes the physicochemical properties of blends. Adding either of these in appropriate proportions improves the engine performance and emission characteristics.[65] Performance and emission tests were carried out on a four-stroke multicylinder CI engine using DMC–EGM–diesel blends. A 5% blend of DMC by volume gives higher brake thermal efficiency than that of petrodiesel. Minimum CO and NO_x emission is found for 10% blend of DMC in petrodiesel. The blends of diesel with 15% DMC and EGM by volume is the best fraction for reduction of smoke.[66]

The above discussion shows that a number of researchers have used different kinds of biodiesel as fuels in CI engines. Not only were they able to run the engine using biodiesel as fuel but also the performance of an engine using biodiesel is almost similar to petrodiesel used as fuel in CI engines. Emissions from CI engines using fuels as a blend of biodiesel and petrodiesel also show good results of lower emissions compared to diesel. Thus, this technology can be explored to reduce the dependence on conventional fossil fuel and utilization of renewable sources of energy as fuels in automobiles.

Conclusions

In this chapter, the current scenario of biodiesel as a fuel is discussed. Biodiesel is basically fuel that can be produced from seeds of edible and nonedible plants, and hence can be classified as renewable sources of energy. The oil from these seeds is extracted by different methods after which the oil undergoes a chemical process known as transesterification. The transesterification process leads to the formation of Biodiesel (Fatty Acid Methyl Ester) and various by-products. Biodiesel has low viscosity and properties similar to conventional petrodiesel; it can be used as fuel in ICEs without any modifications. A number of scientists have done experiments carrying out performance analyses of engines using biodiesel and blends of biodiesel with petrodiesel. They found that engine performance parameters are in line with conventional engines using petrodiesel as fuel. This shows that in the future, biodiesel can fulfill the demand for fuel if conventional fossil fuels are scarce. Also, the engine emissions can meet the various environmental norms set by different countries. There is enough scope in this field for research to investigate new and different seeds from which biodiesel can be produced.

References

1. Mann P, Gahagan L, Gordon MB. Tectonic setting of the world's giant oil and gas fields. In: Halbouty MT, editor. *Giant oil and gas fields of the decade, 1990–1999.* Tulsa (Okla.): American Association of Petroleum Geologists; [accessed 15.05.14]. p. 50.

2. http://www.eia.gov/cfapps/ipdbproject/IEDIndex3.cfm. U.S. EIA International Energy Statistics. [accessed 16.05.14].
3. http://business.financialpost.com/2013/10/17/fossil-fuels-world-energy-council/?_lsa=cf58-ea60 [accessed 10.05.14].
4. Knothe G, Van Gerpen J, Krahl J, editors. *The biodiesel handbook.* (Illinios, USA): AOCS Press; 2004.
5. Agarwal AK, Rajamanoharan K. Biofuels (alcohols and biodiesel) applications as fuels for internal combustion engines. *Prog Energy Combust Sci* 2007;**33**(3):233–71.
6. Singh SP, Singh D. Biodiesel production through the use of different sources and characterization of oils and their esters as the substitute of diesel: a review. *Renew Sustain Energy Rev* 2010;**14**(1):200–16.
7. Lapuerta M, Armas O, Rodríguez FJ. Effect of biodiesel fuels on diesel engine emissions. *Prog Energy Combust Sci* 2008;**34**(2):198–223.
8. Ahmad AL, Yasin NHM, Derek CJC, Lim JK. Microalgae as a sustainable energy source for biodiesel production: a review. *Renew Sustain Energy Rev* 2011;**15**(1):584–93.
9. Patel NK, Nagar P, Shah SN. Identification of non-edible seeds as potential feedstock for the production and application of bio-diesel. *Int J Energy Power* 2013;**3**(4):67–78.
10. Parotta JA. Moringa oleifera *Lam. Reseda, horseradish tree. Moringaceae. Horseradish tree family.* USDA Forest Service, International Institute of Tropical Forestry; 1993.
11. Rashid U, Anwar F, Moser B, Knothe G. *Moringa oleifera* oil: a possible source of biodiesel. *Bioresour Technol* 2008. http://dx.doi.org/10.1016/j.biortech, 2008.30.066.
12. Wickens GE. *Non-timber uses of selected arid zone trees and shrubs in Africa.* Rome: FAO; 1988. p. 98.
13. Radovich T. *Farm and forestry production and marketing profile for Moringa (*Moringa oleifera*).* PO Box 428, Holualoa, Hawai'i 96725, US: Permanent Agriculture Resources (PAR); 2009 [Retrieved 20. 04.14].
14. Ramachandran C, Peter KV, Gopalakrishnan PK. Drumstick (*Moringa oleifera*): a multipurpose Indian vegetable. *Econ Bot* 1980;**34**(3):276. http://dx.doi.org/10.1007/BF02858648.
15. http://en.wikipedia.org/wiki/Soybean [accessed 20.05.14].
16. Shah Shailesh N, Sharma Brajendra K, Moser Bryan R, Erhan Sevim Z. Preparation and evaluation of jojoba oil methyl ester as biodiesel and as blend components in ultra low sulfur diesel fuel. *Bioenergy Res* 2009;**3**(2):214–23.
17. Jhama Gulab N, Moser Bryan R, Shah Shailesh N, Holser Ronald A, Dhingra OD, Vaughn Steven F, et al. Wild Brazilian mustard (*Brassica juncea* L.) seed oil methyl esters as a biodiesel fuel. *J Am Oil Chem Soc* 2009;**86**(1):917–26.
18. Shah Shailesh N, Iha Osvaldo K, Alves Flávio CSC, Sharma Brajendra K, Erhan Sevim Z, Suarez Paulo A. Potential application of turnip oil (*Raphanus sativus* L.) for biodiesel production: physical-chemical properties of neat oil, bio-fuels and their blends with ultralow sulfur diesel (ULSD). *Bioenergy Res* 2013;**6**:841–50.
19. Qi DH, Geng LM, Chen H, Bian YZH, Liu J, Ren XCH. Combustion and performance evaluation of a diesel engine fuelled with biodiesel produced from soybean crude oil. *Renew Energy* 2009;**34**(12): 2706–13.
20. Labeckas G, Slavinskas S. Performance of direct-injection off-road diesel engine on rapeseed oil. *Renew Energy* 2006;**31**(6):849–63.
21. Jeong G-T, Oh Y-T, Park D-H. Emission profile of rapeseed methyl ester and its blend in a diesel engine. In: *Twenty-seventh symposium on biotechnology for fuels and combustions.* ABAB Symposium; 2006. p. 165–78.
22. Zeev W. International publication number WO 2006/137069 A2. 2009. Downloaded from http://www.freshpatents.com 01.13.09.

23. Stephanie RB. *African oils: health and beauty from the motherland.* 2008. Downloaded from, http://www.foa.org/docrep. 23.01.09.

24. EPA-560/6-82-003, PB82-233008. *Test guidelines: chemical fate, aerobic aquatic biodegradation.* 1982.

25. Pinzi S, Garcia IL, Gimenez FJL, Castro MDL, Dorado G, Dorado MP. The ideal vegetable oil-based biodiesel composition: a review of social, economic and technical implications. *Energy Fuels* 2009; **23**:2325–41.

26. Karmee SA, Chadha A. Preparation of biodiesel from crude oil of *Pongammia pinnata. Bioresour Technol* 2005;**96**(13):1425–9.

27. Scott P, Pregelj L, Chen N, Hadler JS, Djordjevic MA, Gresshoff PM. *Pongamia pinnata* an untapped resource for the biofuels industry of the future. *Bioenergy Res* 2008;**1**:2–11.

28. Balat M, Balat H. Progress in biodiesel processing. *Appl Energy* 2010;**87**(6):1815–35.

29. Azam MM, Waris A, Nahar NM. Prospects and potential of fatty acid methyl esters of some non-traditional seed oils for use as biodiesel in India. *Biomass Bioenergy* 2005;**29**(4):293–302.

30. Sanford SD, White JM, Shah PS, Wee C, Valverde MA, Meier GR. *Feedstock and biodiesel characteristics report.* 2011. Available from, http://www.regfuel.com/pdfs/Feedstock%20and%20Biodiesel%20Characteristics%20Re port.pdfS.

31. Demirbas A. Progress and recent trends in biodiesel fuels. *Energy Convers Manage* 2009;**50**(1):14–34.

32. No SY. Inedible vegetable oils and their derivatives for alternative diesel fuels in CI engines: a review. *Renew Sustain Energy Rev* 2011;**15**(1):131–49.

33. Kumar A, Sharma S. Potential non-edible oil resources as biodiesel feed- stock: an Indian perspective. *Renew Sustain Energy Rev* 2011;**15**(4):1791–800.

34. Ragit SS, Mohapatra SK, Kundu K, Gill P. Optimization of neem methyl ester from transesterification process and fuel characterization as a diesel substitute. *Biomass Bioenergy* 2011;**35**(3):1138–44.

35. Jena PC, Raheman H, Prasanna GVK, Machavaram R. Biodiesel production from mixture of mahua and simarouba oils with high free fatty acids. *Biomass Bioenergy* 2010;**34**(8):1108–16.

36. Ghadge SV, Raheman H. Biodiesel production from mahua (*Madhuca indica*) oil having high free fatty acids. *Biomass Bioenergy* 2005;**28**(6):601–5.

37. Balat M. Potential alternatives to edible oils for biodiesel production – a review of current work. *Energy Convers Manage* 2011;**52**(2):1479–92.

38. Achten WMJ, Verchot L, Franken YJ, Mathijs E, Singh VP, Aerts R. Jatropha bio-diesel production and use. *Biomass Bioenergy* 2008;**32**(12):1063–84.

39. Beerens P. *Screw-pressing of Jatropha seeds for fuelling purposes in less developed countries.* Eindhoven: Eindhoven University of Technology; 2007.

40. Forson FK, Oduro EK, Hammond-Donkoh E. Performance of Jatropha oil blends in a diesel engine. *Renew Energy* 2004;**29**(7):1135–45.

41. Tewari DN. *Jatropha and biodiesel.* (India): Ocean Books Ltd; 2007.

42. Rabe ELM. *Jatropha oil in compression ignition engines, effects on the engine, environment and Tanzania as supplying country.* Eindhoven: Eindhoven University of Technology; 2010.

43. Mahanta P, Shrivastava A. *Technology development of bio-diesel as an energy alternative.* 2011. Available from, www.newagepublishers.com/samplechapter/001305.pdfS.

44. At abani AE, Silitonga AS, Badruddin IA, Mahlia TMI, Masjuki HH, Mekhilef S. A comprehensive review on biodiesel as an alternative energy resource and its characteristics. *Renew Sustain Energy Rev* 2012;**1**(1):2070–93.

45. Kislik VS. *Solvent extraction: classical and novel approaches.* Elsevier; 2012.

46. www.gonuke.org, *Regional Center for Nuclear Education and Training*, 3209 Virginia Avenue, Fort Pierce, FL 34981.
47. http://www.epa.gov/osw/hazard/testmethods/sw846/pdfs/3550c.pdf.
48. Christensen FM. Extraction by aqueous enzymatic process. In: *Inform: international news on fat, oils and related materials*; 1991. p. 984–7. New York.
49. Dominguez HM, Núñez J, Lema JM. Enzymatic pre-treatment to enhance oil extraction from fruits and oil seeds: a review. *Food Chem* 1994;**49**:271–86.
50. Rosenthal A, Pyle LD, Niranjan K. Aqueous and enzymatic processes for edible oil extraction. *Enzyme Microbiol Technol* 1996:402–20.
51. Camargo UA, Mandelli F, Antonio F Conceicao M, Tonietto J. Grapevine performance and production strategies in tropical climates. *J Food Agro-Ind* 2012;**5**(4):257–69.
52. Koh MY, Ghazi TIM. A review of biodiesel production from *Jatropha curcas* L. oil. *Renew Sustain Energy Rev* 2011;**15**(5):2240–51.
53. Balat M, Balat H. A critical review of bio-diesel as a vehicular fuel. *Energy Convers Manage* 2008; **49**(10):2727–41.
54. Ma F, Hanna MA. Biodiesel production: a review. *Bioresour Technol* 1999;**70**(1):1–15.
55. Utsav V, Patel Nikul K, Shah Shailesh N. Economical study and estimation for production of biodiesel from Jatropha seeds in Indian context. In: *National seminar on climate change and sustainable development issues and challenges, conference held at the M. S. University of Baroda, Vadodara, India on 23–24 January*; 2013.
56. Shah SN. Sulfur compounds significance in fossil fuels and their impact on the environment. In: Taylor JC, editor. *Advances in chemistry research*, vol. 22. NY (USA): Nova Publication; 2014. p. 203–16.
57. Chung KH. Transesterification of *Camellia japonica* and *Vernicia fordii* seed oils on alkali catalysts for biodiesel production. *J Ind Chem* 2010;**16**(4):506–15.
58. Panwar NL, Shrirame HS, Rathore NS, Jindal S, Kurchania AK. Performance evaluation of a diesel engine fueled with methyl ester of castor seed oil. *Appl Therm Eng* 2010;**30**(2–3):245–9.
59. Vera C, Busto M, Yori J, Torres G, Manuale D, Canavese S. Adsorption in biodiesel refining a review. In: *Biodiesel – feedstock and processing technologies*. Croatia: InTech Europe; 2011.
60. Pugazhvadivu M, Sankaranarayanan G. Experimental studies on a diesel engine using mahua oil as fuel. *Indian J Sci Technol* 2010;**3**(7):787–91.
61. Oza NP, Rahtod PP, Patel NK. A review of recent research on non-edible vegetable oil as fuel for CI Engine. *Tech J Online J Eng Res Stud* 2012;**3**:84–6.
62. Oza NP, Rathod PP, Patel NK. Performance comparison of 4-stroke multi-cylinder CI engine using neem biodiesel and diesel as fuel. In: *Proceedings of the Indian journal of technical education 1st national conference of futuristic trends in mechanical engineering*; 2012. p. 143–8.
63. Hotti SR, Hebbal O. Performance and combustion characteristics of single cylinder diesel engine running on karanj oil/diesel fuel blends. *Sci Res J* 2011;**4**(2):371–5.
64. Sivalakshmi S, Balusamy T. Experimental investigation on a diesel engine fuelled with neem oil and its methyl ester. *Therm Sci* 2011;**15**(4):1193–204.
65. Prajapati S, Rathod PP, Patel NK. Effect of oxygenated fuel additives on performance and emission characteristics of CI engine – a review study. *Tech J Online J Eng Res Stud* 2012;**3**(1):108–10.
66. Prajapati S, Rathod PP, Patel NK. Performance and emission improvements of 4-stroke multi-cylinder CI engine by use of DMC-EGM-diesel blends. *Tech J Online Int J Adv Eng Res Stud* 2012; **1**(4):70–3.

12

Introduction to Solar Photovoltaic Technology

Parimita Mohanty[1], Akshay Tyagi[2]

[1]FELLOW AND TEAM LEADER, TERI, NEW DELHI, INDIA;
[2]RESEARCH ASSOCIATE TRAINEE, TERI, NEW DELHI, INDIA

CHAPTER OUTLINE

Food, Energy, and Water. http://dx.doi.org/10.1016/B978-0-12-800211-7.00012-0

Introduction

A majority of photovoltaic (PV) cells/solar cells are semiconductor-based junction device, which convert solar radiation into direct electricity. PV cells/solar cells show PV effect. PV effect is the phenomenon by virtue of which voltage difference is created across a p–n junction as result of absorption of photons.

Figure 1 shows wavelength, frequency, and photon energy of solar radiation. The entire spectrum of sunlight covers a range of about 0.5 eV to about 2.9 eV (from infrared to ultraviolet). Red light has the energy of about 1.7 eV and blue light has energy of about 2.7 eV.

Unless and until the photon energy is greater than the band gap energy, electrons cannot be removed from the valence band and light energy will pass through the material. This is shown in Figure 2.

Potential Candidate for Solar Cells

One key requirement of an efficient solar cell is to convert as much sunlight into electricity as possible.The power from an electrical device such as a solar cell is equal to the product of the voltage (V) and the current (I).

The current produced by the solar cell by the absorption of photons can be given as

$$I = qNA$$

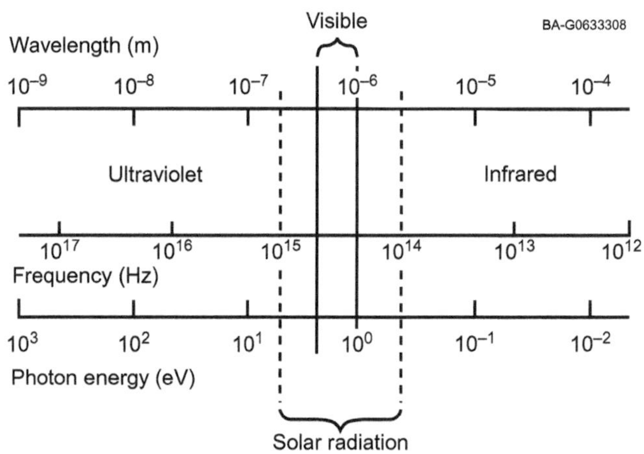

FIGURE 1 Wavelength, frequency, and photon energy solar radiation.

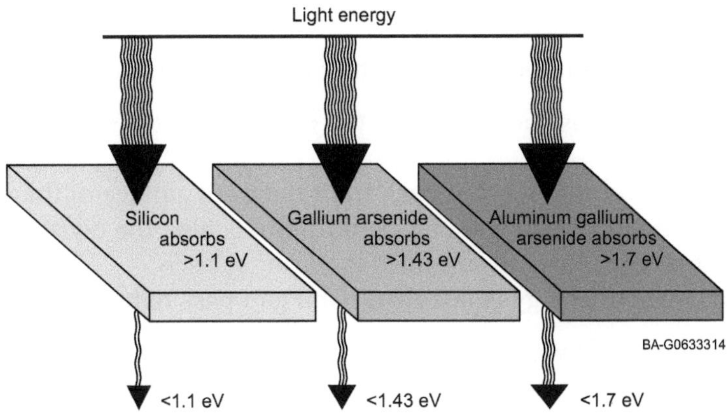

FIGURE 2 Passage of light through different materials.

where

N = number of photons incident upon the solar cell
A = surface area of the solar cell which is exposed to light

Thus materials with lower band gap energies can generate a large number of charge carriers as well as the current. Hence, one can assume that material with the lowest band gap would make the best PV cell, but it is not true.

The band gap energy also influences the strength of the electric field, which determines the maximum voltage the cell can produce (as $V = Eg/e$). The higher the band gap energy of the material, the higher the generated open-circuit voltage (V_{oc}).

To summarize

- Low band gap cells have high current but low voltage;
- High band gap cells have high voltage and low current.

Therefore, a compromise is necessary in the design of solar cells. Cells made of materials with band gaps between 1 and 1.8 eV can be used efficiently in PV devices. Table 1 shows the potential materials for solar cells.

Table 1 Energy Gap of Potential Materials for Solar Cells

Material	Energy Gap (eV)
Crystalline silicon	1.12
Amorphous silicon	1.75
Copper indium diselenide (CuInSe2)	1.05
Cadmium telluride (CdTe)	1.45
Gallium arsenide (GaAs)	1.42
Indium phospite (InP)	1.34

How Solar Cells Work

When light is incident upon the solar cell, the following occurs:

- Photons whose energy is comparable to the band gap of the semiconductor get absorbed.
- Some of the photons are reflected back from the front surface of the cell.
- Some photons are lost as they are blocked from reaching the crystal by the current collectors (metal grid) that cover part of the front surface.
- Some photons whose energy is less than band gap pass right through the solar cell.

When a photon of light with the appropriate amount of energy penetrates the solar cell (e.g., a silicon solar cell) near the junction and encounters one of the silicon atoms, the following occurs (Figure 3):

- It dislodges one of the electrons, which leaves behind a hole.
- The generated carriers will tend to move to the layers where it acts as majority carriers, i.e., electron tends to migrate into the layer of n-type silicon whereas hole tends to move to the layer of p-type silicon.
- Electron can be collected by the current collector on the front surface of the cell, generate an electric current in the external circuit, and then reappear in the layer of p-type silicon.
- It recombines with the hole in the layer of p-type silicon.

Characteristic Curve of a Solar Cell

When the solar cell is not irradiated, it is just like a semiconductor diode. In that case, it follows the ideal diode equation. When a solar cell is irradiated, current flows through the circuit exactly opposite to that of the diode current. The output current (I) is then equal to the difference between light generated current (I_L) and the diode current (I_D). Figure 4 shows the equivalent circuit of the solar cell.

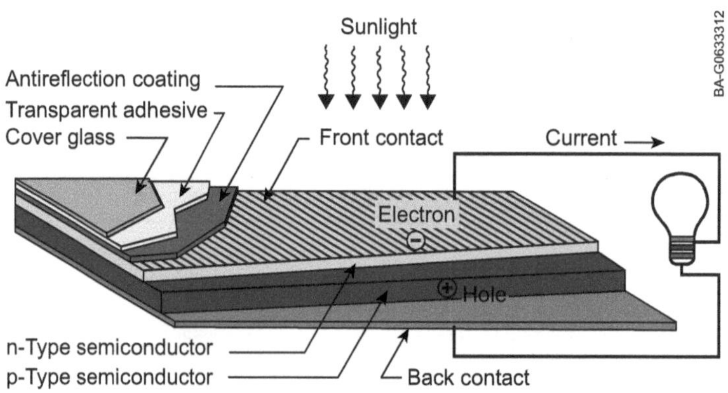

FIGURE 3 How a solar cell works.

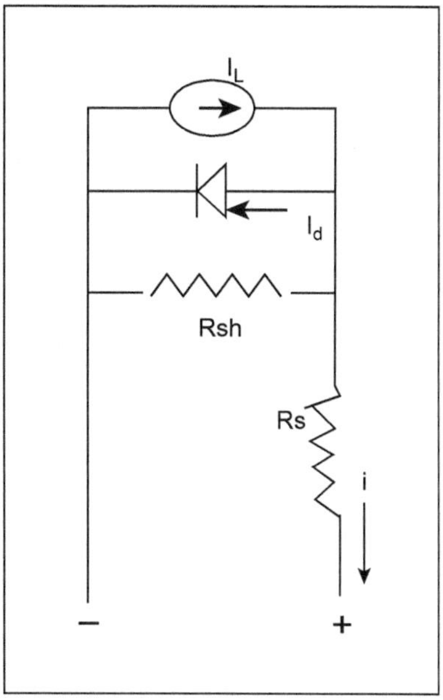

FIGURE 4 Equivalent circuit of the solar cell.

Thus the operating equation of current–voltage characteristic of the solar cell can be derived (by using eq. (1)) as follows:

$$\begin{aligned} I &= I_L - I_D \\ &= I_L - I_O[\exp(q(V + RsI)/\eta k_B T) - 1] - (V + RsI/Rsh) \end{aligned} \tag{1}$$

Where,

I_L = the photo-generated current
I_D = diode current
I_O = saturation current
R_s = Series resistance
R_{sh} = Shunt resistance
T = Junction temperature

If connected to an external source then diode equation would be,

$$I_D = I_O[\exp(q(V + RsI)/\eta k_B T) - 1]$$

The I–V curve of a solar cell is the superposition of the I–V curve of the solar cell diode in dark with the light generated current.

Figure 5 shows typical I–V characteristics.

If the I–V curve is plotted for a solar cell, then the curve passes through three significant points i.e., short circuit current (I_{sc}), V_{oc}, and maximum power point (MPP) (V_{mp} and I_{mp}).

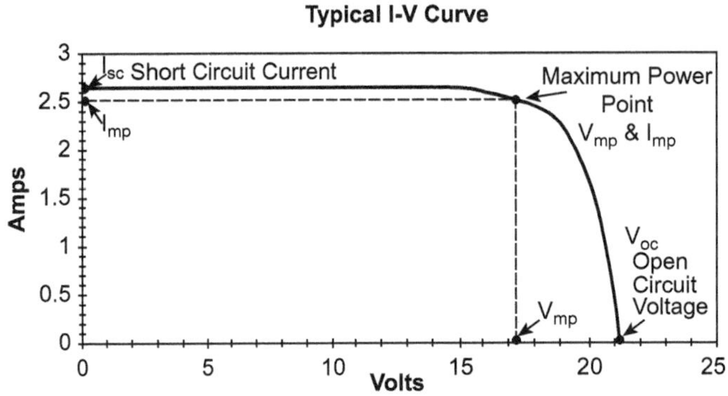

FIGURE 5 Typical I–V curve of a solar cell.

SHORT CIRCUIT CONDITION

The circuit is called a short circuit when $V = 0$ as shown in Figure 6.

In such case $I = I_{SC}$ where I_{SC} is the short circuit current (all the current will pass through the external load) and eq. 1 becomes

$$I_{sc} = I_L - I_O[\exp(qRsI_{sc}/\eta k_B T) - 1] - (RsIsc/Rsh) \tag{2}$$

At normal irradiance, the series resistance is very small and thus in short circuit condition it can be neglected ($R_S \rightarrow 0$). Therefore, eq. (3) reduces to.

$$\begin{aligned} I_{sc} &= I_L - I_O * 0 - 0 \\ I_{sc} &= I_L \end{aligned} \tag{3}$$

I_{sc} depends on a number of factors, which are described below.

- The area of the solar cell.
- The number of photons.
- The spectrum of the incident light. For most solar cell measurements, the spectrum is standardized to the air mass (AM) 1.5 spectrum; maximum value of $I_{SC} \sim 42$ mA/cm^2 for Si.
- Optical properties (absorption and reflection) of the solar cell.
- The collection probability of the solar cell, which depends on the surface passivation and the minority carrier lifetime in the base.

The I_{SC} is equivalent to the light generated current thus proportional to the irradiance.

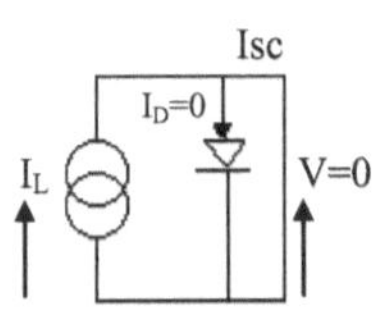

FIGURE 6 Short circuit condition.

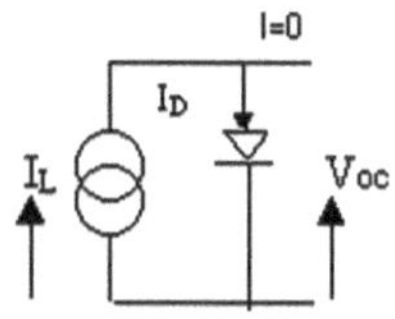

FIGURE 7 Open-circuit condition.

OPEN-CIRCUIT CONDITION

In the open-circuit condition as shown in Figure 7 ($V = V_{oc}$), $I = 0$. In such case R_{sh} tends to ∞ (Figure 7)

Therefore Eqn 1 reduces to.

$$V_{oc} = \eta k_B T/q[\ln((I_L + I_O)/I_O)] \quad (4)$$

The open circuit voltage (V_{oc}) increases logarithmically with increasing irradiance level and decreases linearly with an increase in junction temperature. V_{oc} depends on the saturation current of the solar cell and the light generated current.

Requirement for high V_{oc} is as follows:

- High carrier diffusion length
- Low recombination
- High doping concentration
- Low crystal volume

Laboratory silicon solar cells have V_{oc} of up to 720 mV under one sun and AM 1.5 condition.

MAXIMUM POWER OPERATION

If the cell's terminals are connected to a variable resistance R, the operating point will be determined by the intersection of the I–V characteristics of the solar cell and the load. For a resistive load, the load characteristics are a straight line with a slope 1/R (where $I/V = 1/R$). The maximum power is obtained for $R = R_{opt}$ at point as shown in Figure 8.

$$R_{opt} = V_{max}/I_{max}$$

It is always desirable to connect the load in such a way that it will work always under MPP. Details about maximum power point tracker (MPPT) will be discussed in subsequent sections.

POWER DELIVERED BY THE SOLAR CELL

The electrical power that can be extracted from the solar cell is proportional to its area and to the intensity of the sunlight that strikes the area; it is measured in watt peak (Wp). Under standard test conditions (STCs), a solar cell of 1 Wp will supply a power of 1 W. STCs are defined as when the global irradiation is 1000 W/m^2, AM is 1.5, and the cell operating temperature is 25° C. A typical 4-inch diameter crystalline Si solar cell or a 10×10 cm multicrystalline cell should provide a power of 1–1.5 W under STCs. But these

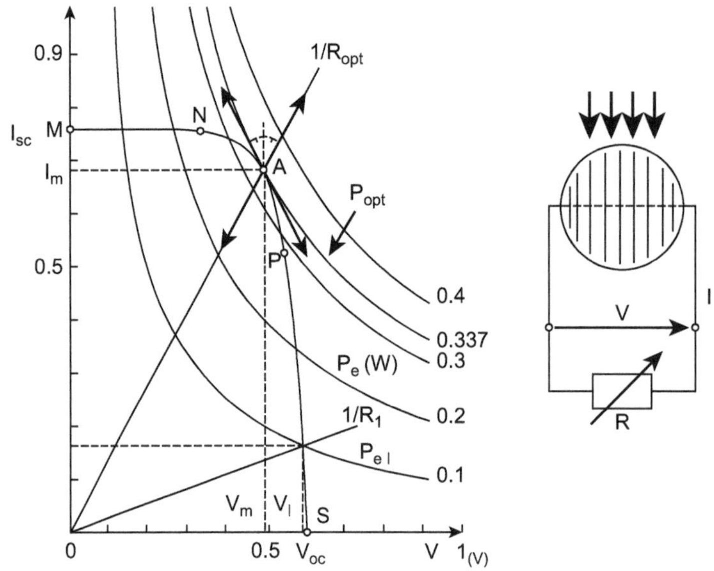

FIGURE 8 Finding the maximum power point.

STCs are rarely found in real-world operating conditions. In real field condition, in a bright sunny day, the global irradiance on the ground is usually 700–800 W/m^2. Again, when the ambient temperature is 45° C, cell operating temperatures go as high as 60–65° C. So in reality, 1 Wp cell never produces 1 W, rather on a good sunny day, it produces 60–70% of that i.e., 0.6–0.7 W.

FILL FACTOR

Fill factor is defined as the ratio of the maximum power (peak power) from the solar cell to the product of the V_{oc} and I_{sc}.

It is expressed as,

$$FF = \frac{V_m * I_m}{V_{oc} * I_{sc}} \tag{5}$$

Graphically (as shown in Figure 9), FF is the measure of the squareness of the I–V characteristic of solar cell and is also the area of the largest rectangle, which will fit in the I–V curve.

Higher voltage will have a higher possible FF. However, parasitic resistance reduces the FF. The maximum FF is for smallest series resistance (R_s) and largest shunt resistance (R_{sh}) and its value is higher than 0.7 for good cells.

CONVERSION EFFICIENCY (η)

Conversion efficiency is the ratio of the optimal power output (P_{opt}) delivered by the PV module to the solar irradiance E_e (input power) received at a given cell temperature (T).

$\eta = P_{opt}/AE_e$; where P_{opt} is in watt, E_e is in watt/m^2, and cell area is in m^2.

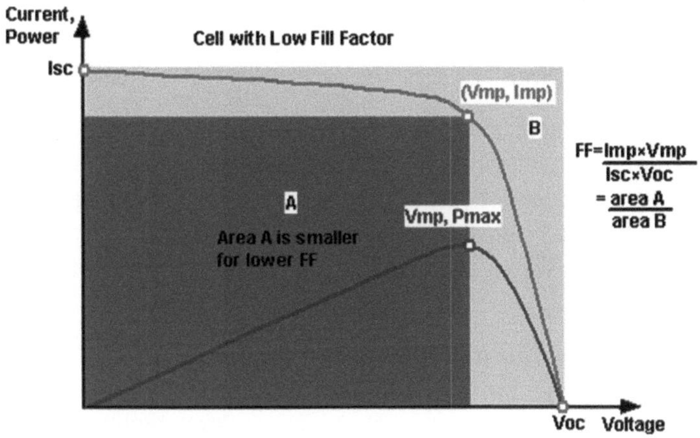

FIGURE 9 Fill factor.

It is expressed as,

$$\eta = \frac{V_{oc} * I_{sc} * FF}{Pin} \tag{6}$$

CELL PARAMETER OF SILICON SOLAR CELL

The theoretical power conversion efficiency for a solar cell based on a single semiconducting material is limited to 31%. One of the most fundamental limitations of solar cell efficiency is the band gap of the semiconductor from which the cell is made. Band gap of the material cannot perfectly match the broad range of solar radiation, which has usable energy in the photon range of 0.4–4 eV. Light with energy below the band gap of the semiconductor will not be absorbed and thus not be converted. Light with energy above the band gap will be absorbed, but the excess energy above the band gap will be lost in the form of heat. Decades of research in developing single-material solar cells has led to cell efficiencies close to the theoretical limit (Table 2). However, different layers could be stacked to receive photons at all energies, reaching efficiencies better than 70%, but in that case many other problems intervene. The most efficient multijunction solar cell yet made—30%, out of a possible 50% efficiency (with double junction).

Table 2 Cell Parameters of Silicon Solar Cells

Cell Type	Efficiency (η) (%)	V_{oc} (V)	I_{sc} (mA/cm^2)	FF
Si (crystalline)	24.7	0.706	42.2	82.8
Si (multicrystalline)	19.8	0.654	38.1	79.5
Si (supported film)	16.6	0.608	33.5	81.5

Solar Cell Technologies

Performance of solar cell depends upon three key factors.

- First, the type of absorbing material used (such as different semiconductors), which absorbs light (photons) and converts it into electron–hole pairs.
- Second, the type of junction (semiconductor junction) formed, which separates the photo-generated carriers (electrons and holes).
- Third, the contacts on the front and back of the cell that allow the current to flow to the external circuit.

Depending upon the type of light absorbing material used, the solar cell technology can be broadly divided into three groups; silicon-based technologies, compound semiconductor-based technologies, and other emerging technologies. However, crystalline silicon is one of the most reliable and most developed PV technologies at the present time. Out of all the solar cell technologies, wafer-based crystalline silicon (indirect band gap material) requires a considerable thickness (several hundred microns) of material, as it is a poor absorber of light. Due to the high cost of crystalline silicon wafers, cheaper materials are being explored. These selected materials (compound semiconductors, amorphous silicon etc.) are all strong light absorbers and only need to be about 1 μm thick, so materials costs are significantly reduced.

The main advantage of employing thin films for solar cells is the fact that a small amount of the material is required for the device to function. One of the most important criteria for thin film solar cell is that the material to be employed is a direct band gap with a high optical absorption coefficient. A typical thickness of an indirect band gap material is 20–50 μm in order to absorb most of the solar spectrum whereas only 1–3 μm of the direct band gap material is required to absorb most of the spectrum. However, the direct band gap material homojunctions are frequently limited by surface recombination. Thus, solar cell technologies can also be classified as wafer-based crystalline silicon solar cell technology, thin film solar cell technology, and other new emerging technologies as shown in Figure 10.

Wafer-Based Crystalline Silicon Solar Cell Technology[4,5,6]

Wafer-based crystalline silicon solar cell manufacturing process involves three stages:

- Material preparation and shaping,
- Cell processing, and
- Cell interconnection and encapsulation.

Wafer-based solar cell manufacturing process is represented in Figure 11.

The three different stages are discussed below.

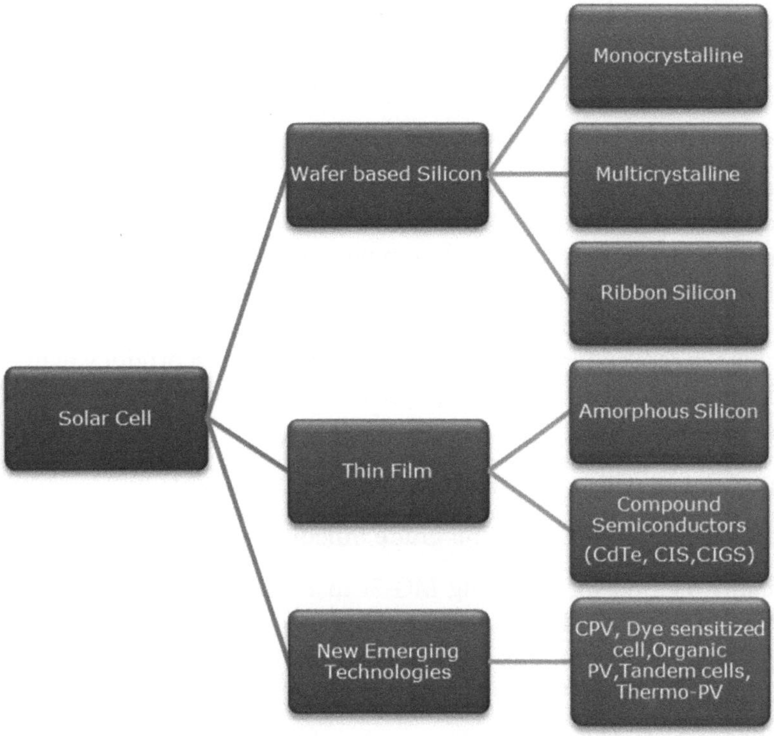

FIGURE 10 Classification of solar cell technologies.

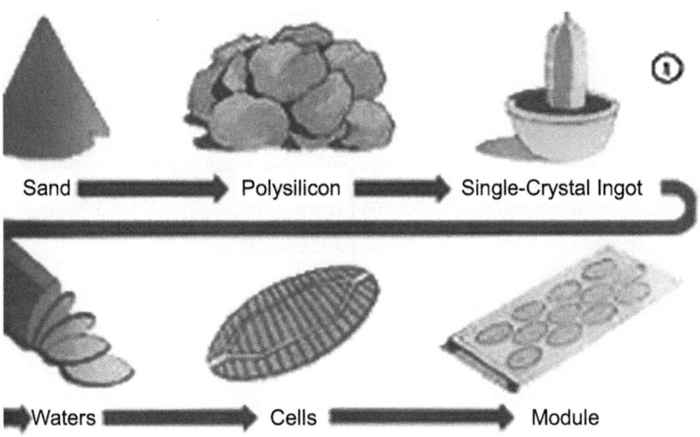

FIGURE 11 Wafer-based solar cell manufacturing process.

Material Preparation and Shaping

The following steps are involved in material preparation and shaping:

Figure 12 shows steps involved in material preparation and shaping.

Step-1: Quartzite to metallurgical-grade silicon

Quartzite rock and special strata of quartz sand are the most suitable raw materials for the production of elemental silicon. Commercial extraction process uses the crystalline form of silicon dioxide, quartzite with 90% or more silica rather than sand whose main constituent is also silica. Sand with its high amount of impurities is too expensive to process and hence economically unviable.

The quartzite is reduced in large arc furnace by carbon to produce elemental silicon. The chemical reaction involved in this process is shown below:

$$SiO_2(s) + 2C(s) \rightarrow Si(l) + 2CO(g) \tag{7}$$

MG-Si is produced with a purity of 98–99% which is low.

Step-2: MG-Si to EG-Si (Electronic grade Silicon)

The standard method for purifying MG-Si to EG-Si is the Siemens C process. The silicon produced by this method is 99.99% pure and is in polycrystalline form.

In this process, a bed of fine MG-Si particles is fluidized and chlorinated with hydrochloric acid (HCl) in the presence of a copper catalyst to start the reaction.

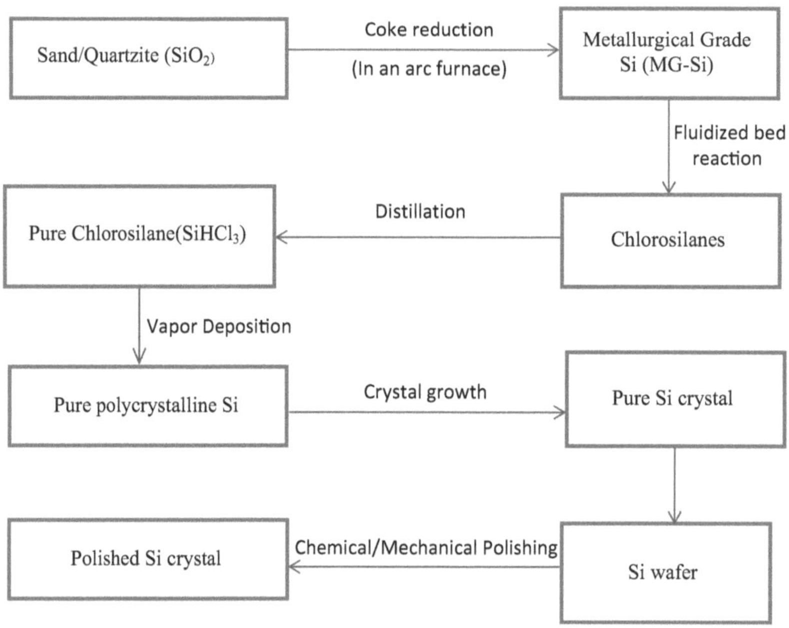

FIGURE 12 Material preparation and shaping.

The trichlorosilane ($SiHCL_3$) thus formed has to be purified through subsequent fractional distillation (to reduce impurities by a factor of 104–105).

$$Si + 3HCl \rightarrow SiHCL_3 + H_2$$
$$SiHCl_3 + H_2 \rightarrow Si + 3HCl$$

A chemical vapor deposition (CVD) method is subsequently used to isolate the pure silicon from the EG-$SiHCl_3$.

Vaporized $SiHCl_3$ is decomposed and reduced with hydrogen at about 1000° C, resulting in the deposition of silicon precipitates on the hot surface of a bridge in the shape of an inverted U. The bridge is heated in a reactor by passing an electric current through it. By this process, polycrystalline silicon rods can be produced.

The deposition process is characterized by a low throughput and a discontinuous operation. Analysis of the present module costs shows that 50% of the manufacturing cost is because of silicon substrate alone.

Solar Grade Silicon (SG-Si) concept: Due to the high processing cost of SG-Si, development of SG-Si had been thought of in PV industry for making solar cell, since Si may not be as pure as EG-Si. It has not been considered a second-rate material but a specific material for efficient solar cell production.

Step-3: Crystal growth

Crystal growth method is broadly divided into two types:

- Bulk growth leading to ingots, which have to be cut into wafers.
- Ribbon pulling.

Though, there is an economic advantage of ribbon growth than ingots (as in case of ribbon pulling technique, silicon consumption is 4–8 times lesser than ingots and there is no need of slicing the ingot into wafers), it is not widely adopted in the industry as the techniques are less mature.

There are two bulk growth technologies for single crystal silicon growth.

- Ingot pulling
- Ingot casting

Single-crystal silicon can be produced in different forms such as ingot, sheet, and ribbon.

Figure 13 summarizes about different processes involved in crystal growth.

The Czochralski (CZ) and floating zone (FZ) techniques use ingot-pulling approach for growing ingot from molten silicon. The ingot grown is then sawed into ingot wafers. However, newer processes such as dendritic web growth and edge-defined film-fed growth (EFG) are used for growing ribbons, which can then directly be used as wafers.

FIGURE 13 Different processes involved in crystal growth.

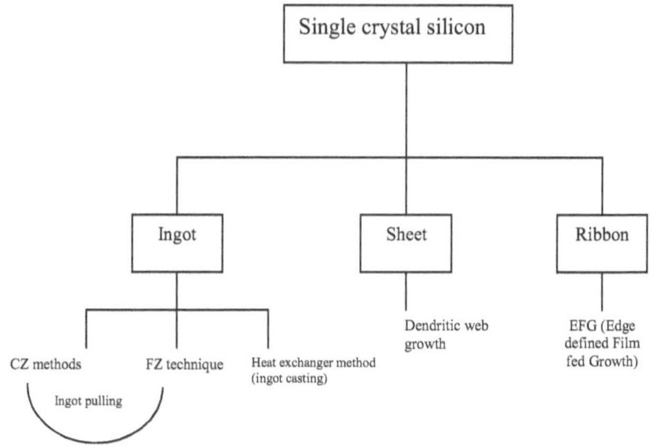

Ingot-pulling technology

CZ method: It is the major commercial method of manufacturing ingots of single-crystal silicon. Figures 14 and 15 shows how a crystal is grown.

To start the crystallization process, a small silicon crystal is properly cooled and used as a seed in molten silicon. When the seed is pulled out slowly, silicon solidifies at the interface with the melt and due to controlled pulling, the silicon atoms arrange themselves according to the crystallographic structure of the seed. This yields an ingot of single-crystal silicon. The cylindrical ingots are typically 1 m long, 7.5 cm in diameter. A p-type crystal is manufactured directly by adding boron into the melt. Crystals with diameter as large as 15 cm and length of 1.5 m can be grown using the continuous process. The rechargeable process also allows the crucible to be reused several times thus reducing the cost of replacement.

In another recent development, a magnetic field is applied to reduce the interaction between molten silicon and crucible thus reducing the usual contamination by carbon and oxygen.

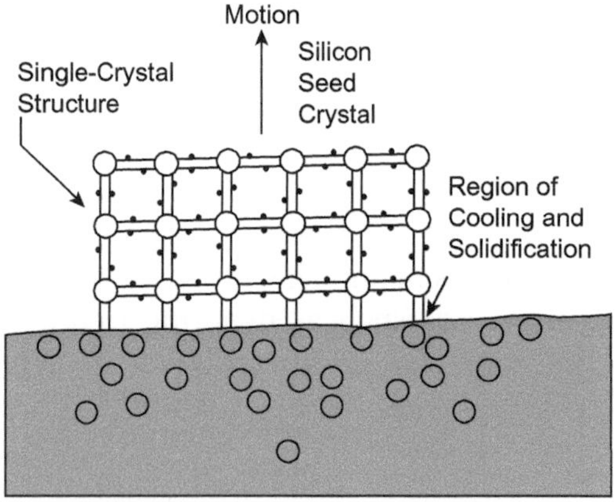

FIGURE 14 Crystal growth by Czochralski.

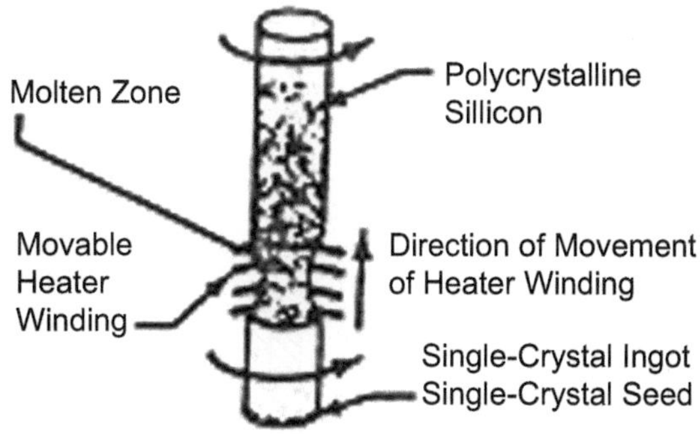

FIGURE 15 Floating zone method.

FZ technique

Here the polycrystalline ingot is placed over the single-crystal seed and the interface between the two is melted by a movable heating coil. As the moving coils go up the stationary ingot, the material below the molten interface gets solidified into single-crystal silicon. The process is called FZ method (Figure 15) as the molten silicon is supported by its own surface tension.

Ingot-casting technology

Heat exchanger method

In Heat exchanger method, the single-crystal seed is placed in a crucible, which itself is placed on a high temperature heat exchanger. The seed is not allowed to melt by passing gaseous helium through the heat exchanger. Molten silicon is allowed to freeze from a seed crystal to form a nearly perfect cubic single-crystal ingot.

RIBBON GROWTH: EFG

A graphite-shaping die is employed to shape the crystal growth to a desired form. When a seed crystal of single-crystal silicon comes in contact (shown in Figures 16 and 17) with the surface of the melt, the melt gets drawn through the die by capillary action. As the crystal is slowly raised (at about 15 cm/min), the silicon cools and freezes into a single-crystal ribbon. The main advantage of ribbon grow is the simultaneous multiribbon growth in one operation and efficient utilization of silicon material with no loss involved in slicing.

A silicon ribbon forms between a pair of dendrites, which provide stability to the edge for width control. It is a slow process without any requirement of die.

POLYCRYSTALLINE SILICON GROWTH

Polycrystalline silicon is made of many grains of single-crystal silicon whose size and quality determines the performance of the solar cell. The grains should be as large as possible so that losses such as recombination losses will be minimal at the grain boundaries. The grain size has major detrimental effect on polycrystalline silicon solar cell characteristic and efficiencies.

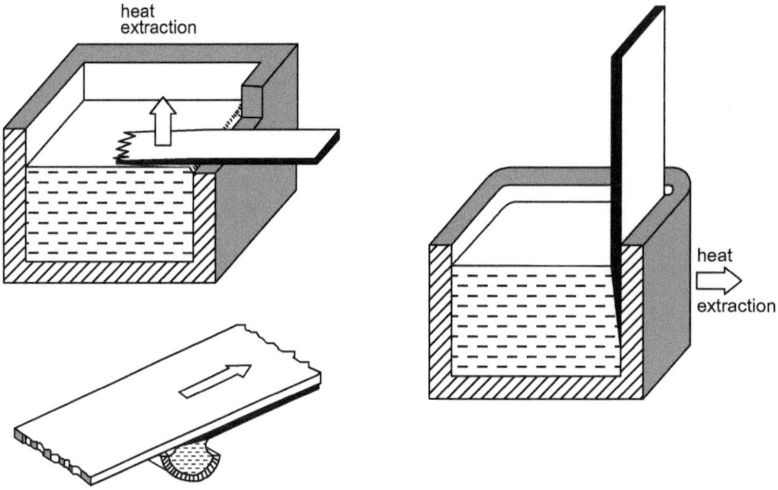

FIGURE 16 Si ribbon growth by edge-defined film-fed growth.

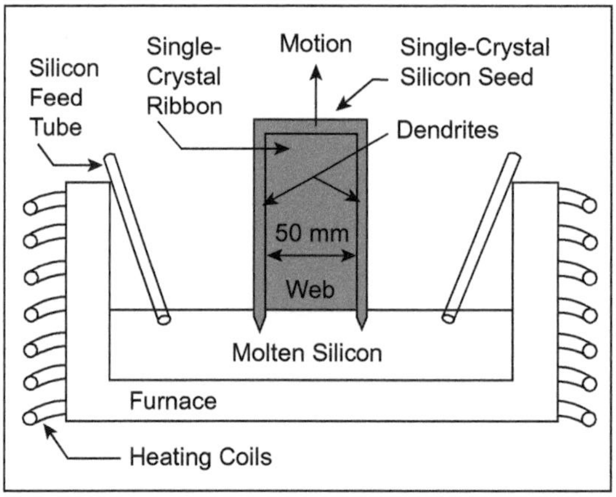

FIGURE 17 Sheet growth: dendritic web growth.

Almost all the methods explained above are used for manufacturing polycrystalline silicon solar cell under less controlled condition. However, ingot pulling is used mainly in single-crystal growth.

Step-4: Silicon crystal to silicon wafer

As the major constraint to crystalline silicon solar cell is its high manufacturing cost, it is understood that low-cost production of crystalline solar cell is possible if efficient cutting and wafering technique is incorporated (50% of the silicon is wasted by the wafering operation itself). There are several ways of slicing the ingot into wafers. The usual way is inner diameter saw as shown in Figure 18.

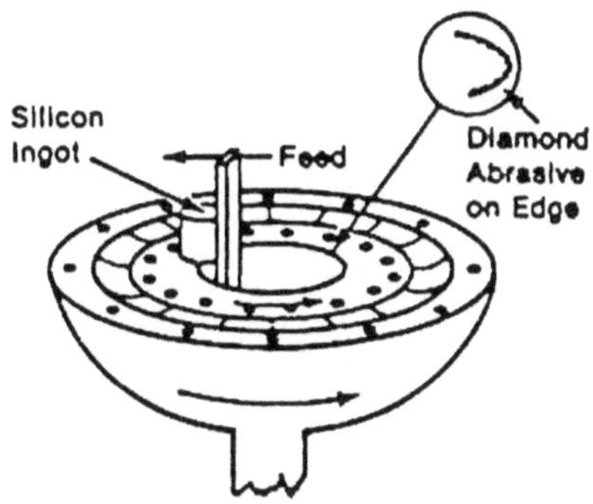

FIGURE 18 Inner diameter saw.

Here, the output is much less because of slow slicing rate (about 30 wafers/h). In addition to slow rate, the main disadvantage of this process is the amount of silicon wasted. Almost half of the silicon gets wasted as kerf loss. However, multiblade slurry saw machine as shown in figure 5.47 can decrease the kerf loss to a large extend. In this case, thinner wafer of 200 μm can be produced as compared to 450 μm in case of inner diameter saws (Figure 19).

There are other material removal process and table shows the comparison of different silicon slicing methods (Table 3).

Step-5: Silicon wafers to polished silicon wafers

After ingot sawing, about 30 μm thick surface layer is damaged. This layer is removed by acid etching (mixture of 1 HF (48%) + 1 CH_3COOH (99.8%) + HNO_3 (65%)) at a rate of 1 μm/min at 20° C.

Surface is also polished with alumina powder.

After polishing, it is cleaned by any of the following procedures.

Simple cleaning (hot chlorethylene + hot acetone + hot methanol + ringing in distilled water).

RCA cleaning (10 min in each of H_2O_2 (30%) + $4SO_4H_2$ (96%), $1H_2O_2$ (30%) + $5H_2O$, $1H_2O_2$ (30%) + 1HCl (37%) + $5H_2O$ and then HF + 1/2 distilled water).

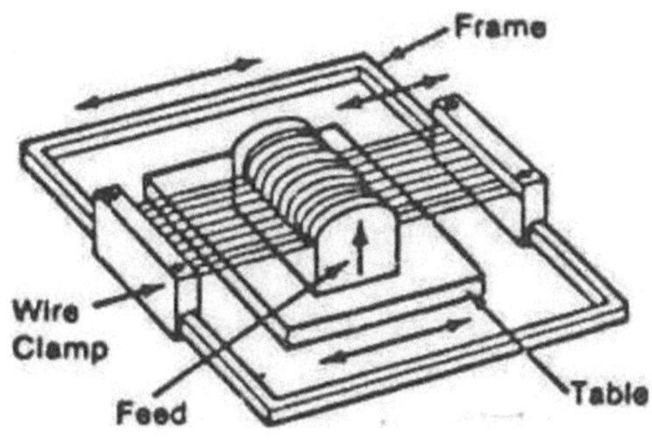

FIGURE 19 Multiblade slurry saw machine.

Table 3 Different Silicon Slicing Method

Slicing Method	Mode of Application	Production Rate (wafers/h)	Yield (%)	Surface Quality	Technical Reliability
ID sawing	Conventional	Average	High	Average	High
ID sawing	Advanced	High	Very high	Good	Very high
MBS	Conventional	Low	Average	Average	Good
MBS	Advanced	Average	High	Good	Fair

ID sawing = Internal diameter sawing, MBS = Multiblade slurry sawing.

Cell Processing (Cell Fabrication Process)

Once silicon wafer is prepared, several fabrication processes are employed to convert it into a solar cell.

The fabrication processes used are as follows

- p–n junction formation,
- Possible back p + region formation,
- Electrical contact formation (front and back metal contact), and
- Antireflection layer deposition.

P–N JUNCTION FORMATION

The wafer so formed is either p-type or n-type. However, to make one solar cell, one p–n junction has to be created to separate the photo-generated carriers. A p–n junction can be created either by a high temperature diffusion process or by an ion implantation process, which is a cold junction process. The advantages of cold junction processing over high temperature processing are lower energy consumption, full automation, and lower overall cost. In both the processes, diffusion can be from a solid phase or from a vapor phase.

Parasitic junctions at the edge of the wafer can be removed by chemical etching, plasma etching or laser scribing.

HIGH TEMPERATURE DIFFUSION To form p-type layer on n-type silicon, boron is commonly used as the impurity from sources such as B_2O_3 (solid), BCl_3 (gas), and BBr_3 (liquid) and to form n-type layer on p-type silicon, phosphorus is used as impurity from the source such as P_2O_5 (solid source) and $POCl_3$ (liquid source). Figures 20 and 21 show the diffusion process for solid, liquid, and gaseous source.

A carrier gas (mixture of N_2 and O_2) passes over the source, which carries the impurity to the wafer surface.

POSSIBLE BACK P+ REGION FORMATION

COLD JUNCTION PROCESSING Though cold junction processing (ion implantation) is at the developmental stage, it could be an alternative to slow, energy-intensive high temperature diffusion process. Here the dopants are ionized and accelerated to penetrate the target through their kinetic energy. Due to their collision, surface of the base material is damaged, which can be rectified through annealing.

Advantages of cold junction processing over thermal annealing are less energy usages, greater control over dopant concentration, and junction depth.

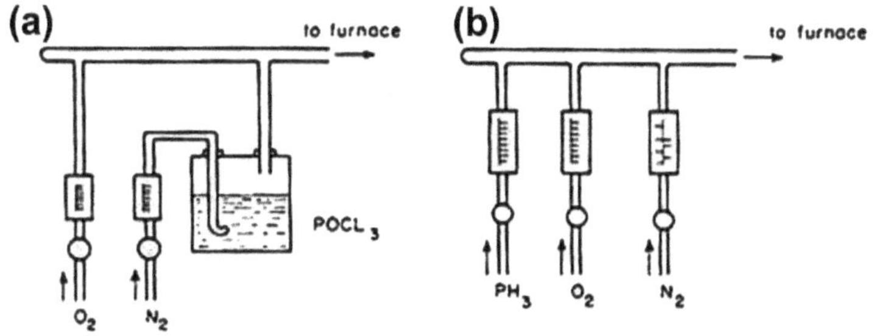

FIGURE 20 Apparatus for phosphorus diffusion (n-type layer).

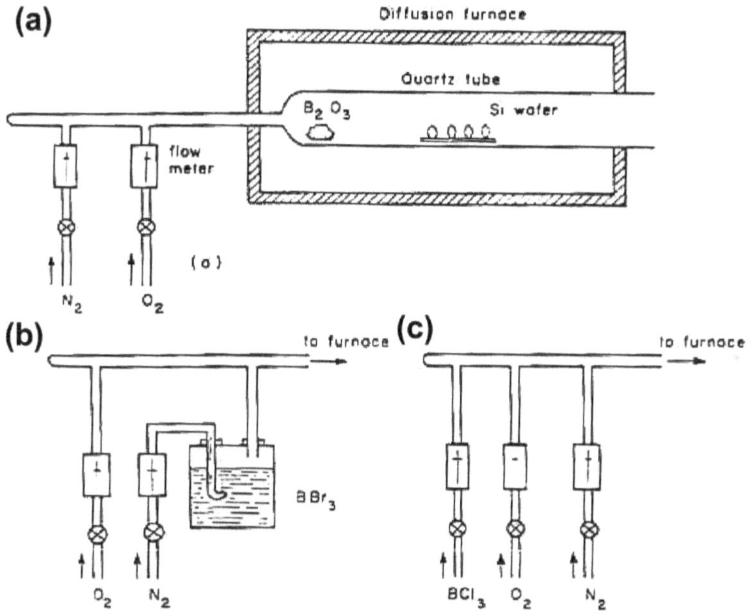

FIGURE 21 Apparatus required for boron diffusion (p-type layer).

A back p+ region may be formed to improve the cell performance by creating a back surface field, which reduces the carrier recombination loss at the back surface.

ELECTRICAL CONTACT FORMATION

Once the photo-generated carriers are separated, they have to be collected to flow in an electrical circuit. Therefore, an electrical contact has to be made. The electrical contact at the front surface (through which light passes) should be made in such a way that it can collect as many carriers as possible and at the same time not shade more than 10% of the cell surface. The back surface contact can simply be metal sheets. However, good ohmic contact has to be created between the metal and semiconductor interface in order to reduce the losses. Aluminium and silver are generally used on the backside. In certain newer cell design (bifacial cells), light enters from both surface of the cell. In that case grid contacts and antireflection coating is required in both the sides of the cell.

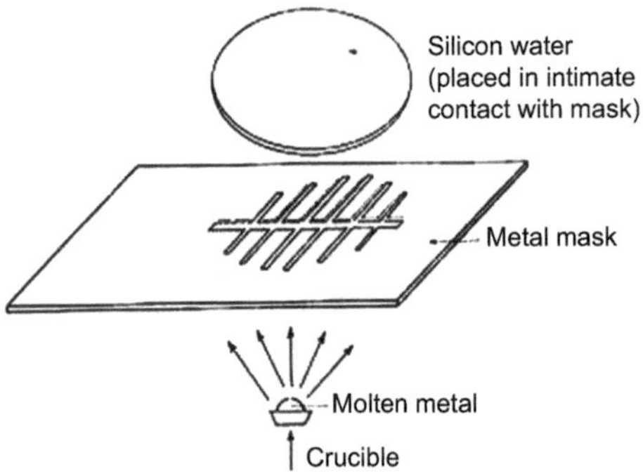

FIGURE 22 Vacuum evaporation.

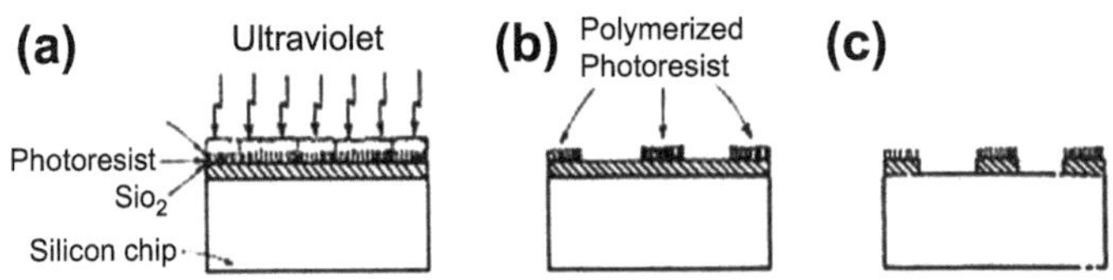

FIGURE 23 Screen printing process.

The following processes are used for electrical contact formation.

- Vacuum evaporation
- RF sputtering
- Electroplating
- Electroless plating
- Screen printing

VACUUM EVAPORATION Metal, which is to be deposited over the cell, is placed in a crucible and heated in a vacuum to a very high temperature to cause melting and evaporation. The metallic vapor is then deposited on a cell through a mask forming a grid pattern as shown in the Figure 22. Titanium, palladium, and silver are used for the front contact with a nickel mask for grid pattern. However, the process is capital intensive.
SCREEN PRINTING Metal is painted over the entire front surface of the cell and subsequently etched away from the unnecessary parts using a photographic technique called photolithography as shown in Figure 23.

There are three separate layers of metal: a thin layer of titanium as a bottom layer for good adherence; a top layer of silver for low resistance and ease of soldering; and a middle palladium layer for restricting the reaction between titanium and silver.

ANTIREFLECTION COATING

Untreated silicon reflects more than 30% of the incident light. However, a thin layer of transparent material can be deposited over the cell before and after the metal contact, which can reduce the reflectivity of the surface and absorb as much incident light as possible. A single layer of antireflection coating will reduce cell surface reflection to 10% where as a double coating will reduce the surface reflection to 3%. Silicon dioxide (SiO_2), titanium dioxide (TiO_2), and tantalum pentoxide (Ta_2O_5) are generally treated as antireflection coating. SiN_x is deposited as an antireflection coating of thickness of about 80 nm. The main advantage of SiN_x is that in addition to acting as an antireflection coating, it also passivates the emitter surface.

Cell Encapsulation and Interconnection (Module Fabrication)

During manufacturing process, there is some variation in the quality of the cells and it is very difficult to maintain the exact quality for all cells. *In general, the current produced by commercial cells suffers a higher degree of dispersion than the voltage.* After measuring the efficiency of each finished cell, they are sorted to minimize mismatch on module interconnection. It is observed that under STC, a typical 4-inch diameter crystalline Si solar cell or a 10 cm*10 cm multi-crystalline cell will provide power of 1–1.5 W at voltage of 0.5–0.6 V. Such less voltage is not sufficient for most of the appliances to work. So, to increase the voltage, the solar cells are connected in series to make one module. Crystalline silicon modules consist of individual cells connected together and encapsulated between a transparent front, usually glass, and a backing material. PV cells are thin and brittle and have to be protected from the weather. For that reason they are encapsulated with a transparent cover on the top. This encapsulation not only provides mechanical strength to the cells but it also protects them from birds, dirt, dust, moisture penetration, etc. It also saves the interconnections and metallic contacts from the weather and wind. The material used for encapsulation should be UV-stabilized, and be able to withstand temperature extremes and thermal shock. The top cover should be transparent to solar radiation, easily cleanable, should not allow the temperature of the cells to go high, etc. The front transparent material is usually made of glass but it is sometimes made of a UV-stabilized plastic sheet. The back of the module is covered with a layer of Tedlar or glass. A frame of aluminium or other composite material provides the necessary mechanical stability to the module (Figure 24).

FIGURE 24 Module structure before lamination as viewed from the rear.

Facts About Silicon Solar Cells

Single Crystalline

- use of single-crystal silicon as early as in 1954,
- well-understood technology,
- semiconductor industry is based on this material,
- made from material that has been purified to one impurity down to one part per billion,
- produced by slicing wafer up to 150 mm diameter and around 350 μm thick,
- highest demonstrated solar to electric conversion efficiency (best single-crystal cells have reached a laboratory efficiency of 25 ± 0.5%),
- single-crystal silicon module yield efficiencies of 16–18% in field,
- cells made of single-crystal silicon are highly stable, and
- major research and development goal in fabricating these cells is to improve the ratio of cost to the total electric energy produced by a cell over its operating life.

Multicrystalline Silicon

- considered as a low cost alternative to single-crystal silicon,
- it is a solid made up of many crystallites or grains simply,
- needs a less pure material,
- automatic mass production techniques in which starting material can be prepared,
- grain boundaries make this material less efficient than single-crystal silicon cells, and
- shares the broad technology base associated with single-crystal silicon

Crystalline silicon represents around 80% of the total PV market today.

Figure 25(a) and (b) show modules made of single crystalline silicon whereas figure (c) shows multicrystalline silicon module.

Module Manufacturing Process

Several steps mentioned below are required to produce a commercial module. The module is also expected to withstand the weather conditions in an outdoor environment. So, a protective covering becomes quite necessary:

LEAD ATTACHMENT AND STRINGING

The leads are attached to each solar cell via a soldering process. The lead attachment is done either manually or with the help of a semiautomatic tabbing machine. Following this, a definite number of cells are joined in a series to form a string of solar cells in a defined pattern. The simple idea is to obtain a sizeable power output. The stringing or cell interconnecting operation is also done by a simple soldering process. Currently, automatic/semiautomatic equipment is also available to carry out these operations. Figure 26 presents a view of single-crystal silicon module available in the marketplace today.

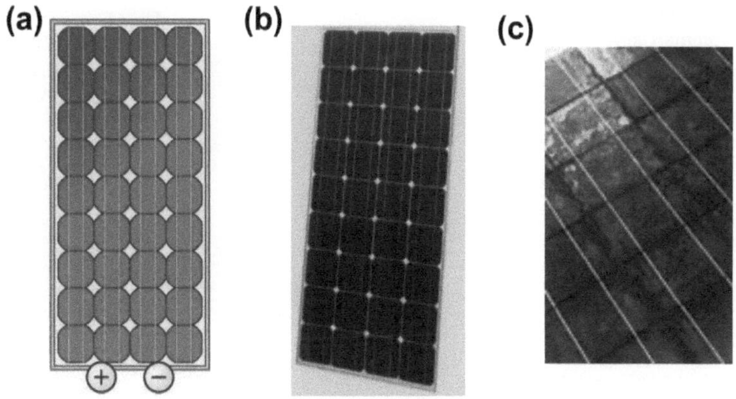

FIGURE 25 Module structure.

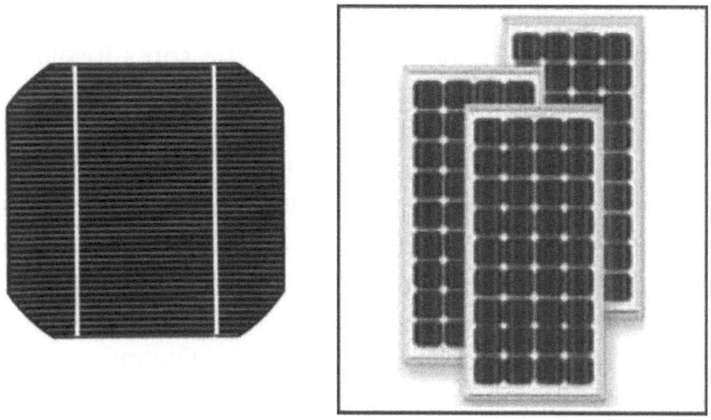

FIGURE 26 View of single-crystal silicon module available in the marketplace today.

LAYUP AND ENCAPSULATION

It is a common feature to treat the string of solar cells to chemical cleaning. Following which, these are placed between a sheet of a soft special polymer known as ethyl vinyl acetate or simply EVA and a sheet of poly vinyl fluoride (PVF). These simply act as encapsulant substrate for the strings. Following which, a sheet of high transmission toughened glass (also known as low iron-content glass) is used as a superstrate. The completed layup is then put in encapsulating equipment known as a laminator. Within which, it is subjected to a specific time–temperature curing cycle to form the laminated module.

FRAMING AND TERMINATION

The encapsulated module is framed using aluminum channels of a special design to protect the module from mechanical damage. It is also for the purpose of mounting the module on a suitable support structure to form a panel. A nylon terminal box is then fixed to provide electrical terminals to extract electrical current from the module. Figure 27 presents a view of multicrystalline silicon module available commercially.

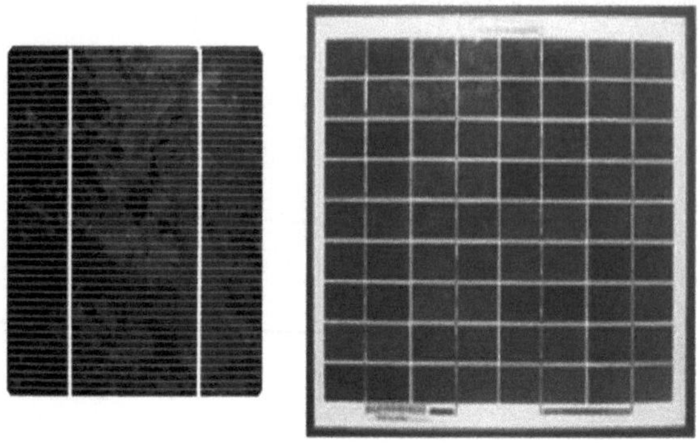

FIGURE 27 A view of multicrystalline silicon module available commercially.

Characterization of Solar Module[3]

Solar modules once laminated are usually tested in a sun simulator. Within it, a module gets exposed to a uniform illumination simulating the sun's insolation on a standard midday (i.e. normally designated as "One Sun" i.e., 100 mW/cm2). Electrodes are then attached to a solar module and the voltage (V) across the module and current (I) from it are measured at different load conditions. These may be open-circuit condition to the short circuit condition. The complete current–voltage curve better known as I–V curve is plotted automatically. Each such I–V curve is stored in the computer memory after suitably tagging it with the module number and other batch data. Figure 28 shows a typical I–V characteristic curve for a solar module.

The power of a solar PV module is measured in Wp. The rating shows the amount of power produced by a solar PV module under STCs. Under STC, a solar cell of 1 Wp will

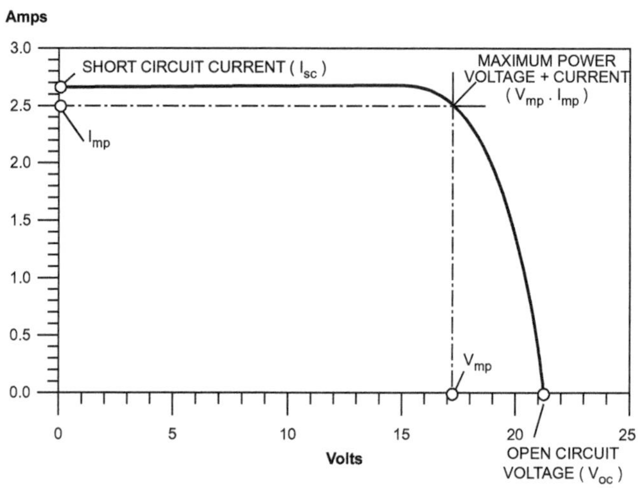

FIGURE 28 I–V Characteristic curve of a typical solar module.

supply a power of 1 W. STCs are defined as when the global irradiation is 1000 W/m^2, AM is 1.5, and the cell operating temperature is 25°C.

This means if you buy a module of rating 50 Wp, then it will produce 50 W power only if the cell operating temperature is 25 °C, global irradiance is 1000 W/m^2. A typical 4-inch diameter crystalline Si solar cell or a 10 × 10 cm multicrystalline cell should provide a power of 1–1.5 W under STCs. But these conditions are rarely found in real-world operating conditions. In the real field condition, in a bright sunny day, the global irradiance on the ground is usually 700–800 W/m^2. Again, when the ambient temperature is 45 °C, cell operating temperatures go as high as 60–65 °C. So in reality, a 1 Wp cell never produces 1 W. Instead on a good sunny day, it produces 60–70% of that i.e., 0.6–0.7 W.

Key Factors Affecting Solar Cell Performance

Sunlight is the basic fuel for a solar module. Any loss of sunlight means a sure loss of power too. This also means that our means of collecting the sunlight should be perfect. There is one more side of the story. A PV module can suffer loss of power due to some other reasons too. It may keep working for around 20–30 years. Ideally, nothing goes wrong with it. Perhaps, the best thing about a module is that it has no moving parts. So, nothing will wear out with time.

The following factors can have some effect on the workings of a solar system.

TILT ANGLE

A solar module should be put up in the proper path of sun. Thus, module is tilted at an angle. It is roughly equal to the latitude of the place where a module is put up. Any error in the tilt angle will lead to loss of some amount of power.

DUST

The modules need to be kept as clean as possible. The dust settles on the module surface mostly in the dry season. So, proper care should be taken to clean the glass surface of a module regularly. Remember, dust may cause energy losses as high as 5–10%.

SHADING

Solar module faces the sun all day long. There should be no shade on it. So, keep on looking for any extended tree branches, plants, TV antennas etc. So, place the panels where they will not be in the shade. Figure 29 shows the net effect of shading on the power output expected from a solar module.

A solar module is made of a string of individual solar cells. These are connected in series with one another. The current output from the whole module is limited to that passing through the weakest link cell. Take, for example, 1 cell from 36 cells in a module. In case, it is fully shaded, the power output from the module will come down to zero. However, if, one cell is 50% shaded, then the power output from the module will reduce by 50% only. The simple solution is to make use of a bypass diode. It can be connected between modules in a system or between groups of cells in a module. The idea is to ensure that power loss takes place from just the shaded unit (cell). Now think of a

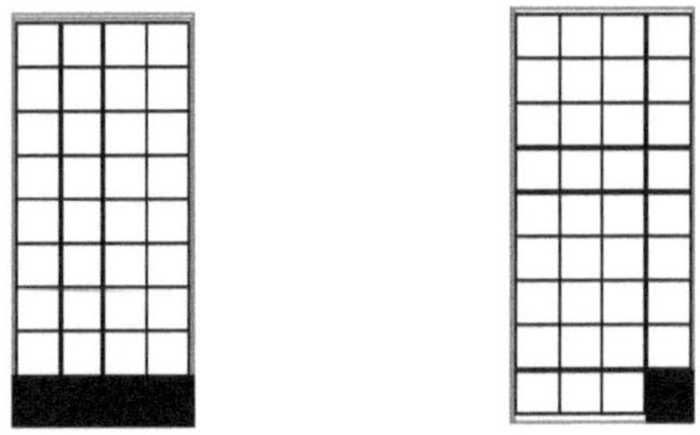

FIGURE 29 Net effect of shading on the power output expected from a solar module.

partially shaded module. The current from the unshaded part of module passes through a diode. It thus bypasses the shaded group of cells.

LIGHT INTENSITY

The brighter the sunlight, the more power the panels will produce. So, if, there is 1000 W/m^2 of sunlight, you will see almost the full rated output of the panel. But, if there is 500 W/m^2 only, you will see half the rated power of the panel. Figure 30 shows a clear effect on the module current as a result of changing values of solar insolation. It goes up in direct proportion with the increasing solar insolation. The same is not the case with the module voltage.

TEMPERATURE

Modules are tested at a standard temperature of 25 °C. The higher the temperature, the lower is the power output of a module. Simply put, a module loses power at higher temperatures.

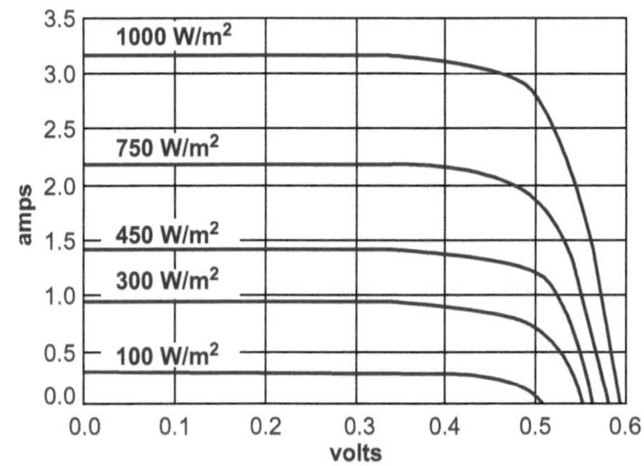

FIGURE 30 I–V curve of a typical solar module under different solar insolation??

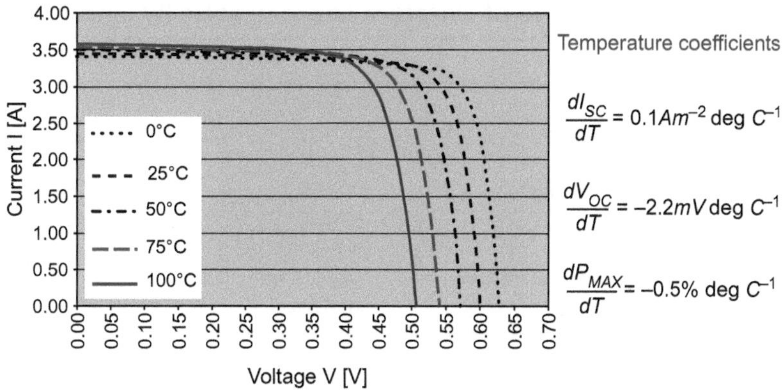

FIGURE 31 I–V Characteristic curve at different temperatures.

The open-circuit voltage V_{oc} of an individual silicon solar cell reduces by 2.3 mV for every one degree rise in temperature T of the solar cell. Therefore, the voltage coefficient is negative and is expressed as,

$$dV_{oc}/dT = -2.3\ mV/°C$$

As a module consists of large number of cells in series, the voltage coefficient can be expressed as,

$$dV_{oc}/dT = -2.3 \times nc\ mV/°C$$

where nc is the number of cells connected in series.

The above equation can be integrated to obtain,

$$V_{oc}(Tc) = V_{oc}(at\ 1kW/m2) - 0.0023.nc(Tc - TSTC)$$

As mentioned, it is important to note that the voltage is determined by the operating temperature of the solar cells, which is not the same as the ambient temperature (T_a).

The cell temperature can reach around 70 °C under the bright sunlight. The power in case of crystalline silicon cells decreases by about 0.4–0.5% per degree C of temperature increase above 25 °C. Amorphous silicon modules have a lower temperature coefficient of about 0.2–0.25% per degree C of temperature increase above 25°C. Figure 31 presents the effect of increasing temperature(s) on the module voltage.

The voltage reduces as the temperature goes up. The same is not the case with module current. Thus, it can be said that power output of a module comes down with reducing values of solar insolation and increasing temperatures.

OTHER LOSSES

Energy consumption in the charge regulator

The charge controller continuously draws a small current of about 5–25 mA. It may amount to a system power loss of around 1%.

Table 4 Typical Range of Different Losses in Solar PV System

Loss Factor	Normal Range	Typical Value
Shading of the module	0–20%	0%
Orientation not proper	5–10%	5%
Dust on the module	5–10%	5%
Temperature effect on the module	0–20%	16% (at 60 °C)
Cable losses	5–10%	5%
Semiconductor energy loss	5–8%	7%
Charge regulator energy consumption	1–5%	2%
Losses inside the battery	10–20%	15%

Source: multiple.

Semiconductor energy loss

The charge controller contains components like the metal oxide semiconductor field-effect transistors and blocking diodes etc. Both these convert a certain amount of energy into heat.

Cabling losses

Some loss of power can take place through cables. The solution lies in choosing a large diameter wire size to bring down the loss.

Improper connections

Poorly made electrical connections produce resistance. It can thus result in less power going to the batteries. Make them tight and keep them clean too (Table 4).

Thin Film Solar Cell Technology

In thin film solar cell, a very thin film of material is sufficient to absorb light and give the desired result. Here, less material is used with relatively easy manufacturing process. So, this technology has the potential of reducing the module cost through lower material and energy consumption. In case of thin film solar cells integrally connected modules are produced directly without the interconnection of cells. Though the module efficiency of thin film solar cell is quite less (around 8–10%) as compared to that of crystalline silicon solar cells, but potentially it is cheaper to manufacture. Figure 32 shows the maximum efficiencies that could be obtained with different materials.

Thus $CuInSe_2$, a-Si:H are potential candidates for the thin film solar cell application. Though GaAs is the material of choice but it is very expensive and used only for space application.

Processing of Thin Film Devices

Following are some of the emerging thin film solar cells.

1. Amorphous silicon (a-Si)

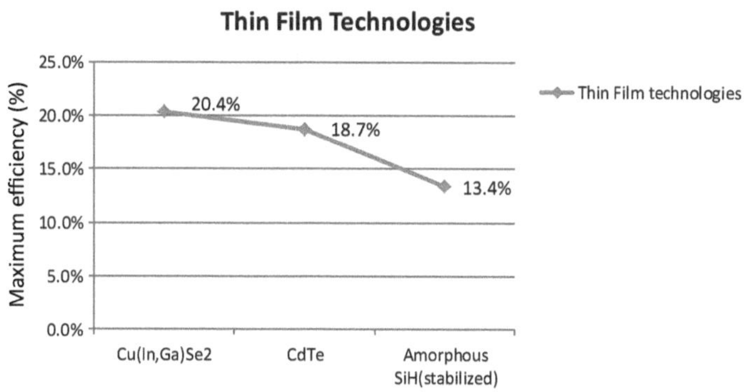

FIGURE 32 Maximum efficiency of different thin film technologies.

2. Copper indium diselenide (CIS), Copper indium gallium diselenide (CIGS)
3. Cadmium telluride (CdTe)

AMORPHOUS SILICON (A-SI) SOLAR CELL

The crystallographic structure of amorphous silicon differs from that of the crystalline silicon. The randomness in the atomic arrangement in amorphous silicon material makes it a direct-gap material with an optical energy gap of about 1.75 eV. However, there exist large number of incomplete bonds (called dangling bond) in a-Si, which could be passivated with the incorporation of hydrogen i.e., a-Si:H solar cells. The advantage of a-Si is that fabrication of a-Si can be fully automated and thin film of a-Si can be deposited over inexpensive substrate such as glass or stainless steel. However, shorter carrier lifetime and so lower efficiency and degradation of cell efficiency under exposure to light are the major drawback of the a-Si technology.

The optimum structure of the a-Si is the p+ -i-n with a transparent conducting contact to the p + layer and an ohmic contact to the n-layer. As both n and p materials have poor transport properties, the incorporation of i layer (hydrogenated amorphous Si) will improve the transport properties.

PRODUCTION OF AMORPHOUS SILICON PV MODULES Figure 33 shows the process followed for the production of amorphous silicon solar cells/module.

Generally, silane (SiH_4) is decomposed in an RF glow discharge reactor near a heated substrate. The exposure time, discharge power, and flow rate all control the film thickness. Diborane (B_2H_6) is added to silane to produce p-type and phosphine (PH_3) is added to silane to make n-type silicon.

A–SI FABRICATION PROCESS Figure 34 shows different fabrication process for a-Si module.

GLOW DISCHARGE METHOD Multichamber deposition system by RF glow discharge as shown in Figure 35 is the widespread approach to the fabrication of high quality a-Si solar cells. It consists of four reactors in a large chamber.

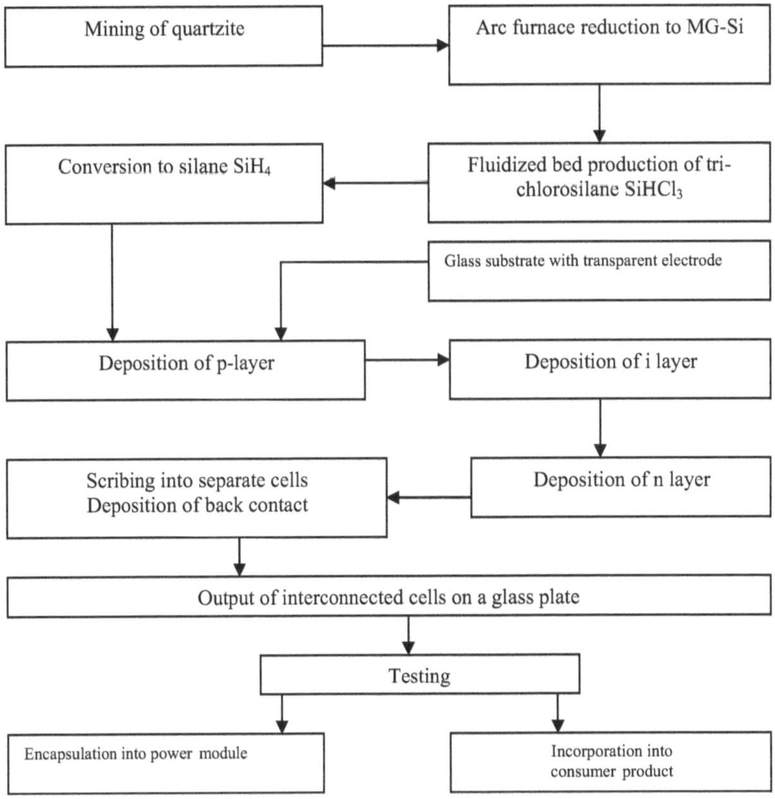

FIGURE 33 Flow diagram for the production of amorphous silicon solar cells/module.

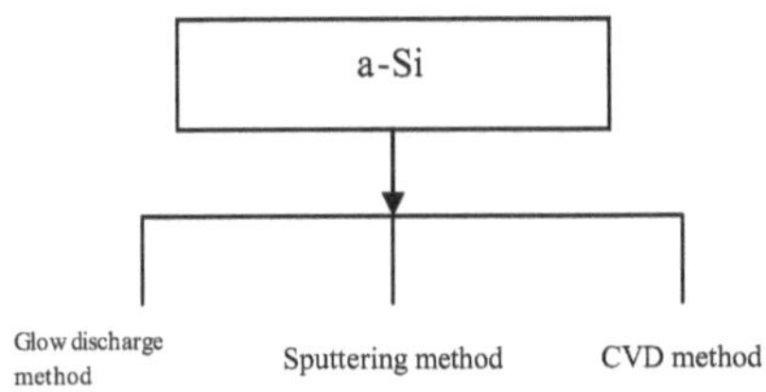

FIGURE 34 Different fabrication process for a-Si module.

The first reaction chamber is for hydrogen plasma etching and other three chambers are for deposition of p + -i-n + layers. Substrate temperature during deposition is 200–250° C.

The main disadvantage of a-Si solar cell is the degradation in performance of silicon solar cell when exposed to light. It is because the light absorption by i layer creates additional defects thus increasing the density of trapping and scattering states. This effect is called *Staebler—Wronski effect*. The main decrease in efficiency occurs in first 100 h of exposure, after which cell parameters remain fairly constant.

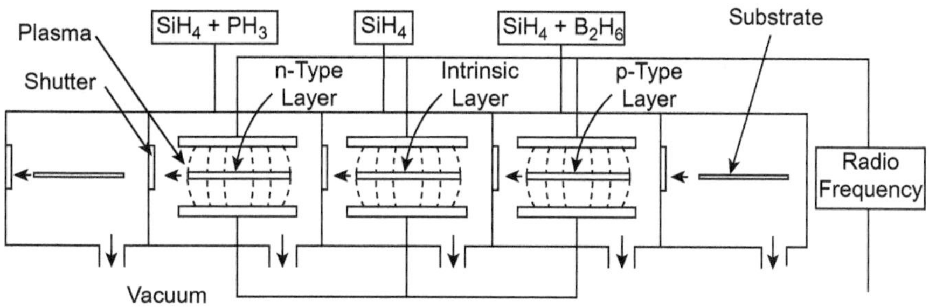

FIGURE 35 Multichamber deposition system by RF glow discharge.

MULTI JUNCTION SOLAR CELL (TANDEM SOLAR CELLS) As Staebler—Wronski effect depends primarily on the thickness of i layer, it can be reduced by creating multijunction structure (tandem structure) where the absorption of light can be split into two or three separate layers. Such cells are more stable with higher efficiency than single junction solar cell. The schematic diagram of the multijunction structure is given in Figure 36.

In a tandem solar cell, higher efficiency can be obtained if a bottom layer of narrower band gap is used.

COPPER INDIUM DISELENIDE CELLS (CIS CELLS)

Copper indium diselenide (CIS) is a direct-gap material and can be either p-type or n-type.The electronic properties of CIS depend on the copper/indium ratio. Though a p–n homojunction of CIS is possible, it is unstable and inefficient. However, it

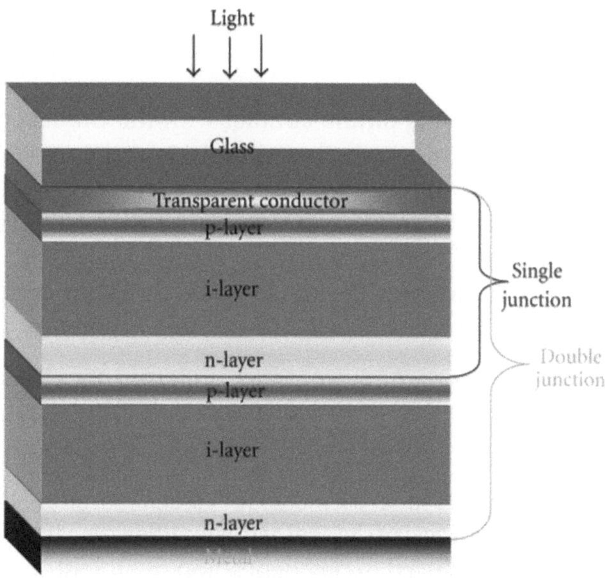

FIGURE 36 shows the configuration of the a-Si:H solar cell.

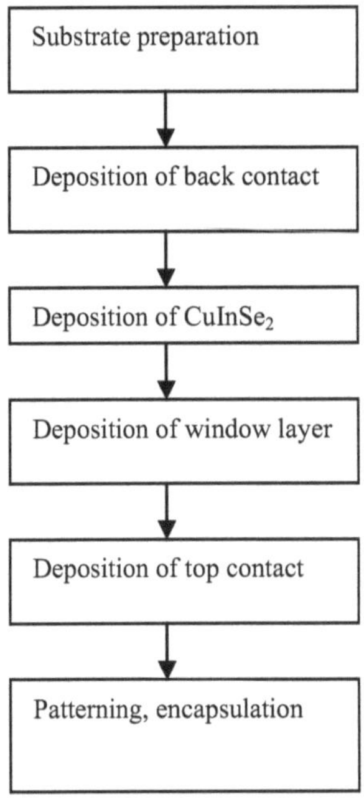

FIGURE 37 A flow chart for production of CIS module.

is observed that heterojunction of CIS with CdS would be the best possible option. Since CdS is only grown as n-type, CIS has to be of p-type in order to create the p–n junction. Figure 5.61 shows the flow diagram for production of CIS module (Figure 37).

- Ohmic contact to CdS can be easily made but contact to p-type CIS is more difficult.
- Molybdenum is used for making ohmic contact with the CIS layer.
- For efficient use CIS has to be heavily doped at the metal contact side where as it has to be lightly doped near the junction.
- Optimum device used a very thin layer of CdS with a wide gap window layer of highly conductive material.
- The incorporation of gallium into CIS improves the fill factor and requires fewer cells per module. Processing is more reliable and less expensive.

Vacuum evaporation is the common approach followed for deposition of different layers in CIS/CIGS solar cell.

CADMIUM TELLURIDE CELLS

Though cadmium telluride has the ideal band gap for solar absorption, but its performance is limited by the presence of high level of defects at the middle of the band gap. These defects (recombination states) can be reduced in postdeposition treatment.

TYPICAL STRUCTURE OF CdTe Solar Cell

Metal back contact
Cadmium telluride
Cadmium sulphide
Transparent front
Glass substrate

Facts About Thin Films

Amorphous Silicon

- An a-Si cell was produced in 1974 in Bell Laboratories and had an efficiency of 0.2%.
- Alloying amorphous silicon with hydrogen increases the efficiency by lowering the band gap.
- Amorphous silicon alloys absorb solar radiation 40 times more efficiently than does c-Si.
- A thickness of only about 1 μm is needed to absorb 90% of the available solar energy.
- The small amount of material needed; saves significantly on cost.
- The material can be produced at low temperatures and deposited on low-cost substrates.
- The materials are amenable to large-scale automated production.

Cadmium Telluride

- has near optimal band gap of 1.47 eV;
- has high absorption coefficient, which enables it to absorb almost all of the incident light greater than the band gap in less than one micron of the absorber layer;
- use of inexpensive manufacturing methods like electro deposition, CVD etc., possible.

Copper Indium Diselenide

- its absorption coefficient is the highest among for semiconductor materials;
- 90% of the sunlight incident on this material is absorbed in the first micrometer of the device;
- has one of the highest current densities of any semiconductor materials;
- amenable to large area automated production.

Other Emerging PV Technology

Emerging solar PV technologies, such as organic PV cells and dye-sensitized solar cells (DSSC) are still under demonstration and have not yet been commercially deployed on a large scale. They are also called third-generation solar PV technology and have been described below:

1. DSSC use photo-electrochemical solar cells, which are based on semiconductor structures formed between a photo-sensitized anode and an electrolyte. In a typical DSSC, the semiconductor nanocrystals serve as antennae that harvest the sunlight (photons) and the dye molecule is responsible for the charge separation (photocurrent) 10. It is unique in that it mimics natural photosynthesis. These cells are attractive because they use low-cost materials and are simple to manufacture. They release electrons from, for example, titanium dioxide covered by a light absorbing pigment. However, their performance can degrade over time with exposure to UV light and the use of a liquid electrolyte can be problematic when there is a risk of freezing. Laboratory efficiencies of around 15% have been achieved due to the development of new broadband dyes and electrolytes, however, commercial efficiencies are low—typically under 4–5%. The main reason why efficiencies of DSSC are low is, because there are very few dyes that can absorb a broad spectral range. An interesting area of research is the use of nanocrystalline semiconductors that can allow DSSC to have a broad spectral coverage. Thousands of organic dyes have been studied and tested in order to design, synthesize, and assemble nanostructured materials that will allow higher power conversion efficiencies for DSSC. DSSC are currently the most efficient third generation solar technology available.[1,2]
2. Organic solar cells are composed of organic or polymer materials (such as organic polymers or small organic molecules). They are inexpensive, but not very efficient. They are emerging as a niche technology, but their future development is not clear. Their success in recent years has been due to many significant improvements that have led to higher efficiencies. Organic PV module efficiencies are now in the range of 4–5% for commercial systems and 6–8% in the laboratory. In addition to the low efficiency, a major challenge for organic solar cells is their instability over time. Organic cell production uses high-speed and low temperature roll-to-roll manufacturing processes and standard printing technologies.

As a result, organic solar cells may be able to compete with other PV technologies in some applications, because manufacturing costs are continuing to decline and are expected to reach USD 0.50/W by 2020. Organic cells can be applied to plastic sheets in a manner similar to the printing and coating industries, meaning that organic solar cells are lightweight and flexible, making them ideal for mobile phones, laptops, radios, flashlights, toys, and almost any hand-held device that uses a battery. The modules can be fixed almost anywhere to anything, or they can be incorporated into the housing of a device. They can also be rolled up or folded for storage when not in use. These properties will make organic PV modules attractive for building-integrated applications as it will expand the range of shapes and forms where PV systems can be applied. Another advantage is that the technology uses abundant nontoxic materials and is based on a very scalable process.

Various Losses in Solar Cells

The power of a solar cell and therefore its efficiency is limited by the following factors.

Fundamental losses such as:

1. Recombination
2. Reflection losses
3. Unabsorbed radiation and excessively strong radiation
4. Shadowing effect
5. Current losses due to collection inefficiency
6. Losses due to series and shunt resistance

Losses Due to Recombination

Recombination (process by which electron combine with hole and the pair annihilated) occurs mostly at semiconductor surface or at the lattice defect or at the impurities site. It also occurs at the ohmic metal contacts to the semiconductor.

Methods of reducing the losses: Surface passivation can reduce largely the recombination losses. For example, in a passivated emitter solar cell, a layer of passivating oxide and in GaAs a top layer of GaAlAs reduces the recombination losses.

Reflection Losses

Some part of the solar radiation is reflected back from the top surface and contributes in overall losses.

Methods of reducing the losses: By texturing, or by applying antireflection coating, reflection losses can be reduced from around 30–4%.

Unabsorbed Radiation and Excessively Strong Radiation

A considerable amount of solar radiation is not absorbed by the semiconductor because of the inability of the semiconductor to absorb the lower band gap light. Again when radiation with higher energy than the band gap of the solar cell is absorbed, in addition to electron–hole pairs, it also generates heat.These two factors contribute maximum to the losses from a solar cell.

Methods of reducing the losses: Tandem structure can reduce the losses. As tandem structure consists of stack of several cells of different band gap, photons of different energy will be absorbed by different layers. The top cell must be made of larger band gap so that light of shorter wavelength can be absorbed first. The transmitted light is then converted by the bottom cell.

Current Losses Due to Collection Inefficiency

All the generated carriers are not able to reach the junction. Some of them recombine before reaching the junction and so do not contribute to current. The ratio of charge carriers that are able to reach the p–n layer to the total number of charge carriers is known as *collection efficiency*. In crystalline silicon solar cell, carrier transport is good because of good transport property of the material. However, in amorphous and polycrystalline thin films, electric fields are required to pull the carriers. Therefore, the collection efficiency can be influenced during the manufacturing process itself.

Comparative Analysis of Various PV Technologies

Table 5 compares the PV technologies and PV devices. The comparative includes the efficiencies of the various PV modules (at an AM index of 1.5), the area required to produce 1 kW of electricity, the stage of commercialization, the advantages, and the drawbacks of the respective technologies.

Solar PV Systems for Water Pumping[8,9]

Solar pumps are used principally for three applications:

- Village water supply
- Livestock watering
- Irrigation

A solar irrigation system needs to take account of the fact that demand for irrigation water will vary throughout the year. Peak demand during the irrigation seasons is often more than twice the average demand. This means that solar pumps for irrigation are underutilized for most of the year. Attention should be paid to the system of water distribution and application to the crops.

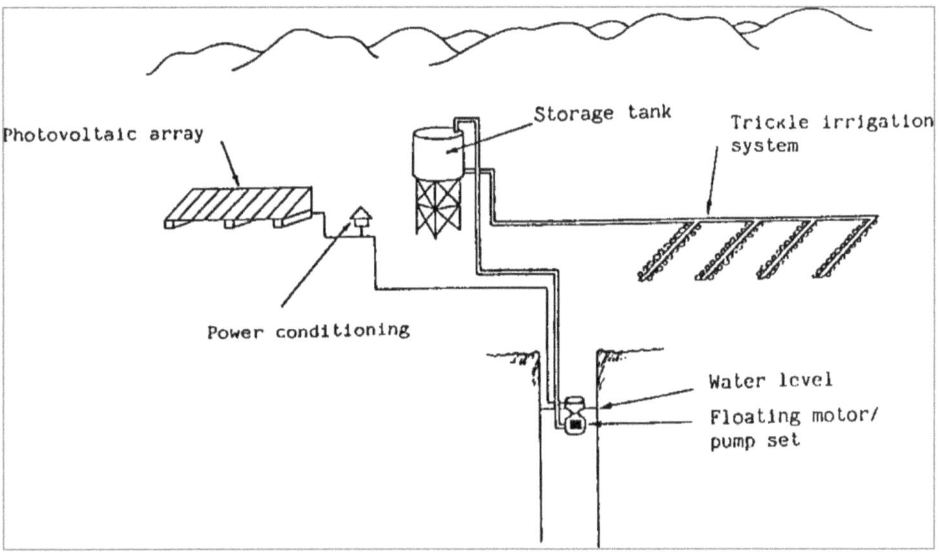

Source: Technical brief of practical action on solar (photovoltaic) water pumping.

Table 5 Summary of Different Solar PV Technology

Technology	Photovoltaic Device	PV Module Efficiency at Air Mass 1.5	Area Required/kW (m^2)	State of Commercialization	Advantages	Drawbacks
Wafer-based silicon	Monocrystalline	15–19%	7	Mature with large-scale production	Suitable for applications in low to warm temperature areas and where there are space constraints.	Operates at decreased efficiencies in higher temperature. Less shadow tolerance.
	Polycrystalline	13–15%	8	Mature with large-scale production	Lower manufacturing cost as compared to monocrystalline. Higher shadow tolerance than monocrystalline.	Less space efficient than monocrystalline. Less durable than monocrystalline. Less temperature tolerance than monocrystalline.
Thin film cells	Amorphous silicon	5–8%	15	Early deployment phase with medium-scale production	High shadow tolerance. Low manufacturing cost. Serial connection of cells can be done during manufacture. a-Si cells dominate the small output market (clocks, calculators etc)	Strong Performance degradation under light exposure. Low-space efficiency.
	Cadmium telluride	8–11%	11	Early deployment phase with small-scale production	Lowest specific cost (cost/watt). Low degradation rate (~0.9%/yr). Serial connection is easy and in an integral manner over a large substrate.	Cd and Te are toxic materials, although the compound is stable and harmless. Low availability of Te.
	CIGS	7–11%	10	Early deployment phase with medium-scale production	Flexible production technology. High scope of improvement in future.	High manufacturing cost compared to the efficiency obtained. Low availability of Indium.

Continued

Table 5 Summary of Different Solar PV Technology—cont'd

Technology	Photovoltaic Device	PV Module Efficiency at Air Mass 1.5	Area Required/ kW (m^2)	State of Commercialization	Advantages	Drawbacks
New emerging technologies	CPV	25–30%	-	Just commercialized, small-scale production	Highest efficiency achieved in PV technology	High manufacturing cost. Requires two axis tracking, which further adds to the cost.
	Organic PV	1%	-	Research and Development stage	Infinite variability of starting materials and low-cost manufacturing processes Energy payback time is expected to be short	Long term stability and protection against environmental influences are significant challenge
	Dye-sensitized solar cell	1–5%	-	Research and Development stage	Less stringent purity requirement of starting material and a simple processing without clean room step is possible as surface/bulk recombination cannot happen.	Low photocurrent efficiency with respect to incident light. Decreased stability with time and over different temperature ranges.

Source: Technical brief of practical action on solar (photovoltaic) water pumping.
A solar PV water pumping system consists of,

1. PV array
2. Motor pump set (surface or submersible): It can be a D.C. motor pump set (with brushes or brush less D.C.) or A.C. induction motor pump set with a suitable inverter
3. Electronics with MPPT controls/protections.
4. Interconnect cables and "On–Off" switch.

Configuration of Various Solar Water Pumping Systems[10,11]

Systems are broadly classified into five categories.

Submerged Multistage Centrifugal Motor Pump Set

This type is probably the most common type of solar pump used for providing drinking and other water supply at a village level. The most commonly employed system consists of an AC pump and inverter with a PV array of less than 1500 Wp. The advantages of this configuration are that it is easily installed, with lay-flat flexible pipe work and the motor pump set is submerged away from potential damage. Either AC or DC motors can be integrated into the pump set. If a brush DC motor is used then the equipment will need to be pulled up from the well (approximately every 2 years) to replace brushes.

Submerged Pump with Surface Mounted Motor

This configuration gives easy access to the motor for brush changing and other maintenance purposes. However, the relatively lower efficiency from power losses in the shaft bearings and the high cost of installation has allowed users to go for other choices/configurations. In general, this configuration is largely being replaced by the submersible motor and pump set.

Reciprocating, Positive Displacement Pump

The reciprocating positive displacement pump is quite popular for high head, low flow-rate applications. The output is proportional to the speed of the pump.

Floating Motor Pump Sets

The floating motor pump set is ideal for irrigation pumping for canals and open wells.

The pump set is easily portable and generally does not go dry. Most of these types use a single-stage submersed centrifugal pump. The most common type utilizes a brushless (electronically commutated) DC motor.

Surface Suction Pump Sets

This type of pump set is not very popular and not recommended, if it is not consistently attended by an operator. Although the use of primary chambers and nonreturn valves can prevent loss of prime, in practice self-start and priming problems are widely experienced.

Conclusions

The PV technology development has come a long way. Further investment in research and development, process innovation, and supportive policies will soon produce solar technology that is competitive with global fossil fuel market prices.

References

1. Lasnier F, Ang TG. *Photovoltaic engineering handbook.* Bristol: IOP Publishing; 1990.
2. Renewable energy engineering and technology–a knowledge compendium"-TERI Publication
3. Davidson J. *The new solar electric home: the photovoltaic how-to handbook.* Ann Arbor (MI): aatec Publications; 1990.
4. Green MA. *Solar cells: operating principles, technology and systems applications.* University of New South Wales; 1986.
5. Komp RJ. *Practical photovoltaics: electricity from solar cells.* Ann Arbor (MI): aatec Publications; 1984.
6. Maycock P, Stirewalt E. *A guide to the photovoltaic revolution.* Emmaus (PA): Rodale Press; 1985.
7. Sandia National Labs. *Stand-alone photovoltaic systems: a handbook of recommended design practices.* Albuquerque (NM): PV Design Assistance Center; November 1991.
8. Strong S. *The solar electric house: a design manual for home-scale photovoltaic power systems.* Emmaus (PV): Solar Design Associates, Rodale Press; 1987.
9. Strong S. *Photovoltaics in the built environment.* Solar Design Associates; 1997.
10. World Bank. *Rural energy and development: improving energy supplies for two billion people.* Washington, DC. 1996.
11. Zweibel K. *Harnessing solar power, the photovoltaics Challenge.* New York, NY: Plenum Press; 1990.

13

Contaminated Irrigation Water and the Associated Public Health Risks

Tsanangurayi Tongesayi[1], Sunungurai Tongesayi[2]

[1]DEPARTMENT OF CHEMISTRY, MEDICAL TECHNOLOGY AND PHYSICS, MONMOUTH UNIVERSITY, WEST LONG BRANCH, NJ, USA; [2]WALDEN UNIVERSITY PUBLIC HEALTH, MINNEAPOLIS, MN, USA

CHAPTER OUTLINE

Food, Energy, and Water. http://dx.doi.org/10.1016/B978-0-12-800211-7.00013-2

Introduction

The issue of food safety is critically important and requires unfettered yet balanced and unbiased evidence-based discourse to avoid unnecessary consternation. It is very easy to overestimate the risk from food contamination as it is equally easy to do the exact opposite. To get the full scope of the impact of contaminated irrigation water on food safety, the contaminants need to be identified and then quantified, as appropriate. Also, the sources of the contaminants must be identified and characterized without ambiguity. If unknown, the health effects of the pollutants have to be evaluated using standardized protocols and then fully characterized with an acceptable degree of certainty. Because there is no risk without exposure, the exposure routes of the identified contaminants will need to be identified, and the risks from the contaminants correctly evaluated and characterized.

Of course, there is no doubt that environmental pollutants, both chemical and biological (pathogens) are getting into agricultural food at levels that may be detrimental to human health through agricultural practices and the general pollution of the environment.[1] What needs to be done is risk analysis, a protocol that involves both science and policy.[2] Contaminated food may well be fast becoming one of the major routes of exposure of environmental pollutants to humans but without fully unveiling and effectively communicating the kind and extent of the risk, we may never know, yet consequences will be real and unrelenting.

The twenty-first century has seen tremendous advances in technology and overall improved quality of life. People are living longer, and more are being born.[3–5] The increase in population is bringing with it boosts in urbanization, industrialization, and agriculture, activities that are the major anthropogenic polluters of the environment in general and water in particular. Ironically, all these activities need unpolluted water to thrive. On the technology front, a new buzz has gripped the world and is revolutionizing applications in all aspects of human life, based on nanotechnology. Unfortunately, ingenious as it may be, the small can get into any space, wanted or unwanted, and may result in unintended consequences. There is evidence of engineered nanomaterials (ENMs) and micromaterials (EMMs) getting into environmental compartments, via both liquid and solid waste, where they can inadvertently come into contact with humans.[6–8] Water contamination by ENMs and EMMs is of particular concern because not only will drinking water be affected, food can also be contaminated through irrigation. In fact, there is evidence of the uptake of ENMs by crops from contaminated soil and water.[9] As a consequence, the technological advances of the twenty-first century, which are meant to improve the quality of human life, may achieve the exact opposite if they are not balanced with pollution prevention and control.

This chapter seeks to provide a balanced assessment of the presence, nature, and the extent of the human health risk that is associated with the use of contaminated water in agriculture. The authors hope to achieve this goal by critically evaluating the current and future global freshwater situation and identifying the major water contaminants and their sources, health effects, and exposure routes to humans, using credible scientific literature. An attempt will be made to identify and discuss the reasons behind the widespread use of contaminated water in agriculture despite the availability of extensive scientific literature that is against this practice.

The Current and Future Water Situation—It Is All in the Water

Imagine waking up one day and the earth is dry. The consequences will be horrendous but, nevertheless, simple to comprehend; it simply means you will no longer exist. The earth may, however, never run dry, but the water may certainly become economically unavailable if the underlying causes of current and projected freshwater stressors are not addressed. All forms of life revolve around water, and without it, life can hardly exist. Water, which by chemical definition is a molecule containing one oxygen atom and two hydrogen atoms, can be classified as fresh or saline depending on the amount of dissolved salts, which are $<1\%$ and 1–3.5%, respectively.[10]

Even though all water can support various forms of life, only freshwater, which can be on surface or underground, is available for use by humans for domestic (drinking and hygiene), agricultural, and industrial purposes. Freshwater contaminants, if they exist at levels that render the water unsuitable for use in any of the three broad categories, have to be removed before it is used. According to the Food and Agricultural Organization (FAO) of the United Nations, the world water reserve is about 1400 million km^3, of which only about 0.003% (about 45,000 km^3) is freshwater.[11] However, only a small fraction of the freshwater (about 9000–14,000 km^3) is economically available for use in domestic, agricultural, and industrial applications.[11] The distribution of water within the three sectors is not uniform, with almost 70%, on average, (or 95% in some developing countries) of all water withdrawals going to agriculture. At continental basis, the current proportions of freshwater withdrawals that go to agriculture are 86.7, 81.4, 68, 38.7, and 32.4% for Asia, Africa, Latin America, North America, and Europe, respectively.[11] In North America and Europe, the highest shares of their total freshwater withdrawals go to industry and are 40.8% and 52.4%, respectively.[11] The differences between developing and developed countries in terms of the amounts of water going to agriculture relative to other sectors could be due to massive industrialization and more efficient water technologies compared to the former where economies are mainly agro-based and water technologies are not as efficient.

The high demand for water in agriculture is hardly surprising given that about 1.5 tons of water (1500 L) are required to grow a kilogram of cereal compared to drinking water that is about 2–4 L per person per day. In fact, it is estimated that a person's daily food requirements need about 3000 L of water to produce.[11] Such demands for water in agriculture, however, put tremendous pressure on water, particularly in poor nations where there are crippling perennial droughts, as well as poor or inadequate water infrastructure. The high demands amid severe shortages have resulted in some nations resorting to unorthodox practices, chief among them being the use of raw wastewater to irrigate crops in agriculture.

There is, however, no doubt that water is critical to food security. This is particularly the case in developing countries where people depend on agriculture for food and income. The water, however, has to meet certain quality standards to make sure human health is preserved. Unfortunately, the lack of capital, water infrastructure, and droughts are stifling agriculture, particularly in the majority of developing countries, to the extent that famine and malnourishment have become the norm, albeit by default. According to FAO, "The right to food is not negotiable. Feeding our growing population and reducing hunger will only be possible if agricultural yields can be increased significantly - and sustainably."[11] The key terms in the quote are "increased significantly" and "sustainably." Can agricultural yields be increased significantly, and sustainably, without sacrificing other needs? The question is quite fair given the current trends in water consumption, scarcity, and contamination. Obviously, water contamination has to be kept in check and water technology has to be significantly improved if the FAO goal has to be realized without causing water shortages in other sectors.

There is also no doubt that the quality of drinking water and water for hygiene and other domestic purposes is critical to public health. It is a known fact that contaminated drinking water is one of the major routes of exposure of toxic chemicals and pathogens to humans, and, because of its mobility and its central role in life, water can easily spread disease-causing agents at unprecedented rates. In fact, it has been reported that the majority of diseases, particularly in developing nations, are the result of lack of clean water for drinking, washing, and getting rid of waste.[12] According to Brian Richter, at least a billion people have no access to clean drinking water, and at least half of hospital beds in the world are occupied by people with preventable waterborne diseases.[12]

Also, without water of the right quality and quantity, most, if not all, industries will grind to a halt. As a consequence, a balance has to be struck between these competing water needs, a feat that many nations are failing miserably to achieve, thanks to the growing national populations amid growing poverty, climate-induced scarcity of water, and the anthropogenic contamination of freshwater, among other factors. Climate change is being projected to become one of the major causes of future water shortages on Earth. According to the International Panel on Climate Change, climate change is going to cause less rain in many regions while increasing evaporation rates on almost the entire planet, thus causing droughts in almost all parts of the world.[12]

Going by current estimates, it seems the water situation is already critical as 80% of the current 7 billion plus world population is reportedly living under a significant threat to water security.[13] Developed countries are said to be mitigating the threat by making extensive investments in water technology and infrastructure, with poor nations remaining vulnerable because of the lack of resources.[13] With the rate at which the human population is growing and the rate of societal advancement in terms of the quality of living, particularly in developed countries, it is inevitable that industrialization, urbanization, and agriculture will also increase. The result is, obviously, more pressure on freshwater reserves. The increased food demand that comes with an increase in human population will inevitably further increase the demand for water in the agricultural sector, which is already very high and skewed in its favor relative to other sectors. As a matter of fact, the Organization for Economic Co-operation and Development, an international organization comprising 34 nations mainly from North America, Europe, and Asia, is projecting a 55% increase in the global water demand by 2050 compared to the demand levels in 2000, which will subject at least 40% of the world population to severe water stress.[14]

Despite shortages of water in certain regions of the world, food production has been on the increase even though shortages in other parts of the world existed and still exist. Going by FAO estimates, food production increased by more than 100% in the last 30 years and is projected to increase by about 60% by 2050, with 70% of gains in cereal production coming from irrigated land, compared to 2006.[11] The overall irrigated food production is projected to increase by 8% by 2050, with only about 10% increase in water that is used in agriculture.[11] However, the marginal increase in agricultural water needs is based on the assumption that technology will improve the efficiency of water utilization, which, unfortunately, can only be realized in the developed world. Poor countries can only achieve the required water utilization efficiencies through technology and capital transfers from developed countries through trade if the current gaps in capital, water infrastructure, and water technology situations between the rich and poor nations are anything to go by in this matter.

In addition to natural and demand-driven water scarcity, the contamination of the economically available freshwater by natural and synthetic pollutants is emerging as one of the major water stressors. Contamination of water reduces the economic viability of its utilization because the contaminants must be removed before it is used, and the cost of removing the contaminants is proportional to the extent of contamination. Contaminated water can be a source of human disease if it is used for any purpose without prior treatment. In agriculture, the water contaminants can be transferred from the water to food crops via irrigation. In fact, the current trend in food contamination from the use of contaminated water, among other sources, has elevated food to one of the principal sources of human exposure to chemical and biological disease-causing agents. The transfer of the chemical and biological disease-causing agents from water to food presents an enormous challenge to modern society because, unlike water, which is locally consumed, food is marketed globally and consumed globally. As a consequence, while water is essential to life, it can also be a cause of human suffering and loss of life if it is abused.

In addition to anthropogenic contamination, water can also be contaminated by natural sources. Humans have no causative effect on natural contamination of water but can have a control effect. Because there is no risk if there is no exposure, human health risk from contaminated water can be avoided by the use of alternative sources that are not contaminated or by treating the contaminated water before using it. However, the current trend, as dictated by human needs, seems to indicate an out-of-control situation where the human causative effect on water pollution is on the increase while the control effect on the same cannot cope. Such a situation is resulting in unprecedented levels of water contamination by chemical and biological agents across the globe. The current level of water contamination is putting enormous pressure on water, a situation that has led to a widespread diminished access to safe drinking water in nations that cannot afford the cost of purifying enough water to meet their basic needs. The widespread shortage of clean water for drinking and sanitation has led the world to direct more attention to the provision of drinking water and proper sanitation.[15]

The focus on domestic water has, unfortunately, removed the needed attention from agriculture, a situation that has contributed to the widespread use of contaminated water and untreated wastewater (domestic, commercial, industrial, storm water, etc.) in agriculture in mid- to low-income countries across Africa, Asia, and Latin America.[15] Scientific research is showing that the practice is causing significant buildup of pollutants in agricultural soils thus exposing humans to occupational, recreational, and dietary risks that are associated with exposure to the pollutants. The phytoavailability of chemical pollutants in the soil is dependent on the chemical load, form, and mobility, among other factors.[16] An increase in the chemical load will saturate the sorption sites in the soil, a situation that will increase the mobility and, hence, availability of the chemicals. The increased mobility and availability of chemicals will also increase the chances of freshwater contamination through leaching and run-off.

Apart from the competing human needs for water, there is also competition from natural ecosystems for the precious commodity. The sharing of water between all these competing needs will be neither equitable nor fair. History informs us that conflicts flourish whenever and wherever there are shortages of essential resources.[11] Given the role that water plays in human life, the possibility and probability of conflicts from water shortages and the resulting famine, disease, and death, are quite significant if the water situation is not remedied. Unfortunately, the water situation seems to be already dire in some poor countries where people are barely getting a liter of water per person per day for drinking, cooking, and washing, an amount that is a far cry of the normal daily requirement of 20 L per person.[11]

Agricultural Water Contamination—Contaminants, Sources, and Health Effects

Water contaminants are unwanted chemical and biological agents that are introduced to the water intentionally or inadvertently through both natural and anthropogenic processes and activities. The contaminants can be natural or synthetic, with the former getting into the water from natural background sources as well as from anthropogenic activities that redistribute and often concentrate them in environmental compartments that include water, while the latter are solely from anthropogenic sources. A large number of the water contaminants can cause harm to the human body. In fact, all chemicals have the potential to cause harm to the human body, with dose providing the fine line between harm and no harm. Most water contaminants, once they get in, cannot naturally disappear from the water. Chemical contaminants can transform from one form to the other, depending on water conditions, but will remain in the water except in cases where the transformation leads to the formation of gases that can then escape into the atmosphere, or solids that get deposited in the sediment. However, the sediment-bound chemicals can be remobilized back into the water column depending on the physicochemical properties of the water at given instances. Biological pollutants, because they are living pathogenic organisms, have life cycles that include reproduction and deaths, and will, therefore, remain in the water and, their populations can grow depending on the availability of nourishment. As a consequence, water contaminants have to be removed by humans, through physical, biological, and chemical means, to render the water safe for use.

As already mentioned, water contaminants are of natural and anthropogenic origin, with the latter being the principal source. The major anthropogenic sources of water contamination include industrialization, urbanization, agriculture, commercial activities, and solid waste management. These anthropogenic sources load contaminants in water through the pumping of untreated effluents (sewage, industrial, commercial, etc.) into urban drainage systems, and surrounding water bodies, and also through storm water, leachates, and run-off. Current practices in a large number of countries have seen sewage, industrial (including mining), commercial (including hospitals) effluents, and storm water being combined into a large effluent flow of wastewater that is then pumped into traditional sources of water for domestic, agriculture, and industrial purposes or directly used for irrigation in agriculture.[17] Such a practice makes wastewater the major source of water contamination. In some countries, the raw wastewater is used directly to irrigate crops in agriculture.[17]

Basing on the nature of wastewater as described above, it is not surprising that wastewater and wastewater-contaminated water can contain all chemicals that are used in all sectors of the economy in modern day society and those that were banned a few decades ago, but are still present in some old products, and in the environment. These chemical pollutants include toxic heavy metal(loid)s, chlorinated dibenzo-*p*-dioxins (CDDs), polychlorinated biphenyls (PCBs), polycyclic aromatic hydrocarbons (PAHs), phthalates, chlorinated organic solvents, steroids, prescription drugs, nonprescription drugs, fragrances and flavors, antibiotics, pesticides, fire retardants, detergents, disinfectants, cosmetics, dyes, resins, and fuels.[18–20] Most of these contaminants appear on the United States Environmental Protection Agency (US EPA)'s list of priority pollutants.[21] Recent advances in nanotechnology have resulted in a large number of consumer products containing ENMs flooding the market. As a result, concerns have been raised about the impacts of ENMs that are getting in wastewater during manufacture and waste management.[6] Also, wastewater contains microplastics that are added to washing and beauty products as exfoliating "microbeads."[7,8] In addition to chemical pollutants, wastewater also contains a whole range of pathogens that are associated with raw sewage and other raw effluents from commercial establishments such as hospitals. The discussion below will focus on examples of heavy metal(loid)s, organic compounds, nanomaterials (NMs), microplastics, and pathogens that are some of the most common yet harmful contaminants that are found in wastewater and their health effects. It has to be pointed out that most chemicals are potentially harmful to the human body and each section below just discusses a few examples in each category.

Heavy Metal(loid)s

Heavy metal(loid)s such as arsenic (As), cadmium (Cd), lead (Pb), and mercury (Hg) and their compounds represent some of the most harmful yet very common chemicals in wastewater. The maximum contaminant levels of these elements in water, as established by the US EPA, are 10, 5, 15, and 2 ppb, respectively.[21] These, together with other metal(loid)s, form the heart of the chemical industry and are, therefore, found in almost all consumer products. As a result, they form a significant constituent of wastewater. They can also leach into both surface and groundwater from solid waste that include electronic waste (e-waste). E-waste includes disused computers, computer monitors, television sets, and cell phones. These electronic gadgets contain toxic and carcinogenic heavy metals such as Pb, Hg, Cd, hexavalent Cr, Ni, and Sb, as well as organic pollutants such as polybrominated di-phenyl ethers and PCBs.[22]

Arsenic is a known human neurotoxin, carcinogen, and cocarcinogen that is linked to lung, bladder, kidney, liver, and nonmelanoma skin cancers, as well as cardiovascular diseases and diabetes.[23,24] It induces tumorigenicity through cellular proliferation, apoptosis, and differentiation as well as by decreasing the expression of the phosphatase and tensin homolog tumor suppressor gene; and via oxidative stress, genotoxic damage, chromosomal abnormalities, and epigenetics.[24] Epigenetics involves the alteration of methylation levels of global DNA and gene promoters; histone acetylation, methylation, and phosphorylation; miRNA expression; and the acceleration and exacerbation of plaque formation in arteries, thus, increasing the risk of cardiovascular diseases.[23,24] At sufficiently high levels, the metalloid can cause dyspigmentation, keratosis, peripheral vascular diseases, reproductive toxicity, and neurological effects.[24,25] Exposure to sufficiently high levels of As in the womb and early childhood has been linked to the development of chronic diseases that include bladder, lung, and skin cancers, and cardiovascular and respiratory diseases in early life or later in life.[26–30] Neurological and cognitive dysfunction, as well as memory and learning impairments, have been reported to result from prenatal exposure to low or moderate levels of the metalloid.[25,31] Some of the symptoms of As exposure include nausea, vomiting, diarrhea, abnormal heart rhythm, blood vessel damage, and a pins and needles sensation in hands and feet.[32]

Cadmium is also a classified human carcinogen that is associated with cancers of the pulmonary system, liver, bladder, stomach, renal, and the hematopoietic system.[23,33] It is also linked to kidney and cardiovascular diseases, as well as fractures following long-term exposure to low levels of the metal.[34] Cadmium exposure is also linked to increased cancer susceptibility in type-2 diabetic patients, with prolonged exposure reportedly leading to prediabetes, diabetes, and overall increased sex-dependent cancer mortality in some types of cancer.[35] The metal has also been linked to decreased performance in attention- and perception-related tasks in adults exposed to low levels over extended periods,[36] while children and teens exposed to Cd levels that are considered safe are reported to have high chances of developing learning disabilities.[33] Cadmium toxicity is thought to involve epigenetic mechanisms that include hypo- and hypermethylation of DNA.[23]

Lead is a neurotoxin that affects the central nervous system, the renal system, skeletal system, and hematopoietic systems. The pre-, peri-, and postnatal exposures to Pb are reported to cause severe irreversible mental retardation, learning disabilities, and in some instances coma and death in children.[23] The heavy metal induces toxicity by causing oxidative stress, disruption of cell signaling and neurotransmission pathways, and influencing the methylation of DNA.[23] Exposure to Pb in early life is reported to induce epigenetic mechanisms that increase susceptibility to diseases such as hypertension, cardiovascular disease, diabetes, schizophrenia, and neurological diseases later in life.[23,37] In adults, Pb exposure is associated with increased chances of hypertension, cardiovascular diseases, reproductive defects, and calcium deficiency. The major target organs for Pb include blood plasma, nervous system, and soft tissues, but it subsequently accumulates in bone. Recent studies have linked the heavy metal to lung and stomach cancers.[38]

Mercury is a well-known neurotoxin.[39,40] It exists in water in various forms that include methyl mercury (CH_3Hg^+) and inorganic mercury (Hg(II)), with the former being the most toxic of all the forms of the heavy metal. Fetuses and children are more susceptible to the neurological effects of CH_3Hg^+ compared to adults. The symptoms of CH_3Hg^+ exposure include impairment of the peripheral vision, disturbances in sensations, lack of coordination of movements, impairment of speech, hearing, walking, and muscle weakness. Prenatal exposure to CH_3Hg^+ is known to lead to impairments in cognitive thinking, memory, attention, language, as well as fine motor and visual spatial skills. Also, prenatal exposures to CH_3Hg^+ at levels that do not induce symptoms to the mother have been reported to cause severe effects to the infants. Inorganic mercury, on the other hand, can cause damage to the gastrointestinal tract, the nervous system, and kidneys. The symptoms of exposure to Hg(II) include skin rashes and dermatitis, mood swings, memory loss, mental disturbances, and muscle weakness. There are no data that currently link Hg to human cancers even though mercuric chloride is reported to have caused tumor development in laboratory mice and rats while CH_3Hg^+ is reported to have induced kidney tumor in male mice.[39,40]

Organics

Organic contaminants that will be discussed include CDDs, PCBs, PAHs, phthalates, and chlorinated organic solvents using examples in each category that are considered more harmful to humans. Research has shown that these organic pollutants are a significant component of wastewater, and, hence, some of the major contaminants in freshwater.

CDDs consist of 75 chemically related compounds that are commonly referred to as chlorinated dioxins.[41] They are known to be human carcinogens that affect the skin, organ development, immune system, and the reproductive system. While CDDs are not manufactured for commercial purposes, they are formed as by-products during some industrial processes and also during chlorination in wastewater and drinking water treatment. For example, 2,3,7,8-tetrachlorodibenzo-*p*-dioxin, one of the most toxic CDD, is formed at pulp and paper mills during the chlorine bleaching process.[41] CDDs can also occur as contaminants in the manufacture of certain organic chemicals.[41] As a result, CDDs can easily get into wastewater and hence freshwater sources and eventually agricultural soils and crops if the wastewater and wastewater-contaminated freshwater is used to irrigate crops.

PCBs are a group of up to 209 chlorinated compounds that have no known natural sources.[42] They were, however, manufactured and extensively used as coolants and lubricants (oils) in transformers, capacitors, and other electrical gadgets before being banned (1977 in the United States) because of their adverse health effects and persistence in the environment.[42] Even though they are no longer being produced, PCBs can still get into freshwater sources from hazardous waste sites and leaks from old electrical transformers. However, only small amounts of PCBs remain dissolved in water, with the bulk getting deposited onto bottom sediments. The PCBs can easily get into the food chain from water and can accumulate in fish and marine animals in levels several times higher than in water.[42] The health effects that are associated with exposure to PCBs include acne like skin conditions in adults and neurobehavioral and immunological disorders in children, as well as liver damage. PCBs are currently not linked to human cancers but are classified as animal carcinogens.[42]

PAHs are a group of more than 100 different chemical compounds that are produced during the incomplete combustion of fossil fuels such as coal, oil, gas, wood, garbage, as well as burning of tobacco and charbroiled meat.[43] Although PAHs can occur naturally, they can also be manufactured for research as well as for use in the manufacture of dyes, plastics, and pesticides, with some reportedly being used in medicine. PAHs are also found in asphalt that is used in road construction, crude oil, coal, coal tar pitch, creosote, and roofing tar, and are found throughout the environment in the air, water, and soil. PAHs can enter surface water through discharges from industrial plants and wastewater treatment plants, and they can be released in soils at hazardous waste sites. Some PAHs in soils can also leach into underground water. These compounds can be accumulated by both plants and animals from contaminated soil or water.[43] The PAHs that have been classified as probably or possibly human carcinogens by both the US EPA and the International Agency for Research on Cancer (IARC) include benzo[a]anthracene and benzo[a]pyrene, benzo[b]fluoranthene, benzo[j]fluoranthene, benzo[k]fluoranthene, and indeno[1,2,3-c,d]pyrene.[43] The IARC also added chrysene and dibenzo[a,h] anthracene to the list. However, the US Department of Health and Human Services classifies all the above PAHs as animal carcinogens.[43]

Phthalates, also known as plasticizers, are aromatic compounds that are used to make plastics more flexible and resilient. They are found in a wide range of consumer products that include toothbrushes, automobile parts, tools, toys, packaging, cosmetics, insecticides, aspirin, wall coverings, floor tiles, furniture upholstery, shower curtains, garden hoses, swimming pool liners, rainwear, baby pants, shoes, sheathing for wire and cable, medical tubing, blood storage bags, carpet backings, paints, glue, and rocket fuel. Because they do not form part of the polymer (plastic) chain, they can easily leach from the products. Examples of phthalates include di(2-ethylhexyl) phthlate, di-*n*-butyl phthalate, and di-*n*-octylphthalate. Even though no harmful effects have been reported in humans, some phthalates have been reported to cause impaired reproduction and developmental effects, as well as deaths in laboratory animals. They are released into the aquatic system from hazardous waste sites and as part of industrial, agricultural, and domestic effluents.[44,45]

As pointed out earlier, the classes of organic water contaminants listed above are not a comprehensive list of organic compounds that can be found in wastewater. They represent some of the common organic water pollutants that are harmful or potentially harmful to humans. There are others not included in the list such as chlorinated organic compounds, which are also very common in wastewater. Typical examples of chlorinated organic compounds include trichloroethene and perchloroethene, which are considered human carcinogens.[46]

Nanomaterials

Nanotechnology is the new buzz and a true whiz-tech of the twenty-first century that is based on a new concept of ultrasmall. It has taken the world by storm and is revolutionizing technological applications in all aspects of the economy from cosmetics to modern medicine. The recent advances in this exciting and promising field have seen an increase in the number of consumer products containing ENMs and large volumes of waste containing ENMs from the manufacturing and usage of the products.[6] As a result, wastewater can contain significant amounts of ENMs of any kind and concerns have been raised about the environmental and public health impacts of these materials in wastewater.[6] Some of the most common ENMs include metals, metal oxides, carbon nanotubes, and quantum dots. The latter are nanoparticles (NPs) or nanocrystals of semiconductor material.

Large chunks of national budgets are being channeled toward research and development in the technological applications of ENMs. Unfortunately, the investments in potential applications are not being matched by investments in the understanding of the fundamental properties of the ENMs, especially environmental impact, ecotoxicity, and cytotoxicity. Significant questions still exist regarding their increased reactivity because of increased surface area and surface properties, increased absorption capacities, ability to reach deep airways, dermal penetration, as well as systemic distribution, penetration of the blood–brain barrier, penetration of the placental barrier, and biological and environmental persistence. Questions also still exist regarding the possibility and the potential risk of ENMs getting in the food chain from contaminated irrigation water, among other sources.

NMs are generally considered safe, but this assertion or perception is largely based on the properties of the bulky counterparts of the ENMs. However, ENMs are known to behave differently from their bulk counterparts, and it is from this property that they owe their novel applications.[47] Ironically, it is also the same property of ENMs that may cause them to elicit adverse effects to humans. For instance, gold NPs (Au-NPs) are considered biocompatible because of the inertness of gold yet inert NPs can produce morphological changes, loss of function, inflammation, or cell damage.[48] Another typical example of NPs that are considered biocompatible yet studies have shown some evidence of dose- and size-dependent cytotoxicity and genotoxicity are the superparamagnetic cobalt ferrite NPs that are used for biosensing, tumor treatment, and as contrasting agents in magnetic resonance imaging.[48]

Microplastics

There have also been recent concerns about the environment and public health impact of microplastics, commonly referred to as microbeads, which are being loaded into wastewater from the use of beauty and cleaning products (shower gels, face washes, toothpaste) that have flooded the markets.[7,8] The microbeads are too small to be filtered out at water treatment plants and end up in wastewater and eventually drain into freshwater sources. Aside from harmful effects that may be inherent, the small sizes and large surface areas of the microbeads may result in them adsorbing and harboring toxic chemicals and pathogens. The microbeads and the adsorbed contaminants may eventually get into the food chain through the use of wastewater in agriculture. The public health concerns about the microbeads have resulted in five US states considering banning their use, with large cosmetics firms such as Johnson & Johnson promising to end their use by 2017.[7,8]

Pathogens

Even though waterborne pathogens can naturally exist in all types of water, untreated wastewater is the major source of these organisms. Health effects caused by pathogens have moved from being just diarrhea and intestinal illnesses to a variety of chronic diseases that include hepatitis, myocarditis, insulin-dependent diabetes, aseptic meningitis, and ocular and respiratory infections, thanks to an increased variety of pathogens that currently exist in wastewater.[49] The most common pathogens that are found in wastewater include (1) *Burkholderia pseudomallei*, which is also known as *Pseudomonas pseudomallei*. It is a gram-negative bipolar bacterium that infects humans, animals, and plants. In humans and animals, it causes melioidosis[50]; (2) *Cryptosporidium parvum*, a protozoan that causes a parasitic disease in the gut called cryptosporidiosis. The parasite also causes respiratory cryptosporidiosis, hepatitis, and cholecystitis when it gets to the lungs, liver, and gall bladder, respectively; (3) *Giardia lamblia*, which is a parasitic flagellated protozoan that causes giardiasis in the small intestines; (4) *Salmonella*, a gram-negative bacteria that causes illnesses such as typhoid fever, paratyphoid fever, and food poisoning; (5) Norovirus, also known as the Norwalk virus, is a single-stranded RNA virus that is the most common cause of viral gastroenteritis in humans, and affects people of all ages. The virus is reported to affect around 267 million people, causing over 200,000 deaths each year mostly in developing countries; and (6) helminths, which are eukaryotic parasitic worms that live and feed on living hosts while causing malnutrition, weakness, and disease.[51]

The chemical and microbial composition of wastewater as described above, clearly shows that wastewater cannot be just dumped into surrounding water bodies, or used untreated in agriculture. The ecological and public health consequences of doing so will be severe and overreaching. Unfortunately, the use of raw wastewater and raw wastewater-contaminated water in agriculture, particularly urban and periurban agriculture (UPA), has become the norm across Africa, Asia, and Latin America.[52] The practice is being fueled by perennial shortages of water, increased contamination of traditional irrigation water sources, and pressure to provide food for the ever-growing human population.[52]

Human Health Risk

Health risk is defined as the likelihood that a negative health outcome will occur in a person or group of people that is exposed to a particular amount (dose) of a hazardous agent. This essentially means that without exposure, there is no risk. The risk associated with the use of untreated wastewater and/or wastewater-contaminated water in agriculture can result from occupational and recreational activities, and diet. Recreational exposure is mainly associated with children as they engage in normal playing activities in areas that are polluted by the contaminated irrigation water, which, in some countries, are not separated from agricultural fields. In some countries, parents work in agricultural fields with babies on their backs or around the fields, with some children providing help. As a result, occupational exposure can occur to both children and adults during normal agricultural activities, via direct contact with the contaminated water (wastewater), and wastewater-contaminated soils, as well as the inhalation of gases from the wastewater, and the inhalation and ingestion of dust from contaminated agricultural soils. Children can routinely, purposefully or accidentally, swallow the wastewater and wastewater-contaminated soils while playing.[16] Wastewater can also inadvertently get into the mouths of adults during irrigation, and can get ingested. Direct contact with the wastewater and the wastewater-contaminated soils can result in certain pollutants, chemical or pathogenic, entering the human body through dermal penetration.

However, the oral route remains one of the major routes of human exposure to environmental pollutants.

Dietary exposure occurs to everyone, but unlike occupational and recreational exposures, it can be both local and global. The dietary exposure emanates from the consumption of contaminated foodstuffs. At local level, food can be further contaminated by washing or processing it with wastewater-contaminated natural water. Unlike the farmers involved and maybe a few locals, most consumers may not know the precise origin and, therefore, safety of foods in their local supermarkets. The consumers, therefore, cannot make informed choices that are based on the safety of the food. Their choices will obviously be based on visual quality, diversity, and advertisements, among other factors, and may get poisoned in the process. The lack of consumer awareness, the widespread and ongoing use of untreated wastewater in agriculture, and the fact that locally produced food can be marketed globally, are some of the principal reasons why agricultural food contamination has become such a huge global challenge.

Some researchers who have attempted to assess the safety, that is, the probability that harm will occur under specified conditions, and risk assessment, which is the probability that a person's health will be adversely affected under a particular set of conditions, of contaminated water and its use in agriculture mostly focused on dietary exposure while ignoring risks from occupational and recreational exposures, and in the process, significantly underestimated the risk that is associated with the use of wastewater or contaminated water in agriculture. Regardless, there is overwhelming and undeniable data available in scientific literature about both chemical and biological contaminants from wastewater and contaminated freshwater getting into the food chain via irrigation and people getting sick through eating of contaminated food, as well as through agriculturally related occupational activities and recreational activities around contaminated soils and water.[18] It has been shown, through research, that exposure to environmental contaminants accounts for 85% of all diseases, particularly chronic diseases such as obesity, diabetes, heart disease, and cancer.[53] And, as mentioned earlier, at least half of hospital beds in the world are occupied by people with waterborne diseases.[12] This makes the risk associated with the use of contaminated water in agriculture quite significant.

The following facts become evident from the foregoing discussions: (1) wastewater, broadly defined, can contain a wide variety of harmful contaminants; (2) wastewater is the principal anthropogenic polluter of freshwater in most countries; (3) untreated wastewater and wastewater-contaminated freshwater are routinely being used to irrigate crops in a large number of countries, some of which are the major producers of the world food reserves; (4) contact with wastewater and wastewater-contaminated water and soils during normal recreational and agricultural activities is causing people to get sick; (5) the contaminants from contaminated irrigation water are getting transferred into food in levels that are not safe and human illnesses have been reported as a result; (6) the use of contaminated water in agriculture is widespread and continuing unabated; (7) farmers have little to no knowledge of the hazards involved; (8) agricultural food products can be marketed globally; and (9) consumers are mostly unaware of the precise origins of the food and cannot make choices based on the safety of the food regardless of their knowledge about the potential dangers of consuming food grown using untreated wastewater or contaminated water. These facts combined, show that the risk associated with the use of contaminated water in agriculture is real, significant, and cannot be overemphasized. Something drastic, at both national and international level, has to happen to preserve public health as well as ecological integrity. Below is a detailed literature-based evaluation of risks associated with the different classes of irrigation water contaminants.

Risk from Exposure to Chemicals

Chemical toxicity can be dose related, and effects can be acute, chronic, or both.[54] One of the key concepts in chemical toxicology is the quantitative relationship between the amount of a chemical pollutant in the human body and the extent of its biological effect. The extent of the effect is usually a function of the amount of the chemical pollutant exposed to an individual. It is worth noting that there is always a range of sensitivities to individual chemical pollutants in any given population and doses quoted in toxicology are averages that take into account the various sensitivities.[54] Chemical pollutants cannot multiply in the human body but can be transformed from one form to the other and can be eliminated from the body usually in a form that is different from the form in which they entered the body. The biotransformation can be a detoxification mechanism or a process meant to convert the chemical to an excretable form. However, the biotransformation can also lead to the human body inadvertently poisoning itself by activating and converting a less harmful chemical to a more hazardous chemical.[54] As a result, it is not only chemicals that are traditionally recognized as toxic that can illicit an adverse health effect to humans; the "harmless" ones can do that as well.

In risk analysis, chemical pollutants can be classified as either non-cancer or cancer causing. For non-cancer-causing chemicals, the potential risk is the ratio of the estimated exposure to the exposure that causes negligible or no risk, and a threshold is assumed to exist, with exposures less than the acceptable daily exposure of toxicity reference dose considered safe.[2] However, for carcinogenic chemicals, the type and severity of cancer are considered to be independent of dose since a single genetic mutation, which can be caused by a single molecule of a carcinogen or a quantum of radiation can initiate the development of cancer.[2] This means there is no threshold, even though the probability of cancer development is considered to be proportional to dose.[2]

As pointed out earlier, the exposure to chemical contaminants in irrigation water can be via contact with and inadvertent ingestion of the contaminated water and contaminated agricultural soils, as well as the consumption of contaminated food. For instance, exposure to organic contaminants such as CDDs, PCBs, PAHs, and phthalates has been shown to occur through contaminated dust in addition to contaminated food and water.[41–45] With respect to chemical contamination of food from the use of untreated wastewater and contaminated freshwater in agriculture, research has largely focused on toxic heavy metal(loid)s, and, recently, on ENMs as well. As a result, risk from toxic heavy metal(loid)s and ENMs will be discussed in more detail. However, it has to be pointed out that all chemicals are potentially harmful.

Risk from Toxic Heavy Metal(loid)s

According to credible scientific literature, the use of raw wastewater to irrigate crops across mid- to low-income countries in Africa, Asia, and Latin America is significantly contaminating natural freshwater sources and agricultural soils with chemical contaminants, chief among them toxic heavy metal(loid)s, which are eventually getting into crops in levels that, in most cases, are way above safe limits.[55] In some crops, the uptake of certain heavy metal(loid)s has been reported to affect crop development and yields, with the accumulation being type and variety dependent in some cases.[56,57] For example, rice has been reported in numerous peer-reviewed research articles to be an effective accumulator of As and other toxic heavy metal(loid)s compared to other food crops.[55,56] Rice has also been reported to be able to accumulate As in the grain to unsafe levels from irrigation water containing the metalloid in levels that are considered safe for irrigation water.[55] Arsenic accumulation in rice has also been shown to interfere with the uptake and accumulation of essential micronutrients such as selenium (Se) and zinc (Zn) in the rice grain. All these findings are quite significant given that rice is the staple food for more than half of the world population.[58] Bhatti et al.,[57] studied the accumulation of As in vegetable species irrigated with As-contaminated water and reported that spinach, in particular, presented a more direct risk to human health based on the As levels it accumulated in the edible parts. Their study also confirmed the species-dependent enrichment and distribution of the metalloid within the plants. Despite the differences in the accumulation rates of metal(loid)s by crops and crop varieties, the overall enrichment levels remain a real threat to public health.

Across Africa, where untreated wastewater is extensively used in UPA, millions of people are being exposed to the risk of heavy metal(loid) toxicity in addition to other associated hazards. In Ghana, for instance, Lente et al. studied the contamination of soils, water, and vegetables in urban vegetables grown with wastewater and reported levels of Pb that were about two to four times higher than permissible limits in the vegetables.[59] In Egypt, a study by Badawy et al. concluded that wastewater irrigation was causing accumulation of heavy metal(loid)s in food to unsafe levels.[60] An earlier study by Shuravilin et al., in the same country, had reported levels of Cd and Pb that exceeded permissible levels in the soils as a direct result of the perennial use of wastewater in agriculture.[61] In Ethiopia, significant enrichment of heavy metal(loid)s in the edible parts of food crops was reported from studies on the impact of the use of untreated wastewater in agriculture.[62] In Nigeria, Adamu et al. analyzed levels of trace heavy metals that included Pb and Cd in sewage water samples from the Kubanni river drainage basin that is used as a source of irrigation water in Zaria City and reported levels that were beyond irrigation water standards established by the US EPA and FAO.[63] They concluded that the use of irrigation sewage water from the drainage basin was potentially contaminating food. In Zimbabwe, a nation whose economy is largely agrobased, the literature has it that raw wastewater is routinely used to irrigate crops in and around some of its major cities. Research conducted in this country has reported severe contamination of freshwater sources, soil, and crops from the practice. For example, Muchuweti et al. (2005) analyzed levels of Cd, copper (Cu), Pb, and Zn in maize and *tsunga* (a green vegetable) as well as in other foods such as beans, pepper, and sugarcane that were under wastewater irrigation at a farm close to the nation's capital and reported levels that were several times higher than the European Union safe limits.[64] Another study in the country by Tandi et al. (2004) at Pension farm, near Harare to determine the effect of the long use of sewage effluent reported levels of Zn and Cu that were as high as 1563.9 and 133.3 mg/kg, respectively, in the top soil (0–20 cm depth), and the subsequent transfer of the metals to the vegetables in levels that were considered detrimental to human health.[65]

In Asia, where on average 87.6% of freshwater drawings go to agriculture, a large number of countries on the continent rely on wastewater in agriculture to ease the pressure on the water demands.[11,17] Unlike in Africa, a lot of work has been done on the impacts of wastewater in this region but the studies have focused largely on contamination by As and its uptake by rice. This may be because rice is the major staple food in the region while As is one of the major chemical pollutants in the region as well. For instance, Khan et al. (2009) reported a substantial buildup of As in some Bangladesh soils from the use of As-contaminated irrigation water, and significant enrichment of the metalloid in the rice grain, which also resulted in a reduction in yield.[66] Another study in Bangladesh by Islam et al. (2012) showed that the concentration of As in irrigation water does not have to be higher than the recommended limits in irrigation water for the metalloid to accumulate to unsafe levels in the rice grain.[55] They concluded that the eating of contaminated rice constituted the major route of As exposure to humans compared to water in that region.[55] This is quite remarkable because, in other regions where there is a risk of As exposure, drinking water is usually the major source. In fact, other researchers in the same country considered the contamination of irrigation water as the major constraint to the country's agriculture because of the impacts of As in terrestrial systems.[67] In Pakistan, the use of contaminated sewage effluent to irrigate crops is reportedly an old tradition, necessitated by the acute shortages of irrigation water, even though the country's regulations prohibit the use of untreated wastewater in agriculture.[17] A study by Ullah et al. (2012) in the country reported higher accumulation levels of toxic metals in the soils that were under raw sewage irrigation and the subsequent transfer of the metals to food.[68] Khan et al. (2013) investigated the health risks associated with the consumption of contaminated agricultural produce from the use of wastewater and reported Cd levels in the soil, as well as Cd, Pb, and Cr levels in the water and food crops that were higher than permissible limits.[69] The accumulation factors for Mn and Pb were greater than one in food crops while the Health Risk Index (HRI) for Pb was greater than one in all food crops that were irrigated with wastewater. The HRI for

Cd and Mn reported were higher than one in some vegetables.[69] The researchers stressed the need for wastewater pretreatment before use in agriculture as well as continuous monitoring to avoid food contamination. However, Al Omron et al. (2012) reported significant buildup of toxic heavy metal(loid)s in soils from long-term impacts of the irrigation of agricultural soils with treated sewage effluent in Saudi Arabia, showing that pretreatment may not offer long-term solution to the problem if contaminants are not completely removed.[70] Another study by Mahmood and Malik (2014) in Pakistan on the human health risks associated with the irrigation of crops with untreated urban and industrial wastewater reported levels of Cr^{2+}, Pb^{2+}, and Cd^{2+} that were higher than European Union permissible limits in vegetables.[71] An earlier study by Shaheen et al. (2011) on the effect of canal water, tube well water, and sewage water irrigation on vegetables collected from all the districts of Pakistan from 2006 to 2008 had reported levels of mercury that was as high as 130.6 ppb.[72] Such levels of food contamination show that fish may no longer be the major source of mercury exposure to humans, at least in this region. Some studies also reported declines in yields of crops other than rice due to the uptake of heavy metals from soils that were under wastewater irrigation. For example, Ahmad et al. (2011) reported a decline in canola yields in Pakistan that was associated with the accumulation of heavy metals such as Pb, Cd, and Cr.[73] They concluded that the municipal sewage water containing high levels of metals was inhibiting canola growth and seed yield.

China is one of the countries where wastewater is routinely used in agricultural crops and one of the countries where there is a widespread condemnation of the practice from researchers and the media.[74–77] Numerous studies in the country have reported significant contamination of agricultural soils and crops with heavy metal(loid)s, with the levels in crops being positively correlated with levels in the soils.[78–81] For instance, a study by Chen et al. (2004) reported the concentration of Pb in rice irrigated with sewage water, in Yinchuan, that was 10 times higher than the national safety standard for food.[82] Wu and Cao (2010) determined the levels of Hg, Cd, Pb, and As in samples of irrigation water, sediment, soil, and groundwater from a field in Tianjin that was irrigated with wastewater and reported elevated levels in river sediments and wastewater-irrigated soils.[83] Wang et al. (2012) reported levels of Cd that exceeded the maximum acceptable limits set by the Ministry of Environmental Protection of China and World Health Organization in soils and vegetables irrigated with raw wastewater.[76] Xue et al. (2012) reported elevated levels of heavy metal(loid)s, which included Cd and Pb, in soils and vegetables that were under sewage irrigation, but concluded that the risk posed by the consumption of the vegetables was not significant based on health risk calculations. It should, however, be pointed that the vegetables are usually part of a meal that may also be contaminated, and the effects of the entire meal will be additive. The fact that the metal(loid)s will accumulate with each application of the sewage water and that crops are accumulating the toxic chemicals in edible parts should be a matter of public concern.[84]

India is reported to receive more than 70% of its annual rainfall during the 3 months of the monsoon, with most of it flooding out to sea with farmers lacking irrigation capabilities having to deal with water scarcity for about three-quarters of the year.[11] Such water shortages have resulted in the use of wastewater in some parts of the country to meet the agricultural water needs.[85–95] The practice has reportedly led to levels of toxic heavy metal(loid)s in agricultural soils and foods that are largely beyond safe limits.[85–95] Studies on crops that were irrigated with uncontaminated water had much lower to undetectable levels of the heavy metal(loid)s showing, beyond reasonable doubt, that wastewater is one of the principal sources of food contamination in the country.[91,94–97] A 20-year study by Bansal (2008) that investigated the long-term effects of the practice reported levels of Zn, Mn, Cu, Ni, Cr, Pb, and Cd in agricultural soil and crops that were higher than levels in control samples and also higher than the permissible limits.[18] Bansal (2008) recommended "proper management of irrigation water and periodic monitoring of soil and plant quality parameters to ensure successful, safe, long term wastewater irrigation."[18] Such recommendations, together with the data on which they are based, as valuable as they are, may remain buried in the scientific literature, and, as a result, may never influence public policy. A more recent study by Rani et al. (2013) on the use of wastewater to grow crops in the country reported levels of heavy metal(loid)s in soils and vegetables that were higher than limits set by FAO, with positive correlations between the levels in plants and the levels in sewage water and soils.[98]

Some studies have directly linked some heavy metal(loid)s to specific diseases from the use of wastewater in agriculture. For example, an extensive survey by the government of Toyama prefecture of Japan that was prompted by an outbreak Cd-induced Itai-itai disease was able to link the outbreak to the use of irrigation water that was contaminated by Cd.[99] After the outbreak of the disease, Cd levels were analyzed in rice and soils of paddy fields that were irrigated with the contaminated water over five decades. The analysis revealed severe contamination of both the soils and rice and it was recommended that the upper soil layer of a total of 1500 ha of paddy fields be replaced. The replacement of the topsoil together with discontinuing of the use of the contaminated water significantly improved the quality of the soils and the agricultural produce.

Nanomaterial Risk

Work that has been done to date regarding the environmental and public health impacts of ENMs point to phytouptake and accumulation, and transfer up the food chain, as well as eco- and cytotoxicity, even though mechanisms remain largely unknown. Kettler et al. (2014) reviewed the literature in an effort to understand the mechanisms of cellular uptake of ENMs and revealed important uptake mechanisms for eukaryotic cells that include macropinocytosis, receptor-mediated endocytosis, and phagocytosis in specialized mammalian cells.[9] Uptake into nonphagocytic cells was reported to depend strongly on NP size, extent of aggregation, and surface charges, with an increase in surface charges (positive or negative) promoting particle uptake. Other studies have also showed that surface functionalization played a role in the uptake of ENMs by cells. For example, Judy et al. (2012) used hydroponic conditions to study tobacco and wheat plants' uptake of Au-NPs of various sizes that were coated with either tannate or citrate and reported uptake and bioaccumulation of the functionalized NPs of all sizes in the study by the tobacco plant.[100] Also, Zhu et al. (2012) showed that uptake and distribution of Au-NPs by rice, radish, pumpkin, and perennial ryegrass depended on both NP surface charge and plant species, with positively charged Au-NPs being most readily taken up by plant roots, while negatively charged Au-NPs were most efficiently translocated into plant shoots (including stems and leaves) from the roots.[101] They also showed that radish and ryegrass roots accumulated higher amounts of the Au-NPs than rice and pumpkin roots, while rice accumulated significant NPs in shoots. A more recent study by Koelmel et al. (2013) also showed that terminal functional groups of the ENMs had a strong influence on the uptake and spatial distribution in the rice plant.[102]

In addition to uptake via plant roots, some researchers have shown that uptake by plants can also occur following leaf exposure to the ENMs. For example, Larue et al. (2012) exposed wheat and rapeseed plantlets to TiO_2-NPs, some via roots and some via leaves and observed accumulation of TiO_2-NP in the plantlets in both cases.[103] They also observed that TiO_2-NP exposure induced increased root elongation but had no effect on germination, evapotranspiration, and plant biomass, showing that the bioaccumulation of the NPs in the plants was not necessarily affecting plant growth.[103] However, other researchers have shown that some ENMs can cause adverse effects to plants upon uptake. For instance, Pokhrel and Dubey (2013) observed structural changes in the primary root cells of maize and cabbage upon exposure to citrate-coated silver NPs and zinc oxide NPs as well as $AgNO_3$ and $ZnSO_4$, even though the observed toxicities in the two plants revealed lower NP toxicity compared to free ions.[104] Zhang et al. (2012) compared the phytotoxicity of Yb_2O_3-NPs, bulk Yb_2O_3, and $YbCl_3 \cdot 6H_2O$ to cucumber plants and reported a concentration-dependent decrease in biomass upon exposure to Yb_2O_3-NPs and $YbCl_3$.[105]

Evidence from the literature has also shown that after uptake, ENMs can be biotransformed during translocation to all parts of the plant, including edible parts, and can be transferred from one trophic level to the other, with biomagnification in some cases. For example, Zhang et al. (2012) reported the biotransformation and translocation of CeO_2-NPs in plant systems.[106] According to a recent review by Gardea-Torresdey et al. (2014), ENMs can accumulate and/or increase levels of the component metal or carbon NMs in the fruits or grains of agricultural crops, with potential detrimental or beneficial effects on the agronomic traits, yield, and productivity of the plants.[107] Judy et al. used model organisms to study trophic transfers from primary producers to primary consumers and observed biomagnifications of the Au-NPs from primary producers (plants) to primary consumers (tobacco hornworm).[6] Pakrashi Sunandan et al. (2014) also reported bioaccumulation of aluminum oxide NPs by a primary producer, which were biomagnified in the primary consumer.[108]

Aquatic organisms, which include fish, have also been shown to take up ENMs directly from water. For example, Joubert et al. (2013) investigated the potential toxicity of Au-NPs using a bivalve model *Scrobicularia plana* and observed particles in the gills, distributed freely in the cytoplasm, or associated with vesicles, and individual or small aggregates of 40-nm-sized NPs within the nuclei, colocalized with DNA in the digestive gland.[109] A more recent study by Garcia-Reyero et al. (2014) explored the potential effects of polyvinylpyrrolidone-coated silver NPs (PVP-AgNPs) as intact NPs or as dissolved silver using females of the model species fathead minnow (*Pimephales promelas*) and reported effects that were consistent with both intact NPs and dissolved silver.[110] Both PVP-AgNPs and $AgNO_3$ were reported to have affected pathways that are involved in Na^+, K^+, and H^+ homeostasis and oxidative stress.[110]

The possible presence of ENMs in wastewater and the ability of plants to take up the ENMs, through both roots and leaves, translocate them to edible parts of the plant, and transfer them, in some cases with biomagnification, up the food chain, coupled with the potential toxicity of the NMs, make the human health risk from ENMs more of a reality than a possibility. Research has also shown that ENMs can be directly taken up by human organs and get translocated to secondary organs, such as the central nervous system, and can be transported across the placenta and the blood–brain barrier.[111] In their work, Englund et al. (2013) reported Au-NPs and Ag-NPs interferences with the growth profile of human embryonic neural precursor cells,[111] while Hanna et al. (2013), using model organisms, reported toxicity of ZnO and CuO ENPs to *Leptocheirus plumulosus*.[112] The colloidal stability of some of the most common ENMs, particularly Au-NPs, means that they persist in the aquatic system and, because of their size and surface reactivities, can adsorb and harbor toxic chemicals and pathogens, which can transfer to humans via the food chain. As a result, even if some ENMs may be safe, they may still elicit adverse health effects from the adsorbed harmful pollutants. Either way more work still needs to be done with respect to the environmental and health impacts of NMs.

Risk from Exposure to Pathogens

The health risk associated with pathogenic pollutants is not necessarily dose related because pathogens have the capacity to grow and multiply in the human body. Their effects can be immediate and fatal if left untreated. A number of studies have reported a range of pathogen-caused diseases that include skin, intestinal, and diarrheal in people within and around farming communities where raw wastewater or wastewater contaminated water is used in agriculture.[113,114] Ensink et al. (2008) reported that intestinal nematode infections, particularly hookworm, *Ascaris lumbricoides* and *Trichuris trichiura*, infection were the main health risk that is associated with the use of sewage water in agriculture.[113] An earlier study in the city of Faisalabad in Pakistan had revealed an increased risk of intestinal nematode and hookworm infections in farmers irrigating with wastewater and their children compared to farmers using regular water.[114] Other researchers reported increased risk of *Salmonella* strains and protozoan infections and helminthiasis in Morocco[115,116]; increased fecal coliforms and giardia cysts, as well as *Shigella* and *Salmonella* giardiasis, amebiasis, and diarrhea, in Eritrea[117] in both children and adults when compared to those in communities where wastewater was not being used for irrigation. Work by Cifuentes et al. (2000) in Mexico showed that children from households where untreated wastewater was being used in agriculture had a 33% higher risk of diarrheal diseases and a fivefold increase in risk of *A. lumbricoides* infection compared to children from households that farmed with rainfall, with the risk for *A. lumbricoides* infection reportedly much higher in older individuals than in children.[118] Their work also showed that the cultivation of vegetables, in addition to the drinking of unboiled water, was associated with a higher risk of diarrheal diseases.[118]

Wastewater Use in Agriculture

The Reasons

In addition to numerous references cited in the sections above, a review by Murtaza et al. (2010) shows that the use of sewage effluent, treated or untreated, is rampant, particularly in the developing world and is ongoing despite the amount of scientific literature that is reporting the environmental and public health ills of the practice.[1] Most, if not all, research is showing significant enrichment of both toxic and essential metals as well as organics and pharmaceuticals in both agricultural soils crops. Such findings call for rigorous treatment of wastewater by subjecting it to physical, chemical, and biological processes to remove contaminants to levels that are not a threat to human health and the integrity of natural ecosystems before it is released into surrounding water bodies or used for any purpose.

The use of properly treated wastewater in agriculture can be an effective and sustainable conservation strategy that ensures nutrient cycling and the preservation of human health and the environment without compromising agricultural production in both quality and quantity. Other obvious benefits of the utilization of treated wastewater in agriculture include the provision of livelihoods and fresh produce to cities, and better quality of living for farming families, especially in developing countries where the generality of citizen subsist on agriculture.[17] Unfortunately, the process of recycling wastewater via treatment is a very costly, undertaking which many poor countries may not afford. Coincidentally, water stress is most severe in a large number of poor countries, and a recent survey in cities across Africa, Asia, and Latin America suggested a positive correlation between the use of untreated wastewater in agriculture and poverty.[17] For example, Africa, a continent with the highest prevalence rates of hunger and one of the poorest continents, is reported to be only second to Oceania in terms of dryness yet its water needs in agriculture stand at 84.1% of its total freshwater withdrawals.[17] As a result, the poverty and scarcity of freshwater are used as reasons for the widespread use of untreated wastewater in agriculture on the continent. The water situation is further worsened by the rate at which freshwater sources are getting contaminated by population-driven anthropogenic activities. Because of unprecedented growth rates of cities across the developing world amid declining national economies and undercapacity wastewater infrastructure, wastewater is being generated in urban centers in quantities and rates that authorities and infrastructure cannot cope with. As a result, the wastewater is pumped, untreated, in surrounding water bodies or directly used in agriculture.[17] According to the International Water Management Institute (IWMI), 85% of cities across the developing world are discharging large volumes of untreated wastewater into natural drainage systems and water bodies around the cities at rates that are overwhelming the dilution capacity of the water bodies.[17] Such a practice makes freshwater in and around the cities essentially raw wastewater. Faced with the increased demand for irrigation water to meet the increasing demand for food, the widespread contamination of traditional irrigation water sources leaves farmers without options but to use the raw wastewater to irrigate their crops.[119] Further pressure is put on farmers because they subsist on agriculture, and market incentives in cities can be hard to resist.[17]

In addition to poverty, freshwater scarcity, contamination of traditional sources of water, and lack of alternative water sources, against unprecedented demands, some farmers are reported to prefer wastewater over traditional irrigation water citing reliability of wastewater and its nutrient content as the reasons for their choice.[1,119,120] All these reasons, unfortunately, make some sense under the circumstances, but the consequences can be costly. The nutrient value of wastewater is undeniable and, in fact, reduces the cost of production because farmers will not need to buy costly fertilizers. Also, unlike natural rainfall that can be erratic and unreliable, wastewater can provide a continuous flow, in more than sufficient volumes, and can be much cheaper, if not free, compared to traditional freshwater. The reduced cost of production will make the practice more lucrative because large profits can be realized. Regardless of the incentive-related factors, the lack of options amid unprecedented demands remains the principal driver for the use of wastewater in agriculture, which in some cases can be the only source available. The major questions are who is supposed to provide safe alternatives and when will that happen.

There is no doubt that UPA, which uses much of the wastewater, can be a vital strategy in mitigating hunger in poor nations if practiced in a safe and sustainable manner. In actual fact, FAO supports UPA under the theme of "growing greener cities" and encourages its recognition by public policy, and its inclusion in urban development strategies and land-use planning.[52] FAO also encourages support of UPA by agricultural research and extension departments, as well as by linking it to sources of technological innovation, investment, and credit, and to urban markets and consumers.[52] It is envisaged that, in addition to improving food security to poor city dwellers, UPA can stimulate local economic growth by creating employment.[52] Unfortunately, because of the reasons given above, untreated wastewater will continue to be used in UPA with obvious detrimental consequences to human health and the environment if the current water situation is not remedied.

Regulations

Despite the number of scientific publications on the human health and ecological hazards of the use of contaminated water, particularly untreated wastewater, in agriculture, the practice remains ongoing with largely no regulations in place to protect both the farmers and the consumers. According to the IWMI, the majority of countries surveyed had no regulations regarding the use of untreated wastewater to irrigate crops by the time of their study, while the few that had their own rules in addition to the FAO or WHO regulations lacked monitoring and enforcement mechanisms.[17,119] For example, in Ghana, regulations on the use of urban wastewater are available but farmers mostly ignore them without any consequences.[119] The enforcement of regulations is reportedly made difficult by lack of alternative sources of irrigation water.

In some countries, the authorities who are supposed to be custodians of the regulations are forced by circumstances to supply untreated wastewater to farmers to meet their obligations to the farmers.[17,119] In Pakistan, for instance, untreated wastewater is usually auctioned for use in agriculture even though the country's laws stipulate that untreated wastewater cannot be used in agriculture.[17] In Bangalore, it is reported that authorities distribute wastewater for use in agriculture, treated or untreated, while, in Nam Dinh, the authorities are reported to pump wastewater, on a regular basis, from the city drainage system directly into the irrigation system in times of scarcity.[17] In India, use of wastewater for irrigation is not encouraged but, as the literature reports, it is used regardless.[17] If truth is to be told, the authorities in all places where untreated wastewater is used in agriculture are aware of the potential hazards of the practice, but because they cannot provide their citizens with uncontaminated water, they pretend all is well.

If regulations cannot be enforced, they are just as good as nonexistent, and enforcing regulations on the use of wastewater without providing farmers with alternative means makes no sense at all. Some developed countries realized the public health hazards of the use of wastewater in agriculture and banned it completely.[19] However, chemical measurements in the soils previously irrigated with sewage before the practice was stopped still show elevated levels of toxic chemicals, particularly heavy metal(loid)s, that are a potential threat to human health.[121] This shows that the practice makes agricultural lands essentially hazardous sites that will need intensive remediation for food safety to be ensured. The longer wastewater is used on the soils, the more difficult and expensive it will likely be to remediate.

Risks Associated with Reuse of Wastewater

Authorities may be aware of some of the risks associated with the use of wastewater, but the farmers may not be as knowledgeable. In fact, Raschid-Sally and Jayakody reported significant evidence of lack of knowledge of the extent of the hazards posed by the use of wastewater by farmers in cities across Africa, Asia, and Latin America.[17] The consumers, regardless of their knowledge about the practice and the resultant health effects, are largely unaware of the exact source of the food, and as studies report, usually make their food selection based on visual quality, cost, and perceived quality.[17] Studies have shown that while consumers would want to avoid food grown using untreated wastewater, about 90% in one study were reported to be unaware of the source of the food and hence could not make choices based on potential contamination.[17] It is also reported that poor consumers are more at risk compared to the affluent because they cannot afford agricultural produce that is sold in up-market supermarkets that get their produce from uncontaminated sources.[17] Up-market supermarkets may, however, provide a false sense of security because they may source cheaper supplies from farmers using contaminated water to boost their profits, or worse still, may not have alternative suppliers or may not be aware their suppliers are using wastewater.

Despite increasing efforts by the FAO and other organizations and a growing number of individual studies and reviews, there are no comprehensive data sets that provide an understanding of wastewater agriculture and related practices across countries and cultures to date.[17] In the absence of complete data about the practice, it will be difficult to pass and enforce regulations, and most importantly, the consumers will never know and will continue to eat food that is potentially contaminated thus increasing the risk of getting sick.

Recommendations—Risk Analysis

The issue of food safety is such an important subject that cannot be left to individual scientists and media outlets alone. It must be incumbent upon governments to ensure food security without compromising public health in their countries. This can only be achieved through government-mandated risk analysis, which must involve scientists, farmers, the media, and most importantly the consumers. Risk analysis will enable nations to formulate practical policy that will result in lasting and sustainable solutions to the problem of the use of wastewater in agriculture. Regulations will be easy to enforce and monitor because everyone will be aware of the risk of not complying and, most importantly, consumers will be aware.

Risk analysis involves the quantification of risk, both in terms of probability and severity, the information that will then be used by risk managers to evaluate the economic viability of available options for alleviating the risk relative to the benefits of risk reduction.[2] The other component of risk analysis, which is particularly crucial with respect to awareness, is risk communication, which involves the dissemination of information about the risk assessment and management to make sure the general public understands the risks, policies, and options available.[2] Without risk analysis data, it will be hard to justify, let alone enforce government regulations. It essentially amounts to regulating the unknown. Facts, not speculation, must be the cornerstones of public policy. Also, without risk analysis data, governments may never feel compelled to invest in water infrastructure and water technology and world citizens will remain in grave danger.

A number of researchers have attempted to perform risk analysis, but their efforts lacked some of the key components of risk analysis and, as a result, the data may not be complete enough to influence public policy. Because of the nature of individual research, particularly in academia, it is rare to find all the elements of risk assessment, i.e., hazard identification, dose–response assessment, exposure assessment, and risk characterization, in a single published journal article. As a result, even though information may be available in the literature, it may be too fragmented to inform public policy. Media outlets in some countries have attempted to publicize the potential hazards associated with the use of wastewater in agriculture in their respective countries. While this is commendable in the absence of coherent government policy and complete data on wastewater, it may cause panic and discontent because of potential inaccuracies. The fact that many farmers in some countries choose to use wastewater instead of uncontaminated alternatives based on minor benefits betrays lack of information. In fact, Raschid-Sally and Jayakody (2008)[17] reported a clear lack of understanding of the types of infections and risks associated with the use of wastewater in agriculture among both consumers and farmers.

We cannot continue to use the excuse of poverty every time they fail to meet their obligations to their citizens. While help should be sought from developed countries in terms of technology, which, by the way, should not be free, national resources must be carefully prioritized to make sure public health is sustainably preserved. After all, some of the countries that are using untreated wastewater in agriculture are not poor by any means. International organizations can help, as they have always tried to do, but they cannot and certainly should not replace national obligations.

Conclusions

The benefits of wastewater use in agriculture are truly insignificant when compared to the risks involved. However, in the absence of complete and accessible data and sound national and international policy to the contrary, those involved will continue using this practice. Risks will be manifesting, of course, but the cause may not be fully understood early enough for corrective measures to be implemented with economic viability. The loading of contaminants in the soil over a long period will essentially turn agricultural lands into hazardous waste sites, which will need extensive remediation to make them usable for food production. Continuing the practice will affect not only the current generation but also the future generations. It is hardly sustainable and should be stopped. Because food is an international commodity, food safety must be given the global attention that it deserves, but ultimately individual nations must do everything in their power to protect their citizens.

References

1. Murtaza G, Ghafoor A, Qadir M, Owens G, Aziz MA, Zia MH, et al. Disposal and use of sewage on agricultural lands in Pakistan: a review. *Pedosphere* 2010;**20**(1):23–34.
2. Hammitt JK. Risk assessment and economic evaluation. In: Rom WN, editor. *Environmental and occupational medicine*. 4th ed. Philadelphia: Lippincott–Raven Publishers; 2006.

3. United Nations Department of Economics and Social Affairs Population Division: 24/2/2005 Press Release: http://www.un.org/News/Press/docs/2005/pop918.doc.htm [accessed 22.10.13].
4. World Population Growth: http://www.worldbank.org/depweb/english/beyond/beyondco/beg_03.pdf [accessed 22.10.13].
5. Cai X, Rosegrant MW. Global water demand and supply projections, Part 2: results and prospects to 2025. *Water International* 2002;**27**(2):170–82.
6. Judy JD, Unrine JM, Bertsch PM. Evidence for biomagnification of gold nanoparticles within a terrestrial food chain. *Environ Sci Technol* 2011;**45**(2):776–81.
7. BBC. *Who, what, why: why do people want to ban 'microbeads'?*. 2014. from: http://www.bbc.com/news/blogs-magazine-monitor-27440188 [accessed 31.05.14].
8. Steinmetz K. *States are cracking down on face wash.* 2014. from: http://time.com/#74956/states-are-cracking-down-on-face-wash/ [accessed 31.05.14].
9. Kettler K, Veltman K, van de Meent D, van Wezel A, Hendriks AJ. Cellular uptake of nanoparticles as determined by particle properties, experimental conditions, and cell type. *Environ Toxicol Chem* 2014;**33**(3):481–92.
10. Perlman H. *Saline water use in the United States.* 2014. from: http://water.usgs.gov/edu/salineuses.html [accessed 31.05.14].
11. FAO. *Water – the most basic resource but also the most essential: how much water do we use for…?.* 2014. from: http://www.fao.org/zhc/detail-events/en/c/231215/ [accessed 31.05.14].
12. Richter B. *Water policies suffer sinking feeling.* 2009. from: http://news.bbc.co.uk/2/hi/science/nature/8205132.stm [accessed 31.05.14].
13. Voeroesmarty CJ, McIntyre PB, Gessner MO, Dudgeon D, Prusevich A, Green P, et al. Global threats to human water security and river biodiversity. *Nat(Lond United Kingd)* 2010;**467**(7315):555–61.
14. *Water chapter of the OECD environmental outlook to 2050: the consequences of inaction.* http://www.oecd.org/environment/indicators-modelling-outlooks/waterchapteroftheoecdenvironmentaloutlookto2050theconsequencesofinaction.htm.
15. Chartres C. *Water everywhere and not a drop to grow.* 2008. from: http://news.bbc.co.uk/2/hi/science/nature/7564140.stm [accessed 31.05.14].
16. Tongesayi T, Dasilva P, Dilger K, Hollingsworth T, Mooney M. Speciation and cysteine-simplified physiological-based extraction technique (SBET) bioaccesibility of heavy metals in biosolids. *J Environ Sci Health Part A* 2011;**46**(10):1138–46.
17. Raschid-Sally L, Jayakody P. *Drivers and characteristics of wastewater agriculture in developing countries: results from a global assessment.* IWMI Research Report 127. 2008. from: http://www.iwmi.cgiar.org/Publications/IWMI_Research_Reports/PDF/PUB127/RR127.pdf [accessed 31.05.14].
18. Bansal OP. Heavy metals in soil, plants as influenced by irrigation with sewage effluents: a 20 year study. *J Indian Assoc Environ Manag* 2008;**35**(3):143–8.
19. Mueller B, Scheytt T, Asbrand M, de Casas AM. Pharmaceuticals as indictors of sewage-influenced groundwater. *Hydrogeol J* 2012;**20**(6):1117–29.
20. Focazio MJ, Kolpin DW, Barnes KK, Furlong ET, Meyer MT, Zaugg SD, et al. A national reconnaissance for pharmaceuticals and other organic wastewater contaminants in the United States – (II) untreated drinking water sources. *Sci Total Environ* 2008;**402**(2–3):201–16.
21. EPA. *Priority pollutants.* 2013. from: http://water.epa.gov/scitech/methods/cwa/pollutants.cfm [accessed 31.05.14].
22. Robinson BH. E-waste: an assessment of global production and environmental impacts. *Sci Total Environ* 2009;**408**(2):183–91.

23. Fragou D, Fragou A, Kouidou S, Njau S, Kovatsi L. Epigenetic mechanisms in metal toxicity. *Toxicol Mech Methods* 2011;**21**:343–52.
24. Ren X, McHale CM, Skibola CF, Smith AH, Smith MT, Zhang L. An emerging role for epigenetic dysregulation in arsenic toxicity and carcinogenesis. *Environ Health Perspect* 2011;**119**(1):11–9.
25. Cronican AA, Fitz NF, Carter A, Saleem M, Shiva S, Barchowsky A, et al. Genome-wide alteration of histone H3K9 acetylation pattern in mouse offspring prenatally exposed to arsenic. *PLoS One* 2013; **8**(2):e53478.
26. Farzan SF, Karagas MR, Chen Y. In utero and early life arsenic exposure in relation to long-term health and disease. *Toxicol Appl Pharmacol* 2013;**272**(2):384–90.
27. Fei DL, Koestler DC, Li Z, Giambelli C, Sanchez-Mejias A, Gosse JA, et al. Association between in utero arsenic exposure, placental gene expression, and infant birth weight: a US birth cohort study. *Environ Health Global Access Sci Source* 2013;**12**:58.
28. Moon K, Guallar E, Navas-Acien A. Arsenic exposure and cardiovascular disease: an updated systematic review. *Curr Atheroscler Rep* 2012;**14**(6):542–55.
29. Liaw J, Marshall G, Yuan Y, Ferreccio C, Steinmaus C, Smith AH. Increased childhood liver cancer mortality and arsenic in drinking water in Northern Chile. *Cancer Epidemiol Biom Prev* 2008;**17**(8): 1982–7.
30. Yorifuji T, Tsuda T, Doi H, Grandjean P. Cancer excess after arsenic exposure from contaminated milk powder. *Environ Health Prev Med* 2011;**16**(3):164–70.
31. Pilsner JR, Hall MN, Liu X, Ilievski V, Slavkovich V, Factor-Litvak P, et al. Influence of prenatal arsenic exposure and newborn sex on global methylation of cord blood DNA. *PLoS One* 2012;**7**: e37147.
32. National Institute of Environmental Health Sciences: Health Homes = health kids. http://www.niehs.nih.gov/health/assets/docs_f_o/health-kids-water.pdf [accessed 22.10.13].
33. Ciesielski T, Weuve J, Bellinger DC, Schwartz J, Lanphear B, Wright RO. Cadmium exposure and neurodevelopmental outcomes in U.S. children. *Environ Health Perspect* 2012;**120**:758–63.
34. Riederer AM, Belova A, George BJ, Anastas PT. Urinary cadmium in the 1999–2008 U.S. national health and nutrition examination survey (NHANES). *Environ Sci Technol* 2013;**47**(2):1137–47.
35. Satarug S, Moore MR. Emerging roles of cadmium and heme oxygenase in type-2 diabetes and cancer susceptibility. *Tohoku J Exp Med* 2012;**228**(4):267–88.
36. Ciesielski T, Bellinger DC, Schwartz J, Hauser R, Wright RO. Associations between cadmium exposure and neurocognitive test scores in a cross-sectional study of US adults. *Environ Health (Lond United Kingd)* 2013;**12**:13.
37. Cory-Slechta DA, Virgolini MB, Rossi-George A, Thiruchelvam M, Lisek R, Weston D. Lifetime consequences of combined maternal lead and stress. *Basic Clin Pharmacol Toxicol* 2008;**102**(2): 218–27.
38. Technology Planning and Management Corporation. *Report on carcinogens background document for lead and lead compounds*. Contract Number N01-ES-85421. Durham (NC): U.S. Department of Health and Human Services; May 8, 2003. http://ntp.niehs.nih.gov/ntp/newhomeroc/roc11/Lead-Public.pdf#search=lead [accessed October 2013].
39. EPA. *Mercury health effects*. 2014. from: http://www.epa.gov/mercury/effects.htm [accessed 31. 05.14].
40. EPA. *Mercury*. 2014. from: http://www2.epa.gov/region8/mercury [accessed 31.05.14].
41. Agency for Toxic Substances and Disease Registry (ATSDR). *Toxicological profile for chlorinated dibenzo-p-dioxins (CDDs)*. Atlanta (GA): U.S. Department of Health and Human Services, Public Health Service; 1998. from: http://www.atsdr.cdc.gov/toxfaqs/tf.asp?id=363&tid=63 [accessed 31. 05.14].

42. Agency for Toxic Substances and Disease Registry (ATSDR). *Toxicological profile for polychlorinated biphenyls (PCBs)*. Atlanta (GA): U.S. Department of Health and Human Services, Public Health Service; 2000. from: http://www.atsdr.cdc.gov/toxfaqs/tf.asp?id=140&tid=26 [accessed 31.05.14].
43. Agency for Toxic Substances and Disease Registry (ATSDR). *Toxicological profile for polycyclic aromatic hydrocarbons (PAHs)*. Atlanta (GA): U.S. Department of Health and Human Services, Public Health Service; 1995. from: http://www.atsdr.cdc.gov/PHS/PHS.asp?id=120&tid=25 [accessed 31.05.14].
44. Agency for Toxic Substances and Disease Registry (ATSDR). *Toxicological profile for Di-*n*-butyl phthalate*. Update. Atlanta (GA): U.S. Department of Health and Human Services, Public Health Service; 2001. from: http://www.atsdr.cdc.gov/toxfaqs/tf.asp?id=858&tid=167 [accessed 31.05.14].
45. Agency for Toxic Substances and Disease Registry (ATSDR). *Toxicological profile for Di(2-ethylhexyl)phthalate (DEHP)*. Atlanta, GA: U.S. Department of Health and Human Services, Public Health Service; 2002. from: http://www.atsdr.cdc.gov/phs/phs.asp?id=376&tid=65 [accessed 31.05.14].
46. Agency for Toxic Substances and Disease Registry, Division of Community Health Investigations. *Possible health risks from exposure to TCE & PCE*. from: http://www.atsdr.cdc.gov/HAC/pha/NavyYardMillsSite/FINALNavyYardMillsfactsheet508.pdf [accessed 31.05.14].
47. Bergamaschi A, Iavicoli I, Savolainen K. Exposure assessment. In: Fadeel B, Pietroiusti A, Shvedova AA, editors. *Adverse effects of engineered nanomaterials*; 2012. p. 25–43.
48. Porredon C, de Lapuente J, Garcia JP, Sendra J, Castano A, Egizabal A, et al. Toxicology review in metal nanoparticles: approximation in gold and cobalt ferrite nanoparticles. *Adv Sci Lett* 2012;**6**: 1–16.
49. Crockett CS. *Cross-cutting drinking water wastewater issues*. 2010. from: http://www.slideshare.net/crockettc/emerging-pathogens-in-wastewater [accessed 31.05.14].
50. Lee YH, Chen Y, Ouyang X, Gan YH. Identification of tomato plant as a novel host model for *Burkholderia pseudomallei*. *BMC Microbiol* 2010;**10**(28). http://dx.doi.org/10.1186/1471-2180-10-28.
51. Gerardi MH, Zimmerman MC. *Rehydration project: focus on diarrhoea, dehydration and rehydration*. 2014. from: http://rehydrate.org/diarrhoea/#Health%20hazards [accessed 31.05.14].
52. FAO (n. d.). *Urban and peri-urban agriculture in Latin America and the Caribbean*. from: http://www.fao.org/ag/agp/greenercities/index_en.html [accessed 31.05.14].
53. National Institute of Environmental Health Sciences. *Child development and environmental toxins*. https://www.niehs.nih.gov/health/assets/docs_a_e/child_development_and_environmental_toxins_508.pdf [accessed 22.10.13].
54. Wilson R, Crouch EAC. *Risk-benefit analysis*. Harvard University Press; 2001.
55. Islam MR, Brammer H, Mustafizur Rahman GKM, Raab A, Jahiruddin M, Solaiman ARM, et al. Arsenic in rice grown in low-arsenic environments in Bangladesh. *Water Qual Expo Health* 2012; **4**(4):197–208.
56. Bhattacharya P, Samal AC, Majumdar J, Banerjee S, Santra SC. In vitro assessment on the impact of soil arsenic in the eight rice varieties of West Bengal, India. *J Hazard Mater* 2013;**262**:1091–7.
57. Bhatti SM, Anderson CWN, Stewart RB, Robinson BH. Risk assessment of vegetables irrigated with arsenic-contaminated water. *Environ Sci Process Impacts* 2013;**15**(10):1866–75.
58. Duan G, Liu W, Chen X, Hu Y, Zhu Y. Association of arsenic with nutrient elements in rice plants. *Metallomics* 2013;**5**(7):784–92.
59. Lente I, Keraita B, Drechsel P, Ofosu-Anim J, Brimah AK. Risk assessment of heavy-metal contamination on vegetables grown in long-term wastewater irrigated urban farming sites in Accra, Ghana. *Water Qual Expo Health* 2012;**4**(4):179–86.

60. Badawy RK, Abd El-Gawadand AM, Osman HE. Health risks assessment of heavy metals and microbial contamination in water, soils and agricultural foodstuff from wastewater irrigation at Sahl El-Hessania area, Egypt. *J Appl Sci Res* 2013;**9**(4):3091–107.

61. Shuravilin AV, Chernukha NI, Saad GB. The influence of long-term irrigation by urban sewage water on pollution of soils with heavy metals in Egypt. *Dokl Ross Akad Sel'skokhozyaistvennykh Nauk* 2010;**6**:42–3.

62. Gebrekidan A, Weldegebriel Y, Hadera A, Van der Bruggen B. Toxicological assessment of heavy metals accumulated in vegetables and fruits grown in Ginfel river near Sheba Tannery, Tigray, Northern Ethiopia. *Ecotoxicol Environ Saf* 2013;**95**:171–8.

63. Adamu H, Uzairu A, Harrison GFS. Assessment of trace metals in sewage water and sludge from River Kubanni drainage basin. *Afr J Biotechnol* 2013;**12**(1):49–55.

64. Muchuweti M, Birkett JW, Chinyanga E, Zvauya R, Scrimshaw MD, Lester JN. Heavy metal content of vegetables irrigated with mixtures of wastewater and sewage sludge in Zimbabwe: implications for human health. *Agric Ecosyst Environ* 2006;**112**:41–8.

65. Tandi NK, Nyamangara J, Bangira C. Environmental and potential health effects of growing leafy vegetables on soil irrigated using sewage sludge and effluent: a case of Zn and Cu. *J Environ Sci Health Part B Pesticides Food Contam Agric Wastes* 2004;**B39**(3):461–71.

66. Khan MA, Islam MR, Panaullah GM, Duxbury JM, Jahiruddin M, Loeppert RH. Fate of irrigation-water arsenic in rice soils of Bangladesh. *Plant Soil* 2009;**322**(1–2):263–77.

67. Ahmed FRS, Alexander IJ, Mwinyihija M, Killham K. Effect of arsenic contaminated irrigation water on *Lens culinaris* L. and toxicity assessment using lux marked biosensor. *J Environ Sci (Beijing, China)* 2012;**24**(6):1106–16.

68. Ullah H, Khan I, Ullah I. Impact of sewage contaminated water on soil, vegetables, and underground water of peri-urban Peshawar, Pakistan. *Environ Monit Assess* 2012;**184**(10):6411–21.

69. Khan MU, Malik RN, Said M. Human health risk from heavy metal via food crops consumption with wastewater irrigation practices in Pakistan. *Chemosphere* 2013;**93**(10):2230–8.

70. Al Omron AM, El-Maghraby SE, Nadeem MEA, El-Eter AM, Al-Mohani H. Long term effect of irrigation with the treated sewage effluent on some soil properties of Al-Hassa Governorate, Saudi Arabia. *J Saudi Soc Agric Sci* 2012;**11**(1):15–8.

71. Mahmood A, Malik RN. Human health risk assessment of heavy metals via consumption of contaminated vegetables collected from different irrigation sources in Lahore, Pakistan. *Arabian J Chem* 2014;**7**(1):91–9.

72. Shaheen Q, Mahmood Z, Imran M, Gillani SR, Faridi IA. Mercury concentration in vegetables of Pakistan irrigated by different water sources. *J Chem Soc Pak* 2011;**33**(4):490–4.

73. Ahmad K, Ejaz A, Azam M, Khan ZI, Ashraf M, Al-Qurainy F, et al. Lead, cadmium and chromium contents of canola irrigated with sewage water. *Pak J Botany* 2011;**43**(2):1403–10.

74. Huang G-X, Chen Z-Y, Sun J-C, Liu J-T, Zhang Y-X, Wang J-C. Chemical fractions and potential mobility of lead in soils irrigated by sewage in Pearl River Delta, South China. *J Central South Univ* 2012;**19**:2620–6.

75. Wang Y, Zhang P, Chen Y. Analysis on content of heavy metal in farmland soil irrigated by sewage from iron-steel factory. *Anhui Nongye Kexue* 2012;**40**:8512–4.

76. Wang X, Nan Z, Liao Q, Ding W, Wu W. Fractions and bioavailability of cadmium and nickel to carrot crops in oasis soil. *Pol J Environ Stud* 2012;**21**(6):1867–74.

77. Chen ZS. Metal contamination of flooded soils, rice plants, and surface waters in Asia. *Biogeochem Trace Met* 1992:85–108.

78. Huang Z-Y, Chen T, Yu J, Qin D-P, Chen L. Lead contamination and its potential sources in vegetables and soils of Fujian, China. *Environ Geochem Health* 2012;**34**(1):55–65.

79. Li JX, Yang XE, He ZL, Jilani G, Sun CY, Chen SM. Fractionation of lead in paddy soils and its bioavailability to rice plants. *Geoderma* 2007;**141**:174–80.
80. Hseu Z-Y, Su S-W, Lai H-Y, Guo H-Y, Chen T-C, Chen Z-S. Remediation techniques and heavy metal uptake by different rice varieties in metal-contaminated soils of Taiwan: new aspects for food safety regulation and sustainable agriculture. *Soil Sci Plant Nutrition* 2010;**56**:31–52.
81. Lai H-Y, Hseu Z-Y, Chen T-C, Chen B-C, Guo H-Y, Chen Z-S. Health risk-based assessment and management of heavy metals-contaminated soil sites in Taiwan. *Int J Environ Res Public Health* 2010;**7**:3595–614.
82. Chen Z, Wang G, Xu S, Zhang M. Analysis on amount of lead in rice planted in sewage area near a paper mill in Yinchuan. *Ningxia Gongcheng Jishu* 2004;**3**(3):285–6.
83. Wu G-H, Cao S-S. Mercury and cadmium contamination of irrigation water, sediment, soil and shallow groundwater in a wastewater-irrigated field in Tianjin, China. *Bull Environ Contam Toxicol* 2010;**84**(3):336–41.
84. Xue Z-J, Liu S-Q, Liu Y-L, Yan Y-L. Health risk assessment of heavy metals for edible parts of vegetables grown in sewage-irrigated soils in suburbs of Baoding City, China. *Environ Monit Assess* 2012;**184**(6):3503–13.
85. Tiwari KK, Singh NK, Patel MP, Tiwari MR, Rai UN. Metal contamination of soil and translocation in vegetables growing under industrial wastewater irrigated agricultural field of Vadodara, Gujarat, India. *Ecotoxicol Environ Saf* 2011;**74**:1670–7.
86. Rana L, Dhankhar R, Chhikara S. Soil characteristics affected by long term application of sewage wastewater. *Int J Environ Res* 2010;**4**:513–8.
87. Sharma RK, Agrawal M, Marshall F. Heavy metal contamination of soil and vegetables in suburban areas of Varanasi, India. *Ecotoxicol Environ Saf* 2007;**66**:258–66.
88. Patel KP, Singh MV, Ramani VP, Patel KC, George V, Zizala VJ. Impact and effect of sewage water on soils and crops in Peri urban areas of Gujarat. *Pollut Res* 2006;**25**:25–30.
89. Khurana MPS, Singh M, Nayyar VK. Assessment of heavy metal contamination in soils and plants irrigated with sewage waters containing industrial effluents in district Amritsar, Punjab, India. *Indian J Environ Ecoplanning* 2004;**8**:221–8.
90. Karthikeyan K, Singh K. Chemical composition of some sewage water used for irrigation purpose. *J Industrial Pollut Control* 2004;**20**:211–9.
91. Madhavi A, Rao AP, Ramavatharam N. Effect of irrigation with groundwater adjoining the effluent streams on essential plant nutrient and pollutant element status in a crop growing area of Hyderabad, India. *Pollut Res* 2004;**23**:433–7.
92. Patel KP, Singh MV, Ramani VP, Patel KC, George V, Zizala VJ. Impact and effect of sewage water on soils and crops in Peri urban areas of Gujarat. *Pollut Res* 2006;**25**(1):25–30.
93. Kamal N, Gupta N, Joshi HC. Impact of use of sewage irrigation on heavy metal accumulation in agricultural field of Keshopur area of national capital territory Delhi. *Environ Ecol* 2010;**28**(1B): 629–34.
94. Sherene T. Heavy metal contamination seen in groundwater of Coimbatore district, Tamil Nadu. *India J Ecotoxicol Environ Monit* 2009;**19**:397–400.
95. Prabhavathi M, Reddy KS, Indirani R. Effect of municipal waste on quality of irrigation water in Tirupati, Andhra Pradesh, India. *Indian J Environ Ecoplanning* 2010;**17**:343–50.
96. Singh J, Upadhyay SK, Pathak RK, Gupta V. Accumulation of heavy metals in soil and paddy crop (*Oryza sativa*), irrigated with water of Ramgarh Lake, Gorakhpur, UP, India. *Toxicol Environ Chem* 2011;**93**:462–73.

97. Raj GB, Patnaik MC, Babu PS, Kalakumar B, Singh MV, Shylaja J. Heavy metal contaminants in water-soil-plant-animal continuum due to pollution of Musi River around Hyderabad in India. *Indian J Animal Sci* 2006;**76**:131–3.

98. Rani KU, Sharma KL, Nagasri K, Chandrika DS, Savithri VL, Lal M. Effect of long-term sewage water irrigation on micronutrient and heavy metal content in soil and plants under Musi River Basin in Hyderabad. *J Industrial Pollut Control* 2013;**29**(2):227–42.

99. Aoshima K. Itai-itai disease: cadmium-induced renal tubular osteomalacia. Current situations and future perspectives. *Nippon Eiseigaku Zasshi* 2012;**67**(4):455–63.

100. Judy JD, Unrine JM, Rao W, Wirick S, Bertsch PM. Bioavailability of gold nanomaterials to plants: importance of particle size and surface coating. *Environ Sci Technol* 2012;**46**(15):8467–74.

101. Zhu ZJ, Wang H, Yan B, Zheng H, Jiang Y, Miranda OR, et al. Effect of surface charge on the uptake and distribution of gold nanoparticles in four plant species. *Environ Sci Technol* 2012;**46**(22): 12391–8.

102. Koelmel J, Leland T, Wang H, Amarasiriwardena D, Xing B. Investigation of gold nanoparticles uptake and their tissue level distribution in rice plants by laser ablation-inductively coupled-mass spectrometry. *Environ Pollut* 2013;**174**:222–8.

103. Larue C, Veronesi G, Flank AM, Surble S, Herlin-Boime N, Carriere M. Comparative uptake and impact of TiO_2 nanoparticles in wheat and rapeseed. *J Toxicol Environ Health Part A Curr Issues* 2012;**75**(13–15):722–34.

104. Pokhrel LR, Dubey B. Evaluation of developmental responses of two crop plants exposed to silver and zinc oxide nanoparticles. *Sci Total Environ* 2013;**452–453**:321–32.

105. Zhang P, Ma Y, Zhang Z, He X, Guo Z, Tai R, et al. Comparative toxicity of nanoparticulate/bulk Yb_2O_3 and $YbCl_3$ to cucumber (*Cucumis sativus*). *Environ Sci Technol* 2012;**46**(3):1834–41.

106. Zhang P, Ma Y, Zhang Z, He X, Zhang J, Guo Z, et al. Biotransformation of Ceria nanoparticles in cucumber plants. *ACS Nano* 2012;**6**(11):9943–50.

107. Gardea-Torresdey JL, Rico CM, White JC. Trophic transfer, transformation, and impact of engineered nanomaterials in terrestrial environments. *Environ Sci Technol* 2014;**48**(5):2526–40.

108. Pakrashi S, Dalai S, Chandrasekaran N, Mukherjee A. Trophic transfer potential of aluminium oxide nanoparticles using representative primary producer (*Chlorella ellipsoides*) and a primary consumer (*Ceriodaphnia dubia*). *Aquat Toxicol Amst Neth* 2014;**152C**:74–81.

109. Joubert Y, Pan J-F, Buffet P-E, Pilet P, Gilliland D, Valsami-Jones E, et al. Subcellular localization of gold nanoparticles in the estuarine bivalve *Scrobicularia plana* after exposure through the water. *Gold Bull Berl Ger* 2013;**46**(1):47–56.

110. Garcia-Reyero N, Kennedy AJ, Escalon BL, Habib T, Laird JG, Rawat A, et al. Differential effects and potential adverse outcomes of ionic silver and silver nanoparticles in vivo and in vitro. *Environ Sci Technol* 2014;**48**(8):4546–55.

111. Englund SE, Johansson F, Klefbohm B, Johansson, Ulrica E. Gold- and silver nanoparticles affect the growth characteristics of human embryonic neural precursor cells. *PLoS One* 2013;**8**(3):e58211.

112. Hanna SK, Miller RJ, Zhou D, Keller AA, Lenihan HS. Accumulation and toxicity of metal oxide nanoparticles in a soft-sediment estuarine amphipod. *Aquat Toxicol* 2013;**142–143**:441–6.

113. Ensink JHJ, Blumenthal UJ, Brooker S. Wastewater quality and the risk of intestinal nematode infection in sewage farming families in Hyderabad, India. *Am J Tropical Med Hyg* 2008;**79**(4):561–7.

114. Ensink JHJ, van der Hoek W, Mukhtar M, Tahir Z, Amerasinghe FP. High risk of hookworm infection among wastewater farmers in Pakistan. *Transact R Soc Tropical Med Hyg* 2005;**99**(11):809–18.

115. Melloul A, Amahmid O, Hassani L, Bouhoum K. Health effect of human wastes use in agriculture in El Azzouzia (the wastewater spreading area of Marrakesh city, Morocco). *Int J Environ Health Res* 2002;**12**(1):17–23.

116. Said EK, Mostafa AE. Prevalence of helminths in a rural population using wastewater for agricultural purposes at Settat (Morocco). *Sante (Montrouge, France)* 2006;**16**(4):245–51.

117. Srikanth R, Naik D. Health effects of wastewater reuse for agriculture in the suburbs of Asmara city, Eritrea. *Int J Occup Environ Health* 2004;**10**(3):284–8.

118. Cifuentes E, Blumenthal U, Ruiz-Palacios G, Bennett S, Quigley M. Health risk in agricultural villages practicing wastewater irrigation in central Mexico: perspectives for protection. *Schriftenr Vereins fur Wasser- Boden- Lufthyg* 2000;**105**:249–56.

119. International Water Management Institute (IWMI). *Wastewater use in irrigated agriculture: confronting the livelihood and environmental realities.* 2004. from: http://www.google.com/url?sa=t&rct=j&q=&esrc=s&source=web&cd=2&ved=0CCoQFjAB&url=http%3A%2F%2Fwww.ais.unwater.org%2Fais%2Fpluginfile.php%2F225%2Fmod_label%2Fintro%2Fiwmi_4.pdf&;ei=pfCHU7DqC_DLsASy6oD4Aw&usg=AFQjCNG1mHha5YLKnLE5pFLc9_OFTxEbuA&bvm=bv.67720277,d.cWc [accessed 28.05.14].

120. Wang Y, Qiao M, Liu Y, Zhu Y. Health risk assessment of heavy metals in soils and vegetables from wastewater irrigated area, Beijing-Tianjin city cluster, China. *J Environ Sci (Beijing, China)* 2012;**24**(4):690–8.

121. Lottermoser BG. Effect of long-term irrigation with sewage effluent on the metal content of soils, Berlin, Germany. *Environ Geochem Health* 2012;**34**(1):67–76.

14

Analysis of Dietary Supplements with a Hand-held XRF Analyzer

Alexandra E. Schroeder, Zachary R. Smith, Mark A. Benvenuto, Elizabeth S. Roberts-Kirchhoff

DEPARTMENT OF CHEMISTRY AND BIOCHEMISTRY, UNIVERSITY OF DETROIT MERCY, DETROIT, MI, USA

CHAPTER OUTLINE

Introduction

Dietary supplements are products taken by mouth that contain a dietary ingredient. These ingredients may be vitamins, minerals, and other substances. These supplements are sold on the market in many forms, commonly as pills or tablets. Dietary supplements can be purchased at many places including health food stores, the Internet, and even dollar stores. In 2007, over 50% of the adults in the United States consumed some form of dietary supplement.[1] The sale of dietary supplements including vitamins is a growing billion-dollar industry. Sales of these dietary supplements was nearly 24 billion US dollars in 2007.[1] Many dietary supplements claim to have various positive effects on the body. Although they may state that the supplement has beneficial effects, the manufacturer cannot legally market the supplement as a cure or treatment for specific diseases. Dietary supplements are regulated by the Food and Drug Administration (FDA).[2,3] One requirement is that the supplement must be labeled as a "dietary supplement." The manufacturer of the supplement does not need the approval of the FDA for the accuracy of the claims made by the manufacturer. It is the responsibility of the FDA to monitor dietary supplements and the reports of serious side effects produced by the supplements.[2]

Food, Energy, and Water. http://dx.doi.org/10.1016/B978-0-12-800211-7.00014-4

A common dietary supplement is calcium, a mineral naturally found in many food products. Calcium is important in muscle movement, maintaining strong bones, and movement of blood through blood vessels. Without calcium, the human body would not be able to function. It is recommended that an adult between the ages of 19 and 50 years should ingest 1000 mg/day. Calcium is found naturally in many foods such as dairy products and most grains.[4] Zinc, another dietary supplement, is found in cells in the body and is known for fighting off viruses through the immune system and plays a role in the senses, taste and smell. Zinc is required for the activity of more than 300 biological enzymes by directly participating in catalysis or helping to maintain protein structure. Adult men should ingest 11 mg/day and women should ingest 8 mg/day. Zinc may be taken naturally in the diet by eating red meat, poultry, nuts, and whole grains.[5, 6] Iron, a very common metal, is also taken as a dietary supplement. Iron plays an important role in the human body by regulating cell growth and differentiation and is mostly found in the protein in red blood cells, hemoglobin, which delivers oxygen to tissues. Iron is found in red meats, lentils, and soybeans. Adult aged between 19 and 50 years should consume 8 mg/day or 18 mg/day, for men and women, respectively.[7] Potassium is commonly consumed as a dietary supplement. Potassium is responsible in regulating blood pressure, muscle contractions, and nerve impulses. It is recommended that an adult consumes 4700 mg/day. Potassium can be found naturally in potatoes, yogurt, and bananas.[8]

X-ray fluorescence (XRF) spectroscopy deals with the emission of inner shell electrons when a material is bombarded with high-energy X-rays. Radiation from the X-ray source ejects electrons from a particular atom's first and second inner orbitals. The empty orbitals then fill with electrons from higher orbital shells. When these electrons enter the lower state, characteristic XRF is emitted. Elements from atomic number 11 to 92 within a sample each have unique emission signatures; making it possible to quantify the presence of an element by the strength of the emission.[9] Portable or hand-held XRF analyzers can be used to scan quickly a variety of samples for a number of different elements. The Bruker S1 Turbo hand-held XRF analyzer used in this study is able to determine the elements in a sample in 30 s, and has sensitivity that matches any of the available nonportable XRF models, which is especially important when examining the lighter elements often found in food supplements, such as potassium and calcium. The analysis of major and trace elements in geological, metallic, ceramic, food and food supplements, and other materials is made possible by the interaction of atoms with this radiation.[10] XRF spectrometry including the use of a hand-held XRF has been used to analyze food and dietary supplements.[11–22] Convenient methods are needed to monitor food contaminants to aid in tracking the sources of contamination.

Experimental Methods

Standard reference materials (SRMs) were obtained from the National Institute of Standards and Technology (NIST). The SRMs analyzed were SRM 2586 and SRM 2587. SRM 2586 is the trace elements in soil standard containing lead from paint (Nominal

500 mg lead/kg).[23] SRM 2587 is the trace elements in soil standard containing lead from paint (Nominal 3000 mg lead/kg) standard.[24] Both of these have reported certified mass fractions for arsenic, cadmium, chromium, and lead. They also have reported reference mass fractions for many elements including calcium, iron, manganese, potassium, titanium, and zinc.[23,24] A Bruker S1 TURBO hand-held XRF analyzer with a silicon drift detector and a resolution of approximately 145 eV at 200,000 cps was used to analyze the dietary supplement samples and SRMs as previously described.[11] For each SRM, three separate samples (0.80 g/sample) were analyzed three times each for 30 s. To ensure each sample was analyzed from an equal distance, the sample cup was placed on the safety platform that would mount to the tip of the instrument. The instrument was mounted on the bench top stand. The instrument used a voltage of 45 kV and a current of 30 μA, with a Ti/Al filter. A Hewlett Packard PDA with XBruker Elemental S1 software was used to control the instrument. The soil FP calibration method in the XBruker Elemental S1 software was used to analyze the samples. This calibration has been optimized for a SiO_2 matrix, is nonnormalized, and relies on using a repeatable distance between the sample and detector. Data files were imported into Microsoft Excel for statistical analysis and results were compared to the SRM reported results with an unpaired *t*-test using GraphPad software. The SRMs, used as provided, were placed into Chemplex Spectrocertified® Quality XRF microsample cups (31.0 × 22.4 mm) and were covered with Chemplex Spectromembrane® perforated thin film mylar polyester sample support carrier films (3.6 μm) with a sealing ring. The concentrations (averages and standard deviations) of various elements for each of the standards were determined and compared to the NIST standard values.[23,24] The limit of detection (LOD) and limit of quantitation (LOQ) for each element analyzed with the soil FP calibration were provided by Bruker.[25] These are shown in Table 1. The LOQ was defined as five times the LOD for this specific method.[11]

Nine dietary supplements obtained from three different dollar stores in the Detroit area were analyzed with a Bruker S1 TURBO hand-held XRF instrument as described for the SRMs. Each of the nine dietary supplement samples were measured out to five equal samples (0.80 g/sample) and analyzed for 30 s. The samples were taken from tablets that were ground using a mortar and pestle. The concentration of selected elements was determined for each sample and compared to values reported by the manufacturer on the product label.

Table 1 Detection Limits for Elements with the S1 Turbo Hand-held XRF Analyzer Using the Soil FP Calibration Method[25]

	K	Ca	Ti	Mn	Fe	Ce	Zn	Pb
LOD[a] (ppm)	450	200	90	18	17	180	3	7
LOQ[b] (ppm)	2250	1000	450	90	85	900	15	35

[a]LOD, Limit of Detection.[b]LOQ, Limit of Quantitation.

Results and Discussion

NIST standards were used to determine the accuracy of the soil FP calibration method on the Bruker S1 TURBO hand-held XRF spectrometer. The values (average ± standard deviation) acquired with the XRF are given as ppm values and also include a percent relative standard deviation (in parentheses) as seen in Tables 2 and 3. The calculated averages were compared to the reference values for two soil SRMs to determine a percent error. SRM 2586 is the trace elements in soil standard containing lead from paint (Nominal 500 mg lead/kg).[23] SRM 2587 is the trace elements in soil standard containing lead from paint (Nominal 3000 mg lead/kg) standard.[24] The percent errors for results from the XRF as compared to the standards were less than 23% for potassium and calcium (Table 2). The values obtained with the hand-held XRF are statistically, significantly different from the SRM values as evidenced by the P values. There was no statistically significant difference between the titanium values measured with the XRF as compared to the reference materials (Table 2). In addition, there were no statistically significant differences between the values determined with the hand-held XRF for manganese, iron, zinc, or lead as compared to the reported values (Table 3). These error values are largest for the lightest elements, potassium and calcium, simply because low X-ray yields for lighter elements tend to produce results that are more difficult for the instrument to quantify.

Nine dietary supplements (Table 4) purchased at three different dollar stores in the Detroit metropolitan area were analyzed with a hand-held XRF instrument using the soil FP calibration method. All of the dietary supplements advertised a specific amount per tablet for each element. The dietary supplements were designated by a sample ID. The results from the XRF analysis of elements that are typically found within dietary supplements are found in Table 5. From the XRF analysis, each of the supplements contained the minerals reported on the label. In addition, many of the supplements contained additional elements as determined by XRF analysis. The products V1, V32, and V35 were all calcium supplements and contained calcium and a smaller amount of potassium. None of the labels for these listed any sources of potassium. The product V16, which is a potassium supplement contained more calcium than potassium by XRF

Table 2 Concentrations of Potassium, Calcium, and Titanium from Analysis with the Hand-held XRF Compared to the NIST Reported Values for the SRMs[a]

	K (ppm)	% error[b]	Ca (ppm)	% error	Ti	% error
2586 XRF[c]	8037 ± 369 (4.6)[d]	17.7 ($P = 0.01$)[e]	17278 ± 533 (3.1)	22.1 ($P = 0.004$)	5714 ± 241 (4.2)	5.55
2586 SRM	9760 ± 180 (1.8)		22180 ± 540 (2.4)		6050 ± 660 (10.9)	
2587 XRF	13567 ± 689 (5.1)	14.3 ($P = 0.01$)	7978 ± 354 (4.4)	17.9 ($P = 0.018$)	3939 ± 180 (4.6)	0.48
2587 SRM	15830 ± 550 (3.5)		9720 ± 200 (2.1)		3920 ± 650 (16.6)	

[a]All values for SRM 2587 and 2586 are reference values.[23,24] [b]Percent error between measured XRF value and NIST reported value. [c]Values (average ± standard deviation) for XRF from analysis with hand-held instrument. [d]Percent relative standard deviation. [e]Statistical analysis using unpaired t-test.[23,24]

Table 3 Concentrations of Manganese, Iron, Zinc, and Lead from Analysis with the Hand-held XRF Compared to the NIST Reported Values for the SRMs[a]

	Mn (ppm)	% error[b]	Fe (ppm)	% error	Zn (ppm)	% error	Pb (ppm)	% error
2586 XRF[c]	908 ± 47 (5.2)[d]	9.20	52111 ± 1510 (2.9)	0.96	379 ± 45 (11.9)	7.12	424 ± 24 (5.7)[d]	1.85
2586 SRM	1000 ± 18 (1.8)		51610 ± 890 (1.7)		352 ± 16 (4.5)		432 ± 17 (3.9)	
2587 XRF	602 ± 22 (3.7)	7.53	29322 ± 1021 (3.5)	4.07	432 ± 60 (13.4)	22.22	3621 ± 255 (5.7)	10.4
2587 SRM	651 ± 23 (3.5)		28130 ± 250 (0.9)		336 ± 8 (2.4)		3242 ± 57 (1.8)	

[a]All values for SRM 2587 and 2586 are reference values except those for lead, which are certified values.[23,24] [b]Percent error between measured XRF value and NIST reported value. [c]Values (average ± standard deviation) for XRF from analysis with hand-held instrument. [d]Percent relative standard deviation.

Table 4 Specifics of Mineral Supplements

Sample ID	Supplement Name	Brand	Supplemental Dosage (mg)[a]	Source[b]
V1	Calcium with D3	Healthy Sense®	600	Calcium carbonate
V16	Potassium	Health Care	99	Potassium gluconate potassium chloride
V18	Zinc	Health Care	50	Zinc gluconate
V28	Iron	Health Care	65	Ferrous sulfate
V32	Oyster calcium	Health Care	500	Calcium carbonate from oyster shell
V33	Calcium, magnesium, and zinc	Family Wellness™	333, 133, and 5	Calcium carbonate magnesium oxide zinc sulfate
V34	Iron	Family Wellness™	65	Ferrous sulfate
V35	Calcium 600 + D	Family Wellness™	600	Calcium carbonate
V37	Zinc	Family Wellness™	50	Zinc gluconate

[a]per tablet. [b]as listed on product label.

Table 5 Mean Concentrations (ppm) ± Standard Deviation of Elements in Mineral Supplements

Sample ID	K	Ca	Ti	Mn	Fe	Zn	Pb
V1	401 ± 18	48600 ± 256	ND[a]	ND	ND	ND	ND
V16	4395 ± 386	10630 ± 313	ND	ND	ND	ND	ND
V18	175 ± 30	16360 ± 164	120 ± 41	ND	ND	12133 ± 690	ND
V28	ND	ND	ND	130 ± 12	24010 ± 2640	120 ± 26	ND
V32	364 ± 26	43370 ± 471	ND	ND	ND	ND	ND
V33	166 ± 19	22270 ± 107	ND	ND	ND	496 ± 86	ND
V34	ND	ND	88.6 ± 39	129 ± 16	25710 ± 3320	133 ± 29	11.9 ± 2
V35	332 ± 19	42850 ± 712	ND	ND	ND	ND	ND
V37	ND	4276 ± 149	ND	ND	ND	15690 ± 787	ND

[a]ND, not detected.

Table 6 Reported Amounts of Minerals in Dietary Supplements as Compared to XRF-Determined Amounts

Supplement ID	Element	Concentration Sample[a] (mg/g)	Concentration XRF (mg/g)	% Difference
V1	Ca	325.7	48.6	85.1
V16	K	167.5	4.4	97.4
V18	Zn	84.9	12.1	85.7
V28	Fe	208	24.0	88.5
V32	Ca	304.9	43.4	85.8
V33	Ca, Mg, Zn	253.4, NA[b], 3.8	22.3, NA, 0.50	91.2, NA, 87.0
V34	Fe	206.4	25.7	87.5
V35	Ca	326.6	42.9	86.9
V37	Zn	77.2	15.7	79.7

[a]Concentration of Sample: Dosage (mg) per mass of pill (g). [b]NA, no analysis since calibration method does not detect magnesium.

analysis. The label for this product does list dicalcium phosphate as one of the "other ingredients." The zinc supplement V18 contains zinc but also contains potassium, calcium, and titanium. As seen before, the label for this product does list dicalcium phosphate as one of the "other ingredients." The iron supplements, V28 and V34, contained iron but also other elements including manganese and zinc. In addition, V34 also contained detectable amounts of lead. The XRF-determined amount of 11.9 ppm is above the LOD for lead but below the LOQ. The calcium, magnesium, and zinc supplement, V33, contained calcium and zinc as well as potassium. The magnesium levels were not determined since this instrumental calibration method does not detect magnesium. The zinc supplement V37 contained zinc and calcium. For this product dicalcium phosphate was listed as an "other ingredient" on the label. The amounts of the element in each tablet (mg element of interest/g total tablet) were calculated using the information from the label and the weight of a tablet. This can be compared to the amount determined from the XRF analysis for each tablet as shown in Table 6.

These values were used to determine the percent difference between the reported amount on the bottle label and the XRF-determined amount (Table 6). The XRF results show that there are lower amounts of the element of interest compared to what the reported dosage is on the bottle.

Conclusions

A hand-held XRF using a soil FP calibration method was used to analyze two NIST SRMS and these results were compared to the XRF recorded values for each. The results from the analysis of the SRMs with the hand-held XRF were within 23% of the reported values for potassium, calcium, and zinc; 11% for manganese and lead; and 6% for titanium and iron. The XRF-determined values for titanium, manganese, iron, zinc, and lead were not statistically, significantly different than those reported for the SRMs. Each dietary supplement was analyzed using the same method. From the XRF analysis, all of the supplements contained significant amounts of the element listed on the label. This amount was determined from the XRF analysis to be less than reported on the label. In addition, many of the supplements contained additional elements. One of the nine supplements contained detectable amounts of lead. This method for analysis is very straightforward, quick, nondamaging, and fairly cost-efficient method for the analysis of these dietary supplements. These results as well as other brand-name supplements could be further investigated to determine the accuracy of the dosages presented on the bottles.

Acknowledgments

Funding for this research was made possible by the University of Detroit Mercy's McNichols Faculty Assembly Internal Research Fund and the Michigan-Ohio University Transportation Center Grant.

References

1. GAO-09–250; United States Government Accountability Office Report to Congressional Requesters *Dietary Supplements: FDA should Take further Actions to Improve Oversight and Consumer Understanding*. Washington, DC: Government Accountability Office; 2009.
2. FDA 101: dietary supplements. U.S. Food and Drug Administration. www.fda.gov [accessed March 2014].
3. Dietary supplement health and education act of 1994. *Public Law* 1994:103–417.
4. Calcium fact sheet for consumers. *Natl Inst Health.* p. 1–4. http://ods.od.nih.gov/ [accessed March 2014] p. 1–4.
5. Zinc fact sheet for consumers. *Natl Inst Health.* p. 1–3. http://ods.od.nih.gov/ [accessed March 2014] p. 1–3.
6. McCall KA, Huang C, Fierke CA. Function and mechanism of zinc metalloenzymes. *J Nutr* 2000;**130**: 1437S–46S.
7. Iron dietary supplement fact sheet. *Natl Inst Health.* p. 1–3. http://ods.od.nih.gov/ [accessed March 2014] p. 1–3.
8. Potassium recommendation fact sheet. *Natl Dairy Counc.* p. 1–2. www.nationaldairycouncil.org [accessed March 2014] p. 1–2.
9. XRF research. *X-Ray Fluorescence.* http://www.xrfresearch.com/technology/x-ray-fluorescence [accessed April 2013].
10. Geochemical instrumentation and analysis. *X-ray Fluoresc Instrum.* http://serc.carleton.edu/research_education/geochemsheets/techniques/XRF.html [accessed April 2013].
11. Labond JL, Stroeters NH, Benvenuto MA, Roberts-Kirchhoff ES. Analysis of nine Edible Clay supplements with a handheld XRF analyzer. In "Chemistry of food and food Contact materials: from Production to Plate" *ACS Symposium Series*, vol. 1159. American Chemical Society: Washington, D.C.; p. 99–111.
12. Anderson DL. Analysis of beverages for Hg, As, Pb, and Cd with a field portable X-ray fluorescence analyzer. *J AOAC Int* 2010;**93**:683–93.

13. Anderson DL. Screening of foods and related products for toxic elements with a portable X-ray tube analyzer. *J Radioanal Nucl Chem* 2009;**282**:415–8.

14. Palmer P, Jacobs R, Baker PE, Ferguson K, Webber S. Use of field-portable XRF analyzers for rapid screening of toxic elements in FDA-regulated products. *J Agric Food Chem* 2009;**57**:2605–13.

15. Hou XD, He YH, Jones BT. Recent advances in portable X-ray fluorescence spectrometry. *Appl Spectrosc Rev* 2004;**39**:1–25.

16. Garshott DM, Macdonald EA, Murray MN, Benvenuto MA, Roberts-Kirchhoff E. Elemental analysis of a variety of dried, powdered, kelp food supplements for the presence of heavy metals via energy-dispersive X-Ray fluorescence spectrometry. In: Benvenuto MA, Roberts-Kirchhoff ES, Murray MN, Garshott DM, editors. *It's all in the water: studies of materials and conditions in fresh and salt water bodies. ACS symposium series,* Vol. 1086. Washington, DC: American Chemical Society; 2011. p. 123–33.

17. Chan JC, Palmer PT. Determination of calcium in powdered milk via X-ray fluorescence using external standard and standard addition based methods. *J Chem Educ* 2013;**90**:1218–21.

18. Gao H, Zhao H, Wang Z, Zhou J, Guan Y, Shi J. Analysis on spirulinas by PXRD and XRF. *Shipin Keji* 2008;**12**:270–2.

19. Pajchel L, Nykiel P, Kolodziejski W. Elemental and structural analysis of silicon forms in herbal drugs using silicon-29 MAS NMR and WD-XRF spectroscopic methods. *J Pharm Biomed Anal* 2011;**56**: 846–50.

20. Saleh-e-in MM, Sultana A, Hossain MA, Ahsan M, Roy SK. Macro and micro elemental analysis of Anethum sowa L. (Dill) stem by X-ray fluorescence spectrometry. *Bangladesh J Sci Indus Res* 2008;**43**: 479–90.

21. Janosova V, Sykorova M, Stroffekova O, Havranek E. Determination of selected elements in dietary supplements by X-ray fluorescence analysis. *Farm Obz* 2008;**77**:177–81.

22. Raggi MA, Schiavone P, Lucchini F, Mandrioli R. Determination of selenium in diet supplement by spectrometry and X-ray fluorescence. *Boll Chim Farm* 1995;**134**:574–9.

23. Wise SA, Watters RL. *Certificate of analysis SRM® 2586 trace elements in soil containing Lead from paint (Nominal 500 mg/kg Lead).* Gaithersburg, MD: National Institute of Standards & Technology; 2008.

24. Wise SA, Watters RL. *Certificate of analysis SRM 2587 trace elements in soil containing Lead from paint. (Nominal 3000 mg/kg Lead).* Gaithersburg, MD: National Institute of Standards & Technology; 2008.

25. *S1 Turbosdr calibrations.* Kennewick, WA: Bruker Elemental; November 2010.

15

Beyond the Brink

Jim Thebaut

THE CHRONICLES GROUP, REDONDO BEACH, CA

CHAPTER OUTLINE

In many parts of the world, water is a luxury and extremely difficult to obtain and the human impact is profound. More than 3,600,000 people die each year from water-related disease, and 4000 children die daily because it is polluted or not existent. In addition, 884 million people on the planet—one in eight—live without safe drinking water and basic sanitation services.

Regrettably, the mainstream of the American public, as well as most members of the US Congress do not seem to comprehend the reality of the global water crisis. Public and private utilities provide inexpensive, free flowing, high quality, clear, clean drinking water to the US consumer and it is available whenever desired. The consequence, water is taken for granted by US citizens and government leadership is slow in confronting the evolving crisis by not implementing critical international and US public policy and strategies to confront the issue globally and domestically.

The planet is in deep trouble and heading for a calamity unless decision-makers begin implementing constructive solutions and recognize the physical evidence. They seem unwilling to accept or recognize, the actuality of significant global population growth, which is projected to exceed 10 billion by 2050, as well as the intense drought conditions and severe weather events that are occurring around the world.

There is energy scarcity, ongoing potential for disease pandemics and epidemics around the world, there is profound absence of sanitation, a growing lack of agriculture land and food supply, severe poverty conditions, degradation of the environment and the lack of education for children, both boys and particularly young girls. These conditions already create local violence with illegal arms. Worse, present circumstances will likely spawn nuclear arms proliferation in those regions, which are only seeing the secure access to scarce resources.

It should be clear the United States is not exempt from the crisis. Its infrastructure is in decay and drought and extreme weather events are threatening water supply, the US economy and national security. The recent West Virginia chemical spill that activated a

Food, Energy, and Water. http://dx.doi.org/10.1016/B978-0-12-800211-7.00015-6

tap water ban for an estimated 300,000 citizens as well as the potential for a public health calamity should more than demonstrate how vulnerable the United Sates is to an environmental disaster.

The ongoing spectrum of disasters, which could impact US infrastructure and its water supply includes earthquakes, tornadoes, monsoon/hurricanes, toxic spills, droughts, etc., as well as the possibility of terrorism and/or cyberattacks.

Southern California could be considered the canary in the mine shaft. Historically, the region has been vulnerable to extreme water scarcity and is at a crossroads in maintaining its glamorous quality of life as well as its economy. The region has depended almost totally on imported water from great distances and across demanding, desolate terrains. It has depended on the depleting Colorado River and the very fragile Sacramento San Joaquin River Bay Delta System for its primary water supply. Consequently, if the flow of water from any of those sources were to dramatically cease to function, Southern California's economy would experience a severe knockout blow. Furthermore, the current water systems for the entire state were designed to serve a population of 25 million and now must support 38 million with future projections of 50 million or more Californians by mid-century.

Following is a list of Water Strategy Principles and a Holistic Environmental Planning Process, which will be a key in implementing vital solutions for 2050.

Water Strategy Principles for 2050

- World governments formulate partnerships with countries, states, cities, provinces, townships, counties, NGO's and the private sector in order to implement integrated water management programs, public policies, and specific strategies in order to confront local and regional water concerns (i.e., infrastructure financing, etc., and efficient water management).
- Implementing a national water strategy and holistic environmental land use planning on a regional and local levels to assure the integrity of watersheds and other natural resources are environmentally safeguarded from drought, severe weather events and the evolving population growth and density.
- Establishing water as the integral component within the global ecological system and utilizing environmental planning as the foundation for all land use decision-making.
- Implementing water education programs for the public and elected officials about uncontrolled growth and its ultimate impact on the economy and supply of natural resources (water, energy, prime agriculture and grazing land, etc.) and to ensure public support for conservation programs and understanding for integrated water management, public policies, strategies, and rate increases.
- Monitoring the quantity and quality of surface water and groundwater in communities, regions and transboundary river systems.

- Create public/private partnerships in order to implement comprehensive rural and urban water infrastructure development projects across the world.
- Allocating specific amounts of water to all urban and rural users.
- Protecting the integrity of water and its overall security.
- Ensuring all people has access to reliable safe drinking water, basic sanitation, and making fundamental hygiene practices the norm.
- Replacing and creating new infrastructure to store, transport, treat, and recycle water and wastewater on a micro and macroscale.
- Ensuring adequate water for future energy, agriculture, and food supply needs.
- Developing programs for managing farming and crop protection, water and climate, water and health, and groundwater management.
- Increasing efficient water use, which includes improving irrigation efficiency.
- Utilizing renewable energy for all brackish groundwater and seawater desalination plants.
- Protecting groundwater resources from drought, overuse, and pollution.
- Implementing common water management rules and establishing an efficient water management system.
- Examining and managing subsidy and pricing in order to assure water efficiency and conservation.
- Developing science and technology programs in order to solve future problems.
- Empowering women to have a voice in the usage of water.

Holistic Environmental Planning Process and Solutions for 2050

The following is a comprehensive environmental planning decision-making tool. This methodical and meticulous approach will provide a "blueprint" for projecting and analyzing future growth trends. The process is functional and can be applied to local, regional, national, and global public policy strategies and importantly in implementing constructive social, economic, and physical planning solutions for 2050.

For example, by overlaying each environmental planning parameter, an integrated water management plan can be implemented in order to confront local and regional growth projections, patterns, and future concerns.

- Begin with the Global trend 2015, 2025 and the evolving 2035 projected study as a baseline.
- Overlay existing global hot spots.
- Overlay global groundwater and aquifer depletion and specifically the 300 transboundary aquifers.
- Overlay transboundary watersheds.
- Overlay projected global drought models.

- Overlay specific global population growth projections through 2050. This should be for cities and countries.
- Overlay global infrastructure needs for cities and countries (To determine if specific cities can accommodate projected populations).
- Overlay existing prime agriculture lands within countries and regions in order to determine projected food supply through 2050 (Drought projections may determine the loss of lands in specific regions).
- Overlay fragile ecosystems and wetland, etc.
- Overlay global areas/regions that are venerable to climate change and extreme whether events, flooding and natural disasters, etc.
- Overlay regions where there is significant unemployment, poverty, crime, conflict, terrorism suicide bombings, etc.
- Overlay existing locations where there is a lack of clean drinking water and sanitation (This should also include regions and countries where this factor effects education and schools).
- Overlay country's which control Weapons of Mass Destruction (WMD). This should also include projections where these weapons programs may exist in the future (i.e., Iran, etc.)
- Overlay regions/countries where there is an existing and evolving mental and physical health crisis.
- Overlay countries and regions where there is extreme illiteracy and lack of educational opportunities and schools.
- Overlay regions of the world where there is extreme obesity.

There are other parameters that can be explored in this planning and decision-making approach. In addition to the above, it would be important to include an impartial and independent multidisciplinary analysis of the findings, which should include a follow-up with recommendations for future strategy planning and solution-oriented decision-making.

Conclusions

Our planet is in deep trouble and heading for a calamity unless decision-makers begin implementing constructive solutions based on the physical evidence. They seem unwilling to recognize the significant global population growth which is projected to exceed 10 billion by 2050, as well as the intense drought conditions and severe weather events that are occurring around the world.

16

Water Scarcity: Global Challenges for Agriculture

Hessy L. Taft

ST. JOHN'S UNIVERSITY, NEW YORK, NY

CHAPTER OUTLINE

Introduction

According to United Nations estimates, the world population is expected to rise above 9,000,000,000 by 2050. This phenomenon will place major demands on increased freshwater supplies for consumer use and irrigated agriculture. Yet our Earth's freshwater resources are finite, comprising only 2.5% of the total freshwater available from both surface water and icecaps. Preserving this precious resource is endangered partly from neglect relating to contamination and partly by the waste that arises from either an absence of adequate infrastructure in developing countries or a failing aging infrastructure in most developed countries. Compounding the problem of freshwater availability is the realization that demands for consumer use will rise as the socioeconomic status

Food, Energy, and Water. http://dx.doi.org/10.1016/B978-0-12-800211-7.00016-8

of world populations attain middle class status. In the past century, the global demand for water use has grown at more than twice the rate of population increase.[1] But perhaps the most overwhelming impact on water resources is the effect of the dramatic climate change that we have witnessed in recent years. According to the Intergovernmental Panel for Climate Change, the average temperature of the Earth's atmosphere has risen by 0.74 °C in the last 100 years.[2] A warmer atmosphere, holding more water vapor, releases more energy during precipitation, intensifying storms and floods, increasing droughts, depleting groundwater levels, and promoting forest fires, which thereby reduce photosynthetic carbon dioxide uptake. Warmed ocean waters are less dense contributing to sea level rise, salinization of river basin deltas and coastal aquifers, and mortality of benthic fauna due to hypoxia/anoxia of dissolved carbon dioxide. Acidification of oceans from absorption of anthropogenic CO_2 disrupts $CaCO_3$ shell formation of marine organisms. These processes contribute to a decline in food supply, which is contraindicated for a rising world population. In addition, melting mountain glaciers threaten much needed water resources for irrigated agriculture as well as limiting the supply of fresh drinking water. The real impact of water scarcity will be food scarcity.

In addition, freshwater has been diverted for uses other than domestic use and agriculture. Water is intrinsically linked with energy production. Demand for electricity has led to development of hydropower worldwide, which in turn has spurred the widespread construction of dams causing deleterious environmental effects to various degrees. Water is being used for cooling of power plants, although increasingly the use of freshwater for these purposes is being limited. More recently, demands for renewable energy sources have resulted in freshwater use in agriculture for the production of crops for conversion to biofuels. Finally, water is currently a major component of the new process of hydraulic fracturing, which holds promise for electric power generation from shale gas but which also poses urgent needs for innovative approaches to safeguard the quality of precious groundwater.

The issues described above define a crisis that merits serious committed attention on a global scale. This chapter is a high-level overview that identifies some trouble spots where such attention is needed and presents potential approaches to resolving the challenges.

Where Are the Crisis Spots Throughout the World?

Figure 1 identifies areas throughout the world where water resources are most at risk. Risk indicators include not only the baseline water stress but also seasonal variability, flood occurrence, and drought severity. The concept of water stress represents an imbalance between water use and water resources. It is noteworthy to mention that more than 25% of the world's agricultural crops grow in water-stressed areas. From this map, it is evident that the areas of midlatitudes, both north and south, are those most susceptible to water stress, whereas the equatorial and high-latitude zones are relatively protected from water stress. This is due to the ocean salinity patterns of the water cycle. At equatorial and high-latitude zones, the rate of precipitation exceeds the rate of

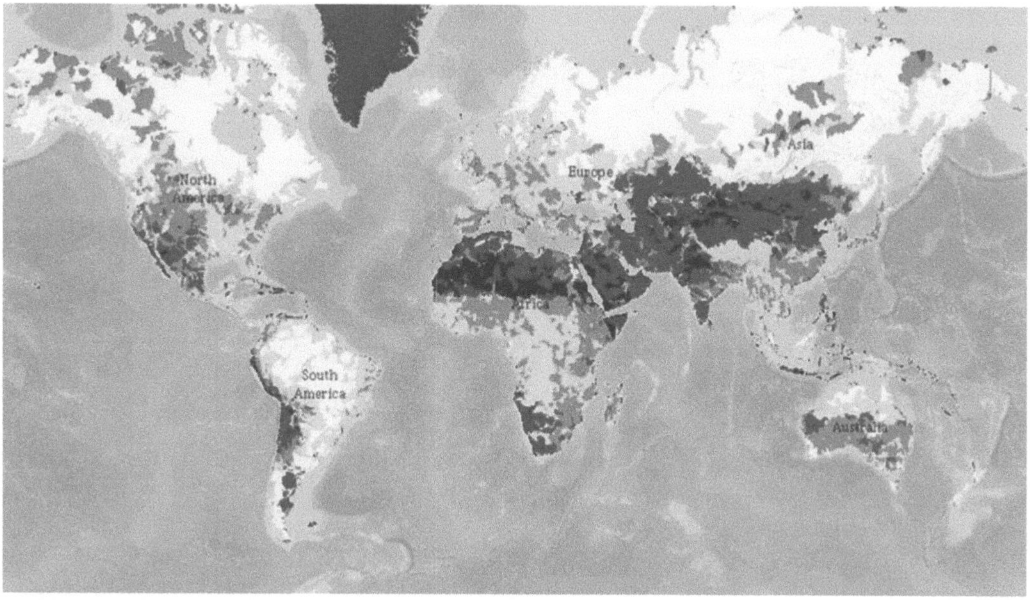

FIGURE 1 Worldwide water risk. White = low risk (0–1), yellow = low to medium risk (1–2), orange = medium to high risk (2–3), red = high risk (3–4), dark red = extremely high risk (4–5). *Aqueduct Water Risk Atlas, World Resources Institute, 2013.* (For interpretation of the references to color in this figure legend, the reader is referred to the online version of this book.)

evaporation, resulting in lower salinity than in central regions of the ocean basis where evaporation is higher than precipitation leading to higher salinity in those regions. The ocean salinity patterns show the fingerprint of an intensified water cycle for the past 50 years. The hydrologic cycle is accelerated at approximately 4% for the 0.5 °C rise in temperature of the past 50 years.[3] This is consistent with what would be predicted by the Clausius–Clapeyron equation, which would predict a 7% rise in moisture per degree of warming. It is important to note that the oceans cover about 71% of our globe, hold 97% of the water, and receive 80% of world precipitation. Thus, the oceans receive a somewhat disproportionately large amount of rainfall, which preferentially adds to salt water rather than terrestrial groundwater.

It is also of interest to examine the changes of ocean and land temperatures that have occurred as the acceleration of anthropogenic CO_2 into the atmosphere has increased. Not unexpectedly, Figure 2 shows that changes in ocean temperatures have been more modulated than land temperatures. The steep uninterrupted rise in temperatures of the last 30 years, particularly for land temperatures, is historically unprecedented.

Thermal expansion of ocean waters mean larger volume of sea water and melting of aquatic ice sheets. Figure 3 records the sea level measured over the past 120 years. A continued rise in sea level, which currently amounts to about 3 mm per year, will have enormous agricultural and economic implications.

There are three different causes that contribute to the rise in sea level. As noted, expansion of sea water due to warming of oceans is one; another is the melting of the

FIGURE 2 Global land–ocean temperature index. *NASA GISS.*

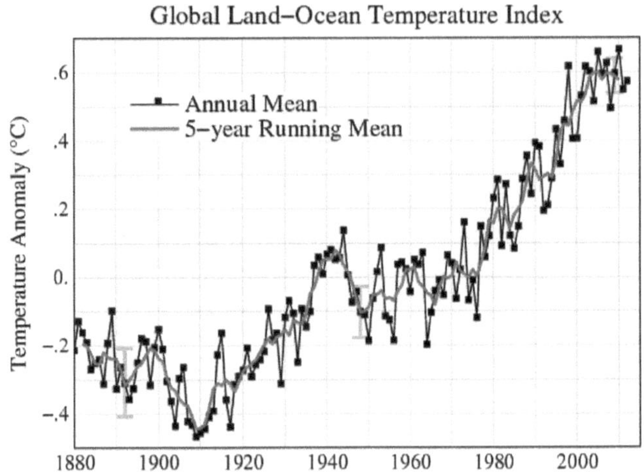

FIGURE 3 Impact of climate change on sea level change.[4]

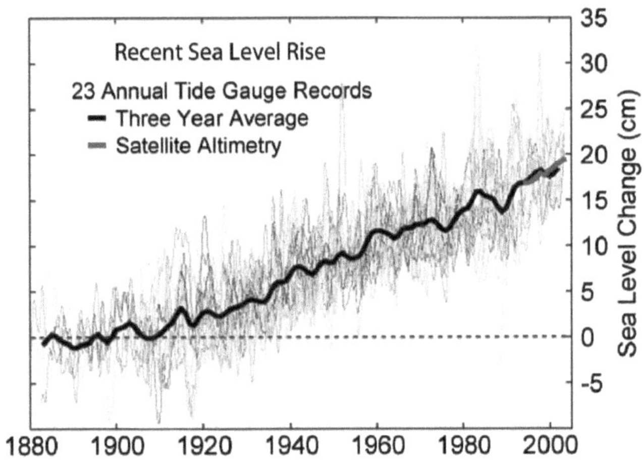

large ice sheets of Greenland and Antarctica, which lost about 263 billion tons and 81 billion tons of ice, respectively, during the period from 2005 to 2010.These effects are bound to have high economic and agricultural costs, as tourism suffers from erosion of beach zones and coastal deltas become salty, impeding the cultivation of food crops. For example, this process has already led to salinization of farmland and wetlands around the mouth of the Yellow River in China, where salt accumulates in topsoil as a result of evaporation. Salt distribution at coastal wetlands is directly affected by human activities. Restoration to improve agricultural management will require committed effort and significant modifications of human activities, primarily in the nature and quantity of energy consumption.

The third cause for sea level rise comes from the melting of mountain glaciers. The ubiquitous melting of glaciers worldwide, from the Himalayas; Tibetan plateau; Alps;

western North America; Ruwenzori range, which includes Kilimanjaro; South American Andes in both Patagonia and tropical regions has been attributed to global rise in temperatures. Millions of people depend on water from glacial flow for their drinking water, their farmland irrigation, and their electricity, which is stored in dams and reservoirs. The snows of Kilimanjaro are now 20% of what they were a century ago. In the Himalayas and Tibetan Plateau, the acceleration of glacier melting is triggered by the accumulation of "black carbon" in the higher altitudes, which results from the burning of wood, dung, and crop residues for cooking and heating by the local poor populations. The persistence of black carbon reduces glacier reflectivity, which, in turn, increases glacier melting and interferes with cloud formation and precipitation. The central and southern glaciers on the Himalayas, which feed the Indus, the Ganges, and other Asian rivers, are disappearing so rapidly that most are likely be totally nonexistent by 2035.[5] The Qori Kalis glacier of the Peruvian Andes has retreated 1.2 km since 1978. This process has reduced the availability of water needed for farming in the Peruvian Andes. The local farmers are implementing the age-old custom of cultivating crops on terraced land, which not only reduces erosion but also is a viable management strategy for minimizing run off. An outreach effort has been made to encourage the farmers to cultivate drought-resistant crops but this approach has been met with considerable resistance due to a reluctance by the local populations to consume genomic modified food. Figure 4 shows the Qori Kalis glacier in 2000. It covers about 84 acres and is now more than 200 ft deep.

Irrigated Agriculture: Water Sources and Conservation Methods

Agriculture consumes about 70% of the world's available freshwater (more in certain regions) and it is thus paramount to engage in practices that will deter waste. Agriculture depends on three sources of the Earth's freshwater for the cultivation of vegetation. The contribution of natural rainfall does not constitute irrigation but few places in the world

FIGURE 4 Qori Kalis glacier, Andes of Peru. *Lonnie Thompson, Ohio State University.*

today can rely on it as the sole source of water. (The Amazon basin in Brazil is one such example.) Irrigated agriculture depends on the use of groundwater and surface water from streams and river. Surface water for irrigated agriculture is increasingly at risk due to multiple causes, among which industrial and agricultural pollution as well as damming and drought feature prominently. Groundwater is drought-proof and currently constitutes 56% of the overall irrigated agriculture worldwide. In Asia, groundwater constitutes 75% of global irrigated agriculture, with India, the front runner, at 37.7% of Earth's land dedicated to groundwater irrigation. Table 1 shows the groundwater extraction in selected countries during 1950–2010.

As is evident from Table 1, the extraction of groundwater has witnessed an explosive growth in the past half century, which is a consequence of the increasing demands for agricultural food production for an expanding population. Hypothetically, one would expect the level of groundwater withdrawal to correlate positively with the size of the country but, in fact, withdrawal is not distributed evenly among nations or even throughout a country; rather it is limited to specific areas where demand for cultivation is intense. The use of groundwater for irrigation is dependent on several key factors, including the availability of surface water for irrigation, the climate variability, the population density, the status of the local economy, and the quality of the groundwater. In Bangladesh, the groundwater has been particularly plagued by contamination with arsenic, a condition that has caused serious health problems for the local population and that has been the subject of much international research and assistance in order to remedy the situation. Arsenic can be typically released into groundwater by the weathering of arsenopyrite, which converts As(III) into soluble As(V) in the presence of oxygen and water. The initial identification of arsenic-contaminated wells for remediation is becoming increasingly ineffective since widespread pumping has facilitated fluctuating water flows in groundwater.[7] Numerous arsenic removal systems based that

Table 1 Groundwater Withdrawal in Selected Countries, 1950–2010

	Groundwater Withdrawal, $km^3 year^{-1}$		
Country	**1950**	**1980**	**2010**
India	14	100	251
USA	42	100	112
China	14	50	112
Pakistan	10	25	65
Iran	10	25	64
Bangladesh	–	10	30
Mexico	10	25	29
Saudi Arabia	0	10	24
Turkey	–	–	13
France	3	5	6

Adapted from Ref. 6.

rely on arsenic-selective adsorbents are presently in use throughout countries in south central Asia but much remains to be done.

A level of groundwater withdrawal becomes unsustainable when the water table is lowered faster than aquifers can be replenished by precipitation, rivers, and streams. The level of abstraction in the alluvial aquifer of Jaipur, Rajasthan, India, for example, has led to the lowering of the water table by 20 m between 1988 and 2003. Continued extraction at the present rate will also upset the natural equilibrium of an aquifer with the ground surface, resulting in subsidence, or the sinking of land, to form craters. As a result of several United Nations Educational, Scientific and Cultural Organization studies, numerous cases of land subsidence have been identified, among which we find the Toluca Valley, Mexico, with a decline of 90 mm/yr between 2003 and 2008 and Yunlin, China, showing a decline of 100 mm/yr during 2002–2007 (IAHS/UNESCO, EISOLS, 2010).[6]

China is particularly vulnerable to the effects of climate change. The country contains approximately 20% of the world's population but has just 7% of its arable land. The North China Plain (Huang Haui Hai), the heartland of wheat production in China, is dependent on irrigation from the Yellow River and the Huang Haui Hai aquifer below. In response to dry conditions, increased pumping from the aquifer has resulted in a decline of the groundwater table of roughly 20 m in the past three decades (Konikow and Kendy, 2005). In response to rising temperatures in the region, the initial stages of wheat growth have accelerated but the later phases of flowering and maturity remain essentially unchanged so that the overall effect is a shorter growing season. This means lower yields due to fewer total grains. According to *Agricultural and Forest Meteorology* (2009), warming is responsible for a 4.7% decline in growth of wheat yields in China from 1979 to 2000.[8]

In the southeastern region of China, where rice is the major agricultural product, warming will also have negative long-term effects. While warming can initially enhance the yield of rice cultivated, prolonged exposure of rice seedlings during pollination to temperatures above 37 °C can sterilize the pollen and thus lead to substantially lower yields in the long run. In addition, the rice-growing eastern region of China could be threatened by a projected rise in sea level by 2050. Compounding the food security problems in China, it is relevant to note that the rising middle class in the country has shown a rising propensity for meat consumption. This translates to diverting about 33% of the country's grain production for the feed of livestock in 2011, adding additional strain to the water allocations.[8]

In the High Plains of the Midwest in the United States, water abstraction from the underlying Ogallala aquifer since the 1940s has converted about 450,000 square kilometers of uncultivated land to one of the most high-quality producers of wheat. The region has become a leading player among bread baskets of the world and has flourished as an economic powerhouse that affects the livelihood of millions of people. However, continued water abstraction at the present rate is clearly unsustainable because the water table is lowered faster than it can be replenished by inflow from rivers and streams or precipitation. Differences in the decline of the water table throughout the High Plains are the result of different local or state-wide regulatory legal limitations. In areas of Texas, the Ogallala Aquifer has declined up to 60 m while in Kansas the decline has been about 50 m in the past half century. Resource management approaches have been stymied to varying degrees by reluctance of the population to accept regulations that could have an immediate adverse economic impact.[6]

When coastal aquifers are overexploited so that the water tables become increasingly depressed, intrusion from sea water occurs, resulting in contamination of the pumping wells from aquifers with brackish water. Limiting the pumping of aquifers in areas where the water table is decreasing faster than it is being replenished is critical for the long-term availability of groundwater. Thanks to the innovative research by National Aeronautics and Space Administration (NASA) (Jet Propulsion Laboratory, Goddard Institute for Space Studies), in collaboration with the University of Texas, Austin, a system of groundwater monitoring was launched in 2002. It is called the Gravity Recovery and Climate Experiment, GRACE, and consists of two satellites orbiting the Earth in the same plane, 220 km apart at a height of 550 km above the Earth. These satellites detect slight variations in the gravity field arising from fluctuations in density due to moving underground water. A microwave ranging system, it detects minute changes in distances between the satellites resulting from different attractions to gravity when the volume of water varies. Global Positioning System receivers can measure the positions of satellites to within 1 cm. Figure 5 is a photograph of the two modules of the GRACE mission as they orbit the Earth. The separation between them, which changes with local variations in gravity, is carefully recorded.

Reliance on surface water is similarly troubled by deterioration that is evident worldwide. The inherent pristine clarity of mountain streams is compromised as the rivers they feed become increasingly contaminated by industrial wastes, fertilizer run-off, pharmaceutical and personal care products (PPCP), and the like. Increasingly,

FIGURE 5 GRACE satellites mapping local variations in groundwater. *NASA, Jet Propulsion Laboratory, Goddard Institute for Space Studies.*

millions of farmers worldwide are irrigating their farmlands with marginal quality water. The rapid proliferation of reservoir dams has alleviated shortages in irrigation water in parched lands afflicted by drought, although the primary purposes of these dams have been the development of hydropower and concomitant flood control. Inevitably, dam construction has been accompanied by varying degrees of erosion due to sediment trapping along river flows and by considerable ecological upheavals. Reduced flows through main river channels into wetlands and lakes or toxic wastes dumped into rivers from industrial effluents affect habitats, and consequently the net productivity of the area. The 2007 report of the World Water Forum identified 10 large river systems that are most at risk for supplying usable surface waters. These are listed in Table 2 together with the sources of problems that could be causing the deterioration of their surface water.

In each case, multiple problems could be afflicting each river. For example, in the case of the Yangtze, increasing population requirements have led to extensive dam construction, the huge Three Gorges Dam, for the extraction of hydroelectric power and the management of the flooding over the river. This has led to wetlands destruction at the mouth of the river resulting in salinization and subsequent intrusion of invasive species. The Yangtze is currently under revised river management policies to remedy some of the consequences incurred. Not listed here are the transboundary issues that affect most of the rivers listed, where, for example, bitter disputes are noted for the countries along the Nile and the Mekong while reasonable agreement is operating among those along the Danube and the La Plata. Although a reasoned approach to resource management is widely recognized as essential to preserving our natural resources and aquatic ecosystems, implementation of such policies has often not been observed, particularly in arid areas with limited water availability where farming is necessary for feeding a growing population or in nations where the economics of export products is a major factor.[9]

Table 2 Ten Rivers Most at Risk, Each Afflicted by Multiple Problems

Rivers	Problems
Danube	Extensive dam construction
La Plata	Overextraction for agriculture
Rio Grande	Arid conditions (climate change)
Nile	Salinization (climate change)
Salween or Nu	Pollution: Fertilizer run-off, industrial and domestic wastes
Ganges	Increasing population needs
Indus	Extraction for hydroelectric power
Mekong	Wetlands destroyed
Yangtze	Invasive species
Murray-Darling	River management

2007 Report of World's Water Forum.

The case of Pakistan, where agriculture is dependent almost exclusively on the Indus River, merits some discussion. Pakistan is a country where 44.7% of the population is employed in agriculture, which accounts for 21.8% of the gross domestic product. The country receives an average rainfall of less than 240 mm $year^{-1}$, which leaves the country prone to severe drought. Agriculture is dependent on the Indus river system, which receives 40% of its flow from snowmelt of the Himalayas. The irrigation infrastructure is strictly a gravity flow irrigation system consisting of two massive dams, three large reservoirs, and 50,000 km of canals that irrigate 36 million contiguous acres of cultivated land. However, due to an aging infrastructure, massive leakage in canals, and siltation in dams, the current water use efficiency is hovering around 45%. In other words, more than half the freshwater available is wasted. Coupling this with a rapidly rising population and an undependable influx of snowmelt from the Himalayas, agriculture in Pakistan is facing an impending crisis. International recommendations have proposed installing a system of drip irrigation but Pakistan has been reluctant to adopt such modification until recent aid from Japan offered help to remedy the situation.[10]

Drip irrigation was developed in Israel about 50 years ago and has since been implemented in more than 30 countries. Its prime advantage is that it allows for major water savings by "drop to crop" irrigation whereby water is delivered directly to the plant surface or plant root via long tubes. This process reduces soil erosion over conventional flood or furrow irrigation and has proven to be 10–100 times as efficient. It requires no major capital investment or infrastructure and, after initial setup, entails significantly reduced labor costs. It has also increased crop productivity since the water is applied directly to the plant thereby avoiding the loss due to evaporation, which is particularly acute in hot dry areas. The success of drip irrigation observed in India for selected crops is exemplified by the results shown in Table 3.

Drip irrigation is usually not practical for enormous stretches of farmland such as the corn or wheat belts cultivated in the United States. Other conservation methods may be more applicable. Making plants resistant to drought is an effective way to cultivate crops when water is scarce. This involves genomic modifications, which have been widely adopted in the United States but have been resisted by many parts of Europe. The

Table 3 Change in Crop Yield and Water Use for Various Crops by Shifting from Conventional Surface Irrigation to Drip Irrigation in India

Crop	% Increase in Yield	% Decrease in Water Application
Bananas	52	45
Cotton	25–27	53–60
Grapes	23	48
Sugarcane	6–33	44–65
Sweet Potatoes	39	60

Comprehensive Assessment of Water Management in Agriculture, IWMI, 2007.

Monsanto Corporation had discovered several years ago that the *Bacillus subtilis* gene codes for a protein that protects a plant from cold shock and has successfully inserted this gene into the plant genome. Its mode of action is to bind to RNA to correct misfolded forms of RNA in afflicted plants. It turns out that this same gene confers upon the plant a resistance to drought, thereby permitting the plant to survive longer with limited water. This gene has now been inserted in corn and tests have shown that the yield has increased by about 10%. Because people have, for years, voiced concern over the practice of genetic engineering, it is important to note that no instance of harm either to people or the environment has been recorded since the technique was introduced more than 15 years ago.

In order to reduce crop losses resulting from drought, it may be advisable to avoid planting thirsty crops in arid or semiarid land. However, such crops typically grow well in hot climates so having in place adequate infrastructure that minimizes waste becomes critical. For example, cotton is a particularly thirsty crop, consuming almost 5000 L of water per kilogram of product obtained and twice as much to produce the woven cotton textile. Rice, which is the staple food and major caloric intake for millions of people in Asia, also leaves an enormous water footprint if planted in the traditional manner of field flooding. Run-off, seepage, and percolation into the soils are water losses at the field level but they are often captured and reused downstream. In such cases, these processes do not represent true water depletion. Growing rice in flooded fields also has multiple negative environmental consequences, one of the primary ones being the release of large quantities of methane, a greenhouse gas several times more powerful than CO_2. Fortunately, recent work has developed a remarkable wet/dry irrigation system, which alternates flooded soils with dry soils, reducing the water footprint by about two-thirds of that from traditional methods. Table 4 gives the global average water footprint of selected crops.[11]

The data for water footprints must be interpreted with care. The values given are simply averages of water footprints exhibited in different countries and reflect numerous

Table 4 Global Average Water Footprint for Selected Crops

Crop	**Global Average Water Footprint, L kg^{-1}**
Sugarcane	210
Maize	1222
Rice (flood irrigation)	5000
Rice (dryland)	1673
Wheat	1827
Oil palm	1098
Soybean	2145
Rapeseed	2271
Cotton seed; fabric	4029; 10,000

From Ref. 11.

local factors including climate, soil fertility, use of fertilizers, oversight of policies for agricultural management, and effective practices to maximize production yields. For example, the average water footprint for maize in the United States is 760 L of water kg^{-1} of maize produced; in China, it is 1160 L kg^{-1}, in Brazil it is 1750 L kg^{-1}, and in India it is 2540 L kg^{-1}.

Water can also be conserved by reducing the rate of evapotranspiration. A sophisticated conservation technique exists that can monitor this phenomenon, thereby saving irrigation water. The Imperial Valley of California receives less than 7.5 cm of rain annually and has a temperature exceeding 38 °C for about 100 days per year. In order for alfalfa fields to grow efficiently, careful monitoring of the rate of evapotranspiration is necessary. An instrument called a scintillometer has been developed that focuses a laser beam on farmland. It records fluctuations in refractive index of air, measuring variations in temperature, humidity, cloud cover, and atmospheric circulation. It measures loss of irrigation water by transpiration from plants and evaporation from the ground over a 3-mile radius, thereby determining precisely the best time of day to irrigate the fields.[12] Of course, when such high-technology equipment is not available, a good conservation technique is to avoid irrigating farmland during the hottest hours of the day in order to minimize evapotranspiration. This practice should be policy that is implemented and enforced globally.

It is now well recognized that plants that engage in a C4 photosynthetic pathway, (such as corn, sugar sorghum) leave a smaller water footprint than those following a C3 photosynthetic pathway (such as rice, wheat, cotton) primarily because C4 plants lose less water from their stomata than C3 plants during photosynthesis. An innovative approach for using genetic engineering to modify rice is currently the subject of intense research. Rice is the predominant food for about half of the world's population. The crop's worldwide yield seems to have reached a peak, that is, it has not increased substantially in recent years but the population that is dependent on rice as its food staple is undergoing particularly rapid expansion. It would therefore be highly advantageous if rice, a plant following the C3 photosynthetic pathway, could be modified to exhibit the more efficient C4 photosynthetic route. To accomplish such major modification, sets of genes that determine both leaf anatomy and biochemical mechanisms as well as their expression in the plant would have to be inserted into the rice genome. The C3 mechanism assimilates CO_2 by the enzyme ribulose 1,5-bisphosphate carboxylase oxygenase (rubisco) but this enzyme also carries out an additional oxygenation reaction, known as photorespiration. At temperatures above 30 °C, typical for rice-growing tropical regions, photorespiration is enhanced considerably, thereby reducing the plant's efficiency in fixing CO_2. In C3 plants, photosynthesis takes place in the mesophyll cells where most of the chloroplasts are located. In C4 plants, both the mesophyll cells and bundle sheath cells possess large numbers of chloroplasts and thus they share photosynthetic functions. The C4 plants have fewer mesophyll cells between their leaf veins and thus their leaf vein density is higher than that of C3 plants. The CO_2 fixing mechanism in the leaves of C4 plants is enhanced, resulting in a much lower rate of photorespiration. The photosynthetic pathway of C4 plants has thus evolved to be more efficient than that in C3 plants. In addition to retaining more water because their stomata exhibit longer closed periods, the photosynthetic activity of C4 plants is not prone to saturation by high-intensity radiation. Furthermore, the C4 plants utilize nitrogen more efficiently than their C3 counterparts do, which translate to reduced fertilizer needs. Thus, the potential for modifying the rice genome to convert it to a C4 photosynthetic pathway has obvious advantages. However, the complex mechanisms controlling the reactions involved are not known at this time and much challenging work remains to be done before such an endeavor can become a reality.[13]

Finally, there are chemical modifications that can be made to a plant to enhance its chance of survival in case of drought. Plant stress occurs under conditions of drought leading to wilting and reduced photosynthesis. Ethylene receptors are then activated to respond to the symptoms of stress. These ethylene receptors are blocked by the compound 1-methylcyclopropene, which, when applied to the plant, allows it to resume photosynthesis with the available water. Another pathway is that of the abcissic acid hormone, the concentration of which rises in case of drought. Again, here is a pathway that allows a plant to slowdown and conserve its own water. A synthetic sulfonamide known as quinabactin triggers the abscisic acid pathway in *Arabidopsis thaliana.* Quinabactin can be applied directly to crops without causing damage.[14]

In areas of the world where sophisticated technology or supplies are not available, rain harvesting is a highly effective way of conserving freshwater. This process involves the collection of rainfall for storage and reuse before it reaches an aquifer. Rain-fed agriculture is, of course, a valuable resource for crop cultivation but vast amounts of rainfall, particularly that which falls in urban areas, is wasted either as evaporation or as run-off that eventually empties into saline waters. Thus, the harnessing of rainfall is a viable conservation method. It has been practiced for centuries throughout the world by a variety of means. Typically, large ponds are constructed whereby the collected water is confined within mud walls or brick walls or even stone walls as in some large urban areas. The quality of collected rainwater is initially higher than that of surface water because it is free of contamination from mud sediments and industrial or agricultural wastes, but it can accumulate debris from airborne particles, insects, or bird feces if allowed to sit in open ponds for storage. More effective means of preserving the fresh water have been devised. In several developing countries, large cisterns used for collecting water have been protected from the environment by installing a rainsaucer, a kind of inverted saucer or wide-open funnel that drains into a narrow pipe at the mouth of a large-capacity storage tank. Harnessing rainwater in storage tanks has numerous advantages besides the obvious ones of preserving a valuable resource for clean drinking water and small farm irrigation water: it requires little capital investment, is easy to install, prevents flooding, and is drought proof.[15] In urban areas, rainwater can be harvested by collection on rooftops of houses and by installing infrastructure for the reuse of greywater (from laundry, bathing, and kitchen uses). These measures, among others, imply lower freshwater withdrawals for municipalities. Harnessing rainwater from storms via various catchment devices is vital not only to collect freshwater but just as importantly to limit run-off from entering streams and rivers, preventing river banks from flooding, and reducing pollutant loads. However, rain harvesting should be limited in farmland areas in order to allow rainwater to recharge aquifers. It should also be noted that the building of large surface water reservoirs has contributed to limit sea level rise by reducing run-off.[16]

Water Resource Management: The Case of the San Francisco Bay Area Delta, California, USA

The Sacramento–San Joaquin River Delta of San Francisco Bay is the largest estuary in the western United States. It is formed by the confluence of the Sacramento River to the north and the San Joaquin River to the south. It is an ecological haven for numerous species of fish and is the lifeblood for the rich Central Valley farmland, which produces about 30% of vegetables and produce in the United States. The Delta comprises an area of about 1100 square miles (2800 square kilometers) and contains around 70 islands surrounded by 700 miles (1100 km) of waterways. It is supported by over 1000 miles of trapezoidal earthen levees, built in the mid-nineteenth century, that have restricted the rivers to their river beds and direct water to San Francisco Bay. Farmers in the Central Valley irrigate about 58,000 km^2 of farmland, producing over 200 different crops and constituting a thriving agricultural economy worth about $17 billion per year. Figure 6 shows the main water flows through the Sacramento–San Joaquin River Delta.

However, problems have ensued in the past century putting at risk the health of the delta. Decomposing peat soils have released CO_2 and islands among the waterways have sunk well below sea level. This allowed for the intrusion of sea water into the delta, resulting in a concomitant reduction in the viability of fish species, notably the smelt. Invasive species are playing havoc with the native ecology. Agricultural pollutants and industrial wastes discarded into the delta have aggravated the problems. Courts have ordered restrictions on the water deliveries extracted from the delta in order to protect the natural ecology and the fisheries that thrive in its natural habitat. On the other hand, millions of cubic meters of freshwater coming to the delta from the Sierra Nevada snowmelt are needed both to serve as irrigation for the Central Valley and to meet the needs of businesses and home residents for 25 million people from San Diego to Silicon Valley. Furthermore, the aging levees sit on six seismic faults and the potential for devastation from earthquakes must be taken seriously. Past attempts to resolve the problems by building peripheral canals from the Sacramento River have met with fierce local opposition and have failed. More recent efforts to restore the vitality of the delta and the reliability of its water supply have spurred several studies involving the collaboration of numerous state and federal agencies that have examined the water use efficiency and ecological health of the delta. The resulting Bay Delta Conservation Plan proposes to reroute freshwater from the Sacramento River via two tunnels linked directly to pumping stations near Tracy, California, that are already delivering water to southern California.

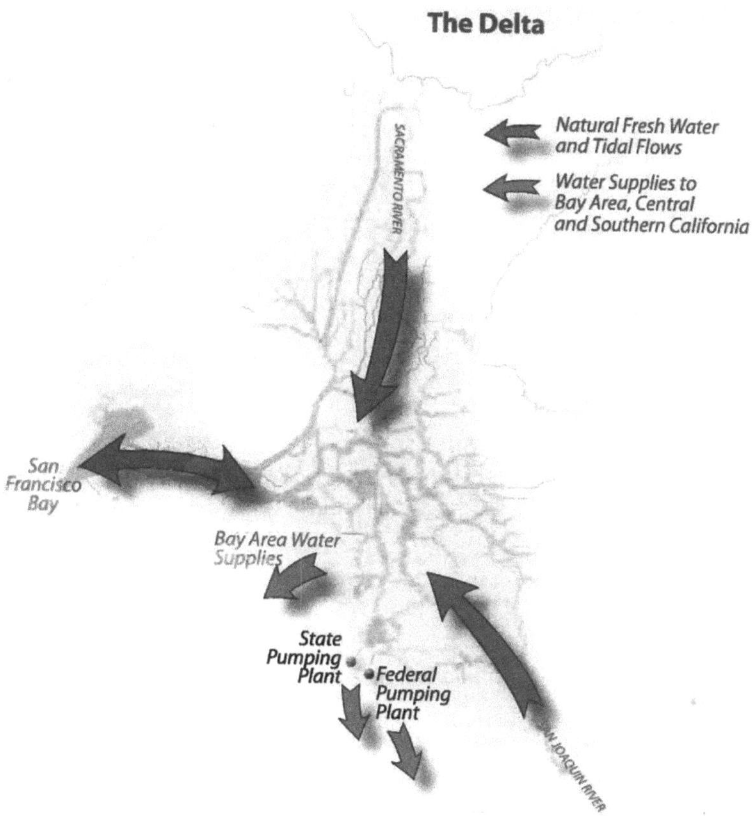

FIGURE 6 Sacramento–San Joaquin delta. *California Department of Water Resources.*

This plan thus alters the way water flows through the delta. It has the added advantage that it will reduce the subsidence of islands in the delta and restore valuable wetlands. Not unexpectedly, this plan has also been subject to considerable public debate, but is likely to be adopted, given the grim realization that the sustainability of the Delta to deliver water supplies and to support the large economy are at stake.[17,18] The enormous amount of work and financial commitments by multiple agencies to address the problems in the Sacramento–San Joaquin River Delta are an example of the kind of concerted cooperation needed in water resource management to correct problems before they attain catastrophic proportions and dissipate essential benefits for all involved.[19]

Yet diversions of water flow in the delta will not solve all the problems in the delta; climate change is showing its effects as well. The Sierra Nevada snowmelt is much reduced. Scientists predict a loss of 25% of Sierra snowmelt by 2050. Already snowmelts as early as late winter have been observed, which creates storage problems and casts doubts on water supply reliability. The consequence of such increased drought is abundantly clear: aquifer pumping now supplies 75% of the farming needs in the San

Joaquin valley, which typically produces cotton, alfalfa, corn, pistachio trees, and raises dairy cows. The Kaweah basin of the San Joaquin basin is 15 m lower in 2013 than it was in 2006. NASA's GRACE satellites have been monitoring the Central Valley aquifer routinely and have recorded a lowering of the water table from 150 m deep when measurements were first started in 2002 to 275–450 m deep in 2013.[20] The long-range reliability of groundwater and reduction of excess withdrawal from groundwater are the bases for a 2013 groundwater management plan that includes storm water capture, storage, and treatment, as well as groundwater use efficiency.

Boosting Our Freshwater Resources

In the last half century, we have witnessed extraordinary technological advances to enhance the Earth's available freshwater supply that is essential to meet the demands of an increasing world population with regard to drinking, sanitation, and food. The major approaches that have been addressed are water reclamation of wastewater and desalinization of sea water. Both have achieved impressive results but both also have incurred environmental and economic costs. Progress to improve cost-effectiveness of both has had considerable success.

Recycling Municipal and Industrial Wastewater

Recognition of freshwater as a vital resource has spurred considerable research in technologies that can provide additional sources of useable freshwater. One of the earliest attempts to augment the supply of freshwater on a large scale was developed in the late 1970s in Israel with the Dan Region Water Reclamation Project. The idea was to capture the water exiting from the metropolitan wastewater treatment plants in Tel Aviv and provide additional purification so the water can be recycled for use on agricultural lands. As of 2000, Tel Aviv recycles 100% of its municipal wastewater from its more than 2 million inhabitants, processing about 350,000 cubic meters of water per day. After emerging from clarifiers in the wastewater treatment plant, the water undergoes soil aquifer treatment (SAT).[21] This process consists of delivering the water to recharge basins where the water is allowed to infiltrate through sand or soil layers into a local aquifer. The recycled water is then is pumped about 100 km to the south to the Negev desert where it serves for agricultural irrigation. Figure 7 shows the basic components of the SAT at a recharge basin.

Surface infiltration through the soil enhances a natural degradation process that reduces organic carbon and nutrients. It is also an additional water purification process via adsorption, ion exchange, sedimentation, and precipitation. The basins are allowed intermittent flooding and drying to provide extra biological treatment by aerobic disinfection with oxygen. The water in the lower saturated zone is maintained there for more than a 300-day retention period in order to destroy microbial pathogens in anaerobic conditions. The quality of the water is monitored via observation wells, which

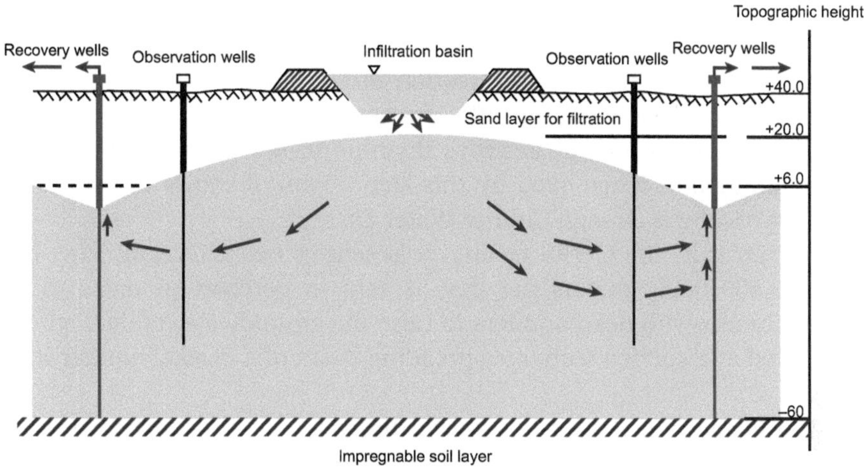

FIGURE 7 Schematic drawing showing the layout of soil aquifer treatment and recovery wells.[22]

determine the time the water is ready for distribution from recovery wells. Testing the water emerging at the observation wells has shown 100% suspended solids, 98% of BOD, 85% of COD, 99% phosphorus, and 57% nitrogen are removed, constituting a level of high-quality water for irrigation. The recharged water, pumped from a depth of 100–200 m, has converted the Negev desert into arable farmland. The sealed recovery wells provide seasonal and multiyear water storage that are not subject to drought or evaporation.

The enormous success of irrigated agriculture with SAT reclaimed water notwithstanding, some concerns regarding the system have become evident of late. One problem is that the increasing population of metropolitan Tel Aviv has increased the quantity of effluent generated, thereby shortening the drying periods between flooding periods at the recharge basins. This deteriorated recharge capacity increases anaerobic conditions, causing clogging from Mn and Fe oxides accumulation and reducing the efficiency of total nitrogen removal. Furthermore, evapotranspiration from cultivated crop fields accumulates salts near plant root zones thereby leading to salinization of irrigated soils and requiring the addition of desalinated water. There is consensus that solutions to the problems cited will necessitate additional technologies such as advanced oxidation processes (AOPs), ultrafiltration (UF), and ultraviolet (UV) irradiation but these all entail considerable economic costs.[23] These will also aid to remove emerging micropollutants, mentioned later.

The semiarid climate of southern California has likewise triggered the need to develop water recycling methods to convert sewage water into fully potable water. The Orange County Sanitation District has built a Groundwater Replenishing System (GWRS) that receives water exiting from standard sewage purification, treats it with NaOCl as a disinfectant that converts ammonia into NH_2Cl, and directs it to its GWRS for additional processing. First, the water is subjected to microfiltration with hollow polypropylene membranes that remove suspended solids, protozoa, and bacteria. This is followed by reverse osmosis (RO) where polyamine membranes remove dissolved salts, viruses, and many pharmaceuticals. The final purification steps involve AOP by treatment with hydrogen peroxide and ozone as an effective disinfectant and irradiation with high-intensity UV radiation. The latter destroys thymine in the DNA of living cells so that microbial contamination is removed by this step. Figure 8 shows the various components of the GWRS of the Orange County Water District.

Once the water exits the GWRS facility, it is sent in two directions: one half of the volume, or 132,500 cubic meters per day, is sent to percolation pools to seep into permeable soil to mix with deep aquifers to raise the groundwater table. Figure 9 shows the SAT delivered at a surface recharge spreading basin of a coastal aquifer in southern California.

As observed in the Negev desert of Israel, the SAT here also provides an additional purification step. As the water percolates into permeable soils, it undergoes a natural biodegradation process that reduces organic carbon within several months of retention in the aquifer. With a 6-month retention time, the organic carbon is effectively reduced from the 5.3 mg C mL^{-1} observed in the effluent that leaves from the GWRS plant to a level of 1.3 mg C mL^{-1} found in the monitoring wells. It is only after this time that pumping from the production wells is permitted.

The remaining 132,500 cubic meters per day is pumped to the coast forming a hydraulic barrier to prevent seawater intrusion into groundwater. The water is added to injection wells that have been built along a16-mile stretch of the coastline at various depths depending on the type of rock formation below. The injected water builds up pressure to overcome intrusion of sea water and, at the same time, also replenishes the aquifer. Figure 10 depicts the process described.

FIGURE 8 Groundwater replenishing system, Orange County, CA.[24]

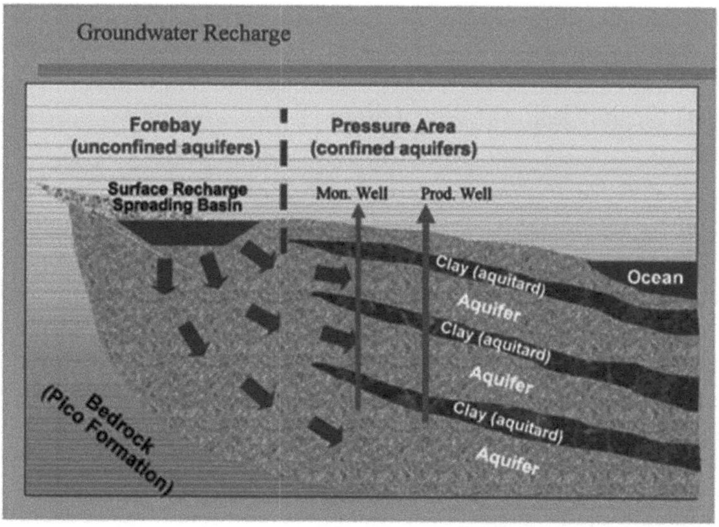

FIGURE 9 Spreading basin for groundwater recharge of coastal aquifer in southern California.[24]

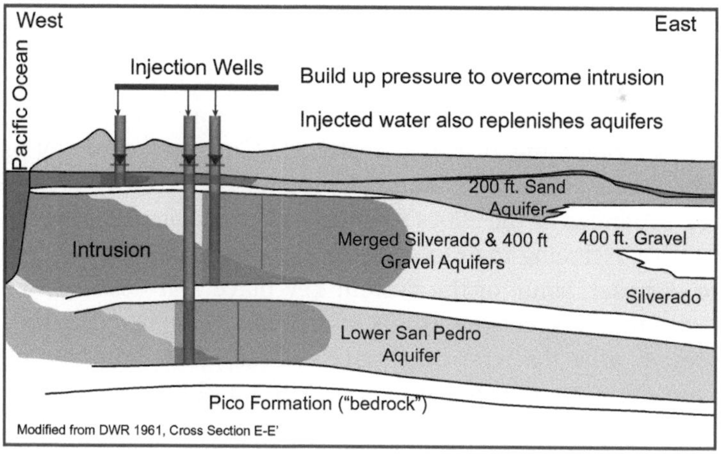

FIGURE 10 Sea barrier injection wells for recycled freshwater.[24]

The water obtained from GWRS is a highly pure source of potable water. It meets the needs of about 10% of the 2.3 million residents of Orange County, California, and costs $0.30–0.40 per cubic meter. It also provides a steady reliable source of freshwater for farming that is independent of drought.

Nevertheless, there is a class of emerging contaminants known as endocrine disruptors and PPCP that are now detectable in water supplies at levels of parts per trillion. Numerous removal methods have been studied that include coagulation; powder activated C treatment; biofiltration; chlorination; membrane treatments such as UF, RO, and nanofiltration; AOP; and UV irradiation. Of all the methods studied, the AOP process, which uses a 1:1 ratio of ozone to hydrogen peroxide, is the most effective, removing 99% of the emerging contaminants.[25] The question to consider is what level of removal is appropriate? An extensive study with highly sensitive instruments conducted by the US Geological Survey in 2007 found that, although 80% of streams examined showed one or more contaminants, only 5% had contaminants ranging above one part per billion. This is well within the federal requirements for safe drinking water. The lengthy retention time in the aquifers replenished by water from GWRS has shown to effectively remove 99% of the PPCP contaminants to a level of nanograms per liter.

In general, the method whereby water is allowed to filter through open ponds is a relatively low-energy approach that can provide much needed nonpotable water to developing countries. The processes of flocculation, gravimetric sedimentation, aerobic and anaerobic microbial action, and exposure to sunlight remove pathogens and organic contaminants that recharge groundwater suitable for irrigation provided a suitable retention time is allocated.

Desalinating Seawater

The oceans contain 97% of the world's water so it would stand to reason that removal of salts from the oceans holds the key to meeting the worldwide freshwater needs. Indeed it does, but at significant economic, environmental, labor, and energy costs. With increasing periods of prolonged drought worldwide, desalination has become a growing reality. By 2005, more than 10,000 desalination plants with a minimal capacity of 100 cubic meters per day were either operating or being installed.[26]The projection for global need of desalinated water is in excess of 38 billion cubic meters per year by 2016. Thus, extensive efforts have been expended in the 2000s to improve the desalination process by making it less energy intensive, reducing maintenance requirements, and providing better mechanisms for brine disposal. Some of the current key players in desalination are shown in Table 5, along with their respective outputs in millions of gallons per day (mgd).

Several processes exist that can desalinate seawater but the ones most extensively utilized by far are multistage flash distillation (MSF) and RO. Even so, MSF is taking a back seat to RO due primarily to the high thermal energy required for its operation. MSF makes use of the property of water to vaporize at different temperatures when under different pressures but the process is now used almost exclusively in the Arabian

Table 5 Key Players in Desalination

Site	Output (mgd)	Comments
Jebel Ali, UAE	220	Multistage flash distillation
Shoaiba, Saudi Arabia	118	Multistage flash distillation
Hadera, Israel	92	RO technology
Barcelona, Spain	53	RO, largest in Europe
Perth, Australia	38	RO, powered by wind
Tampa Bay, Florida, USA	25	RO, largest in USA

mgd, millions of gallons.

Peninsula where energy is plentiful and cheap. In addition, scaling and corrosion is pervasive in the pipes of MSF facilities where temperatures reach above 120 °C. Tons of chlorine used to descale the pipes are released daily into the Persian Gulf, creating an escalating environmental problem for marine life.

The RO process relies on the removal of salt by sending pressurized sea water through semipermeable membranes. These membranes consist of a polyamide composite approximately 100 nm thick that exhibits high water flux and can reject up to 99.6% of salts. The difficulty here is that these membranes are expensive and very sensitive to biofouling and/or breakage so that the RO desalination process requires extensive pretreatment to protect the membranes and careful monitoring of the pressures.

A detailed look at the successful plant in Barcelona (Table 5) will illustrate the scope of the precautions necessary. The seawater open intake consists of two circular concrete towers built below sea level at 2.2 km from the coastline. The raw water from the intake is treated with a hypochlorite for disinfection and carried to the plant via pipes and then undergoes multiple pretreatment steps to prolong the viability of the RO membranes. The water goes from the pumping station through coagulation and flotation chambers and then to the dual media gravity filters, which eliminate suspended solids, algae, and microbial pathogens. Finally, the water is further clarified with pressurized dual media filters and cartridge filters and only then is it ready to pass through the RO membranes for salt removal. The plant is located adjacent to the Baix Llobregat wastewater treatment plant, a situation known as colocation, which enables the brine resulting from desalination to be diluted with cooling water from the treatment plant before being discarded back into the ocean, thereby avoiding environmental hazards to marine life. The heavy energy expenditures normally associated with desalination are minimized with the use of special energy recovery technology known as Energy Recovery PX Pressure Exchangers, which collect energy from the waste stream (brine) of seawater desalination and transfer it back into the low pressure of incoming seawater. Thus, this desalination plant is both an efficient energy saver and an environmentally friendly operation. Figure 11 illustrates the RO building at the Barcelona Llobregat desalination plant.

Similar conditions exist at Tampa Bay desalination plant, which is located next to the Big Bend Power Plant at Tampa Bay, Florida. Here, the water intake from Tampa Bay is also given an initial chemical disinfection treatment with ClO_2 or $NaHClO_3$, which is then sent through mesh screens, particle settlement, sand, and diatomaceous filters before being subjected to the cartridge filters and pumps that are meant to protect the RO membranes. The plant has 10,032 membranes and pumps water at a pressure up to 1000 pounds per square inch. After passage through the RO membranes, turbines collect the water for a second pass through the RO membranes for additional freshwater output. The total brine produced is diluted with cooling waters from the power plant before being released back into the bay, making this plant also ecofriendly. The Tampa Bay Desalination Plant is about to be superseded in size in the United States by the much larger desalination plant now under construction by Poseidon group at Carlsbad in San

FIGURE 11 Osmosis building at the Barcelona Llobregat desalination plant, 2008. *ATTL Concessionaria de la Generalitat de Catalunya, S.A.*

Diego County, which plans to become operational in 2016. It will have an intake pump that will bring in 100 million gallons of ocean water per day of which 50 million gallons will be processed into clean drinking water and the remaining 50 million gallons of brine will be returned to holding channels for dilution with ocean water before being discharged back into the ocean. The Carlsbad plant is expected to operate using approximately 3 KWh per cubic meter of freshwater produced. It is targeted to meet about 7–10% of the region's freshwater needs. The desalination process requires an elaborate infrastructure and maintenance both of which account for the relatively high costs of desalinated fresh sea water. Poseidon estimates that it will be able to sell water at $0.80 m^{-3} in California; in Israel the current price for desalinated water is $0.53 m^{-3} (partially subsidized). Further reductions in price will depend on installation of more efficient pumps and energy recovery devices, and improved membrane techniques that reduce fouling without compromising water flux or salt rejection. One mechanism that has been proposed as a technique to enhance desalination is a systems redesign where two high-pressure pumps and membranes are connected in series. The first stage would operate at a lower pressure than the second stage but a smaller volume would be going through the second stage. This setup would require a net lower overall energy demand for the entire desalination process but it does mean that a higher initial capital investment is needed for the construction of such a facility.[27]

The water that exits from a desalination plant is highly pure, has low pH and salt concentration, and thus a low buffering capacity. In order to be useful for agriculture, the water needs to be remineralized. Calcium, magnesium, and sulfate ions are absent in desalinated water but they are required by plants to avert stunted growth. To remineralize the water, calcite (a source of Ca) and dolomite rock (as a source of Mg) are added after dissolution with sulfuric acid. The low pH increases the corrosion risks in metal distribution pipes and so the alkalinity is stabilized with lime to pH between 8.0 and 8.4. And for countries bordering the Mediterranean Sea (such as Spain and Israel), boron is present in the effluent from desalination plants. The problem arises because the Mediterranean Sea is particularly high in boron (>4.5 mg mL^{-1}) but, it is such a small ion, it typically passes through the RO membranes in acid or neutral conditions. Because agricultural yields are reduced with high boron concentrations, this must be largely reduced from desalinated water to be useful for agriculture in the countries along the Mediterranean.[28]

Desalinating Brackish Aquifers with Solar Energy

In parts of the world where rainfall is plentiful, soluble salts near the ground or surface change in composition and concentration with intense rainfall and are subsequently washed away by good natural drainage. In hot arid climates, water is lost from the surface by evaporation or by plant transpiration in cultivated areas, leaving behind groundwater that is increasingly brackish and unsuitable for further agriculture. For coastal aquifers, the chance of contamination by sea water intrusion is likewise significant. In 2012, a project at Ben Gurion University, Israel, known as the Josefowitz Oasis Project in honor of its sponsor, was begun in the very hot and dry Arava valley of southern Israel with the purpose of growing crops with water obtained by desalination from a brackish aquifer. The key innovation here is that the groundwater used was desalinated using solar panels. Because the brackish groundwater is less salty than sea water, it was possible to use nanofiltration, which is a low-cost technology that permits limited passage of certain ions essential for plant growth. Thus, the feed water was adjusted to have concentrations of Ca^{2+}, Mg^{2+}, and SO_4^{2-} such that the permeate flow contained these ions in optimum concentrations for irrigating water. The desalination could be carried out with lower pressure than sea water RO desalination, requiring less pumping and thus less energy. It was still more expensive to carry out the desalination using the solar panels than using the electrical grid but this price difference is expected to be reduced as solar panels become cheaper and more efficient. Four staple crops were grown: maize, millet, sorghum, and potato. Overall, the crops required 25% less irrigation water and fertilizer and the productivity yields were higher than those obtained under conventional practices. The brine collected from the desalination is a useful waste product: it was used to cultivate red beets, a salt tolerant plant. It has also been proposed that the salty medium be used to grow ornamental fish.[29] Figure 12 shows maize growing on sandy soil of an experimental farm irrigated with solar-powered desalinated groundwater.

FIGURE 12 The solar-powered desalinated facility and agricultural crops in Arava Valley, Israel. *Sam Josefowitz Oasis Project, Zuckerberg Institute for Water Research, Ben Gurion University, Israel.*

The remarkable overall feature of the system described above is that there are no waste products, which makes this system highly ecofriendly. There are approximately 80 million ha of arid land worldwide that are affected by groundwater salinity. The method described above, whereby solar-powered desalination of aquifers is carried out in hot climates where there is plenty of sunshine, has the potential to feed millions of people. This technology should receive the necessary support for the broadest application possible.

Similar efforts to use solar power to desalinate brackish groundwater are underway in the drought-ridden areas of California's Central Valley where climate change has led to undependable surface irrigation from snowmelt and increasingly salty water from pumped groundwater. A new technology, designed by WaterFXAqua4, has been installed as a pilot project in the Panoche water district. A large receiver captures solar energy on a thin reflective film to heat tubes containing mineral oil that are suspended on the receiver. When the temperature reaches 248 °F (120 °C), the heat is pumped to a multistage evaporator system that converts water to steam, which is then condensed to pure water useful for irrigation. The drainage water is collected via tiles below the irrigated crops and is sent for recycling to further concentrate the brine or directed for growing salt-tolerant crops. The system also collects extra heat in the form of molten salt for storage to continue operation when the sun is not shining. This solar thermal desalination system requires an area of 160 × 40 ft for installation on flat level ground and can generate 65,000 gallons of pure water per day. At present, the cost is about 62% higher than the price of water charged by the Central Valley Project but with continuous 24-h operation and dependable supplies, the technology is attempting to become more competitive.[30]

Diverting Water Resources for Energy Production

We have seen that our freshwater resources needed for domestic use and agriculture can be supplemented by various processes, from simple mechanical pumping of groundwater from wells to more complex processes of municipal wastewater treatment by recycling mechanisms to the far more energy-intensive processes of desalination of sea water or brackish water. In each case, energy input is critical to obtain the desired water supply. Likewise, water has been used to increase the ever growing demand for energy by an exploding world population, much of which is rising to middle class socioeconomic status. The nexus of energy security and water security is tightly interwoven, as is of course, food security which depends, in large part, on the accessibility of water for agriculture. We shall examine herein two pathways in which energy demands can be met by water contributions and the consequences that have already arisen from each endeavor.

Dams Across the World's Rivers

Hydropower, or hydroelectricity, is generated in a sustainable fashion by storing water in dams or reservoirs, which are not dependent on seasonal changes of water flow from a stream, as the "run-off-the-river" dams are. Hydropower is a clean, renewable energy source that does not entail carbon emissions. The potential energy of falling water turns turbines, which drive a generator that produces electricity. Hydroelectric plants are capable of up to 90% efficiency (calculated from the total electrical energy produced compared to the potential energy of the water passing through the turbine).

The demand for increased access to electricity in the past three decades has been particularly intense in Asia. When the world's largest hydropower facility, Three Gorges Dam, opened in 2003, it included a 1080 km^2 reservoir that had submerged 13 cities and 466 towns and had displaced 1.3 million people living in the area. Assurances were made that flood control of the Yangtze would be contained, that electricity would meet the supply needed for surrounding industry and urban uses, and that a wild-life reserve of 500 km above the reservoir would provide the much needed habitat of the about 190 native fish species. The first two assurances have certainly been met, the third decidedly not. The huge reservoir reduced fish spawning areas, and created impassable barriers for species like the Chinese river dolphin and the paddlefish. Because sediments accumulate in the reservoir, the water exiting the dam's turbines is relatively clear making erosion of the riverbanks far worse than anticipated. Particularly damaging is the growing realization that the sheer weight of the dam may be contributing to the seismic activity in the region, even if the faults lie below possible contact from leaking water. Nevertheless, four more dams are proposed along the Jinsha River, tributary in the upper Yangtze hoping to counter drought and to produce more power. The further reduction of habitat reserve for endangered species is inevitable and the possibility of seismic aggravation from accumulated stress of multiple stacked dams needs to be carefully scrutinized.[31]

Experience has shown that extensive hydropower development along the Mississippi River in the United States over decades has led to loss of sediment in the Mississippi delta, with accompanying erosion and subsidence of the land. Notwithstanding such negative consequences, the countries in the lower basin of the Mekong delta, which includes Laos, Cambodia, Vietnam, and Thailand, are embarking on major hydropower development projects that would alter the economies of the region. Hydropower can bring energy security to the region. The joint Mekong River Commission formed by the above four countries requires each to consult with the others before launching a major project but approval is not required. In Laos, the proposed Xayaburi Hydroelectric Power Project is planned as the first dam on the mainstream of the lower Mekong River. Set to be completed in 2019, it would be 820 m wide and 49 m high and create a 49-km^2 reservoir, costing $3.8 billion to construct. It is slated to generate 1285 MW of electricity, 90% of which would be exported to Thailand. But the unique ecosystem that nourishes the livelihoods of more than a million people and supplies 30% of the protein nutrition for the local population—a considerable food security bonus—is at risk. In Cambodia, the Tonle Sap Lake swells to more than 10 times its size each year from the monsoon rains pouring in from the Mekong River, washing in organic matter that enhances primary productivity and sustaining the vast number of migratory fish of the world's largest inland fisheries. The Xayahuri Dam could, for example, place an insurmountable barrier to the largest species in the ecosystem, the giant catfish, whose only known spawning area is north of the planned Xayahuri Dam, leading experts to fear the species may go extinct. In addition to retaining sediments that may eventually cause clogging in the dam turbines and erosion of floodplains and agricultural land below it, the dam's design is such that it would retain 40% phosphorus and 30% nitrogen, valuable nutrients essential for productivity. But the Xayahuri Dam is only the first of 12 such dams being planned for the lower Mekong basin, which together could have cascading magnifying effects. Generating electricity and controlling the flood pulse are obvious benefits. But agricultural losses and declines in the abundant fisheries in the Tonle Sap Lake, in the delta, and along riverbeds could have catastrophic consequences. Energy security cannot compromise food security so scrupulous planning to balance the many parameters deserves thoughtful analyses.[32,33]

The quest for surging electrical power demands in India has prompted the government to expand dramatically the construction of dams in the Indian Himalayas, both the reservoir storage and the run-off-the-river kinds. Although India has had a national Environmental Impact Assessment law since 1994, charged with addressing concerns related to endangered species and protected areas, little monitoring to this law is evident in the planning of hydropower. The potential for submergence of undisturbed forests and habitat degradation with concomitant species loss are not being adequately addressed. The needs of populations that will have to be relocated include adequate compensation for unwanted disruptions.[34]

In Iran, the overabstraction of groundwater supplies and climate change are leading factors contributing to the water shortage. The Isfahan River in central Iran and the Ahvaz River near the Persian Gulf have gone dry. But the problems have been exacerbated by the building of dams, particularly in the area surrounding Lake Urmia, once Iran's largest lake and now totally dry. Lake Urmia was a salt lake, somewhat larger than the Great Salt Lake in Utah, and was a thriving tourist center with cruise ships and beach resorts. Although it was not suitable for agriculture, the 11 rivers feeding it were, but these waters have now been diverted to several dams spanning these rivers. Certainly, the dams have brought much needed electricity to previously dark areas, and their water supplies have also contributed to irrigation, but typically in the form of wasteful practices like spraying of water on agricultural fields. With the elevated temperatures observed in most agricultural lands in Iran, the water sprayed on fields is largely lost by evaporation. Meanwhile, experts warn that the valuable agricultural land surrounding the former Lake Urmia could be poisoned by the dried salt.

In Africa, hydropower is still largely unexplored, although the potential is vast. Many rural areas currently obtain energy for heating and cooking by burning biomass, releasing large quantities of CO_2 to the atmosphere. In order to implement responsible hydropower, sustainable and dependable sources of water are needed. The water should be managed to meet the needs of renewable energy, flood control, irrigation, industry, drinking water, and recreation. We have noted above that responsible hydropower also needs to minimize the disruption of ecosystems. Regional hydropower programs in Africa will attain cost benefits if they are designed to share responsibilities for joint projects. For example, the Manantali Dam across the Senegal River is an integrated project that serves Senegal, Mali, and Mauritania. It consists of a run-of- the-river dam that provides irrigation for 255,000 ha of land; has navigation along 900 km; is a reliable water supply for cities like Dakar, St Louis, and Nouakchott; and provides power at \$0.06 kWh^{-1} as opposed to \$0.22 kWh^{-1} for thermally derived electricity.[35]

Water Consumption in Biofuel Production

The rude awakening that our planet's climate is being inextricably changed by the relentless anthropomorphic burning of fossil fuels has produced multiple attempts by policy makers to curb this practice with limited success. One such attempt is to provide more access to electricity to developing countries so that the age-old practice of burning wood and shrubs for cooking and heating can be abandoned. Another effort is for developed countries to adopt legislation for higher energy efficiency in transportation. Yet another approach is to develop technology for introducing fuels into transportation that are derived from biomass, which release lower levels of greenhouse gases to the atmosphere than fossil fuels. The past 20 years have seen major advances in producing ethanol to mix with gasoline and biodiesel to mix with, both of which reduce the amount of CO_2 released in transportation. In 2010, the world produced 105 billion gallons of ethanol, 90% of which was produced in the United States and Brazil, while the nations of the European Union produced 53% of the world's biodiesel.

The basic problem is that these alternative and renewable biofuel energy sources are derived from crops. Ethanol is produced from corn grown in the United States and from sugarcane in Brazil. Soybeans are the main source of biodiesel in both these countries. In the European Union, rapeseed and, increasingly, the high-oil-yielding palm oil are the major sources of biodiesel. But diverting agricultural land from food crops to filling fuel tanks runs counter to the prevailing need for food. The inevitable results are higher food prices and hunger for an expanding population. For example, in Guatemala, plantation owners have been buying farmland cheaply from local farmers to make room for growing palms and the local population now finds itself squeezed on marginal hillside lands to grow the needed food crops. The price of tortillas in Guatemala has quadrupled in the 2000s. Furthermore, when palm oil plantations clear forests, they instill greater danger to the environment than they are trying to relieve: clearing forests has the potential for net global increase in greenhouse gases by reducing the uptake of CO_2, precisely the opposite of the desired goal.

The food–energy nexus is further strained by addition of the water component. The water footprint for several biofuels is dependent on the type of crop, soil type, and irrigation needed. The water demand for corn-based ethanol in the USA is typically 2500–2800 gallons of water per gallon of ethanol produced, the higher values predominating when significant irrigation is needed, as in the case in the Midwest. Water demand for sugarcane-based ethanol production in Brazil is about 2100 L of water per liter of bioethanol produced. Sugarcane in Brazil is grown largely with just natural rainfall except in the northeast where supplemental input with irrigated water is needed. Irrigation is essential in African countries such as Kenya, Malawi, and Zimbabwe, which are currently diversifying their sugarcane production to ethanol, so their practices involve pumping aquifers, often at unsustainable rates. For soybean-based biodiesel in both the United States and Brazil, the typical use of water for soybean cultivation ranges from 11,400 to 14,000 L of water per liter of biodiesel obtained. (A side benefit of soybeans is that their plants can fix nitrogen directly from the atmosphere, thus not requiring fertilizers.) Biodiesel obtained from palm oil or rapeseed has considerably lower water footprints than that from soybean, typically requiring 4975 L and 4823 L of water, respectively, per liter yield of biodiesel.[11] Overall, diverting large quantities of water from food crops is simply not sustainable. And the energy gains are questionable. One gallon of ethanol contains only 66% of the energy of 1 gallon of gasoline. Moreover, the potential yield from crop biofuels is generally poor. Sugarcane is the most productive crop source for ethanol and palm oil for biodiesel. Table 6 shows some potential fuel yields from selected crops.

Table 6 Potential Fuel Yields from Selected Crops

Source	Ethanol (US gallons of fuel acre^{-1} year^{-1})	Biodiesel (US gallons of fuel acre^{-1} year^{-1})
Corn	250	
Sugarcane	450	
Soybean		50
Rapeseed		104–127
Castor beans		140–151
Palm oil		635–640
Jatropha		200
Corn stover	203	
Microalgae		>4000

It follows logically that alternatives to crop biofuel production must be developed. As can be seen in Table 6, use of nonfood crops are being explored with the rationale that they can be grown on marginal land and require less irrigation. But this argument is fortuitous since nonarable land can be made arable, as we have seen in the Negev desert of Israel, and, in any case, restrictions of this kind would never approximate the required demands for renewable energy. The problems described herein would be more realistically reduced, either by oil extraction for biodiesel from cultivation of microalgae or alcohol extraction from cellulosic biomass, but applications of new technologies that make these economically viable are required. Fortunately, recent efforts in these directions have made great strides.

Microalgae have characteristics that make them particularly attractive as fuel feedstock. First and foremost, they do not compete with food production. Microalgae require little land use, can grow in saline or wastewater systems, consume CO_2 from wastewater treatment plants, purify wastewater by assimilating nitrogen and phosphorus to grow biomass, have an approximate 10- to 14-day rapid cycle for harvesting, and give a high mass yield of about 35% lipid content, which can be augmented with selective breeding. The oil extracted as triglycerides then undergoes various processes to produce biofuels, biodiesel and bio jet fuel among others. The big threat to their cultivation, typically in open raceway ponds, is the risk of contamination from predators such as rotifer or zooplankton, which could destroy an entire culture. The open raceway ponds are equipped with paddle wheels to agitate the culture, a considerable energy expense necessary to ensure that all cells are exposed to sunlight for photosynthesis. In addition, large energy expenses can be averted by avoiding the need for drying of the raw biomass before extraction. Strategies where wet algal biomass can be processed directly to produce high-value products will make the technology of algal fuel economically feasible and environmentally sustainable. One company that seems to have addressed most of these limitations is Univerve in Israel, which recently began constructing its first farm. Instead of open raceway ponds, they have developed batch processing, which reduces evaporation and provides culture circulation by bubbling air, a considerable energy saver. Contamination is managed by strict management oversight so that a batch is harvested before contamination completely destroys the affected biomass or infects other batches. The special strain of algae used is ideally suited for optimal oil yield and, importantly, leaves other biomass that is readily processed into products with high market value. The omega-3-type fatty acid obtained is sold as food additives and the proteins and carbohydrates recuperated are valuable to the animal/fish feed industries. Globally, biofuels from algal oil have already been successfully pilot tested as aviation and shipping fuel both by commercial airliners and by the US Navy. This represents a huge potential market but the fact that an entire algal biomass can be commercialized is the key ingredient for ensuring profitability.

Producing ethanol from cellulosic biomass is also a process that does not compete with food crops. It has the attractive characteristic that the feedstock consists of waste materials: agricultural wastes such as corn stover, wheat straw, sugar bagasse, and forest wastes such as wood chips, sawdust, and paper. In the United States, the raw material is plentiful: 105 million tons of corn stover and wheat straw are produced annually. Its use as biomass feedstock could have potential environmental benefits for mitigating CO_2 emissions and reducing wastes in landfills. However, the overarching barrier to implementation of this technology is that platforms for ethanol production and extraction, either by biochemical or thermochemical means, require elaborate procedures and highly intensive energy input. Profitability of the process of converting celluloses or hemicelluloses into ethanol depends on a production design with a high ratio of energy output to input. In 2014, the US Environmental Protection Agency set the goal of 17 million gallons of cellulosic-derived ethanol to enter the fuel supply in the United States, representing 0.01% of the fuel market. In Straubing, Bavaria, Germany, a pilot plant is using an innovative Sunliquid technology developed by the Swiss company Clariant to produce 1000 t of cellulosic ethanol from 4500 t of wheat straw, with government financed support. Yet, in Germany the feedstock supply could, at a maximum, produce enough ethanol to meet just 25% of its fuel transportation needs.

Promoting Transboundary Cooperation

About 270 river basins worldwide flow through two or more countries and roughly 40% of the world's population lives in shared basins. With freshwater resources becoming ever more scarce and sought after, the imperative for transboundary cooperation is becoming increasingly urgent if the explosion of conflict is to be averted. The closely linked issues of water, energy, and agriculture have promulgated sharp debates in a wide spectrum of instances that only the demonstration of real political will can hope to resolve. It will suffice here to illustrate the complex issues of potential conflict with just two examples.

For example, Egypt and the upstream countries of the Nile river basin are engaged in continuous arguments over what constitutes fair share use of the Nile's waters and what problems are caused by wasteful practices. The riparian countries involved (notably Ethiopia, Sudan, Eritrea, Burundi, Uganda, Tanzania, and Kenya) want to abrogate past treaties, which currently give Egypt powerful dominance in the Horn of Africa, and are claiming more equitable benefit sharing. The fierce competition for freshwater creates havoc and raises tensions in a region prone to political and socioeconomic upheavals.

Another example where careful management of transboundary water resources is critical is among the Central Asian countries of Tajikistan and Uzbekistan. Energy-starved Tajikistan wants to attain energy security and growth by building the Rogun Hydroelectric Dam across the Vakhsh River and thereby become an exporter of electricity. The massive Rogun Dam would be 335 m high, with a 13.3-km^3 reservoir capable of generating 13.3 kWh $year^{-1}$. Neighboring Uzbekistan is dependent on the Vakhsh River for irrigation of its cotton fields, producing its prime crop. The World Bank, supported by a European Commission initiative, has currently placed the entire project on hold pending its own environmental assessment and water management study. To date, the region of Central Asia has the highest per capita water withdrawal rate in the world due mainly to inefficient irrigation techniques and poor infrastructure maintenance. If regional cooperation instead of conflict is the outcome along the Vakhsh River, much will have been accomplished.[36]

The nations of Africa are in early stages of regional development and are proceeding with considerable understanding of the benefits that pooling of shared resources can bring. As of 2014, only 5% of crops cultivated in Africa are irrigated with rain-fed subsistence supplying most of the remaining agricultural needs. Less than 10% of the hydropower of 1750 TWh $year^{-1}$ $capita^{-1}$ potential available for Africa has been tapped. In Europe, the level of hydropower potential explored is 75%, and in South America it is 30%. Investments and development are restricted by wide climate variability and risks associated with political instability. Yet the African countries need such investments urgently to boost water and energy security and alleviate endemic poverty. The World Bank has estimated that the minimum consumption of electricity per capita to sustain a reasonable quality of life is about 500 kWh $year^{-1}$ $capita^{-1}$, yet the consumption for sub-Saharan Africa ranges from 21 to 204 kWh $year^{-1}capita^{-1}$. (The electricity consumption in the United States is 11,994 kWh $year^{-1}$ $capita^{-1}$.) As already stated, the construction of transboundary dams and reservoirs can improve energy security, mitigate losses from run-off and drought, provide water storage for irrigation relief, and thus enhance food security and economic stability. But these efforts must be carried out so as to minimize ecosystem upheavals, including forest degradation, and negative impacts on local populations. The largest such project, planned with the intervention of the African Development Bank and other financing institutions, is the proposed Inga I and II Rehabilitation Project in the Democratic Republic of Congo. It would generate 44,000 MW of electricity, sufficient power to supply the entire African continent with adequate electricity. But engineering concerns, availability of technical expertise to guarantee operational sustainability, low-cost operational requirements so as to reach the poor, appropriate environmental precautions, coordination by policy makers among the countries involved, and limited financial capacity are all key components that must be resolved before implementation of this plan becomes feasible.[35]

While the uneven international allocations of water in shared basins has the potential for resolution given the political will and realization of benefit sharing that can be derived, the unequal distribution of freshwater resources throughout the world cannot be remedied by the long-distance transport of water since this process is so energy intensive as to be prohibitive. However, the problem can be alleviated by the trade of virtual water, which refers to water embedded in foods and goods, that is, the amount of water consumed to produce the product. To the extent that a country imports wheat to meet its carbohydrate staple, it can save its scarce water resources for other purposes and can effectively trade by offering other goods for export. The world's current net exporters of virtual water in the form of agricultural products are United States, Brazil, Canada, Australia, and Argentina and the top net importers are Japan, Russia, United Kingdom, Germany, China, and countries of the Middle East.[37]

An exemplary model of transboundary cooperation is the Red Sea–Dead Sea Water Conveyance Program, which received approval for implementation of the first phase in December 2013. The proposal for this project was initiated by Jordan and Israel, with agreement of the Palestinian Authority. It was submitted to the World Bank in 2005, seeking financial support for feasibility studies, with the multiple purposes of saving the Dead Sea from continued desiccation, of providing much needed energy in the form of hydroelectricity for the participants at affordable prices, and of meeting local water demands via sea water desalination. This initial action prompted two major studies, one regarding engineering feasibility and the other concerned with environmental and social assessment.

The Dead Sea is a unique environment about 400 m below sea level and 10 times more saline than ocean water. Its shoreline supplies minerals for numerous health products and supports industries for potash mining and processing, as well as other chemicals. In recent years, it has become a magnet for tourism on which border countries are economically dependent. However, during the past 50 years, the water level has been declining steadily as a result of reduced inflow from the Jordan River whose waters are being extracted to support the local populations and economies. The water level has fallen from 394 m below sea level in the 1960s to about 423 m in 2012. Consequently, the Dead Sea's water surface area has declined by one-third. The water level continues declining at the staggering rate of 0.8–1.3 m per year while the degradation of the environment includes river bed erosion and hundreds of sink holes. The Red Sea–Dead Sea Conveyance Program proposes the intake of water from the Red Sea at the Aqaba location and will take advantage of the natural gravitational flow to redistribute water. This descending flow is an energy-saving setting to develop hydropower for local energy needs and for operating a desalination plant that will provide drinking and irrigating water for a water-starved region. Figure 13 shows a satellite image of the Dead Sea and its surroundings.[38]

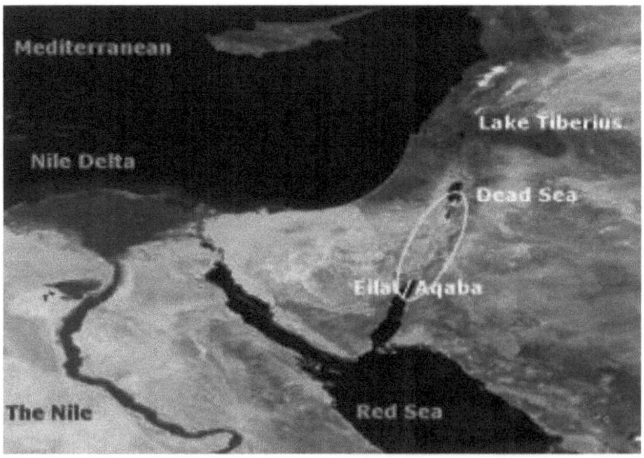

FIGURE 13 Satellite image of the Dead Sea and surrounding areas. *World Bank, Red Sea–Dead Sea Water Conveyance Program, 2013.*

The project envisions a pipeline 180 km long, built entirely in Jordan along the Israeli-Jordanian border. In the first phase of the project, an intake and pumping station at the Aqaba juncture with the Red Sea will withdraw 200 million cubic meters of water annually. The hydropower plant along the pipeline's route will generate electricity for the Wadi Arabah/Arava regions and power the high-level desalination plant that will supply potable water to be shared by Israel and Jordan. In return, Israel will release additional water to Jordan's northern region from Lake Tiberius and will also sell, at a cost, an extra 20–30 million cubic meters of water to the Palestinian Authority for use in the West Bank. The brine resulting from desalination will be directed downward to the Dead Sea and the effects of mixing these waters will receive careful scientific scrutiny. If indeed, this project restores the Dead Sea to stable conditions without negative consequences, the larger final phase of the project will be launched whereby 2 billion cubic meters of water will be withdrawn annually from the Red Sea. Barring unforeseen circumstances, the initial smaller project is expected to be operational within 5–6 years. This Red Sea–Dead Sea Conveyance Program is a historic agreement of cooperation in transboundary water management in a troubled region and represents a measure of strategic importance for environmental restoration and economic well-being of the participating nations.

Conclusions

Water scarcity is a global reality. So are climate change and an expanding world population. Although the impact of water scarcity is felt unequally throughout the world, sufficient attention is being devoted to this problem internationally so that there is cause for optimism: creative innovations have made possible new technologies for conservation; interdisciplinary scientific approaches have discovered processes to boost the availability of freshwater resources and continue to develop mechanisms to make these more affordable; policy makers have become aware of the economic dimensions associated with adequate water access for the poor and of the potential benefits of transboundary cooperation. Responsible water management in agriculture is critical to minimizing water wastes in irrigation practices and requires the financial and economic capacity to oversee appropriate operation and maintenance.

Attempts to mitigate climate change have so far proven to be more elusive, primarily because they require global cooperation to be genuinely effective. Because of social, political, legal, and ethical dimensions and competing priorities among the world's nations, mitigating the emissions of greenhouse gases has been difficult to achieve, much less to enforce.[39] Developing alternative energy sources has attracted extensive research efforts, such as solar, wind, nuclear, and biofuel approaches. Each has had some encouraging success but issues of scalability and associated costs have limited further implementation. Crop biofuels, in particular, are being reconsidered if their production competes for land and water with agricultural products and food. Recent efforts in the United States to extract natural gas reserves contained in shale could make vast reductions in the levels of CO_2 emitted but similar vast amounts of water required for such extraction has led to concerns for legal regulations to address potential groundwater contamination. The expanding world population is likely to be disproportionally poor, which means that biomass burning for cooking and heating would increase and contribute to massive increases in greenhouse gas emissions. Thus, making electricity available to millions of people who currently lack access to it could have a positive impact on mitigating climate change. But development of hydropower, which is a clean energy source of electricity, is usually accompanied with some environmental impact such as upheaval of fish ecosystems that represent food security for millions of people.

Thus, we see that water resources, energy security, and food security are all intrinsically linked. Preserving them all will require a delicate balancing act requiring utmost cooperation of our global and local leaders to sustain the viability of our planet.

References

1. UN Water. www.unwater.org/statistics; 2014.
2. Intergovernmental Panel on Climate Change. *IPCC fourth assessment report: climate change 2007.* 2007. https://www.ipcc.ch/publications_and_data/ar4/wg1/en/tssts-3-1-1.html.
3. Durack PJ, Wijffels SE, Matear RJ. Ocean salinities reveal strong global water cycle intensification during 1950 to 2000. *Science* 2012;**336**:455–8.
4. Douglas B. *Sea level rise.* International Geophysics Series; 2001.
5. Glick D. The big thaw. In: *National Geographic Magazine*; 2007. http://ngm.nationalgeographic.com/2007/06/big- thaw/big-thaw-learn.
6. Margat J, Van der Gun J. *Groundwater around the world: a geographic synopsis.* CRC Press; 2013.
7. Ahuja S, Malin J. The problem of arsenic contamination in water. In: *Chemistry for water, CHEMRAWN XV, Perspectives and recommendations.* Maison de la Chimie; 2006. p. 86–90.
8. Larson C. Losing arable land, China faces stark choice: adapt or go hungry. *Science* 2013;**339**:644–5.
9. Molden P. *Water for food, water for life: comprehensive assessment of water management in agriculture.* International Water Management Institute, Earthscan; 2007.
10. Couton A. Tackling the water crisis in Pakistan: what entrepreneurial approaches can add, in water and agriculture: implications for development and growth. In: *Essays from the CSIS and SAIS year of water conference.* Center for Strategic and International Studies; 2009.
11. Mekonnen MM, Hoekstra AY. The green, blue, and grey water footprint of crops and derived crop products. *Hydrol Earth Syst Sci* 2011;**15**(5):1577–600.
12. Hidalgo H, Cayan D, Dettinger MD. Sources of variability of evapotranspiration in California. *J Hydrometeorol.* February 2005;**6**.
13. Karki S, Rizal G, Quick WP. Improvement of photosynthesis in rice (*Oryza sativa* L.) by inserting the C4 pathway, from the 10th international symposium on rice functional genomics. *Rice* 2013;**6**:28. http://dx.doi.org/10.1186/1939-8433-6-28 [Springer Open Journal].
14. Okamoto M, Peterson FC, Defries A, Park SY, Endo A, Nambaca E, et al. Activation of dimeric ABA receptors elicits guard cell closure, ABA-regulated gene expression and drought resistance. *Proc Natl Acad Sci USA* 2013;**110**(29):12132–7.
15. Kumar R. *Collect up to 10 gallons of water per inch of rain with rainsaucers.* LocalBlu; 2012. http://localblu.com/2012/05/collect-up-to-10-gallons-of-water-per-inch-ofrain-with-rainsaucers-latest-stansalone-catchment/.

16. *State rainwater harvesting statutes, programs and legislation.* NCSL; 2013. http://www.ncsl.org/issues-research/env-res/rainwater-harvesting.aspx.
17. State Water Contractors. http://swc.org/issues/sacramento-san-joaquin.
18. Bay Area Conversation Plan. http://bayareaconseervationplan.com/About The Delta/About the problem.aspx; 2014.
19. Integrated Regional Water Management Plan. http://www.water.ca.gov/delta .
20. Famiglietti JS, Rodell M. Water in the balance. *Science* 2013;**340**:1300–1.
21. Idelovitch E, Medy M. Soil-aquifer treatment – a new approach to an old method of wastewater reuse. *J Water Pollut Control Fed* 1984;**56**(8):936–43.
22. Bixio D, Wintgens T, editors. *Water reuse system management manual, AQUAREC-integrated concepts for reuse of upgraded wastewater.* Directorate-General for Research, European Commission; 2006.
23. Brenner A. Water and wastewater treatment alternatives and quality standards toward a comprehensive reuse policy, *Water supply and sanitation for all,* The IWA Water & Environmental Management Series 14: *Obligation of the water professionals for our common future,* Huber H, Wildener P, Paris S, editors; 2007. pp. 327–37. ISBN:1843395142 and 9781843395140.
24. Johnson T. *Recycled water for groundwater recharge.* 2008. www.epa.gov.
25. Westerhoff P. Removal of endocrine disruptors, pharmaceuticals & personal care products during water treatment. *Southwest Hydrol* 2003.
26. Glieck PH. *The world's water 2006–2007, the Biennial report on freshwater resources.* Island Press; 2006.
27. Elimelech M, Phillip WA. The future of seawater desalination: energy, technology and the environment. *Science* 2011;**333**:712–7.
28. Yermiyahu U, Tal A, Ben-Gal A, Bar-Tal A, Tarchitzky J, Lahav O. Rethinking desalinated water quality and agriculture. *Science* 2007;**318**:920–1.
29. Ghermandi A, Messalem R, Offenbach R, Cohen S. An engineered oasis for solar desalination and dryland agriculture. *Renew Agric Food Syst CJO2013* 2013:1–10. http://dx.doi.org/10.1017/S1742170513000082.
30. WaterFX. *A modular solar-thermal water system to treat and desalinate any water source.* 2014. http://WaterFX.com/Aqua4/.
31. Qiu J. Trouble on the Yangtze. *Science* April 2012;**336**:288–91.
32. Stone R. Mayhem on the Mekong. *Science* 2011;**333**:814–8.
33. Brown FZ. *The price of power on the Mekong delta.* SAISPHERE, The Paul H. Nitze School of Advanced Studies, Johns Hopkins University; 2008.
34. Grumbine RE, Pandit MK. Threats from India's Himalaya dams. *Science* 2013;**339**:36–7.
35. Bahri A. *Water resource management and hydropower in Africa, oral communications IUPAC Istanbul Congress.* Tunisia: African Development Bank; 2013.
36. Central Eurasia Standard. *The Rogun dam: regional conflict and opportunity.* 2013. http://cestandard.files.worldpress.com/2013/05/rogun-dam-a-nexus-of-conflict.
37. Bottelier P. *Getting to the bottom of water scarcity.* SAISPHERE, The Paul H. Nitze School of Advanced Studies, John Hopkins University; 2008.
38. The World Bank Group. Red sea–dead sea water conveyance study program, overview, p. 1, http://www.worldbank.org/rds.
39. US Geological Survey. *Facing tomorrow's challenges—US geological survey science in the decade 2007–2017.* US Geological Survey Circular1309; 2007. http://pubs.usgs.gov/circ/2007/1309/.

17

The Need for Water Reuse

John W. Finley[1], James N. Seiber[2], Satinder Ahuja[3]

[1]SCHOOL OF NUTRITION AND FOOD SCIENCES, LOUISIANA STATE UNIVERSITY, BATON ROUGE, LA, USA; [2]DEPARTMENT OF ENVIRONMENTAL TOXICOLOGY, UNIVERSITY OF CALIFORNIA, OAKLAND, CA, USA; [3]AHUJA CONSULTING, CALABASH, NC, USA

CHAPTER OUTLINE

Water is critical for both food and energy production. Our needs for water continue to grow as the world population approaches 9 billion. In order to meet both food production and energy needs it is imperative that we make more efficient use of our water resources by increasing the reuse of water. Traditionally, energy production from hydroelectric power generation returned up to 98% of the water used for other purposes. Agriculture returned much less. New means of energy production including fracking to recover natural gas and oil, and production of ethanol and biodiesel from agricultural crops add significant strain on our water supply. The influence of climate change further complicates the water availability. Global warming is resulting in a redistribution of rainfall, which leads to greater needs of water for irrigation in our major food-producing regions. In some areas, the river flows are low enough that there may not be sufficient flow for hydroelectric generation. The need for irrigation water requires deeper and deeper drilling to find aquifers for water. More efficient use through improved crops and better cleanup of water from agricultural and industrial use are critical to meeting our current and future needs.

Food, Energy, and Water. http://dx.doi.org/10.1016/B978-0-12-800211-7.00017-X

Introduction

In a recent review, the complex interactions among water, energy, and food were shown to be inseparable. It becomes clear that water is the limiting factor for the sustainable production of food and energy.[1] The greatest demand for water worldwide is for agriculture and food production. Energy production also uses large amounts of water for the generation of electricity, mining, extraction, refining, residue disposal, and processing of fossil fuels, and as a feedstock for biofuels. Most of this water is not available for reuse. However, water used for cooling and generation of hydroelectric power is generally returned to the environment and can be reused for agricultural or municipal applications. Agricultural food and fiber production uses about 70% of the fresh water withdrawn from surface or underground sources in the United States. The need to produce food and energy stresses the limited supply of fresh water on the planet. Some of the major issues surrounding energy, water, and food security are as follows:

- Food production and agriculture are the largest water users
- Food production processing and distribution require major amounts of water and energy
- Agricultural crops including food crops are being used for fuel in addition to food
- Water is required to generate electricity and produce transportation fuels
- Energy is required to transport and treat drinking water and dispose of waste
- Food and energy production can adversely affect water quality
- Excess use of fossil fuels can increase global warming
- All of these have an impact on the environment including wildlife habitat

Water and Food

In this chapter, we will consider the water system as an inclusive term that encompasses all water utilized for human purposes. This includes surface or underground water and treatment or disposal of industrial and municipal wastewater to protect public health and the environment. Adequate supplies of water do not always exist where they are required to meet local needs; thus, water frequently must be moved considerable distances. Movement of large volumes of water from natural basins requires considerable energy input. The movement of water, however, allows us to meet agricultural, municipal, industrial, energy, and recreational requirements. For example, the California Water Project moves water from the Sacramento and San Joaquin rivers to southern California to provide irrigation for one of the most productive agricultural areas in the world, as well as municipal water for the greater Los Angeles area.

Water withdrawal can be defined as water that is diverted from the surface or groundwater. Consumed water is water that is no longer available because it has evaporated, or has been transpired by plants, drunk by animals, incorporated into

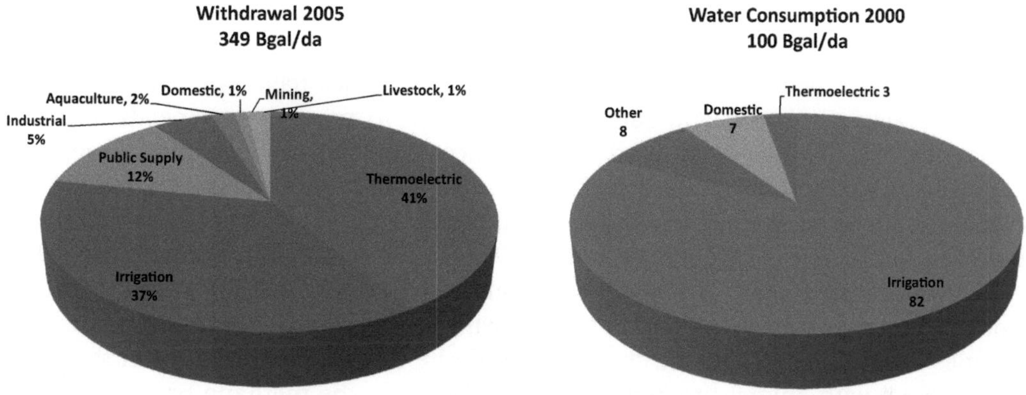

FIGURE 1 Water withdrawals and water consumption in the United States. *Adapted from Faeth Ref 3.*

products, or otherwise transformed.[2] From Figure 1 it can be seen that approximately 30% of the water withdrawals are consumed, as a result it is not available for reuse.[3] Irrigation for agriculture is the largest single consumer of water in the United States.[4] In order to meet the needs for irrigation, we require deep wells to reach aquifers with abundant high-quality water. Repeated droughts and global warming continue to increase our demand for irrigation water. Irrigation uses vast quantities of water while directly competing with other critical water uses such as power plant cooling, municipal needs, fossil fuel production, and transportation. During drought conditions, municipalities and states place restrictions on water use. Restrictions on irrigation can negatively impact food production and the economy.

Water that is returned to the environment may not always be suitable for reuse for agriculture, municipalities, or industrial applications. Agricultural use of water also can have an impact on water quality because of runoff from fertilizers, pesticides, and manure from agricultural production sites. Processing wastewater limits the ability to reuse the water. Downstream runoff of water containing fertilizers, phosphates, manure, and other agrochemicals is among growing concerns. There is a clear need to develop more efficient use of water, through drip irrigation and other sustainable practices that are needed in agriculture. Electric generation accounts for the second largest quantity of water withdrawals. Most of these withdrawals are of water for hydroelectric electric generation and most of this water is returned to the environment in a usable form.

Global climate change has been associated with repeated droughts in the world, including in the United States over the last several years. In order to maintain crop production for fiber, food, and animal feed, sustainable water supplies are critical. To meet our water needs, wells have been drilled deeper and deeper; this approach is not sustainable and will not provide an adequate and safe supply of water in the future. Higher temperatures will also increase evaporation, thus requiring more irrigation. The increasing use of biomass crops, such as corn for production of ethanol, places strains on the already limited water supply and diverts crops away from food and feed production.

Food security would exist "when all people at all times have physical, social, and economic access to sufficient, safe, nutritious food to meet dietary needs and food preferences."[1,5] The need for food raises concerns of sufficient water to meet our increasing food demands. Table 1 summarizes the quantities of water required to bring selected foods to market. The water demand for animal protein foods is much higher than that for plant-based foods, because of water needs for feed production as well as water needs of the animal. This raises questions as to the sustainability of animal protein production under increasing population growth. In addition to population growth, as societies become more affluent, they tend to desire more animal protein in their diet. The problem with animal production is exacerbated by the methane production from terrestrial animals, contributing substantially to greenhouse gases and potential global warming.

All food production requires significant amounts of water. The quantities are altered by growing conditions and the nature of the crop. Animal production, because of the inherent inefficiencies, utilizes relatively large amounts of water, compared to cereals and pulses (Table 1).

The large water demand of agriculture and food production also impacts the environment because of land degradation, salt buildup in the irrigated soil, changes in runoff, disruption of groundwater recharge, water quality, and availability of water and land for other purposes such as natural habitats and aquaculture.

Postharvest processing of foods also requires large amounts of water. Processing wastewater can present environmental concerns because the water may not be suitable for reuse. Table 2 contains examples of water requirements to produce some processed foods. The values include production as well as processing requirements.

It is also important to recognize that production of packaging materials for food and beverages also can require large amounts of water. For example, production of a 1-l polyethylene terephthalate water bottle requires approximately 3 l of water.[5] This

Table 1 Water Requirements for Food Commodities[5]

Product	Unit	Equivalent Water in m^3/unit
Cattle	Head	4000
Sheep and goats	Head	500
Fresh beef	kg	15
Fresh lamb	kg	10
Fresh poultry	kg	6
Cereals	kg	1.5
Citrus fruits	kg	1
Palm oil	kg	2
Pulses, roots, and tubers	kg	1

Table 2 Water Requirements for Processed Foods[5]

Product	Quantity	Equivalent Water in Liters
Glass of beer	250 ml	75
Glass of wine	125 ml	120
Glass of milk	200 ml	200
Glass of apple juice	200 ml	190
Cup of coffee	125 ml	140
Glass of orange juice	200 ml	170
Cup of tea	250 ml	35
Bag of chips	200 g	185
Slice of bread	30 g	40
Egg	40 g	135
Hamburger	150 g	2400

raises serious concerns about the sustainability of continued use of disposable bottles for water and beverage consumption.

Water and Energy

Freshwater withdrawals for irrigated agriculture and thermoelectric energy production in the United States each utilize approximately 345 billion gallons of water per day. Thermoelectric water withdrawals are mostly returned in a form that they can be reused for agricultural or municipal uses. Approximately 70% of the water used in thermoelectric plants is fresh water; the remainder is saline in coastal areas. Of the fresh water withdrawn, 99% of it is surface water.[6]

For domestic use, potable water must be pumped from its source (underground or surface) to the treatment plant, treated, and then pumped to consumers. In areas without adequate fresh water it may be necessary to bring water from very long distances, making the energy footprint for this water extremely high.

Irrigation for agriculture uses 40% of the water withdrawals; however, it comprises more than 80% of the 100 billion gallons of water consumed daily in the United States. Irrigation water losses are from evapotranspiration. Water returned from irrigation (40%) is much less than that from thermoelectric uses (98%).[7] The Electric Power Research Institute projects that by 2025, thermoelectric power production will be negatively affected in all states and 14% of all countries throughout the world because of insufficient available water.[8]

Generation of nuclear power requires about 2.5 times the water per unit of electricity than gas and 25% more than coal. As a result, nuclear power plants are more susceptible to restricted water availability as can occur during droughts, when fresh water is being used for cooling. In 2007, 24 of the 104 nuclear power generators in the United States were in danger of shutdowns because of water limitations.[9] Nuclear power plants in coastal regions are less affected because they can use salt water for cooling.

Global Climate Change

Although the role of fossil fuels in global warming remains controversial, there is clear evidence that global warming and redistribution of rainwater is occurring. Increases in atmospheric carbon dioxide and temperature may result in faster plant growth and more respiration; however, more severe warming, floods, and drought may result in lower overall net yields. Drought reduces both the quality and quantity of forage for grazing animals. Plant breeders are challenged to develop more drought- and heat-resistant forage crops. Heat waves are projected to increase with the current observed climate change, which will influence livestock production. Several states have reported losses of more than 5000 animals from just one heat wave.[9]

It is projected that in the twenty-first century, most dry subtropical areas of the United States will experience significant reductions in surface water and groundwater as a consequence of climate change. In currently dry regions such as the Great Plains, drought frequency will likely increase by the end of the twenty-first century. Conversely, in higher latitude areas water resources are projected to increase.

In many areas the changes in climate-related water distribution will reduce raw water quality and pose risks to drinking water quality because of interacting factors: increased temperature; increased sediment, nutrient, and pollutant loadings from heavy rainfall; coastal flooding; increased concentration of pollutants during droughts; and disruption of treatment facilities during floods. Water management efforts including scenario planning, learning-based approaches, and flexible and low-regret solutions, can help decrease the impact of hydrological changes and climate change.[1,9]

Most aspects of food security are likely to be affected by climate change, including food access, utilization, and consumer cost of food. Temperature increases of ~ 4 °C or more above late-twentieth-century levels, combined with increasing food demand from greater populations and increased demand for animal foods, could pose large risks to food security globally and regionally. Risks to food security are generally greater in low-latitude areas.[1,9] The nature of pests and disease threats to crops will also change, in type and frequency, with as yet unknown consequences.

Energy

Modern agriculture, which is required to feed the current population, uses approximately 30% of total global energy. Fuels are required for land preparation, fertilizer production, irrigation and sowing, harvesting, transportation of crops, and processing and distribution of foods.[1]

The "energy system" is defined as the energy operations necessary to generate, transport, and distribute electricity and the steps required to produce and distribute fuels required for transportation and industry.[10]

Most of our electrical energy is generated at a variety of power plants that operate on combustion of fossil fuels (coal or natural gas) or nuclear fission. The energy sources generate heat that converts water to steam to turn turbines, ultimately generating electricity. Electricity is also generated from flowing water (hydroelectric), wind, sunlight, and Earth's heat. Transportation fuels are those that are used for transportation of goods, automobiles, trains, and airplanes. These fuels are obtained by recovery, production, refinement, and distribution of oil or natural gas including feedstocks for biofuels such as sugar cane, corn, or alternative biofuel crops.

Large quantities of water are required in conventional power production, primarily for cooling. It is estimated that nearly half of the fresh and ocean water withdrawals in the United States are for cooling thermoelectric power plants, including nuclear power plants.[11]

Water is a key component in energy production; it is used in power generation, for cooling thermal power plants, and in the extraction, transportation, and processing of fuels. Water also is required for irrigation and for processing of biomass crops that are dedicated to energy production. Energy is also required for the collection, transport, treatment, and distribution of irrigation and fresh potable water. Water is required for energy extraction of oil and gas, particularly where hydraulic fracturing (fracking) is employed to mobilize the oil or gas from rock. Water is the key requirement for hydroelectric power production. Outflow from hydroelectric power can, however, be used for other applications such as irrigation or municipal water supplies.

Droughts in major agricultural areas greatly increase the demands for irrigation water for food and bioenergy production, placing greater reliance on underground water aquifers. Recovery of groundwater can require considerable amounts of energy. The deeper we drill for groundwater, the greater the demand for energy to bring the water to the surface, with potential impacts on the stability of the overlying geology. Current and growing populations increase the demand for clean water and sewage treatment, which will require energy to move water to the needed sites economically, without adding more environmental stress.

Global warming is resulting in much more glacial and polar ice cap melting. Depending on location, the meltwater could be valuable, but frequently it is not close to agricultural or urban areas where it can be effectively utilized. In addition, most of the polar ice cap melts goes into sea water. Desalination becomes an attractive source of water for coastal regions but does not solve the agricultural irrigation needs in areas like the midwest of the United States. Desalination capability has improved dramatically in the last 10 years and, as a result, the capacity has grown rapidly. Despite improvements, cost remains a significant issue in desalination. In large-scale desalinization operations, the cost of desalinated water ranges from US\$0.50/m^3 to US\$1.00/m^3, depending on location. Recent developments, including the increase in unit capacity, process design and materials, and hybrid systems, have potential for decreased cost and reduced energy consumption in the desalination processes. The developments of low-energy desalination technologies, such as adsorption desalination, indicate good possibilities for low cost and energy demand in the future.[12,13]

Table 3 Average Water Consumption for Production of Transportation Fuels[14]

Source	Raw Material Recovery gal per million BTUs	Transformation gal per million BTUs	Total gal per million BTUs
Traditional oil	1.4	12.5	13.9
Conventional natural gas	0	2	2
Shale natural gas	12.5	2	14.5
Oil sands	260	12.5	272.5
Enhanced oil recovery	1257	172	1429
Irrigated corn (ethanol)	15,750	9	15,759
Irrigated soy biodiesel	44,500	9	44,509

The energy sector has an increasing need for water to meet global energy demands. The relative amounts of water used in the energy sectors differ among the various sectors. Table 3 provides a comparison of the amounts of water required to produce a million British Thermal Units (BTUs) of energy. The data provide interesting comparisons of the amounts of water required for enhanced recovery techniques. It also must be kept in mind that currently it is difficult to recycle that water. The biofuels comparisons are important to deliberate, but it must be recognized that rainwater is not considered in the estimates.

Traditional oil and gas production uses less water than that used for oil and gas production from tar sands and fracturing. Developing agriculture-based energy substitutes produces additional stress on water resources. The need for water may serve as a significant constraint on corn-based ethanol production in the United States. Water demand for use in ethanol fuel production ranges widely. For example, to produce 1 gallon of ethanol, 10–323 gallons of water is required. These differences are because of variations in climate and irrigation requirements. Switch grass requires from 1.9 to 9.8 gallons of water per gallon of ethanol. Gasoline production ranges from 2.8 to 6.6 gallons of water per gallon of gasoline.[14]

Table 4 provides cost and environmental factors for various electricity-generating technologies. It shows that natural gas is the cheapest source of electricity and can be produced with low use of water. CO_2 emissions are also low. This suggests that natural gas could be a fuel of choice in the future, if we handle it properly.

To put water use and energy use in perspective, it requires about 2 million BTUs to drive round trip from New York City to Washington DC. Table 5 presents the amount of water necessary to produce the various fuels required for such a trip.

In 2001, 707 million bushels of grain were used for ethanol production. By 2012, this had grown to 5011 million bushels of the total production of 12,358 bushels of corn produced, which equates to approximately 40% of the corn crop.[3,15] Considering that

Table 4 Cost and Environmental Factors for Various Electricity Generating Technologies[1,26,27]

Energy Technology	**Average Levelized Cost of Electricity USD per MWh**	**Median Water Withdrawal by Cooling Type gal per MWh**	**Median Water Consumption by Cooling Type gal per MWh**	**CO_2 lb per MWh**
Existing technology				
Conventional coal	62	531–36,350	250–471	1886
Nuclear	59	1,101–44,350	269–672	–
New technology				
Conventional coal*	95	531–17,914	493–779	1886
Advanced coal	109	390	372	1755
Advanced coal + carbon capture and sequestration (CCS)	126	596	540	206
Natural gas combined cycle (NGCC)	66	253–5950	198–240	797
NGCC + CCS	89	496	378	86
Advanced nuclear	104	1101–7050	610–672	–
Hydroelectric	86	–	4491	–
Wind. onshore	97		0	
Wind, offshore	243		0	
Solar thermal	312		8–1000	
Photovoltaic	211		26	
Energy Source	**Cost 2011 USD per MWh**	**Water Withdrawals gal per MWh**	**Nonrenewable Water Consumption gal per MWh**	**CO_2 Emissions tons CO_2 per GWh**
Coal	100.1–135.5	390–36,000	42–942	888
Natural gas	67.1–130.3	2–35,000	2–826	499
Nuclear	108.4	1101–44,350	269–672	29
Solar (photovoltaic)	144.3	0	0	85
Geothermal	92	–	0–1406	
Hydroelectric				26
Wind	86.6 land 221.5 offshore	–	0	26

*Reuse of cooling water or dry cooling in power plants.

over 35,000 gallons of water are required to produce a gallon of corn-based ethanol fuel, corn ethanol production cannot be considered sustainable. The water withdrawals for corn ethanol production will deplete aquifers to unacceptable levels.

Hydraulic fracturing, or "fracking," is used to extract gas and oil from rock formations where oil and gas are not normally accessible. However, after fractures are developed in the rock formations using horizontal drilling and high levels of hydraulic pressure, gas

Table 5 Gallons of Water Required to Produce Sufficient Fuel for a Round Trip Between New York City and Washington, DC[14]

Fuel	Gallons of Water to Produce Fuel
Natural gas (on land)	5
Shale natural gas	33
Traditional oil	32
Oil sands mining	616
Irrigated corn (ethanol)	35,616
Irrigated soy	100,591

and oil can move through the fractures and become accessible. These more accessible oil and gas reserves can then be exploited by wells.[16]

Fracking has resulted in vast new areas of potential energy recovery. The Marcellus shale basin, one of the largest, extends from northern Georgia to New York State. Various other formations are found in Texas, Oklahoma, Indiana, Michigan, Arkansas, Louisiana, and the Rockies.[16] Shale gas will make up 46% of the total supply in 2035. Fracking raises a major concern because of the volumes of water required and the risk of contamination of aquifers used for food production. The average shale gas well requires 3–4 million gallons of water and will yield 800 million cubic feet of gas over the 7.5-year lifetime of a well.[17]

Hydraulically fractured wells typically require about 12,000 l/m, or 10 million to 20 million liters of water per well. In many regions, this demand for water will be a minor fraction of the local water consumption. Hydraulic fracturing operations introduce new water demands that could result in competition for water, price increases, aquatic biodiversity loss, and accelerated groundwater and surface water depletion.

The arid Eagle Ford area in south Texas has experienced extended droughts and continued population growth and now large amounts of water are being used for fracking. This has raised concerns over fresh water consumption for energy production. The water use in the Eagle Ford area is a small fraction of the total annual water consumption of the region; however, in some rural low-population areas it accounts for the majority of water consumed. In some parts of the Eagle Ford region, hydraulic fracturing water requirements are expected to exceed sustainable groundwater withdrawal rates. Similar concerns occur in other areas such as the Gulf of San Jorge Basin of Argentina, the Cooper Basin of Australia, the Monterrey Basin in California, and the Denver–Julesburg Basin in Colorado. In these regions, fracking may have to be restricted because of lack of water.

Water and Energy Interactions

Water reuse in fracking can help alleviate some of the stress on available water. Reuse of fracking-produced water is the practice of using treated or untreated produced water as the make-up water for subsequent hydraulic fracturing fluid. In the Marcellus, high disposal costs of used fracking water have resulted in nearly 90% reuse of Marcellus-produced water. In other areas, the majority of produced water is disposed of in deep injection wells. Constraints to full reuse include water production volume that may exceed the injected volume, issues with scale formation, biofouling, and limited performance of the fracking fluid additives, because of the accumulation of dissolved solids as the water is recycled. Research has improved the understanding of the operational and logistical constraints of fracking water reuse. Improvements such as well placement and drilling schedule can maximize reuse potential.[18]

Fracking fluid contains a number of chemical compounds. Companies involved in fracking reported to a congressional committee that up to 750 different chemicals are used and/or formed in the fracking process. They range from the innocuous materials such as sodium chloride, citric acid, coffee, and walnut hulls extractives, to toxic ones such as lead, benzene, and methanol. Twenty-nine known or possible carcinogens have been reported in fracking wastewater.[19] It is not clear at this time, whether this water can be reclaimed for reuse without additional large expenditures.

Thermoelectric power consumption, whether coal, oil, or gas, requires cooling water, which comes from surface water. Currently, cooling for thermoelectric power withdraws up to 75% of the surface water in some areas.[20] Sequestering greenhouse gases from power plants can double the cooling water demand because of the energy required for compressing the carbon dioxide.[21]

The needs for cooling water frequently are in competition with agricultural and municipal demands. In areas where heavy irrigation is required, up to 40% of the available water is used for irrigation, which results in rivers running low in the summer, putting additional strain on power production[22] and waterway transportation and wildlife habitat.

In thermoelectric power plants in the United States, 1 kWh of electricity requires the evaporation of about 2 US gallons (7.6 l) of water to eliminate waste heat from the process. In hydroelectric plants, an average of 18 gallons (68 l) of water is evaporated to generate 1 kWh of electricity. In practical terms, it means that 8–16 gallons of water are required to burn one 60-W light bulb for 12 h. This would mean that annually one incandescent light bulb would consume 3999–6300 gallons of water. If it is assumed that the 111 million occupied housing units (US Census Bureau 2005) burned one 60-W light bulb for 12 h a day for 1 year, it would require between 336 and 656 billion gallons of water use.[23]

Energy production requires large quantities of water, but bringing water from its source to usable supply and disposal of sewage requires energy. Extraction and pumping from ground and surface sources to the site of use requires electrical energy. The energy intensity of treatment required for different types of water sources varies depending on water sources. Fresh surface waters use energy mainly for pumping; however, the use of seawater adds the energy requirements for desalination. The energy required for residential end use is very high relative to other parts of the water supply cycle. Processes such as heating water, washing clothes and dishes, and cooking all require intense energy input. Hot water usage is responsible for making end use the most energy-intensive stage of the water cycle.[24]

Depending on the depth of the well, pumping groundwater requires from 537 kWh to 2270 kWh per 1 million gallons of water. Southern California receives potable water from the Sacramento–San Joaquin delta, which requires pumping water over hundreds of miles over low-lying mountains using 9200 kWh per 1 million gallons.[25] In California, the conveyance, storage, treatment, distribution, wastewater collection, treatment, and discharge of water consumes 19% of the state's electricity, 30% of its natural gas, and 88 billion US gallons (330,000,000 m^3) of diesel fuel annually.[24] The River Network reports that the United States consumes at least 521 million MWh a year for water-related purposes, which is equivalent to 13% of the nation's electricity consumption.[25]

Solutions Must Be Sustainable

The solutions to the water, energy, and food crisis must be sustainable solutions for all three sides of the triangle. Without integrated solutions, the entire system will fail. Sustainable food production includes optimizing health, safety, and quality and consumer appeal of sufficient foods to meet the needs of the growing world population, in part a result of increasing life expectancy in large part because of more abundant, safe, and healthier food alternatives. Over the last 75 years the current world population has increased to approximately 8 billion, and the growth is projected to continue for the next 25 years, reaching 9 billion. Agricultural food production has helped sustain this growth by increasing food production 20% from 1990 to 2010. If we are to meet the food needs of the anticipated population growth, this rate of increased production must continue for the next 25 years.[26] These increases must continue in the face of limited energy supplies, limited water supplies, and increased global warming, which results in lower production using current practices. In effect, we must become much more efficient in food production, generating less waste. (It is estimated that up to 25% of all food produced is wasted, and another 25% is lost as a result of postharvest pests, diseases, and spoilage.)

In order to meet the growing need for food production, we must face the following considerations:

1. Production of more food on less desirable land with less water
2. Decreasing energy and water availability at a low cost
3. Increased globalization of the best measures to provide energy and water resources for food
4. Globalization of compliance regulations

Within the food system, sustainability is regarded quite differently, depending on the particular segment. For example, for food producers sustainability relates to land, water, and energy management. In addition to these macro concerns, producers must consider fertilization and pest management and focus on the sustainable environmental factors as well as energy resources for production of the chemicals. To help address these concerns in recent years the Environmental Protection Agency registration of new pesticides is now largely devoted to biopesticides, rather than further modification of synthetics from the twentieth century.[27] When food sustainability is considered from the food, health, and nutrition viewpoint, consumer health and nutrition are particularly important (e.g., vitamin, antioxidant, minerals balance and avoiding undue levels of fats, carbohydrates, sodium, and synthetic additives[28–32]). Food processors' sustainability issues are focused on preserving and distributing foods, including maximizing shelf life for quality and safety.

There is growing momentum demonstrating the benefits of new biopesticides, biofuels, and biobased products used in the production, processing, and distribution of foods.[33] Food wastes need to be minimized. Capturing and using the energy from processing the food wastes using anaerobic fermentation to biogas, recovering the energy as electricity, and utilizing the by-product of compost as field dressing are steps to minimize wastes. (These steps are important in ensuring that all parts of the food supply are beneficially utilized).[34]

What Are We Doing About the Problem?

The "9 Billion Problem" has implications for the way research, education, and outreach are supported at the National Institute of Food and Agriculture (NIFA). The science, policies, and regulations must align with the steps necessary for addressing the global challenges in a sustainable manner, including food security, hunger, food safety, health and nutrition, childhood obesity, energy, water, and climate change.

The nexus among food, water, energy, and health is critical to help us determine how we should deploy the resources to address the same, because of the need to consider them collectively rather than individually. The increase in population will contribute to escalating scarcity of land, water, energy, and food.[35]

Addressing these complex issues will take a commitment to seeking collaborative efforts among federal and state agencies, private enterprises, and nongovernmental organizations. Investing in enabling sound science and policies that address these complex issues and investments in ensuring food security and sustainable bioenergy and water will be a significant part of NIFA's portfolio. NIFA must continue to support the best and brightest scientists at academic institutions and in private and nongovernmental sectors to find innovative solutions to these challenges. By making the right investments in science, our expectation is that NIFA will enable the scientific community to make great discoveries that can be translated into innovations beneficial to lives of the American people, and for that matter, globally.[35] The same consideration apply to support from other offices within USDA, from DOE, EPA, and other federal and state agencies supporting energy and/ or agricultural research.

The trilemma of food, energy, and water present us with a worldwide multidimensional problem that links resources, politics, and quality of life for all on the planet now, and for the 9 billion that will inhabit the plant in the next 40 years.

Avoiding Disastrous Side Effects: The Impact of Arsenic-Contaminated Groundwater on Food

Groundwater polluted by arsenic has been reported in over 70 countries, posing a health hazard to an estimated 150 million people worldwide.[36–38] Around 110 million of those people live in 10 countries in South and Southeast Asia: Bangladesh, Cambodia, China, India, Laos, Myanmar, Nepal, Pakistan, Taiwan, and Vietnam. In the US Geological Survey studies in the United States, over 31,000 samples analyzed over almost a 30-year period revealed that a large number of states are affected by arsenic in groundwater. Consequently, the maximum contaminant level of 10 ppb was instituted in the United States just a few years ago.

Bangladesh has historically had high arsenic levels in groundwater, which becomes a problem when food crops are irrigated with contaminated water. Arsenic in rice seems to vary widely. Speciation of Bangladesh rice shows the presence of As(III), DMAA, and As(V); more than 80% of the recovered arsenic is in the inorganic form. It has been reported that more than 85% of the arsenic in rice is bioavailable, compared to only about 28% of arsenic in leafy vegetables.

Rice is the staple food for Bangladeshi people. There are two seasons for rice culture: aman and boro. The aman culture period is during the rainy season, when no irrigation is required; but the boro cultivation phase (dry season) is completely dependent on irrigation. So, it can be predicted that arsenic-contaminated irrigation water could easily increase the arsenic level in rice grain, straw, and other parts of rice plants. Arsenic contents in boro rice could be 1.3 times higher than for aman rice.[39]

An average Bangladeshi citizen consumes about 500 g rice/day. The maximum allowable daily limit (MADL) of arsenic ingestion without harm is 0.22 mg/kg, or 220 μg/day limit. Assuming that 3 l/day consumption of drinking water with an arsenic content of 0.05 mg/kg and consumption of rice with an average arsenic content of 0.437 mg/kg, simulations show a mean load of 304 μg/day, far exceeding the 220 μg/day limit. Rice contributes 144 μg/day, approximately 65% of the daily limit.[40] It has been observed that even if a raw rice sample does not contain any detectable amount of arsenic, the cooked rice contains a substantial amount of the element when it is cooked with arsenic-contaminated water. To summarize, the amount of arsenic in cooked rice, plus an average consumption of 3 l/day of drinking water with the Bangladesh standard of 50 μg/l arsenic, along with arsenic-rich vegetables, is sufficient to bring the value of daily arsenic ingestion higher than the MADL.[41]

The Food and Drug Administration (FDA) has increased its testing of arsenic in rice and rice products to determine whether action is needed to reduce long-term exposure to the element. The agency released initial data showing total inorganic arsenic levels in about 200 products. The levels were between 3.5 and 6.7 μg per serving, consistent with data published in *Consumer Reports*. The FDA hopes to better understand how arsenic levels vary by rice type, geographic region in which the rice is grown, and the type of food product.

The arsenic example may be replicated with cadmium, chromium, mercury, lead, and other metallic contaminants, and for some organic contaminants like dichlorodiphenyltrichloroethane (DDT), and for nitrate and perchlorate. We need to anticipate potential problem areas by developing and employing predictive tools and better integration of information from all sources, including targeted monitoring of contaminants in air, water, soil, biota, and all foods.

Conclusions

We must adopt a nexus approach to deal with the relationships of food, energy, and water. None of these issues can be effectively resolved in isolation. Development of new technology in any of the sectors must consider the impact on the other sectors as well as the environment. For example, a new crop that could be used as an energy or food crop, the water and energy, and waste or by-product implementations for production must be considered. Government and international policy need to be developed to assure integrated implementation of technologies considering growing population stresses, the associated energy and food needs and how these impact the environment and potential for global warming. We must avoid isolated silo approaches and consider broad effects of new technologies and their impact on the food, water, and energy systems. The nexus must involve individuals, policy makers, and thought leaders. As scientists, we need to take leadership roles in communicating the issues clearly and bring forth new technologies that solve problems, while considering the broad implications of new technologies. Technology to deliver new sources of energy must be carefully assessed for unintended consequences. For example, fracking is delivering abundant levels of natural gas; however, the influence on groundwater is raising concerns. Fracking also uses large amounts of water and generates large volumes of wastewater that cannot be used for municipal use or agriculture.

References

1. Seiber JN, Finley JW. The nexus of food Energy and water. *J Agric Food Chem* 2014;(27):6255–62.
2. Vickers A. *The handbook of water use and conservation*. WaterPlow Press; 2001. ISBN:1-931579-07-.
3. Faeth P. U.S. Energy security and water: the challenges we face. *Environ Sci Policy Sustain Dev* 2012; **54**(1):4–19. http://dx.doi.org/10.1080/00139157.2012.639595.
4. Barber NL. *Summary of estimated water use in the United States in 2005*: U.S. Geological Survey Fact Sheet 2009. 2009–3098, 2 p.
5. Use of water in food and agriculture. http://www.lenntech.com/water-food-agriculture.htm; [accessed 15.06.14].
6. Climate Change. *Impacts, Adaptation, and Vulnerability*. Summary for Policymakers. IPCC; 2014., ipcc-wg2.gov/AR5/images/uploads/IPCC_WG2AR5_SPM_Approved.pdf.
7. Barber NL. *Summary of estimated water use in the United States 2005. U.S. geological survey fact sheet 2009–3098*. U.S Geological survey; 2009. https://www.google.com/webhp?source=search_app#q=summary+of+estimated+water+use+in+the+united+states+in+2005; [accessed 14.06.14].

8. Use of water in agriculture. http://www.lenntech.com/water-food-agriculture; [accessed 06.06.14].
9. Drought Could Shut Down Nuclear Power Plants. http://www.msnbc.msn.com/id/; [accessed 06.06.14].
10. Hanlon P, Madel R, Olson-Sawyer K, Rabin K, Rose J. *Food, water and Energy: Know the nexus*. New York: Grace Communications; 2013. programs@gracelinks.org.
11. Macknick J, Newmark R, Heath G, Hallet KC. *A review of operational water consumption and withdrawal factors for electricity generating technologies*. Golden, CO: National Renewable Energy Laboratory; 2011. Technical Report NREL/TP-6A20-50900.
12. Ghaffour N, Missimer TM, Amy GL. Technical review and evaluation of the economics of water desalination: current and future challenges for better water supply sustainability. *Desalination* 2013; **309**:197–207.
13. Subramani A, Badruzzaman M, Oppenheimer J, Jacangelo JG. Energy minimization strategies and renewable energy utilization for desalination: a review. *Water Res* 2011;**45**(5):1907–20.
14. Mckenna P. *Measuring Corn Ethanol'S Thirst for Water; MIT Technology Review; MIT: Cambridge*. MA, USA, April 14, 2009.
15. Ethanol usage projections & corn balance sheet (mil. bu.). https://www.extension.iastate.edu/agdm/crops/outlook/cornbalancesheet.pdf; [accessed 15.06.14].
16. Hydraulic fracturing 101. http://www.earthworksaction.org/FracingDetails.cfm; [accessed 06.06.14].
17. *Modern shale Gas development in the United States: Primer*. April 2009. http://www.netl.doe.gov/technologies/oil-gas/publications/EPreports/Shale_Gas_Primer_2009.pdff.
18. Shaffer DL, Arias Chavez LH, Ben-Sasson M, Romero-Vargas Castrillón S, Yip NY, Elimelech M. *Environ Sci Technol* 2013;**47**(17):9569–83.
19. U.S. House of Representatives Committee on Energy and Commerce. *Minority staff, chemicals used in hydraulic fracturing*. April 2011. http://democrats.energycommerce.house.gov/sites/default/files/documents/Hydraulic%20Fracturing%20Report%20 4.18.11.pdf; [accessed 06.06.14].
20. U.S. Energy Security and Water. The challenges we face, http://www.netl.doe.gov/energy-analyses/pubs/Bituminous_Performance_Tool_Rev4.swf; [accessed 06.06.14].
21. Energy-Water Nexus Overview. http://www.sandia.gov/energy-water/nexus_overview.htm; [accessed 06.06.14].
22. Water use, withdrawl and consumption. http://gracelinks.org/1249/water-use-withdrawal-and-consumption-what-does-it-all-mean; [accessed 14.06.14].
23. Hill R, Younos T. *Water cooler, the intertwined tale of energy and water*. Blacksburg VA: Virginia Water Resources Research Center Virginia Tech; 2008.
24. Plappally AK, Lienhard JH. *Energy requirements for water production, treatment, end use, reclamation, and disposal. Renewable and sustainable energy reviews*, vol. 16; 2012. 4818–4848.
25. Griffiths-Sattenspiel B, Wilson W. *River network. The carbon footprint of water*. 2009. http://www.rivernetwork.org/resource-library/carbon-footprint-water.
26. Cobb L. World population from 1800 to 2100, based on UN 2010 projections and US Census Bureau historical estimates. http://simple.wikipedia.Org/wiki/World_population; http://commons.wikimedia.org/wiki/File%3AWorld-Population-1800-2100-es.png [accessed 06.06.14].
27. Cantrell CL, Dayan FE, Duke SO. Natural products As sources for new pesticides. *J Nat Prod* 2012;**75**: 1231–42.
28. Seiber JN. *JAFC selects virtual issue: food quality traits for sustaining agriculture*. American Chemical Society; 2011. http://pubs.acs.org/page/vi/2011/food-quality.html; [accessed 06.06.14].
29. Frankel EN. Nutritional and biological properties of extra virgin olive oil. *J Agric Food Chem* 2011; **2011**(59):785–92.

30. Hafer TJ, Howard LR, Prior RL. Processing and storage effects on ellagitannin composition of processed blackberry products. *J Agric Food Chem* 2010;**58**:11749–50.

31. Sharma GM, Su M, Joshi AU, Roux KH, Sathe SK. Functional properties of select edible oilseed proteins. *J Agric Food Chem* 2010;**2010**(58):5457–64.

32. Argov-Araman N, Smilowicz JT, Bricarello DA, Barboza M, Lerno L, Froehlich JW, et al. Lactosomes: structural and compositional classification of unique nanometer-sized protein lipid particles of human milk. *J Agric Food Chem* 2010;**2010**(58):11234–42.

33. Orts WJ, Holtman KM, Seiber JN. Agricultural chemistry and bioenergy. *J Agric Food Chem* 2008; **2008**(56):3892–9.

34. Zhang R, Zhang Z. Biogasification of solid waste with an anaerobic-phased solids-digester system. US Patent 6342378 B1 2002.

35. Ramaswamy S. Nexus between food, water and energy. Abstracts of Papers, 245th ACS National Meeting & Exposition. New Orleans, LA, United States; 2013.

36. Ravenscroft P, Brammer H, Richards KS. *Arsenic pollution: a global synthesis*. U.K: Wiley-Blackwell; 2009.

37. Ahuja S. Remediation of arsenic contamination of groundwater in Asia and USA. Abstracts of Papers, 245th ACS National Meeting & Exposition. New Orleans, LA, United States 2013.

38. Ahuja S. *Arsenic contamination of groundwater: mechanism, analysis, and remediation*. NY: Wiley; 2008.

39. Ahsan DA, Del Valls TA. Impact of arsenic contaminated irrigation water in food chain: an overview from Bangladesh. *Int J Environ Res* 2011;**5**(3):627–38.

40. Correll R, Huq SMI, Smith E, Owens G, Naidu R. Dietary intake of arsenic from crops. In: Naidu R, smith E, Owens G, Bhattacharya P, Nadebaum P, editors. *Managing arsenic in the environment: from soil to human health*. Melbourne, Australia: CSIRO Publishing; 2006. p. 255–71.

41. Huq SMI, Al-Mamun S, Joardar JC, Hossain SA. Remediation of soil arsenic toxicity in Ipomea aquatica using various sources of organic matter. *J Land Contam Reclam* 2008;**16**(4):333–41.

Index

Note: Page numbers followed by "f" and "t" indicate figures and tables respectively.

Kolibri 360 GmbH
Ferdinand-Jühlke-Straße 7
99095 Erfurt, Deutschland
produktsicherheit@kolibri360.de